CIRCUIT ANALYSIS

ELIAS M. SABBAGH
PROFESSOR OF ELECTRICAL ENGINEERING
PURDUE UNIVERSITY

THE RONALD PRESS COMPANY · NEW YORK

2

Library of Congress Catalog Card Number: 61–6789
PRINTED IN THE UNITED STATES OF AMERICA

PREFACE

The science of engineering is based on results obtained through experimentation. From these results theories are developed. By logical reasoning and application of the theories, new conclusions are forecast and other results are predicted. If these are verified by further experimentation, the theories are accepted and adopted. If at a later time some experimental results, obtained by the use of more accurate instruments or by newer methods of testing, show discrepancies with the theoretical results, either the theories are modified to account for the discrepancies or newer theories are developed.

Oftentimes newer and different processes of thinking result in other ways of interpreting the original experimental results. The newer methods of attack may be better adapted to the analysis and synthesis of some types of systems than the older ones. The engineer should learn when and where to use one method in preference to the other.

Electrical engineering is based on six experimental discoveries and one postulate; namely Coulomb's forces between charged bodies at rest, Oersted's forces between current carrying conductors, Faraday's induced voltage, Maxwell's displacement current postulate, Ohm's law, and Kirchhoff's laws. The first two discoveries led to the assumption of the existence of electric and magnetic fields in the same way that Newton's law of attraction between masses led to the assumption of the existence of a gravitational field. No one, however, has as yet proved that these fields have a real physical existence. For this reason they are considered as hypothetical means for the visualization of forces. Whether these fields do physically exist or not, the assumption of their presence gave a great impetus to the advancement of electrical engineering in all its branches.

The circuit theory provides a logical foundation for the analysis of a large number of physical phenomena in general and of nearly all electrical devices and systems in particular. While it is recognized that wave guides and other devices working at very high frequencies cannot be represented by lumped circuit elements, the circuit theory is, nevertheless, essential for the engineering analysis and synthesis of most electrical systems and devices. The circuit can be electric, dielectric, or magnetic in nature.

In this book the electric circuit is considered in detail and the concepts of dielectric and magnetic circuits are introduced. The book progresses step by step from an introduction to electric and magnetic fields—a must for many students and a valuable review for others. Inductance, capacitance, and resistance are first defined in terms of field quantities and later in terms of their volt-ampere relationship. In the consideration of the capacitor and the inductor, a simple treatment of electrostatics and magnetostatics is deemed necessary to clearer understanding. The fundamental laws of circuit theory are explained and the student is introduced in a logical sequence to more sophisticated tools of circuit analysis such as transformation theories, complex algebra, the complex frequency method, and the Fourier series. Special techniques of analysis are also discussed.

Topics of importance not usually found in an introductory text include equilibrium equations set up from the topology of the circuit, use of the signal flow diagram, and equilibrium equations of magnetically coupled circuits.

With this treatment the author hopes to provide a firm foundation for the student of electrical engineering. The student will be fully prepared to move progressively to more complex application of principles to dielectric and magnetic circuits, the theory of energy converters, electromechanical coupling, and simple magnetic amplifiers. Topics met in subsequent courses such as transients, various methods of synthesis, operational calculus, and functions of complex variables will fall logically into place. The theories thus developed can be reinforced through application to systems such as those involving servomechanisms.

The contents of this book have been developed and tested with Purdue University sophomores over the last five years. Although it bears the name of one author, in reality it is the contribution of several men. Part of it was edited by Professor Robert N. Faiman from notes given him by the author. The chapters on complex frequency were written by Professor George R. Cooper. Professors Harry W. Hale and Stephen Freeman, Jr., are responsible for useful suggestions and for the collection, editing, and solving of the problems following each chapter. Special acknowledgment must be given to the people in charge of the sophomore classes for suggestions and corrections made in the original notes.

ELIAS M. SABBAGH

West Lafayette, Indiana
February, 1961

CONTENTS

SYMBOLS AND NOTATION

The following symbols and notational forms are in use throughout this material. Other specialized notational forms and symbols of limited use are defined at the place of use. AIEE standard symbols have been used where available except those indicated by *.

Symbol	Quantity
$\mathbf{a}$	Acceleration
$\mathbf{B}$	Magnetic flux density
B	Susceptance
B_C	Capacitive susceptance
B_L	Inductive susceptance
C	Capacitance
$\mathbf{D}$	Electric flux density
e	Electric potential—instantaneous value
$\mathbf{E}$	Electric field intensity
E	Electric potential—rms or d-c value
E_{av}	Electric potential—average value
E_m	Electric potential—maximum value of a sinusoid
f	Frequency
$\mathbf{F}$	Force
$\mathcal{F}$	Magnetomotive force
G	Conductance (or the conductive component of an admittance)
$\mathbf{H}$	Magnetic field intensity
i	Electric current—instantaneous value
I	Electric current—rms or d-c value
I_{av}	Electric current—average value
I_m	Electric current—maximum value of a sinusoid
Im	Imaginary part of a quantity
$\mathbf{j}$ or $\mathbf{J}$	Electric current density

Symbol	Quantity
j	Imaginary or 90° rotative operator ($\sqrt{-1}$)
L	Inductance
N	Number of turns
p	Power—instantaneous
P or P_{av}	Power—average
$\mathcal{P}$	Permeance
q	Electric charge—instantaneous value
q*	Electron charge
q_T	Electric charge—total of a system
Q	Electric charge—average
Q	Q factor of a reactance or selectivity factor for a tuned circuit
Q_{av}	Reactive power—average
R	Resistance or the resistive component of an impedance
$\mathcal{R}$	Reluctance
Re	Real part of a quantity
t	Time
T	Period of a periodic function
T	Temperature
T*	Time constant of an exponential quantity
v	Electric potential—instantaneous value
$\mathbf{v}$	Velocity—instantaneous
$\mathbf{v}_d$	Electron drift velocity
V	Electric potential—rms or d-c value
W	Energy or work
W_C	Energy in an electric field or stored in a capacitance

Symbol	Quantity	Symbol	Quantity
W_L	Energy in a magnetic field or stored in an inductance	ϵ_0	Permittivity of vacuum
X	Reactance	ϵ_r	Relative permittivity
X_C	Reactance—capacitive	θ^*	Phase angle—sinusoidal steady state
X_L	Reactance—inductive	λ	Flux linkage
Y	Admittance	μ	Permeability
Z	Impedance	μ_0	Permeability of vacuum
α	Resistance–temperature coefficient	μ_R	Relative permeability
ϵ	Base of Naperian system of logarithms	ρ	Resistivity
ϵ	Permittivity	σ	Conductivity
		ϕ	Magnetic flux
		ψ	Electric flux
		ω	Angular velocity

A bold roman capital letter such as **A** indicates a vector; a bold italic such as ***A***, a complexor or phasor quantity.

A bold capital letter followed by an asterisk indicates the conjugate of the vector, complexor, or phasor quantity denoted by the letter, as ***A**** is the conjugate of the phasor ***A***.

(t) sometimes follows a lower-case letter to emphasize that the letter represents an instantaneous quantity (a quantity that may vary with time).

CIRCUIT ANALYSIS

1

INTRODUCTION

The circuit concept is extremely useful in the solution of problems in all fields of engineering. This is particularly true in electrical engineering because the circuits in this area are usually so well defined that no doubt exists as to just what the circuit consists of. A study of circuit theory, therefore, might be considered as a prerequisite to the study of engineering systems in general and to electrical engineering systems in particular.

In any circuit study it is essential that something be known about the nature of the components of which the circuit is composed. This becomes even more evident if it is recognized that the circuit concept is actually just an engineering approximation of more fundamental concepts. For example, in the case of the electrical engineering circuit all of the behavior can be explained completely and accurately on the basis of electric and magnetic field theory, but the labor involved in carrying out the computation on this basis is so formidable that it is foolish to do so whenever the circuit approach is sufficiently accurate.

A thorough understanding of electrical phenomena must be based upon a knowledge of electric and magnetic fields, but it is not necessary to understand this theory completely in order to use circuit theory effectively. It is, however, desirable that an appreciation of the relationships between the two be obtained in order to determine the situations in which each is appropriate. In this text a brief introduction to electric and magnetic fields is followed by an analysis showing how circuit elements may be derived from these basic field concepts under conditions in which the field effects are restricted to a limited region. The method of presentation combines physical reasoning and mathematical development so that the electrical engineering student may obtain a clear picture of the basic aspects of engineering science.

1–1. The Electrical Structure of Matter. Before an attempt is made to understand electric and magnetic fields some knowledge must be secured of the electrical properties of materials. The physical properties of materials, such as mass, length, surface texture, and so on, are familiar to the student, but in addition to these we must also be concerned with the electrical properties of the material. The first knowledge of electricity was obtained when early investigators discovered certain phenomena that could not be described in terms of the then known physical properties of the system under consideration. For example, it was noted that if certain bodies were rubbed by other materials these bodies either attracted or repelled each other. These bodies were said to be "electrified" or "charged." As a result of these studies it was deduced that two types of electric charge existed, and at a later date the now used positive and negative designations were given to the two types. In general, electrical properties of materials may be described as caused by the effect of an additional physical property known as electric charge. In addition to the dimensions of mass, length, and time we now consider electric charge (usually symbolized by q or Q) as a basic dimension in the description

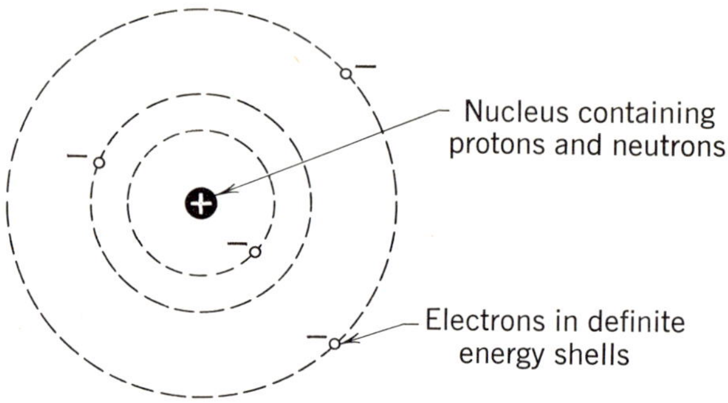

Fig. I–I. Simple atom model.

of a physical system. An understanding of this effect may be obtained from a brief review of some atomic and nuclear theory.

The theories of physics describe all matter as formed of molecules, which in turn are made up of atoms. The atom is a basic physical and electrical structure used to describe the properties of an element. Modern physics and chemistry recognize the existence of about 100 elements, which differ from each other by the basic composition of their atoms. For the purpose of this study, a very much simplified model of the atom will be described that permits a convenient explanation of the electrical properties of materials. It must be recognized that this picture of the atom is not complete, but is satisfactory for the purposes of this discussion.

Figure 1–1 illustrates this simplified model; it shows a number of electrons revolving in orbits (or shells) around a nucleus consisting of protons and neutrons. The electron is the smallest known electric particle with a mass at rest of 9.12×10^{-31} kg, a negative charge of 1.602×10^{-19} coulomb, and a radius in the order of 1.9×10^{-15} m. The proton is positively charged, also with 1.602×10^{-19} coulomb. The masses of the proton and neutron are nearly equal and the mass of each is approximately 1840 times that of the electron. Effectively, therefore, all the mass of the atom is concentrated in the nucleus. The atomic radius is very large compared with the size of the nucleus (for most metals the atomic radius is approximately 2×10^{-10} m or about 10 times as large as the nucleus).

The atom as a whole is normally electrically neutral; therefore, the number of protons of the nucleus must be equal to the number of electrons revolving around it. This number of protons, which in any element is fixed, characterizes the element and gives its atomic number. The remaining mass of the atom is made up of the neutrons in the nucleus; the number of neutrons for any one element varies and determines the isotopes of the atom. The combination of neutrons and protons gives the element its atomic weight. In the actual atom the spacing between the nucleus and the orbital electrons is very large in comparison with the size of the electrons and nucleus, and, although we may think of most materials as solid, they actually consist chiefly of space, certain points of which are occupied by nuclei and electrons. The configuration is often considered to be similar to that of the solar system.

The electron in its orbit around the nucleus may be described by principles of mechanics as having kinetic energy of rotation. For any given element and this postulated model of the atom, only certain discretely separated energy levels can exist. The electrons revolving in their orbits can therefore possess only a discrete amount of energy, and the orbit is said to exist in an energy shell. These electron shells are numbered outward from the nucleus in the order of increasing energy 1, 2, 3, $\cdots$, N, with the number of electrons per shell not exceeding a maximum of $2N^2$ electrons. There may exist in the first shell, therefore, 1 or 2 electrons, in the second shell up to 8 electrons, in the third shell up to 18, and so on.

The electrical characteristics of materials may be partially accounted for in terms of the distribution of electrons in these energy level shells. If the number of electrons in the outer shell is the maximum allowable, these electrons are closely bound to the atom by the intra-atomic forces. These bound electrons are practically the only type which exists in electrical insulating materials, commonly called *dielectrics*. If the outer shells, on the other hand, contain only one or two electrons, these are bound loosely to the atomic structure and are called *free electrons*. Materials having this arrangement are classified as conductors. Between the two extremes of strongly bound and free electrons (dielectrics or conductors) there exist materials which may be neither one nor the other, and whose characteristics may be controlled by their physical state. Semiconductors are one category of current interest. Although conduction, semiconduction, and insulation are explained in terms of shells and electrons around an atom, materials should be treated from a macroscopic point of view. When atoms are close together their combined fields alter the position of the orbital shells and cause them to form energy bands.

From the band point of view materials can be divided into valence regions, forbidden regions, and conduction regions as shown in Fig. 1-2. In insulators the valence region is separated from the conduction region by a wide forbidden region. An electron cannot move from the valence region into the conduction region unless it receives enough energy (about 7 ev) to take it across the forbidden region. The insulator then breaks down. In semiconductors the width of the forbidden region is small (0.7 ev for germanium and 1.1 for silicon). In conductors there is no separation between the valence region and the conduction region. Electrons can move from the valence to the conduction region with little or no energy applied. The electrons in the conduction band move at random when no outside field is applied.

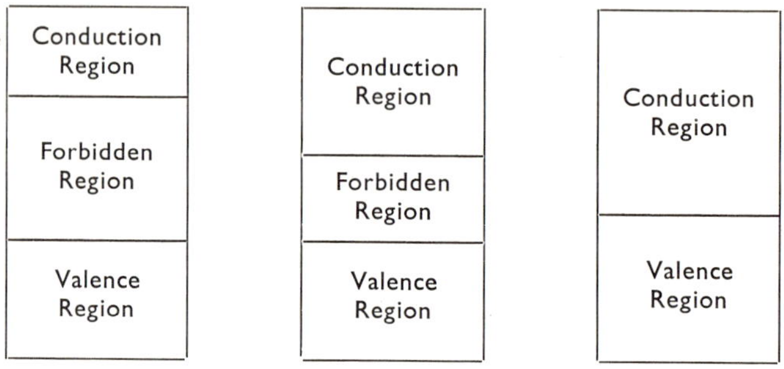

Fig. 1–2. Band regions.

1–2. Electric Charge Effects. If equal and opposite charges were to occupy the same space position, the net effect would be one of neutralization. Microscopically, that is, within the atom or molecule, this cannot occur, but an over-all view of the effect of many charges in a finite-sized material portion, such as is of interest to the engineer, indicates that this seems to be the normal state of matter. The engineer is primarily interested in a macroscopic view in which the dimensions involved are very, very large compared with the atomic dimensions.

A structural study of most solid matter shows that the atoms are arranged in a regular geometric pattern known as a *lattice structure*. If the matter under consideration is a conductor, after the valence bonds are formed electrons are left over and are free to move. These free electrons are continually in random motion within the conductor, but they also may be made to assume a net direction of motion by the application of external electric forces. Usually the electrons removed from a region will be replaced by electrons from an adjacent region; this movement of electrons is termed an *electric current*. If the net number of electrons leaving a given region exceeds the number of electrons arriving, however, the net charge in the region is positive and that portion of the conductor is termed *positively charged*. Similarly, a region in which the net number of electrons arriving is greater than the number normally present is termed *negatively charged*. In a dielectric the electrons are not free to move, but the dielectric may be made to assume a charged condition by distortion of the orbital electron paths caused by the application of an external electric force. The time-averaged positions of the electrons are then such that there may appear to be a net positive charge at one surface of the dielectric and a net negative charge at the other surface. A dielectric may also be forced to assume a permanently charged state by bringing another insulating material in rubbing contact with it (such as fur to rubber), which results in a spontaneous transfer of charge from one surface to the other. One dielectric will then have an excess of electrons and the other a deficiency after separation.

The electric charges described up to this point have been free electrons, bound charges, or positively charged atoms that have a shortage of free electrons. However, negative electric charges are sometimes found not associated with the atomic structure of a conductor or dielectric. Under certain circumstances electrons may become completely separated from any atomic structure. For example, in the electron tube there exists a charge flow through the tube consisting entirely of free electrons. These electrons have been emitted from a conducting surface after an amount of energy has been supplied to permit them to acquire kinetic energy to overcome the intra-atomic and intramolecular forces that normally bind them to the atoms of the material. The additional energy required to free these electrons may be provided in the form of heat energy, light radiation, or bombardment by other sources of electric charge.

In a gas an initially neutral atom may be caused to free electrons by a bombardment process involving other free charges. As a result, free negative charges and atoms or molecules having a net positive charge are obtained and the gas is said to be *ionized*.

Chemical action such as exists in a battery may also cause charge separation with a resultant accumulation of negative charges at one terminal and positive charges at the other. Electric generators employ magnetic forces to separate positive and negative charges in the generator conductors, and a result similar to that of the battery is obtained. The connection of external circuit elements to these charged terminals will then permit the flow of charges (current) through these circuit elements and through the source.

In summary, it may be said that the electrical engineer is concerned primarily with the total effect or action of electric charges appearing in several modes: as free charges in conductors, as bound charges in dielectric materials, and as free charges in a gas or vacuum in the form of electrons or positively charged atoms. The individual actions of the electric charges in all cases are governed by the same basic laws; the average effect differs, however, and these dissimilarities in the over-all character of the charge action are the basis for the different electrical phenomena. A conducting material has properties that are determined by the number of free electrons and the amount of energy dissipated as heat as these electrons are moved by external electric forces. In a dielectric the electrons in the atom shells are closely bound to their atomic nuclei over a broad range of electric forces, temperature, and conditions of mechanical stress, and the actions of these bound electrons determine the electrical properties of the material as an insulator. The free charges obtained by the dissociation of electrons from solids or liquids by heat, light, or electric or magnetic forces are necessary to the operation of devices such as electron tubes. These charge effects will be accounted for first in terms of electric field and magnetic field relationships and these relationships utilized to obtain the characteristics of the circuit elements.

1-3. Units and Dimensions. A unit is a standard for the numerical evaluation of the magnitude of some physical quantity. The measurement of any electrical or mechanical quantity involves a specified unit, and the result is expressed in terms of this unit together with a number indicating how

TABLE 1-1

Mks Dimensions and Units for Some Quantities Useful in Electrical Engineering

Quantity	Symbol	Unit	Dimensions
Mass	M	Kilogram	M
Length	L	Meter	L
Time	T	Second	T
Charge	Q, q	Coulomb	Q
Force	$\mathbf{F}$	Newton	MLT^{-2}
Energy	W	Joule	ML^2T^{-2}
Power	P	Watt	ML^2T^{-3}
Current	I, i	Ampere	QT^{-1}
Current density	$\mathbf{J}, \mathbf{j}$	Ampere/meter²	$QT^{-1}L^{-2}$
Potential	V, v, E, e	Volt	$ML^2T^{-2}Q^{-1}$
Resistance	R	Ohm	$ML^2T^{-1}Q^{-2}$
Resistivity	ρ	Ohm-meter	$ML^3T^{-1}Q^{-2}$
Capacitance	C	Farad	$M^{-1}L^{-2}T^2Q^2$
Permittivity	ϵ	Farad/meter	$M^{-1}L^{-3}T^2Q^2$
Inductance	L	Henry	ML^2Q^{-2}
Permeability	μ	Henry/meter	MLQ^{-2}
Electric field intensity	$\mathbf{E}$	Volt/meter	$MLT^{-2}Q^{-1}$
Electric flux density	$\mathbf{D}$	Coulomb/meter²	$L^{-2}Q$
Electric flux	ψ	Coulomb	Q
Magnetic field intensity	$\mathbf{H}$	Ampere/meter	$L^{-1}T^{-1}Q$
Magnetic flux density	$\mathbf{B}$	Weber/meter²	$MT^{-1}Q^{-1}$
Magnetic flux	ϕ	Weber	$ML^2T^{-1}Q^{-1}$
Reluctance	$\mathscr{R}$	Ampere/weber	$M^{-1}L^{-2}Q^2$
Magnetomotive force	$\mathscr{F}$	Ampere	QT^{-1}

many of the units exist in the measured quantity. A unit system is a collection of units that may be used together with a specified set of equations to indicate the magnitudes and the relations between the physical quantities involved in an engineering problem.

The dimensions required for this consistent set of units are expressed in terms of mass, length, time, and, in some systems, electric charge. This text will use the rationalized meter-kilogram-second (mks) system, unless otherwise noted. Table 1-1 tabulates some of the units and dimensions for the mks system that are useful in physics and engineering.

2

THE ELECTROSTATIC FIELD AND CAPACITANCE

The discussion of the preceding chapter brought out the fact that the electrical properties of a material are determined by the existence of electrically charged particles within its atomic and molecular structure. This internal behavior of any material was considered in a microscopic sense and found capable of being conveniently explained by a model atom structure. In this chapter the interest is in the external behavior of such materials when considered macroscopically, that is, the characteristics of a finite amount of material.

Although charge-bearing particles within the materials are moving in random motion, a statistical average may be taken and the average position of the charges considered for the material as a whole. If this average position is fixed with regard to an outside observer, the charges are said to be static and the study of the effects of charges on this basis is termed *electrostatics*. In later chapters the effect of the gross motion of charges will be separately considered during the study of resistance and the magnetic field.

2-1. Coulomb's Law. Coulomb experimentally discovered that the force of attraction or repulsion between two point charges at rest when in a homogeneous, isotropic medium is given by

$$\mathbf{F} = \frac{q_1 q_2}{4\pi\epsilon_r\epsilon_0 r^2} = \frac{q_1 q_2}{4\pi\epsilon r^2} \tag{2-1}$$

where $\mathbf{F}$ is the force in newtons

q_1 and q_2 are the charges in coulombs

r is the distance between charges in meters

ϵ_0 is the permittivity constant of free space (or air) and has a value of

$$\epsilon_0 = \frac{1}{36\pi \times 10^9} \quad \text{coulombs}^2 \text{ per newton-meter}^2$$

ϵ is the permittivity of the medium in coulombs^2 per newton-meter^2, and is also called the *dielectric constant*

ϵ_r is the relative permittivity of the medium to the permittivity of free space

$$\epsilon_r = \frac{\epsilon}{\epsilon_0}$$

Equation 2–1 is called *Coulomb's law* and is applicable to situations in which the charge dimensions are very small (sometimes called point charges) compared with the spacing between them.

The force **F** is a vector, that is, it has a magnitude and a direction. The magnitude of the force **F** is given by Eq. 2–1. The direction of the force is along the line joining the two charges. The force is a force of attraction if the

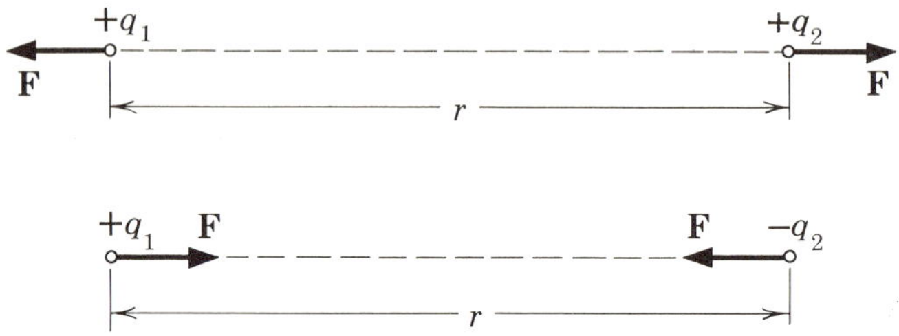

Fig. 2–1. Forces between charged bodies.

charges are dissimilar, namely, one positive and the other negative. It is a force of repulsion when the two charges are of similar nature, namely, both positive or both negative (see Fig. 2–1).

Example 2–1. Find the force between two positive charges of 5×10^{-5} and 15×10^{-7} coulomb when placed 30 cm apart in air (ϵ_r for air is 1).

By Eq. 2–1 the force is found to be:

$$\mathbf{F} = \frac{5 \times 10^{-5} \times 15 \times 10^{-7}}{4\pi[1/(36\pi \times 10^9)] \times 1 \times (0.3)^2}$$
$$= 7.5 \text{ newtons}$$

The direction of the force is along the line $q_1 - q_2$ and is a force of repulsion.

2–2. Forces on More Than Two Charges. If more than two charges are present, the total force may be found by considering the charges in pairs as if the other charges did not exist. The force of attraction or repulsion between each pair is determined from Coulomb's law, and the total force on each charge is determined by adding vectorially the forces due to all the other charges on the charge under consideration. This is an application of the principle of superposition, which will be considered in detail at a later time.

Example 2–2. Three similar concentrated charges, each of 10^{-6} coulomb, are placed at the vertices of an equilateral triangle of 0.03 m side as shown in Fig. 2–2. Find the magnitude and direction of the force exerted on one of the charges.

The force on q_3 caused by charge q_1 is:

$$\mathbf{F}_{3-1} = \frac{10^{-6} \times 10^{-6}}{4\pi[1/(36\pi \times 10^9)](0.03)^2} = 10 \text{ newtons}$$

This force is a repulsion force along the line $q_1 - q_3$. The force $\mathbf{F}_{3-2}$ on q_3 caused by q_2 is also 10 newtons. It is a force of repulsion along the line $q_2 - q_3$. The total force $\mathbf{F}_T$ is the *vector* sum of $\mathbf{F}_{3-2}$ and $\mathbf{F}_{3-1}$:

$$\mathbf{F}_T = \sqrt{(10 + 10 \cos 60)^2 + (10 \sin 60)^2} = 17.32 \text{ newtons}$$

The direction of the force is along a line 30° below the horizontal.

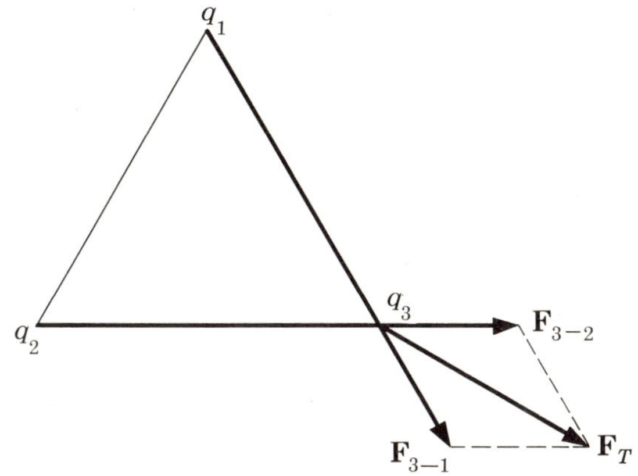

Fig. 2-2. Forces exerted on charge q_3 by charges q_1 and q_2.

2-3. Field Concept. Equation 2-1 gives the value of the force between two concentrated charges. As stated previously, this equation is obtained from results of experimentation. No one knows why a force exists, nor has anyone proved the physical existence of the mechanisms by which this force is transmitted. However, to explain the method of transmission of the force the electric field concept was "hypothesized."

By *definition* an electric field is said to exist in a region if a force is experienced by a static test charge when introduced in the region. Usually the test charge is a small positive charge of such magnitude that it has no effect on the field. From this explanation we see that the electric field is something created by definition.

The field intensity is defined as the ratio of the force experienced by the test charge to the test charge. Mathematically the field intensity is given by

$$\mathbf{E} = \frac{\mathbf{F}}{q_T} \tag{2-2}$$

where $\mathbf{E}$ is the electric field intensity in newtons per coulomb or volts per meter

$\mathbf{F}$ is the force in newtons

q_T is the charge in coulombs of the test charge

As the force $\mathbf{F}$ has magnitude and direction, the field intensity $\mathbf{E}$ is also a vector. It has a magnitude given by Eq. 2–2. Its direction is the direction of the force on the test charge.

EXAMPLE 2–3. A positive charge of 10^{-10} coulomb when placed at a certain point experiences a force of 10^{-8} newton. The direction of the force in cartesian x, y, z coordinates is along the x axis. Find the value of the field intensity at the point under consideration.

By Eq. 2–2,

$$\mathbf{E} = \frac{10^{-8}}{10^{-10}} = 100 \text{ v/m or newtons/coulomb}$$

The direction of $\mathbf{E}$ is along the x axis.

2–4. Field Intensity of a Point Charge. The combination of the experimental results expressed by Eq. 2–1 and the field intensity defined by Eq. 2–2 gives:

$$\mathbf{E} = \frac{q_1 q_2}{4\pi\epsilon r^2 q_2} = \frac{q_1}{4\pi\epsilon r^2} \quad \text{volts per meter} \tag{2–3}$$

2–5. Field Intensity of Several Point Charges. The field intensity at a point caused by several point charges is the vector sum of the field intensities, at the point, of the several charges considered individually.

EXAMPLE 2–4. Two charges, each of 10^{-9} coulomb, are placed at the two vertices of an equilateral triangle of 10 cm side, as in Fig. 2–3. What is the field intensity at the third vertex if the charges are positive?

$$\begin{aligned}\mathbf{E}_{3-1} &= \frac{10^{-9}}{4\pi[1/(36\pi \times 10^9)](0.1)^2} \\ &= 900 \text{ along a line making } -60^\circ \text{ with } x \text{ axis}\end{aligned}$$

$$\begin{aligned}\mathbf{E}_{3-2} &= \frac{10^{-9}}{4\pi[1/(36\pi \times 10^9)](0.1)^2} \\ &= 900 \text{ along } x \text{ axis}\end{aligned}$$

$$\begin{aligned}\mathbf{E} &= \sqrt{(900 + 900 \cos 60)^2 + (900 \sin 60)^2} \\ &= 1558.8 \text{ v/m and makes } -30^\circ \text{ with } x \text{ axis}\end{aligned}$$

2–6. Electric Flux and Flux Density. The field intensity as given by Eq. 2–3 depends upon the medium of permittivity. It is desirable to have an entity that is independent of the medium. To that end the terms *electric* or *dielectric flux* and *flux density* are introduced. It is assumed that a positive charge q emanates q flux lines or tubes. The flux is uniformly distributed around the charge when isolated. The flux lines or tubes start from a positive charge and end on a negative charge. An isolated point charge emanates q

lines ending at infinity. The flux density is the number of flux lines per square meter.

Assume a point charge q. The flux density at a point r meters from q is by definition

$$\mathbf{D} = \frac{q}{4\pi r^2} \tag{2–4}$$

where $\mathbf{D}$ is the flux density at a point r meters from q in coulombs per square meter

r is the distance in meters

From Eqs. 2–3 and 2–4 we get

$$\mathbf{D} = \epsilon_r \epsilon_0 \mathbf{E} = \epsilon \mathbf{E} \tag{2–5}$$

The flux density is also a vector having the same direction as $\mathbf{E}$.

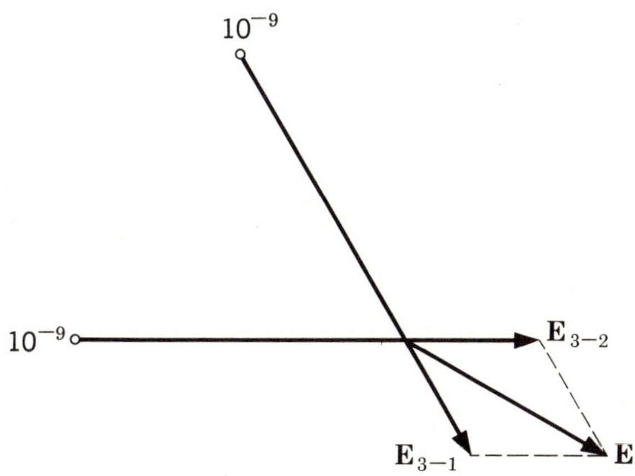

Fig. 2–3. Field intensity at one vertex of an equilateral triangle caused by charges on the two other vertices.

Example 2–5. Find the flux density and the field intensity around a point charge of 5×10^{-9} coulomb at a distance of 10 cm in air.

From Eq. 2–4

$$\mathbf{D} = \frac{5 \times 10^{-9}}{4\pi(0.1)^2} = \frac{5}{4\pi} 10^{-7} \text{ coulomb per square meter}$$

Since ϵ_r for air is 1, the field intensity is:

$$\mathbf{E} = \frac{5 \times 10^{-7}}{4\pi[1/(36\pi \times 10^9)]} = 4500 \text{ v/m}$$

2–7. Conductor in an Electric Field. If a conductor is placed in an electric field, as in Fig. 2–4, the free electrons move toward one end of the conductor, leaving the other end positively charged. The motion of the electrons ceases when inside the conductor the electric field produced by the positive charges and the separated negative charges is equal and opposite to the applied field. Under this condition no field exists inside the conductor although the applied field outside the conductor still exists.

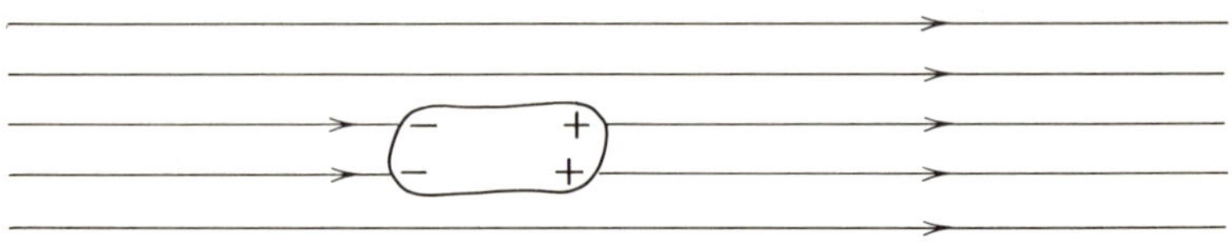

Fig. 2–4. Disruption of electric lines by a conductor.

2–8. Insulator in an Electric Field. When an insulator is placed in an electric field there is no separation of charges in the body of the insulator. A few flux lines of the field may stop at the free electrons on the surface of the insulator but an electric field will exist inside the insulating material. The molecules of the insulator, however, rotate slightly under the influence of the applied electric field. The rotation is in such a direction as to produce a field inside the insulator opposing the applied field. The field inside the insulator is less than the applied field.

Assume a charge q in air or free space. The field intensity at a point r meters away is given by:

$$\mathbf{E}_0 = \frac{q}{4\pi\epsilon_0 r^2} \tag{2–6}$$

Assume the same charge imbedded in a homogeneous dielectric. The field intensity at a point r meters from the charge is given by

$$\mathbf{E} = \frac{q}{4\pi\epsilon r^2} \tag{2–7}$$

where ϵ is the permittivity of the dielectric. The ratio of $\mathbf{E}$ to $\mathbf{E}_0$ is

$$\frac{\mathbf{E}}{\mathbf{E}_0} = \frac{\epsilon_0}{\epsilon} = \frac{1}{\epsilon_r} \tag{2–8}$$

where ϵ_r is the relative permittivity of the dielectric.

The field intensity $\mathbf{E}$ in a dielectric material is the resultant of the applied field intensity $\mathbf{E}_0$ and the field $\mathbf{E}_d$ caused by the partial alignment of the molecule to oppose the applied field, or

$$\mathbf{E} = \mathbf{E}_0 - \mathbf{E}_d \tag{2–9}$$

This is a vector subtraction.

If $\mathbf{E}_0$ is normal to the surface of the dielectric, then:

$$\mathbf{E} = \frac{\mathbf{E}_0}{\epsilon_r} \tag{2-10}$$

From Eqs. 2–9 and 2–10 we get

$$\frac{\mathbf{E}_0}{\epsilon_r} = \mathbf{E}_0 - \mathbf{E}_d$$

and

$$\mathbf{E}_d = \mathbf{E}_0 - \frac{\mathbf{E}_0}{\epsilon_r} = \mathbf{E}_0\left(1 - \frac{1}{\epsilon_r}\right) \tag{2-11}$$

The value of ϵ_r for some materials is given in Table 2–1.

2–9. Gauss's Theorem. If the field intensity at a point is known the flux density at that point is obtained from Eq. 2–5. The field intensity may be the result of several charges; nevertheless Eq. 2–5 gives the flux density due to the several charges.

If the field intensity at every point on a closed surface is known, it is possible to find the net electric charge enclosed by the closed surface. The value of the net charge inside the surface is given by*

$$q = \oint_s \mathbf{D} \cdot \mathbf{dA} = \oint_s D_n \, dA \tag{2-12}$$

where $\oint_s$ means integration over all the closed surface

D_n is the value of the flux density normal to the surface in coulombs per square meter

dA is an element of area in square meters, and is a vector normal to the surface

Equation 2–12 states that the total flux emanating from any closed surface is equal to the *net* charge enclosed by the surface. This is Gauss's theorem.

Caution should be exercised in the application of Eq. 2–12. A positive **D** is directed from the surface to the outside. A negative **D** is directed from the outside to the surface.

Example 2–6. Find the field of an infinitely long straight conductor charged uniformly with λ coulombs per meter of length, assuming the diameter of the wire to be very small.

Imagine a cylinder L meters long with a radius r around the wire, as in Fig. 2–5. The charge on the wire enclosed by the cylinder is λL coulombs. This charge emanates λL flux lines. The flux density at the surface of the cylinder is

$$\mathbf{D} = \frac{\lambda L}{2\pi r L} = \frac{\lambda}{2\pi r}$$

* $\mathbf{D} \cdot \mathbf{dA}$ is called a dot product. Its value is given by the product of D, dA, and the cosine of the angle between **D** and **dA**.

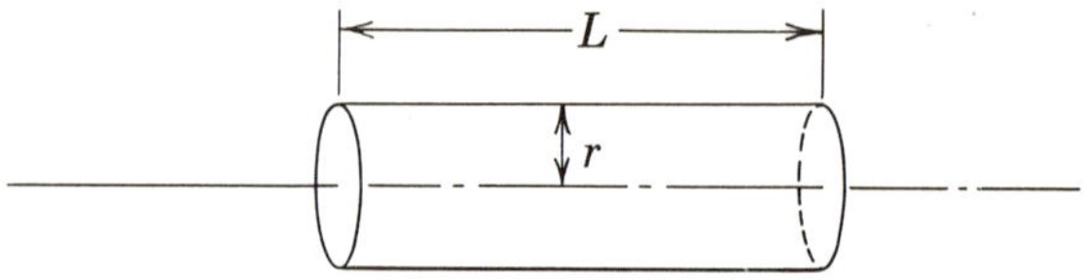

Fig. 2–5. Cylinder around a charged wire.

The field intensity is

$$\mathbf{E} = \frac{\mathbf{D}}{\epsilon} = \frac{\lambda}{2\pi r \epsilon_r \epsilon_0}$$

2–10. Potential Energy. To introduce the concept of electric potential energy in a simple manner, consider a positive isolated point charge of q coulombs set at a fixed point in space. If a positive test charge q_T is introduced in the field of q a force is exerted on q_T. This force tends to push q_T out of the field of q.

To bring q_T to a distance r from the charge q work must be done *on* q_T. If q_T is permitted to move freely it tends to move away from the field of q and goes to infinity where the field intensity of q is zero. The work done on q_T to bring it to a distance r from q is stored as potential energy.

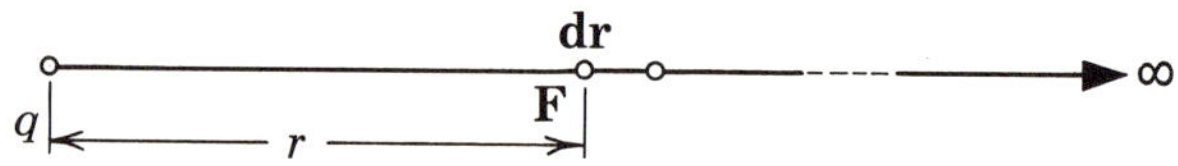

Fig. 2–6. Figure to determine potential energy at a point.

To determine the value of the potential energy at any point r meters from q, proceed as follows: Let the force on q_T at r be $\mathbf{F}$. Assume that q_T moves an infinitesimal distance $\mathbf{dr}$ in the direction of $\mathbf{F}$ under the influence of $\mathbf{F}$, as shown in Fig. 2–6. The work done by the system is then:

$$dw = \mathbf{F} \cdot \mathbf{dr} \tag{2–13}$$

If $\mathbf{dr}$ is positive the work is done *by* the system. This amount of work is equal to the *decrease* in potential energy of the system. Therefore

$$-du = \mathbf{F} \cdot \mathbf{dr}$$

or

$$\mathbf{F} = -\frac{du}{\mathbf{dr}} \tag{2–14}$$

where u is the potential energy of the system at r.

The total potential energy of the system at r can be determined by finding the amount of work that q_T can perform in going from r to a point where the field intensity is zero and thus **F** is zero. This is obtained by integrating Eq. 2–13 thus:

$$w = \int_r^\infty \mathbf{F} \cdot \mathbf{dr} \tag{2–15}$$

But

$$\mathbf{F} = \frac{q_1 q_2}{4\pi\epsilon r^2}$$

Therefore:

$$w = \frac{q_1 q_2}{4\pi\epsilon} \int_r^\infty \frac{dr}{r^2} = \frac{q_1 q_2}{4\pi\epsilon r} \tag{2–16}$$

The work done is equal to the drop in potential energy. Therefore the potential energy at r is:

$$U_r = \frac{q q_T}{4\pi\epsilon r} \quad \text{newton-meters or joules} \tag{2–17}$$

Notice that potential energy is *not a vector*.

2–11. Potential. The potential energy per unit test charge is called the *electric potential at a point.* This is mathematically expressed as:

$$V_r = \frac{U_r}{q_T} \quad \text{joules per coulomb} = \text{volts} \tag{2–18}$$

For a point charge this gives:

$$V_r = \frac{q}{4\pi\epsilon r} \tag{2–19}$$

From usage the electric potential at a point is called its *voltage.* Notice that the potential at a point *is not a vector.*

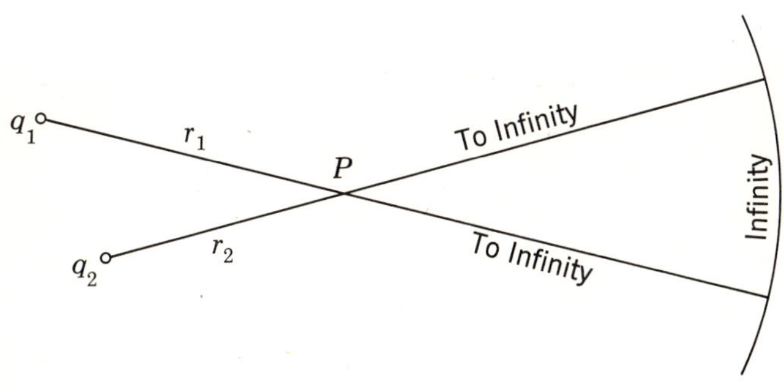

Fig. 2–7. Potential energy of two charges.

2–12. Potential of Several Charges. Assume two positive charges q_1 and q_2 and let it be desired to find the potential at P, r_1 meters from q_1 and r_2 meters from q_2 (Fig. 2–7).

The potential at P due to q_1 alone is:

$$V_{1P} = \frac{q_1}{4\pi\epsilon r_1} \tag{2–20}$$

The potential at P due to q_2 alone is:

$$V_{2P} = \frac{q_2}{4\pi\epsilon r_2} \tag{2–21}$$

The total potential at P is therefore:

$$V_P = V_{1P} + V_{2P} = \frac{q_1}{4\pi\epsilon r_1} + \frac{q_2}{4\pi\epsilon r_2} \tag{2–22}$$

If there are n charges the potential at P will be:

$$V_P = \sum_{i=1}^{n} \frac{q_i}{4\pi\epsilon r_i} \tag{2–23}$$

EXAMPLE 2–7. Assume two positive charges q_1 and q_2, each of 10^{-10} coulomb, placed 0.05 m and 0.10 m respectively from P. Find the potential at P.

The potential at P due to q_1 is:

$$V_{1P} = \frac{10^{-10}}{4\pi[1/(36\pi \times 10^9)](0.05)} = 18 \text{ v}$$

The potential at P due to q_2 is:

$$V_{2P} = \frac{10^{-10}}{4\pi[1/(36\pi \times 10^9)](0.1)} = 9 \text{ v}$$

The total potential at P is:

$$V_P = 18 + 9 = 27 \text{ v}$$

2–13. Field Intensity and Potential. From Eq. 2–14 we get

$$\frac{\mathbf{F}}{q_T} = -\frac{du}{q_T\,dr}$$

or

$$\mathbf{E} = -\frac{dV}{dr} \tag{2–24}$$

Notice that dV/dr is a vector.

From Eq. 2–24 we see that if the potential at a point is given, the field intensity at that point can be found. As a matter of fact, by the use of Eq. 2–23 we get

$$\mathbf{E}_i = -\frac{\partial V_P}{\partial r_i} \tag{2–25}$$

Assume, for instance, that V_P is given by Eq. 2–22 for two charges; then:

$$-\frac{\partial V_P}{\partial r_1} = \frac{q_1}{4\pi\epsilon r_1^2} = \mathbf{E}_1 \text{ in direction of } r_1 \tag{2–26}$$

$$-\frac{\partial V_P}{\partial r_2} = \frac{q_2}{4\pi\epsilon r_2^2} = \mathbf{E}_2 \text{ in direction of } r_2 \tag{2–27}$$

The total field intensity is:

$$\mathbf{E} = -\frac{\partial V_P}{\partial r_1} - \frac{\partial V_P}{\partial r_2} \quad \text{vectorially} \tag{2-28}$$

2-14. Difference in Potential. The potential at a point is not as important as the difference of the potential between two points. If the potential at a is V_a and at b is V_b, then:

$$V_a - V_b = V_{ba} = -V_{ab} = -\int_b^a \mathbf{E} \cdot \mathbf{dr} \tag{2-29}$$

However, because of usage a voltage drop is taken as positive, hence:

$$V_a - V_b = V_{ab} = \int_b^a \mathbf{E} \cdot \mathbf{dr} \tag{2-30}$$

It should be kept in mind that for an electrostatic field

$$V_{ab} + V_{ba} = 0$$

or

$$\oint \mathbf{E} \cdot \mathbf{dr} = 0 \tag{2-31}$$

where $\oint$ means contour integration, and $\mathbf{E} \cdot \mathbf{dr}$ a dot product.

From Eq. 2-31 we see that the potential between two points in an electrostatic field does not depend upon the path of integration but upon the value of the integral at the end points. In other words, it is a function of the points a and b, the starting and end points.

2-15. Storage of Electric Potential Energy. Mechanical potential energy can be stored in springs under tension or compression, in lifted weights, in gases under pressure, and so forth. Electric potential energy can be stored by bringing two similar charges near each other or by separating two dissimilar charges. To recover the potential energy and convert it into work the charges should be permitted to move away from each other when they are of the same sign or toward each other when they are of opposite sign.

A better and more practical way to store potential energy is by the use of two fixed conductors with insulation between them. If some of the electrons are taken from one conductor and added to the other, an electric field will be established between the conductors and potential energy will be stored in the electric field.

Assume two parallel plate conductors. Let them be separated by some insulating medium. If some electrons are taken away from one plate and added to the other, the plates are said to be charged. Let one plate have a positive charge of q coulombs and the other a negative charge of q coulombs. Let the distance l between the plates be small compared with their areas. Electric flux lines of q coulombs emanate from the positive plate and terminate on the negative plate. The flux density at any point between the plates is then:

$$\mathbf{D} = \frac{q}{A} \quad \text{coulombs per square meter} \tag{2-32}$$

The field intensity is:

$$\mathbf{E} = \frac{\mathbf{D}}{\epsilon} \quad \text{volts per meter} \tag{2–33}$$

The total voltage between plates is then:

$$V = \mathbf{E} \cdot \mathbf{d} = \frac{ql}{A\epsilon} \quad \text{volts} \tag{2–34}$$

Let a small charge δq be transferred from one plate to the other. The work done on δq is

$$\delta w = V\, \delta q \tag{2–35}$$

But

$$V = \frac{ql}{A\epsilon}$$

Therefore:

$$\delta w = \frac{lq\, \delta q}{A\epsilon} \tag{2–36}$$

The total work done in charging the capacitor* from 0 to q coulombs is:

$$w = \frac{l}{A\epsilon} \int_0^q q\, dq = \frac{l}{A\epsilon} \frac{q^2}{2} \tag{2–37}$$

This is the amount of stored potential energy in the field between the plates.

2–16. The Capacitance C. Equation 2–34 gives the value of the voltage of a parallel plate capacitor. The ratio of the charge to the voltage is called *capacitance*. For the parallel plate case:

$$C = \frac{q}{V} \quad \text{farads} \tag{2–38}$$

$$= \frac{A\epsilon}{l} \quad \text{farads} \tag{2–39}$$

In general, the value of the capacitance depends upon the permittivity of the medium between the conductors and the geometric configuration of the conductors. Any two conductors separated by an insulating medium form a capacitor and serve as a means to store potential energy.

To find the capacitance of a capacitor, assume a charge q on one conductor and $-q$ on the other. Find the flux density at any point r between the conductors. From the flux density at r find the field intensity $\mathbf{E}$ at r. Assume an infinitesimal distance $\mathbf{dr}$ along r. The potential difference along dr is $\mathbf{E} \cdot \mathbf{dr}$. The total voltage across the capacitor is:

$$V = \int \mathbf{E} \cdot \mathbf{dr}$$

Knowing q and V, we can find the capacitance C.

* Also called condenser.

EXAMPLE 2–8. Find the capacitance per unit length of a concentric cable.

Let the radius of the inner conductor be a and that of the outer one be b, as in Fig. 2–8.

1. Assume a charge q per unit length on one conductor and $-q$ on the other.
2. The flux between conductors is q lines.
3. Assume a point in the dielectric at a radius r.
4. The flux density at r is:

$$\mathbf{D} = \frac{q}{2\pi r \times 1}$$

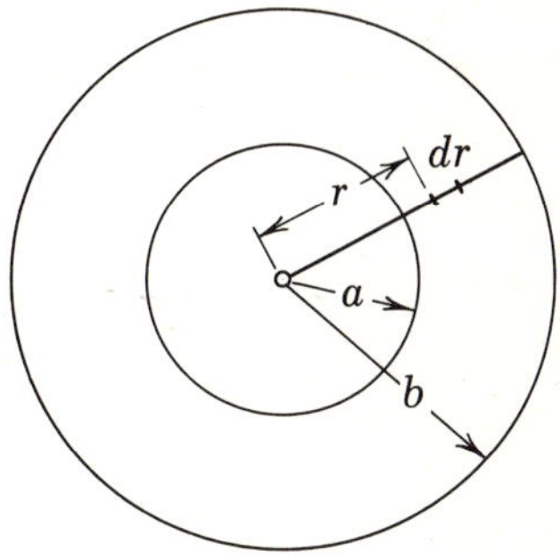

Fig. 2–8. Concentric cable.

5. The field intensity at r is:

$$\mathbf{E} = \frac{\mathbf{D}}{\epsilon} = \frac{q}{\epsilon 2\pi r \times 1}$$

6. The voltage drop across a small increment dr is:

$$dV = \mathbf{E} \cdot \mathbf{dr} = \frac{q\,dr}{\epsilon 2\pi r}$$

7. The voltage difference between the conductors is:

$$V = \int \mathbf{E} \cdot \mathbf{dr} = \int_a^b \frac{q}{\epsilon 2\pi} \frac{dr}{r}$$

$$= \frac{q}{\epsilon 2\pi} \ln \frac{b}{a}$$

8. The capacitance per unit length is:

$$C = \frac{q}{V} = \frac{\epsilon 2\pi}{\ln (b/a)} \quad \text{farads}$$

2–17. Field Intensity and Breakdown. If the field intensity across any part of a dielectric exceeds a certain value called *dielectric strength*, the insulation breaks down and becomes a conductor. There is therefore a limit to the amount of potential energy that can be stored in a given size capacitor. The dielectric strengths of some common dielectric materials are found in Table 2–1.

2–18. The Dielectric Circuit. When the electric flux is confined within a bounded path, that path becomes the dielectric circuit. Under this condition the charge q on the conductors sets q lines of force or coulombs through the circuit. From the configuration of the circuit the flux density **D** can be computed. The field intensity **E** can be determined and the voltage between the conductors can thus be computed.

TABLE 2–1

Dielectric Constants and Dielectric Strengths for Some Typical Engineering Materials

Material	Relative Permittivity	Dielectric Strength (kv/m)
Air	1.0	3,000
Bakelite	5.0	21,000
Glass	5.4–9.9	35,000
Gutta-percha	2.78–4.9	14,000
Mica	4.5–7.5	50,000
Nylon	3.7	16,000
Polyethylene	2.26	48,000
Polystyrene	2.6	30,000
Porcelain	4.5–6.9	10,000
Pyanol	5.0	20,000
Rubber	2.3–4.0	70,000
Teflon	2.1	60,000
Transil oil	2.5	10,000

Frequently, however, for such circuits the capacitance can be found from the configuration of the circuit and its dielectric material by formulas derived in handbooks. Such devices are called capacitors.

PROBLEMS

2–1. If two point charges of 1 coulomb each are placed 1 m apart, what will be the force between them?

2–2. If two point charges of 0.05 and 0.04 coulomb are placed 1 ft apart, what will be the force between them in pounds?

2–3. A charge of 10^{-6} coulomb is in a field of 10,000 v/m. What force does it experience?

2–4. An isolated copper sphere of radius 0.1 m is deprived of 10^{12} electrons. Find the field intensity (a) at its surface and (b) at a point 1 m distant from its center.

2–5. Two charges placed at A and B are 10 in. apart. Charge A is positive and has 10^{-6} coulomb; B is negative and has 2×10^{-6} coulomb. At point C 8 in. from A and 6 in. from B a positive charge C of 10^{-8} coulomb is located. Find the force and direction of the force on charge C, on charge A, and on charge B.

2–6. Find the field intensity at a distance of 1 ft from a 1-μcoulomb charge. What is the force on an electron placed at this point?

2–7. An electron placed in a uniform electric field is acted upon by a force of $(1.602)^2 \times 10^{-16}/5$ newton. (a) Calculate the field intensity. (b) If the charge is moved against the field 5 m, what is the potential change between the end points of the path?

2–8. The potential of an isolated sphere of 5-cm radius is 10,000 v. Find the charge on the sphere, the electric field intensity near the surface of the sphere, and the electric field intensity and the electric potential at a distance 20 cm from the center of the sphere.

2–9. A hollow conducting sphere has a radius of 5.0 cm and is uniformly charged with -20×10^{-6} coulomb. What is the value of the electric flux density inside the sphere, expressed as a function of the radius from the center of the sphere?

2–10. The potential field distribution for a given electrode arrangement may be expressed as

$$V = \frac{3Q}{4}(e^{-x} + e^{-y})$$

where Q is the system charge and x and y are the general dimensions. What is the capacitance between electrodes?

2–11. A parallel plate of area 100 sq cm for a certain capacitor has two dielectrics having constants of 4 and 8. The thickness of each dielectric is the same. One hundred volts potential is applied to the combination and a charge of 10^{-4} coulomb appears on the plates. What is the electric flux density through each dielectric and how much voltage appears across each?

2–12. A parallel plate capacitor has plates of 400 sq cm in area which are 0.5 cm apart. Find its capacitance. If the capacitor is charged to 1000 v, find the charge in coulombs and the energy stored.

2–13. A parallel plate capacitor has plates of 100 sq cm area which are 0.5 cm apart. Find its capacitance. To what voltage can the capacitor be charged before the air breaks down?

2–14. Bakelite is substituted for air in the capacitor of Prob. 2–13. Find its capacitance and breakdown voltage.

2–15. The distance between the plates of the capacitor of Prob. 2–14 is increased to 0.6 cm. The dielectric material consists of Bakelite with 0.5-cm thickness placed adjacent to one plate, and air of 0.1-cm thickness. To what voltage can the capacitor be charged without a breakdown?

2–16. Two capacitors of 5×10^{-6} farad and 1×10^{-6} farad are put in parallel across a 100-v source. What is the total capacitance of the system? What is the charge on each capacitor? *Hint:* Note that the capacitance is proportional to the surface area of the plates.

2–17. Two parallel plates are 1 in. apart. A positive charge of 3×10^{-9} coulomb is uniformly spread over one plate. An equal negative charge is on the other plate. (a) If all the electrostatic flux lines are straight across the 1-in. spacing and uniformly spread out over the area, what is the flux density? (Each plate is 1 ft by 2 ft.) (b) What is the electric field intensity? (c) What is the force on an electron halfway between the plates? (d) Compute the capacitance.

3

ELECTRIC CURRENT AND RESISTANCE

The preceding chapter dealt with the behavior of static (stationary average position) charges and the resultant static electric field within dielectric materials. The bound charges in the dielectric material were displaced from their neutral position, but were not free to move. If, on the other hand, such an electric field is established within a conducting material, the free electrons in the conductor will move under the influence of the field, and the resultant electron flow is termed *electric current.* (The atomic and molecular structure of the solid conductor through which the free electrons travel is fixed in position and therefore only the free electrons can move and constitute conduction current.) This movement of free electrons is from point to point within the conducting material, and the obstructions to this motion will be discussed as determining the resistance of the conductor.

3–1. Electric Current. If a solid conductor, such as a length of wire, is connected to a potential source (that is, a battery or electric generator), an electric field is established along the wire. The free electrons of the conducting material are moved toward the positive terminal of the energy source because of the force exerted on them by the electric field. (The positive charges in the conductor are also influenced by a force toward the negative terminal of the source but cannot move because of their fixed position in the lattice.) This motion of free electrons is defined as current by the relationship

$$i = \frac{q_T}{t} \tag{3–1}$$

where q_T is the total charge passing a chosen point in the time t.

In differential notation

$$i = \frac{dq_T}{dt} \tag{3–2}$$

or, electric current is the rate of flow of charge with respect to time. The unit of current in the mks system is the ampere. An ampere is the flow of charge at a rate of 1 coulomb/sec.* In terms of the foregoing discussion, it has been

* It is commonly stated in the literature that current "flows through a conductor." Of course, it is actually the charge that flows and the current that exists within the conductor. The "current flow" notation is, however, so commonly accepted that it may be used in this material from time to time.

noted that the current in a conductor is actually the flow rate of negative charges. Since, however, a knowledge of electric current preceded an understanding of the mechanism of current flow, an early assumption was made that current was a movement of positive charges. This is the basis for the assigned direction of *conventional* current, where the current is considered to flow externally in the circuit from the positive terminal to the negative terminal of the potential source. The concept of the electron current direction has been adopted in some modern literature, but as long as it is clearly understood which convention is being used it is unimportant which is employed for the purpose of calculation.

3–2. Current Density. Consider a solid conductor of uniform cross section of A square meters having n free electrons per cubic meter of material uniformly distributed throughout the volume. If the charge per electron is q coulombs, then the charge density of these free electrons per cubic meter is nq. In the absence of an applied external electric potential the motion of the electrons in this charge region will be purely random in nature, and no net flow of charge will be observed. If, however, an electric field is established along the conductor, an additional component of electron motion will be induced, and the conductor charge flow may be considered to have an average velocity (or drift velocity) of v_d meters per second along the conductor. The charge in coulombs passing through the cross section A per second is then nqv_dA. By definition the rate of charge flow is the current in amperes, or

$$i = nqv_dA \tag{3–3}$$

where i is the symbol designating current.

Current density is defined as the amount of current per unit of cross-sectional area. Since the average electron velocity induced by the external electric field is in the direction of the conductor, the current density through a cross-sectional area A, perpendicular to the conductor, may be expressed as

$$\mathbf{j} = \frac{i}{A} = nq\mathbf{v}_d \tag{3–4}$$

where $\mathbf{j}$ is the symbol for current density. The mks units for current density are amperes per square meter.

3–3. Resistivity. The free electrons within the conductor were considered to have an average drift velocity caused by an electric field $\mathbf{E}$ established within the conductor. In addition to this drift velocity the electron also experiences a random motion as a result of intra-atomic and interatomic and molecular forces within the conducting material. The net effect of the external electric force, however, is a motion of the electron in the direction of the conductor. The path of this electron is impeded by other electrons, atomic nuclei, and so on, and the electron motion in the direction of the conductor is inhibited by many collisions. The actual distance traveled before such a collision is in all probability very small. The energy which an electron gains through acceleration to a certain velocity will be partially or wholly surrendered to the particle with which it collides, and the energy given up will appear in the form of heat. After the collision the electron will again be accelerated and continue to move along the conductor. The actual length of path traveled may not be specifically defined for one electron, but a statistical average may be obtained for the conducting material as a whole.

The conductor characteristics may be established as follows: If the force on an individual electron of charge q in an electric field of value $\mathbf{E}$ is expressed as

$$\mathbf{F} = \mathbf{E}q \tag{3–5}$$

the electron may be considered to be accelerated according to Newton's second law

$$\mathbf{F} = m\mathbf{a} \tag{3–6}$$

where m is the mass of the electron in kilograms and $\mathbf{a}$ is the acceleration in meters per second squared caused by the electric field. Combining Eqs. 3–5 and 3–6,

$$\mathbf{a} = \frac{\mathbf{E}q}{m} \tag{3–7}$$

This acceleration is constant whenever $\mathbf{E}$ is constant. The average drift velocity $\mathbf{v}_d$ previously described may be written in meters per second as

$$\mathbf{v}_d = \frac{\mathbf{a}\tau}{2} \tag{3–8}$$

where τ is the time in seconds between collisions. If Eq. 3–7 is substituted for the acceleration in Eq. 3–8 and both sides are multiplied by nq, the free charge density in coulombs per cubic meter, the following expression is obtained:

$$nq\mathbf{v}_d = \frac{nq^2\tau\mathbf{E}}{2m} \tag{3–9}$$

An expression for current density was previously developed as

$$\mathbf{j} = nq\mathbf{v}_d \tag{3–4}$$

and therefore from Eqs. 3–4 and 3–9 current density may now be written as

$$\mathbf{j} = \frac{\mathbf{E}}{2m/(nq^2\tau)} \tag{3–10}$$

The quantity $2m/(nq^2\tau)$ is a constant for any chosen material at a fixed temperature.

Since the quantity $2m/(nq^2\tau)$ is the function of the chosen material at a given temperature, it may be used to describe the characteristics of the material as a conductor. The name given to this constant is *resistivity* and it is symbolized by ρ, where

$$\rho = \frac{2m}{nq^2\tau} \tag{3–11}$$

A common form for the current density expression may then be obtained as

$$\mathbf{j} = \frac{\mathbf{E}}{\rho} \tag{3–12}$$

3–4. Ohm's Law. The only quantities in Eq. 3–11 accurately known for a given material are m and q. It is impossible to compute the specific value of the electric field $\mathbf{E}$ at a given point in the conductor since the internal field characteristics are not specifically known. Therefore, instead of writing our expressions in terms of $\mathbf{E}$ it is preferable to work with the voltage drop v, which can be measured for any given length of conductor. Equation 3–12 may be combined with the first equality of Eq. 3–4,

$$\mathbf{j} = \frac{i}{A}$$

and the definition of electric field intensity

$$\mathbf{E} = \frac{dv}{dl}$$

to obtain

$$i = \frac{A}{\rho}\frac{dv}{dl} = \frac{A}{\rho l}v \tag{3–13}$$

where l is the total length of the conductor in meters and v is the potential applied across the ends of the conductors.

The term $(\rho l)/A$ is a constant for any material at a constant temperature and is called *resistance*. It is usually symbolized by R, and written

$$R = \frac{\rho l}{A} \tag{3–14}$$

The relationship between current and voltage for the conductor may then be written as

$$i = \frac{v}{R} \tag{3–15}$$

This expression is known as Ohm's law. The lower case symbols used for current and potential imply that they are instantaneous quantities; that is, the current i exists instantaneously in response to the application of a potential v across a conductor having a resistance R.*

* Magnetic field effects as discussed in Chapter 4 may qualify this statement, but are neglected here.

The unit of resistance is named in honor of the experimental discoverer of this relationship and is termed the ohm. In terms of mks units the ohm is volts per ampere.

3–5. The Calculation of Resistance. Equation 3–14 expresses the resistance quantity in terms of the resistivity of the material and the physical dimensions of the conductor. The resistance depends upon the nature of the material making up the conductor and the chosen geometrical form for that conductor and is independent of the magnitude of the voltage applied across it or the current within it. (It must always be remembered that the value of ρ varies with the temperature.)

In the mks system, R is given in units of ohms, l in meters, and A in square meters. The dimensions of resistivity, ρ, are ohm-meters squared per meter or ohm-meters. The values of resistivity for some common conductor materials are given in Table 3–1.

TABLE 3–1

Conductor Material Resistivities and Temperature Coefficients at 20°C

Material	ρ_{20} (ohm-m)	ρ_{20} $\left(\frac{\text{ohm-cir. mils}}{\text{ft}}\right)$	α_{20} (per °C)
Aluminum	2.63×10^{-8}	15.55	0.0039
Copper	1.72×10^{-8}	10.3	0.00393
Iron	10.0×10^{-8}	60.2	0.0050
Lead	22.0×10^{-8}	130.0	0.0043
Nichrome	100.0×10^{-8}	602.0	0.0004
Silver	1.47×10^{-8}	8.7	0.0038
Tungsten	5.51×10^{-8}	32.6	0.0045

The reciprocal of resistivity is termed *conductivity* and is expressed in units of mho-meters per square meter or mhos per meter. Resistivity describes the opposition of a given material to the flow of charge; conductivity, its reciprocal, describes the willingness of material to permit charge to flow.

EXAMPLE 3–1. CALCULATION OF RESISTANCE USING MKS VALUES. Compute the resistance of 10 m of aluminum wire having a diameter of 5 mm. Temperature is 20°C.

From Table 3–1

$$\rho_{20} = 2.63 \times 10^{-8} \text{ ohm-m}$$

Using Eq. 3–14 with this value,

$$R = \frac{\rho l}{A} = \frac{2.63 \times 10^{-8} \times 10}{\pi \times (2.5)^2 \times 10^{-6}} = 1.34 \times 10^{-2} \text{ ohm}$$

To be consistent with the mks system, the cross-sectional area of the conductor must be expressed in square meters. However, this is often considered an impractical unit for dealing with the usual sizes of electrical conductors. Therefore, a system of units describing the cross-sectional area in terms of circular mils rather than square meters and length in feet rather than meters is commonly employed in engineering work to calculate the resistance of conductors. The area of a conductor in circular mils is arbitrarily defined as the square of the diameter of a circular conductor, where the diameter is expressed in thousandths of inches (1 mil $= 10^{-3}$ in.). Figure 3–1 illustrates this concept.

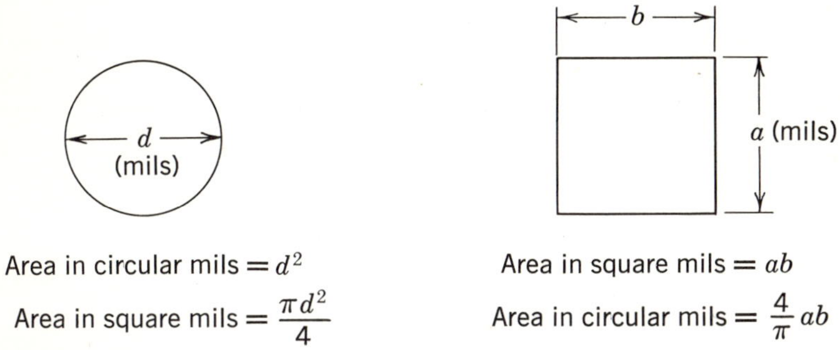

Fig. 3–1. Illustration of the circular mil.

The circular mil unit of area is less than the square mil unit of area, and, therefore, the area in circular mils may be obtained for any area in square mils by multiplying the area in square mils by $4/\pi$. Since the relation between the area in circular mils and the area in square mils is a constant, this constant may be absorbed in the resistivity coefficient and Eq. 3–14 for the resistance of a conductor written in terms of this new resistivity. The conductor length l is measured in feet and A, the cross-sectional area, in circular mils, where $A = d^2$ (d is the diameter of the circular conductor in mils). Table 3–1 also presents a tabulation of the resistivity of the same materials in terms of ohm-circular mils per foot.

Example 3–2. Resistance Calculations Using Circular Mil-Feet Dimensions. Compute the resistance of a copper conductor whose diameter is 0.1 in. and of length 100 ft. Temperature is 20°C.

$$0.1 \text{ in.} = 100 \text{ mils}$$
$$\text{Area in circular mils} = (100)^2 = 10^4$$

From Table 3–1

$$\rho_{20} = 10.3 \frac{\text{ohm-cir. mils}}{\text{ft}}$$

Using Eq. 3–14,

$$R = \rho \frac{l}{A} = \frac{10.3 \times 10^2}{10^4} = 0.103 \text{ ohm}$$

TABLE 3–2

Complete Wire Table, Standard Annealed Copper, AWG

AWG No.	Diameter (mils at 20°C)	Area (cir. mils)	Area (sq in.)	Ohms/1000 ft, 20°C (= 68°F)	Lb/1000 ft
0000	460.0	211,600.0	0.1662	0.04901	640.5
000	409.6	167,800.0	0.1318	0.06180	508.0
00	364.8	133,100.0	0.1045	0.07793	402.8
0	324.9	105,500.0	0.08289	0.09827	319.5
1	289.3	83,690.0	0.06573	0.1239	253.3
2	257.6	66,370.0	0.05213	0.1563	200.9
3	229.4	52,640.0	0.04134	0.1970	159.3
4	204.3	41,740.0	0.03278	0.2485	126.4
5	181.9	33,100.0	0.02600	0.3133	100.2
6	162.0	26,250.0	0.02062	0.3951	79.46
7	144.3	20,820.0	0.01635	0.4982	63.02
8	128.5	16,510.0	0.01297	0.6282	49.98
9	114.4	13,090.0	0.01028	0.7921	39.63
10	101.9	10,380.0	0.008155	0.9989	31.43
11	90.74	8,234.0	0.006467	1.260	24.92
12	80.81	6,530.0	0.005129	1.588	19.77
13	71.96	5,178.0	0.004067	2.003	15.68
14	64.08	4,107.0	0.003225	2.525	12.43
15	57.07	3,257.0	0.002558	3.184	9.858
16	50.82	2,583.0	0.002028	4.016	7.818
17	45.26	2,048.0	0.001609	5.064	6.200
18	40.30	1,624.0	0.001276	6.385	4.917
19	35.89	1,288.0	0.001012	8.051	3.899
20	31.96	1,022.0	0.0008023	10.15	3.092
21	28.46	810.1	0.0006363	12.80	2.452
22	25.35	642.4	0.0005046	16.14	1.945
23	22.57	509.5	0.0004002	20.36	1.542
24	20.10	404.0	0.0003173	25.67	1.223
25	17.90	320.4	0.0002517	32.37	0.9699
26	15.94	254.1	0.0001996	40.81	0.7692
27	14.20	201.5	0.0001583	51.47	0.6100
28	12.64	159.8	0.0001255	64.90	0.4837
29	11.26	126.7	0.00009953	81.83	0.3836
30	10.03	100.5	0.00007894	103.2	0.3042
31	8.928	79.70	0.00006260	130.1	0.2413
32	7.950	63.21	0.0000496	164.1	0.1913
33	7.080	50.13	0.0000394	206.9	0.1517
34	6.305	39.75	0.0000312	260.9	0.1203
35	5.615	31.52	0.0000248	329.0	0.0954
36	5.000	25.00	0.0000196	414.8	0.0757
37	4.453	19.83	0.0000156	523.1	0.0600
38	3.965	15.72	0.0000123	659.6	0.0476
39	3.531	12.47	0.0000098	831.8	0.0377
40	3.145	9.888	0.0000078	1049.0	0.0299

Except for some specialized bus structures and other conductor types designed for special applications, almost all conductors used in electrical systems are wires having circular cross-sectional area. For economy, wire sizes have been standardized. A number of standard systems exist throughout the world, but the basic standard in this country is the Brown and Sharpe gauge, usually called the American wire gauge (AWG). The sizes in this gauge system are based on a geometric progression and bear numbers such that the area, the resistance, and the weight per foot approximately double or halve every three numbers and the diameter approximately doubles or halves every six numbers. The larger numbers refer to the smaller conductor sizes. Table 3–2 is a wire table giving dimensions and resistance for annealed copper conductors.

3–6. Temperature and Resistance. The expression for resistivity has been given as

$$\rho = \frac{2m}{nq^2\tau} \tag{3–11}$$

where n and τ have not been specifically determined. Over the usual range of temperatures encountered in normal conductor operation the number of free electrons per unit volume may not be expected to increase with temperature,* but the energy of these free electrons and the energy of the system do increase. Thus, the random motion of the electrons becomes more spirited and the period of time between collisions with other particles within the conductor decreases. A decrease in the value of τ results in an increase in ρ. This is confirmed for the majority of conductors which are commonly used in electric circuits; the resistance of most metallic conductors has been determined by measurement to increase with temperature.

An expression for the resistance R_T of a conductor at temperature T in terms of a base value (usually chosen as resistance R_0 at 0°C) may be written as follows

$$R_T = R_0(1 + \alpha_0 T + \beta_0 T^2 + \cdots) \tag{3–16}$$

For the usual range of operating temperatures encountered in conventional electrical work the coefficients of the T^2 and higher order terms are very small and may be neglected. Equation 3–16 may then be rewritten as

$$R_T = R_0(1 + \alpha_0 T) \tag{3–17}$$

A useful relationship for this range of temperature, based on the resistance at 20°C for convenience, is written from Eq. 3–17 as

$$R_T = R_{20}[1 + \alpha_{20}(T - 20)] \tag{3–18}$$

The linear relationship between R and T expressed by Eq. 3–18 is useful for the usual range of operating temperatures but cannot be considered valid at very high or very low temperatures.

* This is not valid for materials other than metallic conductors. For example, in semiconductor materials the number of free electrons increases sharply with temperature and a reduction of resistance results from an increase in temperature.

If Eq. 3–18 is plotted as shown in Fig. 3–2, a useful expression may be easily derived for the calculation of the resistance at any temperature if the resistance at any other temperature is known. It is seen that the straight line representing the variation of resistance with temperature will intersect

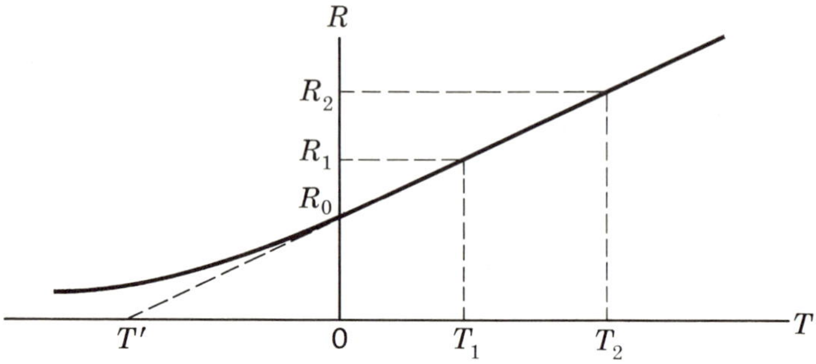

Fig. 3–2. Typical resistance-temperature plot for a conductor.

the horizontal axis at some value T', called an *inferred zero-resistance value* of temperature for a given material. This inferred zero-resistance temperature is a function of the slope of the linear resistance plot. Using the concept of similar triangles, several relationships may be developed for the calculation of resistance at various temperatures; for example, from Fig. 3–2,

$$R_2 = R_1 \frac{T' + T_2}{T' + T_1} \tag{3–19}$$

where T' is the inferred zero-resistance temperature

T_1 is the temperature corresponding to resistance R_1

T_2 is the temperature associated with resistance R_2

The value of T' may be calculated by geometric considerations as shown in Fig. 3–2, utilizing a known value of α, or obtained from engineering handbooks where its value is tabulated for various materials. (The value of T' for commercial annealed copper is $-234.5°C$ and for commercial grade aluminum is $-237.4°C$.)

Example 3–3. Calculation of Resistance After Temperature Change. Determine the resistance of the copper conductor of Example 3–2 if the temperature of the conductor is changed to 100°C. T' for copper is $-234.5°C$.

Using Eq. 3–19,

$$R_2 = R_1 \frac{T' + T_2}{T' + T_1}$$

$$R_{100} = R_{20} \frac{234.5 + 100}{234.5 + 20} = 1.31 R_{20}$$

and from Example 3–2

$$R_{20} = 0.103 \text{ ohm}$$

therefore

$$R_{100} = 1.31 \times 0.103 = 0.135 \text{ ohm}$$

It was previously noted that the free electrons moving under the influence of the electric field collide with the bound electrons and other atoms, giving up kinetic energy in the form of heat. The energy lost in this manner by the charge flow through the conductor will produce a temperature increase of the conductor material. For relatively small currents the heat produced is small and conductor cooling by radiation, convection, and conduction to adjacent objects or to the air limits the temperature rise to small values. On the other hand, if the current density within the conductor is large the amount of heating may become appreciable and the temperature of the conductor may rise to high values.* If the conductor is enclosed by insulating material, an excessive conductor temperature may weaken or destroy this insulation and physically damage the electric circuit. As a result, the allowable current density for a given conductor material establishes the maximum current which may be safely carried by any chosen conductor size. Standards defining the usual maximum current-carrying capacity of various sizes of conductor with different insulation materials have been established and are available in electrical engineering handbooks.

3–7. Energy Loss in a Conductor. The qualitative discussion of electron drift along a conductor describes the electron's path as consisting of a series of motions, each terminated by a collision with some obstruction. The kinetic energy possessed by the free electron was described as being converted to heat energy upon collision. Although the amount of energy possessed and released by any specific electron is not known, the over-all effect for a conductor section as a whole may be evaluated in terms of externally measurable quantities.

Consider a wire conductor connected across the terminals of a potential source of value v volts. Equation 3–15 states that a current $i = v/R$ amperes will flow in the conductor. From a previous definition, the potential difference v between the ends of the conductor may also be defined as the work done in transferring 1 coulomb of charge through the conducting path. If a differential amount of charge dq_T coulombs is transferred through the conductor, the amount of work done in joules is

$$dw = v\, dq_T \tag{3–20}$$

Equation 3–2 defined current as

$$i = \frac{dq_T}{dt}$$

* Conductor temperatures may become so high as to melt the conductor material. The common fuse is an example of deliberate use of this characteristic to provide protection for electric circuits and machines.

or

$$i\,dt = dq_T$$

from which

$$dw = vi\,dt$$

and

$$w = \int vi\,dt \qquad (3\text{–}21)$$

If a constant current of I amperes exists for a time T seconds caused by a constant potential difference of V volts, the energy in joules dissipated in the conductor is

$$W = VIT \qquad (3\text{–}22)$$

Power is defined as the rate of doing work. In the mks system energy is expressed in joules and power therefore in joules per second, also called watts. By rewriting the first of Eqs. 3–21, an expression for power loss in watts for the conductor is obtained as

$$\frac{dw}{dt} = p = vi \qquad (3\text{–}23)$$

Again for a constant current of I amperes and a constant potential difference of V volts, the expression for power loss in watts in the conductor is

$$P = VI \qquad (3\text{–}24)$$

Introducing the relationship between current and voltage in terms of resistance as defined by Eq. 3–15, Eq. 3–23 may be written in either of the following two forms:

$$p = i^2R = \frac{v^2}{R} \qquad (3\text{–}25)$$

It must be remembered that all of the energy supplied to the conductor is dissipated as heat; there is no storage of energy in an electrical form within a conductor.

EXAMPLE 3–4. POWER LOSS IN A CONDUCTOR. One thousand feet of AWG No. 10 copper wire are used to connect a load drawing a constant current of 30 amp to a generator. How much power is lost in the lines if the conductor temperature is 40°C? The diameter of No. 10 wire is 101.89 mils.

From Table 3–1

$$\rho = 10.3 \text{ at } 20°\text{C}$$

Using Eq. 3–18, ρ at 40°C may be computed as

$$\rho_{40} = 10.3 \times \frac{234.5 + 40}{234.5 + 20} = 11.11$$

Using Eq. 3–14,

$$R = 11.11 \times \frac{10^3}{(101.89)^2} = \frac{11.11 \times 10^3}{10{,}380} = 1.07 \text{ ohms}$$

The power lost in the conductor from Eq. 3–24 is $P = VI$ and, using Eq. 3–15, $V = IR$ for constant V and I,

$$P = I^2R = (30)^2 \times 1.07 = 964 \text{ w}$$

PROBLEMS

3–1. What is the drift velocity of the moving electrons in a metal conductor if the current density is 200 amp/sq cm? There are 8.5×10^{28} free electrons per cubic meter.

3–2. If ρ for copper is 1.7×10^{-8} ohm-m, what is the resistance of a copper conductor 1000 ft long and 10 mils in diameter?

3–3. The resistivity of copper at 20°C is given as 1.724×10^{-8} ohm-m in mks units. Calculate the value in (a) ohm circular-mils per foot and (b) ohm-mils^2 per foot.

3–4. Aluminum may be assumed to have 60% of the conductivity of copper. Find the values of Prob. 3–3 for aluminum.

3–5. Find the values of Prob. 3–3 at a temperature of (a) 100°C and (b) −40°C.

3–6. Forty feet of No. 20 AWG, 120 ft of No. 24 AWG, and 50 ft of No. 12 AWG copper wire are connected as shown. What is the circuit resistance?

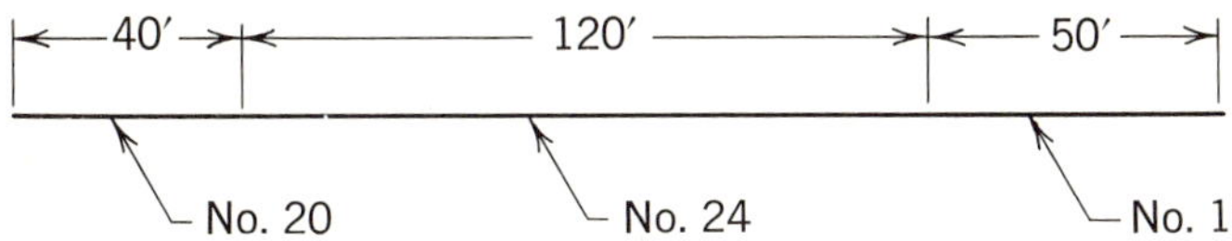

Prob. 3–6.

3–7. When an aluminum conductor and a copper conductor each of length 100 ft and equal cross-sectional area are connected as shown, their resistance is measured as 0.1 ohm. Temperature is 20°C. What is the cross-sectional area of each conductor in circular mils and what is the resistance of each conductor?

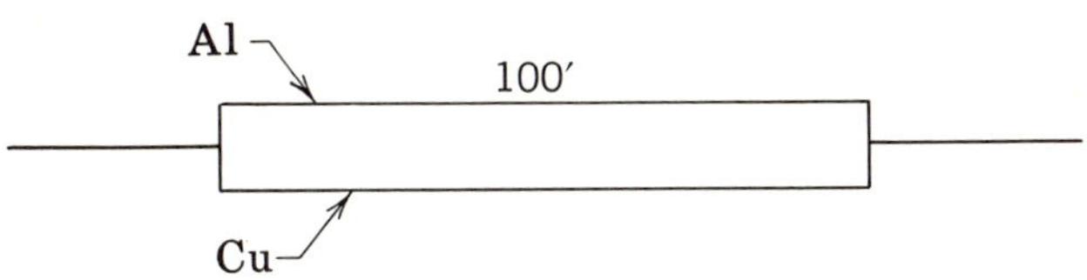

Prob. 3–7.

3–8. The temperature of the system of Prob. 3–7 is increased to 100°C. What is the new resistance?

3–9. A potential of 6 v is connected across the conductor system of Prob. 3–7. What current will flow in each conductor when the switch is closed? Will this be the current value after some time has passed?

3–10. An aluminum bar has dimensions as follows: length = 24 in., width = 2 in., and height = 1 in. Assuming the conductivity of aluminum to be 60% that of copper, calculate the resistance of the bar at 80°C.

3–11. A hollow rectangular copper bus conductor has outside dimensions of 4 in. by 6 in. and a wall thickness of 0.25 in. Calculate the resistance per foot at 100°C.

3–12. If the conductivity of copper is 6×10^7 mhos/m and the field intensity along a conductor is 1.5×10^{-2} v/m, what is the current density?

3–13. A water heater supplies water at 80°C at a rate of 80 gal/hr. The water intake is at 30°C. An electric heating element with an efficiency of 90% is used. If the kilowatt-hour of energy costs 7¢ and the heater is used for 4 hr a day, what is the monthly bill for the water heater for a month of 30 days? (1 kg-cal = 4200 j.)

3–14. The resistance of the armature of a machine was measured when the temperature was 20°C and found to be 0.005 ohm. After the machine was run for a certain time its resistance was measured and found to be 0.0055 ohm. What is the temperature rise and the hot temperature of the machine?

3–15. Fifteen hundred feet of No. 12 AWG were used for a certain continuous circuit which carries a current of 20 amp. What is the energy loss in this circuit if the temperature is 20°C and the current flows for 6 hr?

4

THE MAGNETIC FIELD, MAGNETIC INDUCTION, AND INDUCTANCES

In Chapter 2 the effects of stationary charges were considered in terms of the electrostatic field. Chapter 3 discussed charges in a motional state with uniform drift velocity within a conductor and termed this phenomenon electric current. Also associated with this motion of charges in a conductor is an additional phenomenon called *magnetism*. Oersted discovered that when a compass was brought close to a conductor carrying current and placed so that its axis was perpendicular to the conductor, the compass needle was deflected in a manner similar to the deflection of the needle when the compass was brought close to a bar magnet. Further investigations revealed that too closely spaced parallel conductors carrying current will either attract or repel each other, as do magnets when unlike or like poles are brought together.

Later on Faraday discovered that when two stationary coils were placed close to each other an electromotive force was induced in one coil when the current in the other changed.

These discoveries are two of the main bases upon which the science of electrical engineering is built.

4–1. Magnetic Field. Because of Oersted's discovery it was assumed that an electric current establishes around it a *magnetic* field. No one as yet has really proved that this field exists physically. However, the hypothesis of a magnetic field helps us to visualize and explain the phenomena which are to be discussed in this and the following chapters. Yet we should never lose sight of the fact that the magnetic field is only a *hypothesis*. This hypothesis in essence states that whenever a current flows in a conductor a magnetic flux is set around that conductor. This flux can be represented by lines or tubes called *magnetic lines of force*. Each line forms a loop closed on itself and each loop surrounds the conductor and forms a link as shown in Fig. 4–1. The magnetic flux is usually denoted by the Greek letter ϕ.

Unlike the electrostatic line of flux, which begins at a positive charge and ends on a negative charge, the line of magnetic flux has no beginning and no end. However, the magnetic line of flux, like the electrostatic one, has direction. This direction is found in several ways, the simplest of which is the following: Grasp the conductor with the right hand with the thumb pointing in the direction of the current; the fingers will then point in the direction of the flux lines.

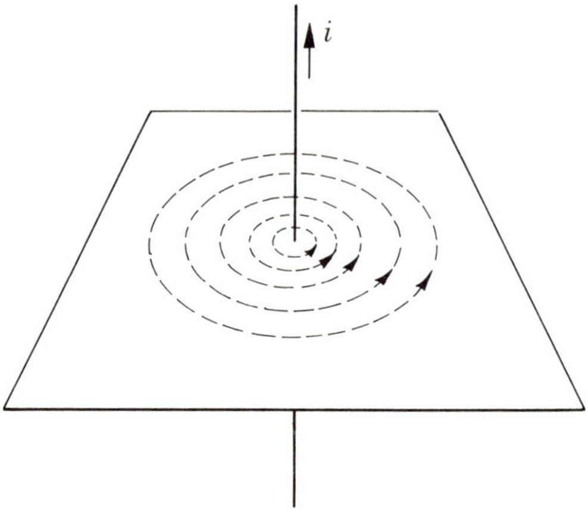

Fig. 4–1. Magnetic field around a conductor carrying current.

4–2. Magnetic Flux Density. The magnetic flux lines are more numerous (more dense) close to the conductor carrying current than at other points. In order to show the value of the magnetic flux at different points in space, namely, how thick or sparse are the flux lines, the flux density, usually denoted **B**, is introduced.

By flux density at a point P is meant the ratio of the flux $d\phi$ in an incremental area dA perpendicular to the flux lines with the point P as center, to the area dA. Mathematically this is expressed as

$$\mathbf{B} = \frac{d\phi}{dA} \tag{4–1}$$

where **B** is the flux density in webers per square meter

$d\phi$ is the amount of flux in webers in the area dA

dA is the incremental area in square meters

If the flux is evenly distributed through the area the flux density will be constant through that area.

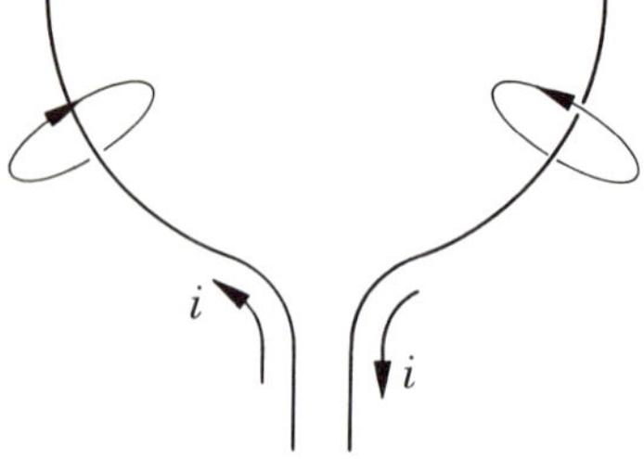

Fig. 4–2. Flux linkages.

4–3. Flux Linkages. As the electric current is a vector closing on itself, a current flowing in a conductor has to return to the source from which it originates. It can be said that the current (or the conductor carrying the current) forms a closed loop.

It was mentioned that the flux lines also form closed loops around the current or the conductor carrying the current. Therefore the conductor and the flux lines form linkages. This is shown in Fig. 4–2. To have a flux linkage, the flux line has to link the circuit. Figure 4–3 shows one case of flux linkage and three situations where there is no flux linkage.

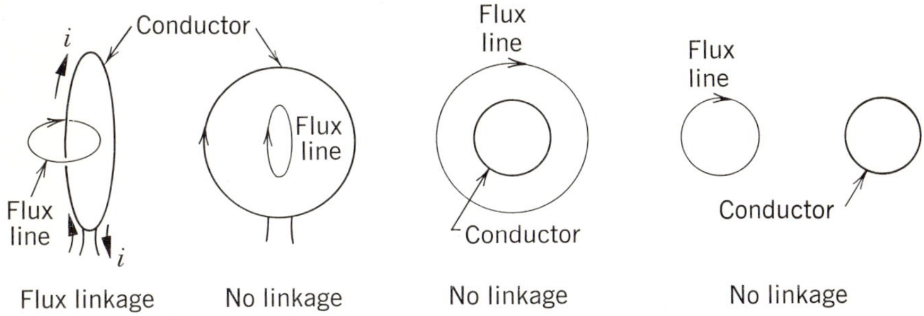

Fig. 4–3. Situations of flux linkage and no flux linkage.

4–4. Faraday's Discovery. Faraday discovered that whenever a magnetic field changes so that the flux linkages of a certain loop are changed an electric field is induced in that loop. The value of the electric field is such that:

$$\oint \mathbf{E}_i \cdot \mathbf{dl} = e = -\frac{d\phi}{dt} \tag{4–2}$$

where $\oint$ denotes a contour integral or integral around the loop surrounding the field

$\mathbf{E}_i$ is the field intensity in volts per meter

e is the induced electromotive force (emf) in the path linking the flux, in volts

$d\phi/dt$ is the rate of change of flux linkages in webers per second

If the loop is a conductor the emf will be induced in the conductor.

Equation 4–2 may be the basis of two definitions. First, it gives a definition for magnetic flux thus:

Magnetic flux is that quantity which, linking a closed path, induces in it an emf whenever its value changes.

Second, it gives the definition of a weber thus:

A weber is the amount by which the magnetic flux linking a circuit of one turn must change in 1 sec to induce an emf of 1 v in the circuit, the change being at a uniform rate.

The induced electric field intensity $\mathbf{E}_i$ caused by the change of magnetic flux linkages does not obey the inverse square distance law and therefore has different properties from the electrostatic field intensity $\mathbf{E}_s$. It was seen that the line integral of the electrostatic field intensity is always equal to zero, or

$$\oint \mathbf{E}_s \cdot \mathbf{dl} = 0$$

The line integral of the induced electromagnetic field intensity $\mathbf{E}_i$ is not zero but gives the value of the rate of change of the flux in the area bounded by the path of integration, or

$$\oint \mathbf{E}_i \cdot \mathbf{dl} = -\frac{d\phi}{dt} = e$$

Furthermore, $\mathbf{E}_s$ starts at positive charges and ends on negative charges whereas $\mathbf{E}_i$ forms a closed path with no beginning or end.

4–5. The Direction of the Electromotive Force. The direction of e, the emf in the closed path around the changing field, depends upon the rate of change of ϕ. If $d\phi/dt$ is positive, meaning the flux is increasing, then e is negative. If $d\phi/dt$ is negative, meaning the flux is decreasing, then e is positive.

These two conditions mean that e is of such polarity that it tends to send a current around the loop of such direction as to *oppose the change of* ϕ. This is called Lenz's law. In reality it is Newton's law that states that the reaction is equal and opposite to the action.

4–6. Multiturn Circuits. If the flux ϕ links N turns in series then the total flux linkage is $N\phi$. The induced emf is then:

$$e = -N\frac{d\phi}{dt} \quad \text{volts} \tag{4–3}$$

Usually the symbol λ is given to $N\phi$, thus

$$\lambda = N\phi \tag{4–4}$$

Equation 4–3 is then written

$$e = -\frac{d\lambda}{dt} \tag{4–5}$$

Care must be exercised when substituting Eq. 4–4 in Eq. 4–5. We should remember that N is a constant and therefore

$$\frac{d\lambda}{dt} = N\frac{d\phi}{dt} \tag{4–6}$$

and

$$e = -N\frac{d\phi}{dt} \quad \text{and not} \quad -\frac{dN}{dt}\phi - N\frac{d\phi}{dt}$$

4–7. Current-Carrying Conductor in a Magnetic Field. Assume a magnetic field as shown in Fig. 4–4. Let a conductor carrying current be placed in this field so that the conductor is perpendicular to the field. In Fig. 4–4 the conductor is perpendicular to the plane of the paper. The current in the conductor sets up a magnetic field of its own. The flux lines of this field form loops around the conductor as shown in the figure. These flux lines will merge with the lines of the applied field and the combined fields will be

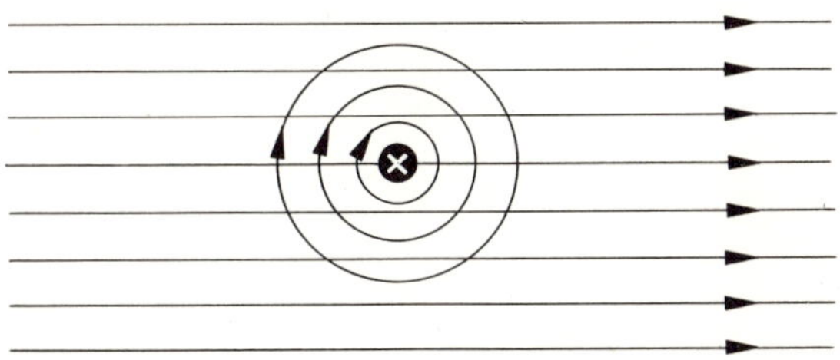

Fig. 4–4. Current-carrying conductor in a magnetic field.

somewhat distorted as shown in Fig. 4–5. The distorted field acts like rubber bands and exerts a force on the conductor. This force is downward in the figure and is perpendicular to both the original field and the conductor.

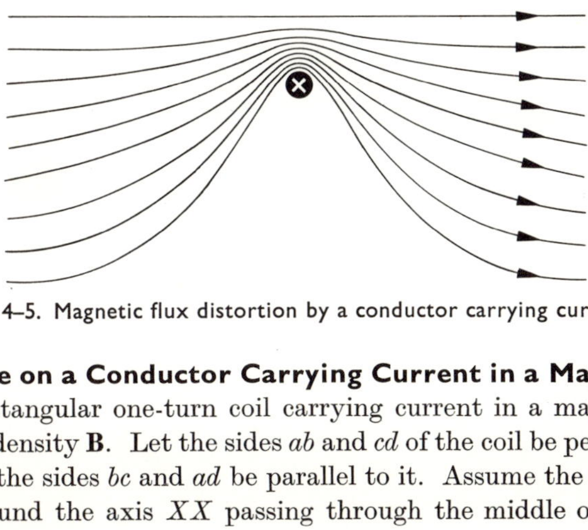

Fig. 4–5. Magnetic flux distortion by a conductor carrying current.

4–8. Force on a Conductor Carrying Current in a Magnetic Field. Assume a rectangular one-turn coil carrying current in a magnetic field of uniform flux density **B**. Let the sides *ab* and *cd* of the coil be perpendicular to the field and the sides *bc* and *ad* be parallel to it. Assume the coil to be able to rotate around the axis XX passing through the middle of *bc* and *da* as shown in Fig. 4–6.

Let θ be the angle between the plane of the coil and the direction of the field. The amount of the applied flux linking the coil depends upon θ. When θ is zero no lines of the applied field pass through the coil. When θ is 90° a maximum applied flux of magnitude $B2RL$ links the coil. The amount of applied flux linking the coil is given as a function of θ by

$$\phi = B2RL \sin \theta \tag{4–7}$$

where ϕ is in webers

B is in webers per square meter

R and L are in meters

Let a constant current i flow in the coil as shown. This current causes a magnetic field ϕ_i of its own to link the coil. However, the flux linkages ϕ_i due to the current in the coil are constant as long as the current i is constant. The total flux linkages of the coil are then

$$\phi_T = B2RL \sin \theta + \phi_i \tag{4–8}$$

where ϕ_i is constant and independent of θ.

From Art. 4–7 we see that a force **F** downward is exerted on ab and an equal force **F** upward is experienced by cd. These two forces exert a torque on the coil, tending to rotate it in a counterclockwise direction.

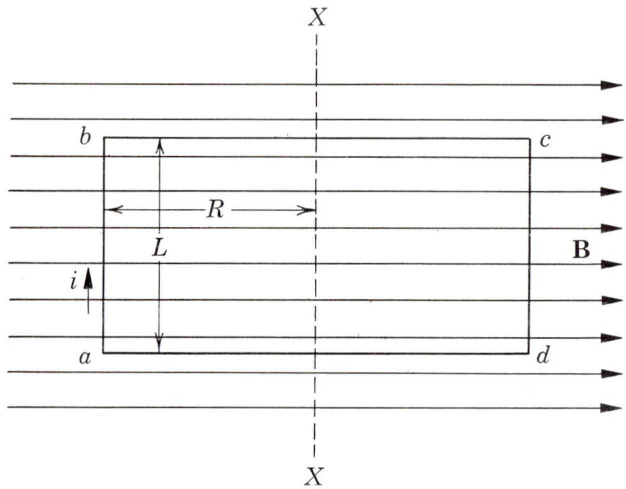

Fig. 4–6. Rectangular coil in a magnetic field.

Let the coil under the influence of this torque move through an angle $d\theta$ in dt seconds. The amount of change of flux linkages in the coil is obtained by differentiating Eq. 4–8 with respect to θ, or

$$\frac{d\phi_T}{d\theta} = B2RL \cos \theta \tag{4–9}$$

and

$$d\phi_T = B2RL \cos \theta \, d\theta \tag{4–10}$$

Because of the change of flux linkages in the coil an emf is induced in the coil. The value of this emf is:

$$e = -\frac{d\phi_T}{dt} = -B2RL \cos \theta \frac{d\theta}{dt} \tag{4–11}$$

In order to keep the current i constant and flowing in the coil circuit an equal and opposite emf must be applied to the coil, thus:

$$e_c = B2RL \cos \theta \frac{d\theta}{dt} \tag{4–12}$$

The current i flows through this emf and thus the source of this emf must supply an amount of power equal to:

$$p = e_c i = B2RLi \cos\theta \frac{d\theta}{dt} \tag{4–13}$$

The amount of energy supplied electrically by the source to the coil is

$$p\,dt = dw = B2RLi \cos\theta\, d\theta \tag{4–14}$$

where dw is the electrical energy in joules supplied in dt seconds. But the mechanical work dw is

$$dw = T\,d\theta \tag{4–15}$$

where T is the torque in newton-meters. Therefore:

$$T\,d\theta = B2RLi \cos\theta\, d\theta \tag{4–16}$$

From Eq. 4–16

$$T = B2RLi \cos\theta \tag{4–17}$$

But from Eq. 4–9

$$B2RL \cos\theta = \frac{d\phi_T}{d\theta}$$

Therefore:

$$T = i\frac{d\phi_T}{d\theta} \tag{4–18}$$

where T is the torque in newton-meters
i is the current in amperes

When $\theta = 0$ the sides bc and da are parallel to the field and therefore have no force exerted on them. Under this condition the forces are on ab and cd. From Eq. 4–17 for $\theta = 0$

$$T = B2RLi \tag{4–19}$$

But

$$T = 2RF \tag{4–20}$$

Therefore from Eqs. 4–19 and 4–20 we get

$$F = BLi \tag{4–21}$$

The force is always in such a direction as to cause the total flux linkages ϕ_T of the coil (owing to its own ϕ_i and the applied field) to become maximum. The force per unit length of the conductors ab and cd is

$$F = Bi \tag{4–22}$$

and for an infinitesimal length dl it becomes:

$$dF = Bi\,dl \tag{4–23}$$

If conductors ab and cd make an angle α with the plane perpendicular to the direction of the field, the force on the conductor becomes:

$$F = BLi \cos\alpha \tag{4–24}$$

4–9. Forces Between Parallel Conductors Carrying Current. The forces between parallel conductors carrying currents can now be explained (see Fig. 4–7). Let the two conductors lie in the plane of the paper. The current i_1 in conductor 1 establishes a magnetic field around it. Let the flux density at a distance R from conductor 1 be B. The direction of this field at conductor 2 is perpendicular to the plane of the paper and therefore is perpendicular to conductor 2. Conductor 2 with its current lies in the magnetic field of the current in conductor 1 and therefore experiences a force.

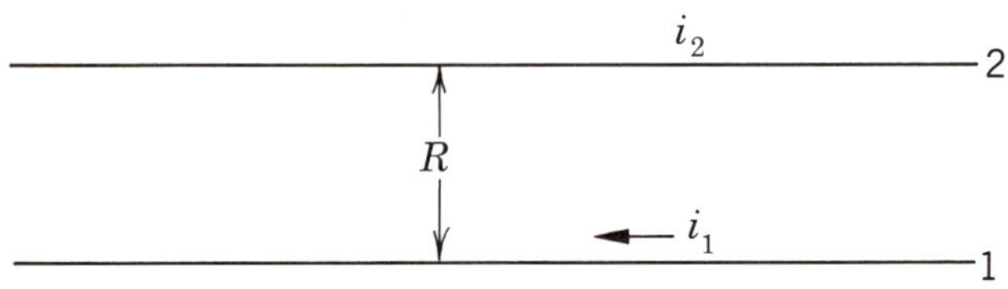

Fig. 4–7. Parallel conductors carrying currents.

The direction of the force is either away from conductor 1 or toward conductor 1, depending upon the direction of i_2.

4–10. Motional Electromotive Force. An emf will be induced in a conductor which is moving through a magnetic field so as to cut the lines of flux.

This principle was first enunciated as a discovery by Faraday, but it will be shown here that it is a consequence of Oersted's discovery. Consider Fig. 4–8, where two conducting rails are separated by a distance L. Assume a uniform flux density B directed into the paper.

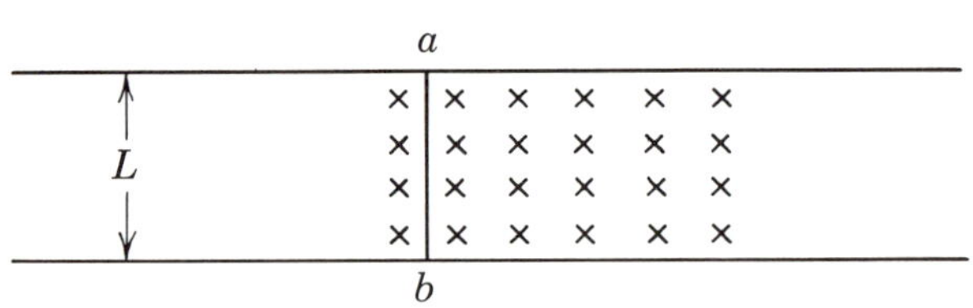

Fig. 4–8. Motional emf in a conductor cutting flux.

Let a conductor ab touching the rail be moved with constant velocity v so as to cut the magnetic flux. The structure of the conductor has previously been described as consisting of positive and negative charges. These charges in the conductor are moving with velocity v, and therefore a magnetic force will act upon them. Although the positive charges and some electrons are bound in the lattice of the conductor, the free electrons may move under the influence of this force. Since the free electrons in the conductor are being moved from left to right, the force due to the magnetic field will cause their motions within the conductor to be toward terminal b. This motion will create a deficiency of electrons at a and an excess of electrons at b. An electric field will then be established within the conductor between b and a, having a direction that opposes this electron motion. When the electron accumulation and deficiency establish an electric field in the conductor whose force on the electrons is equal to that of the magnetic field, electron motion will cease.

The magnetic force per unit length is

$$F_m = Bi = Bqv \tag{4-25}$$

as

$$i = qv \tag{4-26}$$

But the electrostatic force per unit length is

$$F_e = qE \tag{4-27}$$

Therefore

$$E = Bv \tag{4-28}$$

and

$$e = \int_b^a E\,dl = Blv \tag{4-29}$$

where e is the potential difference between a and b in volts

B is the flux density in webers per square meter

l is the length of the conductor in meters

v is the velocity of the conductor in meters per second

Notice that inside the conductor there is no electric field whereas outside the conductor between points a and b there is a field from a to b.

4-11. Self- and Mutual Inductances. So far nothing has been said about the source of the flux ϕ. This flux may be created by the current in the coil itself as in Fig. 4-9a, by the current in another coil as in Fig. 4-9b, or by a

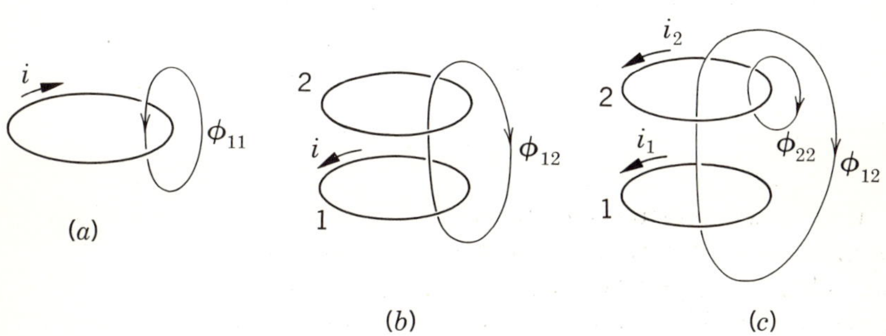

Fig. 4-9. Self- and mutual inductances.

combination of the current in the coil itself and the current in another coil as in Fig. 4-9c.

When the flux linkages of a coil are created by the current in the coil itself, the emf can be written

$$e_1 = -N_1 \frac{d\phi_{11}}{di_1} \frac{di_1}{dt} \tag{4-30}$$

The term $N_1(d\phi_{11}/di_1)$ is called *self-inductance* and is symbolized by the letter L, usually with two subscripts thus:

$$L_{11} = N_1 \frac{d\phi_{11}}{di_1} \tag{4-31}$$

where L_{11} is the self-inductance of coil 1 in henrys

ϕ_{11} is the flux linking N_1 turns of coil 1 and produced by i_1. This flux is in webers

i_1 is the current in coil 1 and is given in amperes

N_1 is the number of turns of coil 1 linked by ϕ_{11}. The turns are in series

Equation 4–30 is then written as:

$$e_1 = -L_{11} \frac{di_1}{dt} \tag{4-32}$$

Notice that, except in ferromagnetic cores, L is a constant and Eq. 4–31 may be expressed as:

$$L_{11} = \frac{N_1\phi_{11}}{i_1} \tag{4-33}$$

Therefore the self-inductance of a coil is defined as the number of flux linkages in a coil set by 1 amp in that coil.

If the flux ϕ_{12} of coil 1 links N_2 turns of coil 2, then the flux linkages in coil 2 are given as $N_2\,\phi_{12}$. Equation 4–30 under this condition becomes:

$$e_2 = -N_2 \frac{d\phi_{12}}{di_1} \frac{di_1}{dt} \tag{4-34}$$

The term $N_2(d\phi_{12}/di_1)$ is called the *mutual inductance* between coils 1 and 2 and is denoted either as M_{21} or M_{12} or L_{12} or finally L_{21}. Therefore:

$$L_{21} = N_2 \frac{d\phi_{12}}{di_1} \tag{4-35}$$

and

$$e_2 = -L_{21} \frac{di_1}{dt} \tag{4-36}$$

The mutual inductance between two coils is defined as the flux linkages in coil 2 caused by a current of 1 amp in coil 1 (or vice versa). In Fig. 4–9c the flux linkages $N_2\phi_2$ are caused partly by ϕ_{22} (the flux in coil 2 produced by the current i_2 of coil 2) and partly by ϕ_{12} (the flux in coil 2 caused by the current in coil 1). For this case:

$$\begin{aligned} e_2 &= -N_2 \frac{d\phi_2}{dt} \\ &= -N_2 \frac{d\phi_{22}}{di_2} \frac{di_2}{dt} - N_2 \frac{d\phi_{12}}{di_1} \frac{di_1}{dt} \\ &= -L_{22} \frac{di_2}{dt} - L_{21} \frac{di_1}{dt} \end{aligned} \tag{4-37}$$

4–12. Energy Stored in Inductances. Assume a voltage e applied across an inductance. The applied voltage is equal and opposite to the voltage induced in the inductance, or

$$e = -e_i = L\frac{di}{dt} \tag{4–38}$$

If a current i flows in the inductance, the instantaneous power supplied by the voltage source to the inductance is:

$$p = ei = Li\frac{di}{dt} \tag{4–39}$$

If the current in the inductance changes from an initial value of i_0 at $t = 0$ to a value i at t, then the increase of energy in joules supplied to the inductance is:

$$\Delta w = \int_0^t p\,dt = \int_{i_0}^{i} Li\,\frac{di}{dt}$$

$$= \frac{L}{2}(i^2 - i_0^{\,2}) \tag{4–40}$$

If the initial current through the inductance at $t = 0$ is zero, then the amount of energy supplied to the inductance is:

$$w = \frac{Li^2}{2} \tag{4–41}$$

The inductance stores this energy in its magnetic field. Therefore the stored energy in an inductance is:

$$w = \frac{Li^2}{2} \tag{4–42}$$

4–13. Displacement Current. Before concluding this chapter on magnetic fields produced by conducting currents it is important to introduce another kind of electric current that also produces a magnetic field, namely the displacement current.

The capacitance of a capacitor is given by

$$C = \frac{q}{V}$$

or

$$q = CV \tag{4–43}$$

If q is a function of time, then V will be a function of time. If both sides of Eq. 4–43 are differentiated with respect to time, then:

$$\frac{dq}{dt} = C\frac{dv}{dt} \tag{4–44}$$

dq/dt is the current flowing in the external or conductor circuit connecting the plates of the capacitor, that is, it is the motion of free electrons in the conductor. Since the plates of the capacitor are separated by a dielectric that does not conduct current, the free electrons must stop at the negative plates. However, the continuity of current, insofar as its magnetic effect is concerned, is a fundamental electrical concept. This reasoning prompted Maxwell to postulate that $C(dV/dt)$ constitutes in effect a current, which he termed "displacement current." Through the capacitor, therefore, an equivalent current exists which is written:

$$i_d = C\frac{dV}{dt} \tag{4-45}$$

The displacement current concept assures that the magnetic effects of the circuit containing a capacitor will be the same as if the current in that circuit were flowing through a closed loop.

PROBLEMS

4–1. A conductor carries current in the direction shown. Find the direction of the flux lines at the conductor and away from it.

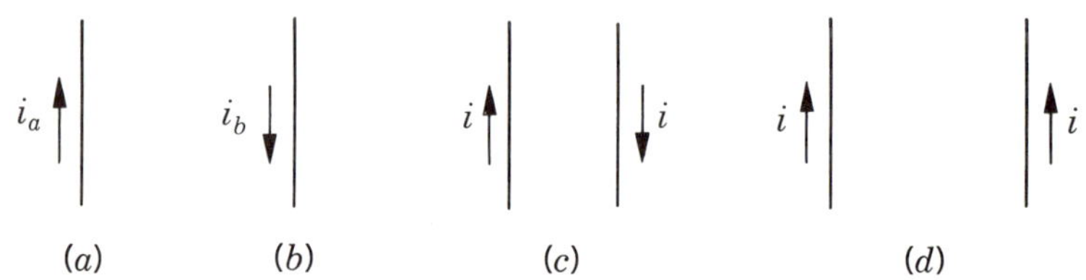

Prob. 4–1.

4–2. In (c) of the figure for Prob. 4–1 does the increase in the separation of the wires increase or decrease the amount of flux between them?

4–3. In Prob. 4–2 does the increase of the separation increase or decrease the flux linkages of the circuit, assuming that the current in one conductor is the return of the current in the other? How about the self-inductance?

4–4. In (c) of the figure for Prob. 4–1 what is the direction of the force between the wires?

4–5. From Probs. 4–3 and 4–4 can you draw a conclusion about forces and self-inductance?

4–6. Repeat Probs. 4–2, 4–3, and 4–4 for (d) of the figure for Prob. 4–1, and check whether the answer of Prob. 4–5 is still true.

4–7. You are given a wide horseshoe magnet in which the flux density between its opposing forces is uniform. You are given a linear torsional spring and an armature that fits between the opposing faces of a horseshoe magnet. Can you make an ammeter or a voltmeter out of these components?

4–8. Assume a circular turn carrying a current of 1 amp and placed in a uniform magnetic field. Find the torque if the turn can rotate on an axis along its diameter.

4–9. Two coils are put close to each other. When the current in one coil changes at a uniform rate of 50 amp/sec a voltage of 1 v is induced in the other coil. Can you explain why? What is the value of the mutual inductance between the coils?

4–10. A current of 10 amp in coil 1 sets up a flux linkage of 0.05 weber-turn in coil 2. If the current in coil 1 is stopped at a uniform rate in 0.05 sec, what is the voltage induced in coil 2?

4–11. The self-inductance of a coil is 2 h. If a current of 10 amp is set in that coil in 1 sec, what is the *average* induced voltage in the coil? (Why is the word average used here?)

5

MAGNETIC CIRCUITS

The computation of magnetic fields in general is difficult, if at all possible. In certain cases, however, when the magnetic flux is constrained to lie in a certain bounded path, the amount of flux can be found from a knowledge of the magnetomotive force (mmf) and the nature of the path by means of the circuital law. The bounded path is the magnetic circuit.

5–1. The Magnetomotive Force $\mathscr{F}$. It was seen in Chapter 4 that magnetism or magnetic flux is a manifestation of the presence of electric current. As current flows in a closed path and as the path links the field produced by the current, it may be said that the current flowing in a turn sets the flux passing through the area bounded by the turn. If the current flows in two turns in series and these turns are concentrated with each other, the amount of flux will be doubled and the flux will link with the two turns. It can then be assumed that flux is created by the mmf $\mathscr{F}$ and that this mmf can be expressed in terms of ampere-turns. To elucidate: 1 amp flowing in five concentrated turns linking a magnetic path sets in that path the same amount of flux as 5 amp flowing in one turn.

5–2. The Circuital Law. In Chapter 4 it was mentioned that the flux lines have direction and the flux density is a vector. The flux density at a point depends upon the nature of the path. A vector independent of the property of the path is the so-called magnetizing force, mmf gradient, or magnetic field intensity. It is a vector in the same direction as the flux density and is usually denoted by **H**. It is related to the mmf thus:

$$\mathscr{F} = \oint H\, ds \cos\theta = \oint \mathbf{H} \cdot \mathbf{ds} \qquad (5\text{–}1)$$

where $\mathscr{F}$ is the mmf in ampere-turns

H is the magnetizing force in ampere-turns per meter

ds is an incremental length of the magnetic path which is the path of integration and is in meters

θ is the angle that the vector **H** makes with **ds**

$\oint$ denotes that integration is being carried out as a line integral along a closed path

H is then the mmf per unit length, hence the name *magnetizing force*. If the flux density over the whole magnetic circuit is constant and the magnetic circuit is of the same material and shape throughout its length, the value of H is given by dividing the mmf by the length of the magnetic path, or

$$H = \frac{\mathscr{F}}{l} \tag{5–2}$$

where l is in meters

H is in ampere-turns per meter

Equation 5–1 states that the line integral of $H\,ds \cos\theta$ around any closed path is equal to the net ampere-turns through the surface bounded by the path. For instance, the magnetic field intensity H for a long conductor carrying current I at a point R meters from the conductor is given by:

$$H = \frac{I}{2\pi R}$$

H forms a closed path around the conductor, as shown in Fig. 5–1. Assume an element of the path to be ds. Then:

$$ds = R\,d\alpha, \qquad \cos\theta = 1$$

and $$\oint H\,ds\cos\theta = \int_0^{2\pi} \frac{I}{2\pi R} R\,d\alpha$$

$$= \frac{I}{2\pi}\alpha\Big|_0^{2\pi} = I$$

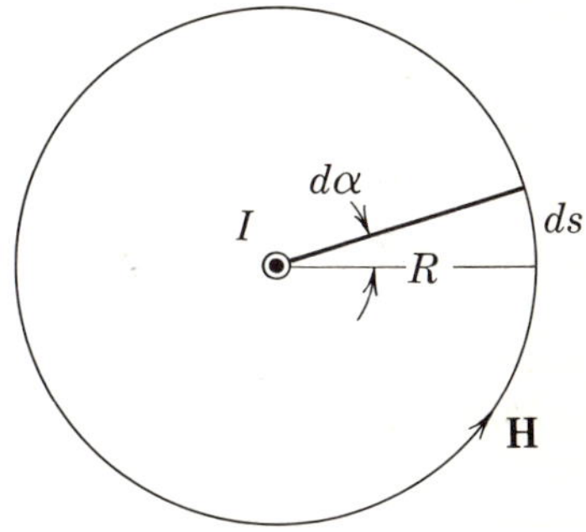

Fig. 5–1. Closed path around a current element of **H**.

which is the current coming out of the surface bounded by the path of integration.

5–3. Flux Density B and Mmf Gradient H. The relationship between **H** and **B** is always given by

$$\mathbf{B} = \mu\mathbf{H} = \mu_r\mu_0\mathbf{H} \tag{5–3}$$

where **B** is the flux density in webers per square meter

μ is the permeability of the core material in webers per ampere-turn-meter

μ_r is the relative permeability and equals μ/μ_0

μ_0 is the permeability of free space or air and equals $4\pi \times 10^{-7}$ weber per ampere-turn-meter

5–4. Applications of the Circuital Law. The foregoing circuital law permits the computation of the magnetic flux in several electric devices and machines such as toroids and transformers.

Let it be desired to compute the flux in a toroid, with N turns closely wound on a core of circular cross section and evenly distributed as shown in Fig. 5–2. Assume the core to be of a material such as wood with relative permeability of unity. The radius of the cross section R_c is very small compared with the radius of the toroid R_t.

The reason why R_c is assumed small compared with R_t is that at different points inside the core there are different **H**'s. Assume, for instance, that points A, B, and C of Fig. 5–3 are inside the core of the toroid. The value of **H** at these points is given by:

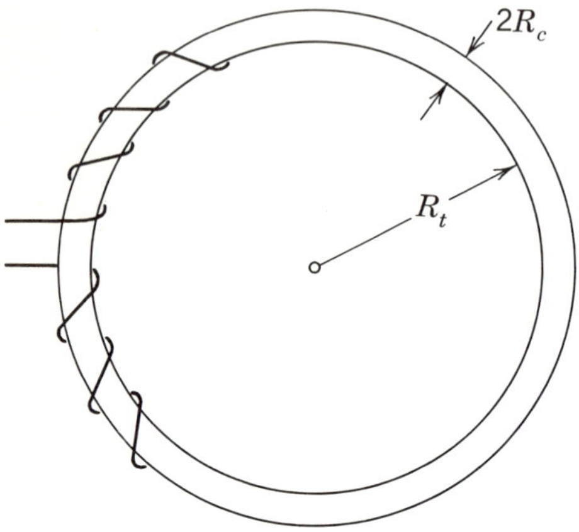

Fig. 5–2. A toroid with closely wound turns.

$$H_A = \frac{Ni}{2\pi R_1}$$

$$H_B = \frac{Ni}{2\pi R_2}$$

$$H_C = \frac{Ni}{2\pi R_3}$$

The three values of H are not the same and consequently the flux density inside the core is different for various points at different distances from the center of the toroid. For engineering purposes the H at the center of the core is taken. Its value is given by:

$$H = \frac{Ni}{2\pi(R_t + R_c)} \tag{5–4}$$

The flux density at that point is:

$$B = \mu H \tag{5–5}$$

$$= \mu \frac{Ni}{2\pi(R_t + R_c)} \tag{5–6}$$

This flux density is assumed constant throughout the cross section of the toroid. The total flux in the toroid is therefore:

$$\phi = BA \tag{5–7}$$

$$= \frac{\mu A N i}{2\pi(R_t + R_c)} \tag{5–8}$$

where A is the cross-sectional area of the core and is given by

$$A = \pi R_c^2 \tag{5–9}$$

From Eqs. 5–8 and 5–9 the total flux in the core is given by:

$$\phi = \frac{\mu \pi R_c^2}{2\pi(R_t + R_c)} Ni = \frac{\mu R_c^2}{2(R_t + R_c)} Ni \tag{5–10}$$

The value of the inductance of the toroid is:

$$L = \frac{N\phi}{i} = \frac{\mu R_c^2 N^2}{2(R_t + R_c)} \tag{5–11}$$

From Eq. 5–11 it is seen that as long as μ is constant the self-inductance is proportional to the square of the number of turns. Furthermore, the self-inductance depends upon the geometry or configuration of the magnetic

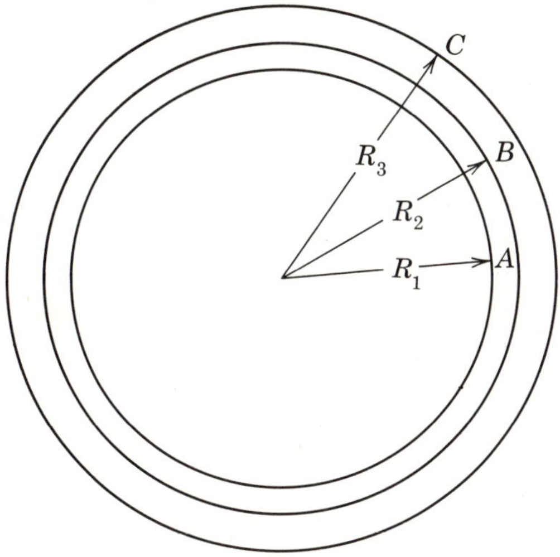

Fig. 5–3. Magnetic field intensities in a toroid.

circuit and the permeability of the material forming the circuit. The permeability μ of most materials except ferromagnetic materials and ferrites is the same as μ_0. Therefore for most materials:

$$\mu = 4\pi \times 10^{-7}$$

Because ferromagnetic materials and ferrites have high permeability, the magnetic circuits for high inductances are wholly or partly made up of these materials. However, the permeabilities of ferromagnetic substances and ferrites are functions of flux density or of the magnetizing force **H**. For low magnetizing forces the permeability may be assumed constant. However, for higher **H**'s the permeability decreases and may even become equal to that of air or free space. For the present the permeability will be assumed constant and no hysteresis assumed. Later on, studies of ferromagnetic materials will be made when energy converters are studied.

5–5. Reluctance and Permeance. The ratio of mmf to flux is called the *reluctance* of the magnetic circuit. From Eq. 5–10 the value of the reluctance of a toroid is equal to

$$\mathscr{R} = \frac{Ni}{\phi} \tag{5–12}$$

$$= \frac{2(R_t + R_c)}{\mu R_c^2} \tag{5–13}$$

The ratio of flux to mmf of a magnetic circuit is called *permeance*. The permeance of a toroid is:

$$\mathscr{P} = \frac{1}{\mathscr{R}} = \frac{\mu R_c^2}{2(R_t + R_c)} \tag{5–14}$$

PROBLEMS

5–1. Assume a toroid with $R = 10$ cm and a cross section of 0.8 sq cm. If the flux density B is assumed uniformly distributed through the core and if $B = 10^{-3}$ weber/sq m, find the ampere-turns necessary to set the flux in the core, knowing that the core is wood with $\mu_r = 1$.

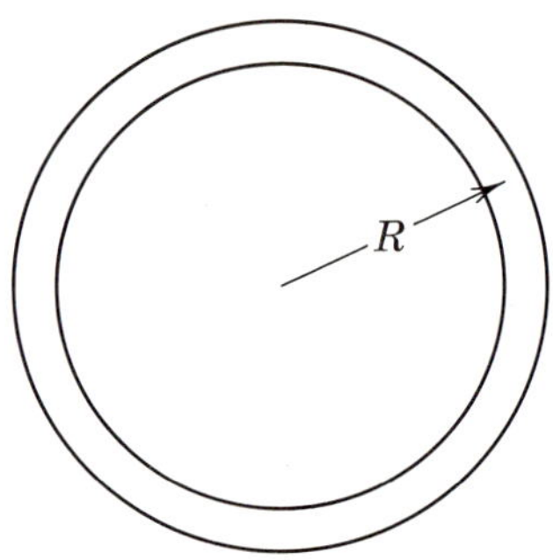

Prob. 5–1.

5–2. If in Prob. 5–1 a current of 4 amp flows in the coil, find (a) the number of turns; (b) the flux linkages; (c) the inductance in henrys.

5–3. If in Prob. 5–1 the core has a relative permeability $\mu_r = 100$ and the flux density $B = 10^{-3}$ weber/sq m, find (a) the ampere-turns necessary to set up the flux. (b) Assuming the turns to be the same as in Prob. 5–2, what is the value of the current? (c) What is the value of the inductance?

5–4. Assume the toroid as shown in Prob. 5–1 with ampere-turns of 6000 and a current of 1 amp. Find (a) H, the magnetizing force; (b) B if $\mu_r = 100$; (c) the flux in the core; (d) the flux linkages; (e) the inductance of the coil; (f) the energy stored.

5–5. A magnetic circuit as shown has a depth d of 3 cm and a width w of 1 cm. The flux density is assumed to be uniform throughout the core. The length of the magnetic path l is 50 cm. A coil is uniformly wound around the core. The flux density in the core is assumed to be 2×10^{-3} weber/sq m. The relative permeability of the core material is $\mu_r = 1$. Find the ampere-turns on the core.

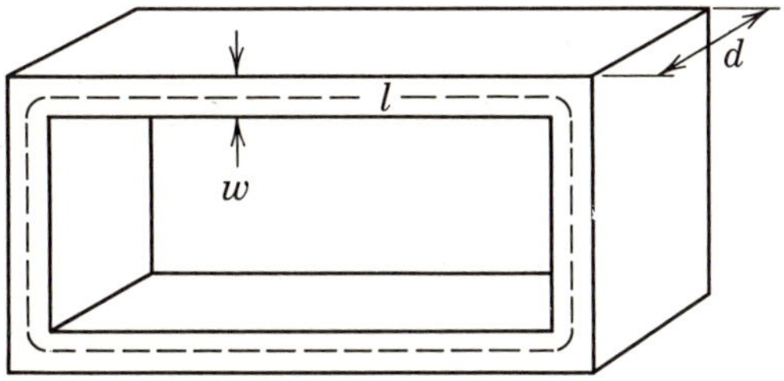

Prob. 5–5.

5–6. If in Prob. 5–5 the current in the coil is 1 amp, find (a) the flux linkages; (b) the inductance of the coil; (c) the energy stored.

5–7. If in Prob. 5–5 the turns are as in Prob. 5–6 and the flux density is still 2×10^{-3} weber/sq m but the relative permeability μ_r is 1000, find (a) the current in the coil; (b) the flux linkages; (c) the inductance; (d) the energy stored.

6

CIRCUIT ELEMENTS

The preceding chapters have introduced three basic concepts: capacitance, resistance, and inductance. Purely electric circuits are made up of combinations of these together with two additional elements termed *voltage sources* and *current sources*. The capacitance element, the resistance element, and the inductance element are termed *passive*, since they either store or dissipate energy. The latter two elements, voltage sources and current sources, are called *active* elements because they supply energy to the circuit.

The names *capacitor*, *resistor*, and *inductor* are given to physical elements which exhibit the properties of *capacitance*, *resistance*, and *inductance*. In circuit terminology the designations capacitance element, resistance element, and inductance element are commonly used; the word "element," however, is often omitted. In this introduction to circuit analysis, the elements will be assumed *linear*, *bilateral*, and *lumped*. Linear means that the value of the element is independent of the magnitude of the current in or the voltage across it. Bilateral means that the value of the element is independent of the sense of the current in or the voltage across it. Lumped means that the instantaneous current is the same in all parts of the element. Circuit analysis is the study of the current in the circuit as a whole and also in the various elements or combinations of elements when a known voltage is applied to the circuit or, conversely, the study of the voltage across the circuit and also the voltage across the various elements when a known current is supplied to the circuit.

From a knowledge of current and voltage the amount of energy consumed or stored may be found and the behavior of the circuit predicted or explained. This chapter will show the volt-ampere relationships for the various individual elements, where this volt-ampere relationship defines the relation between current in the element and the potential across it.

6–1. Resistance or Conductance. The simplest volt-ampere relationship is that of the resistance element. The relation between current and voltage in the resistance was first developed as Ohm's law, Eq. 3–15, and is here rewritten as

$$e(t) = Ri(t) \tag{6–1}$$

or

$$i(t) = Ge(t) \tag{6–2}$$

where R is the resistance of the element in ohms

G is the reciprocal of the resistance, termed *conductance*, in mhos

$e(t)$ and $i(t)$ indicate that the potential and current may be functions of time, that is, their instantaneous value at any time is dependent upon the time of investigation

Equation 6–1 states that if a current $i(t)$ exists in a resistance element R the voltage established across the element is R times that current. Equation 6–2 states that if the voltage across a resistance is $e(t)$ the current established in the element is G times that voltage. Figure 6–1 shows the graphical

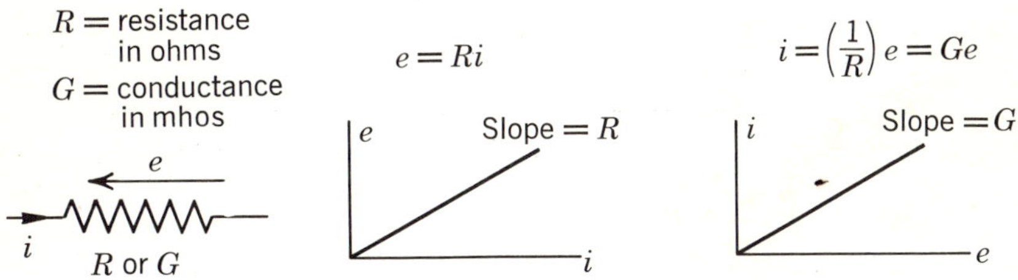

Fig. 6–1. The volt-ampere characteristics of a resistor.

relationship between $e(t)$ and $i(t)$ as well as a conventional symbol for resistance on which the direction of the current and the resultant sense of the voltage for Eqs. 6–1 and 6–2 are indicated by arrows.*

6–2. Inductance. The volt-ampere relation for an inductance element was established as Eq. 4–32† and is rewritten here as

$$e(t) = L\frac{di(t)}{dt} \tag{6–3}$$

* Many notational systems for indicating the sense (direction) of voltage and current have been used. This text will usually employ an arrow notation as shown below, where

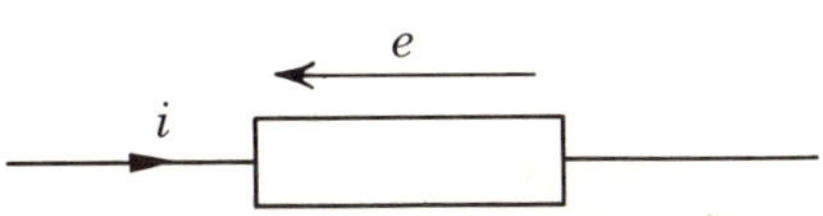

the *closed* arrowhead on the connecting line denotes the direction of current in the element or branch and the *open* arrowhead beside the element indicates the direction (sense) of the voltage. The head of the voltage arrow corresponds to the positive terminal of the element.

Another common representation employs plus and minus signs as illustrated below.

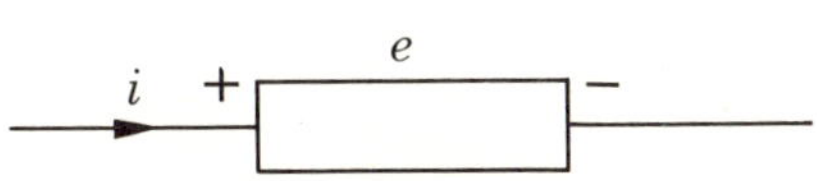

Since in general e and i are time-varying quantities, the relationship between e and i as shown is an instantaneous one.

† Equation 4–32 carried a negative sign which has been omitted here by considering the magnitude of the voltage only. The inclusion or omission of this sign has been a matter of strong debate; however, if the e and i sense arrow notation is followed for this or any other passive element the matter becomes unimportant since for "meeting" arrowheads a positive e to i relationship exists. Chapter 7, in which combinations of elements are considered, will clarify this problem.

or

$$i(t) = \frac{1}{L}\int e(t)\,dt \tag{6–4}$$

where L is the inductance in henrys, and $e(t)$ and $i(t)$ are the instantaneous values of voltage and current as previously defined.

Equation 6–3 states that if the current in an inductance is known the voltage across the inductance is found as L times the derivative of that current with respect to time. The derivative of $i(t)$ with respect to time is the rate of change of current with respect to time or the slope of a plot of current with respect to time. For example, if the current through the inductance is a constant with respect to time, the voltage across the inductance will be zero.

From Eq. 6–4 it is seen that the current through an inductance is equal to $1/L$ times the integral with respect to the time of the voltage across it. For example, if the voltage across the inductance is constant and different from zero, the current in the inductance is increasing linearly with time.

Figure 6–2 is a graphical representation of the relationship between voltage and current in the inductance along with a conventional symbol for

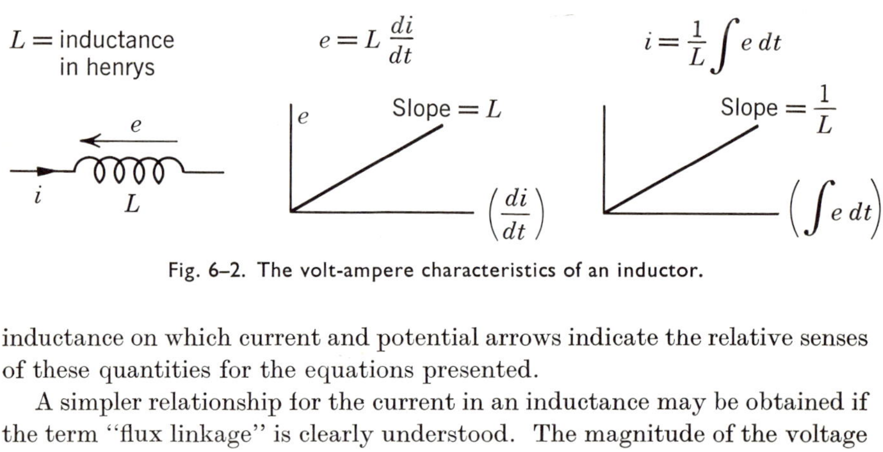

Fig. 6–2. The volt-ampere characteristics of an inductor.

inductance on which current and potential arrows indicate the relative senses of these quantities for the equations presented.

A simpler relationship for the current in an inductance may be obtained if the term "flux linkage" is clearly understood. The magnitude of the voltage across an inductance is

$$e = N\frac{d\phi}{dt} \tag{4–3}$$

which may be rewritten as

$$N\,d\phi = e\,dt \tag{6–5}$$

or

$$N\phi = \int e\,dt \tag{6–6}$$

Dimensionally, it may be seen that the product of turns and flux in units of weber-turns is equivalent to the integral of voltage with respect to time which has the units of volt-seconds.

By combining Eqs. 6–4 and 6–6, the following expressions are obtained:

$$i(t) = \frac{1}{L} N\phi \tag{6–7}$$

or

$$Li(t) = N\phi \tag{6–8}$$

Equation 6–8 indicates that the relationship between current and flux linkage is one of direct proportionality.

6–3. Capacitance. The volt-ampere relation for a capacitance element is established from Eq. 4–45 and is rewritten here as

$$e(t) = \frac{1}{C} \int i(t)\, dt \tag{6–9}$$

or

$$i(t) = C \frac{de(t)}{dt} \tag{6–10}$$

where C is the capacitance in farads

$e(t)$ and $i(t)$ are instantaneous values of voltage and current as previously defined

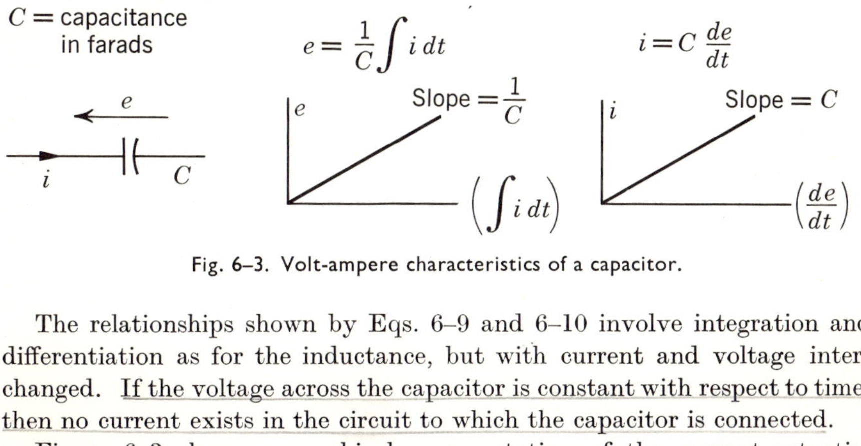

Fig. 6–3. Volt-ampere characteristics of a capacitor.

The relationships shown by Eqs. 6–9 and 6–10 involve integration and differentiation as for the inductance, but with current and voltage interchanged. If the voltage across the capacitor is constant with respect to time, then no current exists in the circuit to which the capacitor is connected.

Figure 6–3 shows a graphical representation of the current-potential relationship for the capacitance along with a conventional symbol for capacitance on which are indicated current and potential senses corresponding to the equations shown.

Equation 6–9 may also be written as

$$e(t) = \frac{1}{C} \int i(t)\, dt = \frac{1}{C} q(t) \tag{6–11}$$

This returns to the basic definition of capacitance, which indicated that the voltage across a capacitor is directly proportional to the charge on that capacitor.

6–4. Energy. A basic physical hypothesis is that energy is neither created nor destroyed but may be converted from one form into another.

In its atomic behavior mass appears to be a form of electrical energy in nature. However, until such time as scientists and engineers are able to obtain electrical energy in usable form directly from the atomic masses of matter, it is necessary to depend on the conversion of other energy forms, such as chemical, thermal, mechanical, and radiant, to electrical energy. At the present time radiant energy may be converted to electrical energy at low power levels only, although the amount of energy released by the solar system makes it highly desirable that in the future some means be found to harness and convert this radiant energy directly to electrical energy. Chemical energy is convertible into electrical energy by means of batteries.

The most common form for the conversion of mechanical to electrical energy is by means of electromagnetic machines. Oersted's and Faraday's discoveries made possible the invention of the electromagnetic dynamo. The mechanical energy necessary to operate the dynamo may be obtained from a potential energy source such as a waterfall or from thermal energy which in turn is obtained from chemical energy. To illustrate one case of this conversion let us reconsider Fig. 4–8, where a conductor ab is moved with a velocity v in a direction that cuts the lines of magnetic flux. Figure 6–4 shows

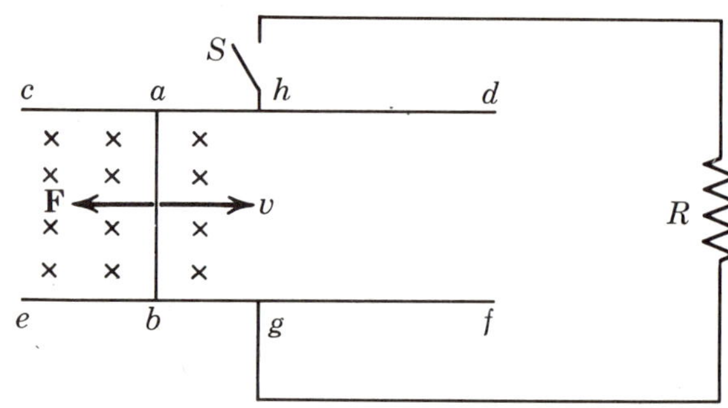

Fig. 6–4. Illustration of electric generator principle.

Fig. 4–8 with an external circuit added which includes a switch S and a resistor R. On this figure the conductor terminals a and b slide on two conductors cd and ef. Conductor cd will be positively charged and ef negatively charged owing to the motion of the conductor ab from left to right. If switch S is closed the potential difference $E = Blv$ is applied across R and the free electrons move from g to h through R. The electrostatic field across ab is momentarily disturbed and weakened. However, the magnetic force immediately pushes an equal amount of electrons toward b to re-establish the electric field and a conventionally directed current exists in the circuit $bahRgb$. The current i_{ba} flowing in the moving conductor ba is in a magnetic field and a force F of magnitude Bli is exerted on the conductor ba, opposing its motion. Note that this force is zero if the current i_{ba} is zero. In order to keep the conductor moving with the velocity v, the force F has to be counterbalanced by a mechanical force applied so as to move the conductor from left to right. The mechanical energy to be supplied by the driving device is

$$W = FS \tag{6–12}$$

where W is the energy in joules
F is the force in newtons
S is the distance in meters through which the force has acted

Since power is the rate at which work is done, differentiation of Eq. 6–12 yields

$$P = \frac{dW}{dt} = F\frac{dS}{dt} = Fv \tag{6–13}$$

But it has been previously stated in Eq. 4–21 that the force on a conductor moved through a magnetic field is

$$F = Bli$$

From this and Eq. 6–13 the power required to move the conductor is

$$P_m = Bliv \tag{6–14}$$

It has also previously been shown in Eq. 4–29 that the magnitude of the voltage induced in a conductor under these conditions of movement is

$$e = Blv \tag{4–29}$$

and therefore the electrical power P_e generated is the same as the mechanical power P_m supplied, or

$$P_m = ei = P_e \tag{6–15}$$

The mechanical power has been converted into electrical power and this power supplied to the circuit.

6–5. Energy Sources: Definition and Types. An energy source is a circuit element which converts energy from another form to electrical energy or vice versa. It may supply energy to the circuit or absorb energy from it. Examples of energy sources are batteries, d-c generators and motors, a-c generators and motors, vacuum tubes, microphones, electroacoustic motors and generators, photoelectric cells, thermocouples, and so on.

Fortunately, all electrical energy sources may be classified as voltage sources or current sources. In turn, these may be subclassified as d-c sources, a-c sources, or sources having some other function of time depending upon the behavior of the voltage (in a voltage source) or the current (in a current source) with respect to time.

6–6. Ideal and Practical Voltage Sources. An ideal voltage source is defined as one whose terminal voltage is independent of the current at that pair of terminals. Figure 6–5 shows a representation for this ideal voltage source together with a plot of its terminal voltage with respect to current variation.

The lower case symbol $e(t)$ beside the voltage source indicates that this voltage may be a function of time. A reference arrow is included to designate that when the potential is acting in the direction of the arrow it is to be

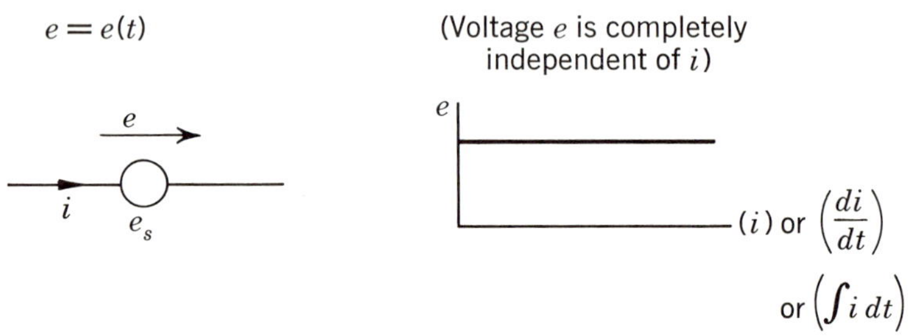

Fig. 6–5. Ideal voltage source.

considered a positive quantity. Also, on the voltage source a plus and minus indicator pair are shown as another method of representing the sense of the voltage.

In all practical cases, however, there exists an internal element (here considered to be a resistance) in series with the source of voltage. Figure 6–6 shows a schematic representation of this situation along with a typical plot of terminal voltage versus current for such a generator. Current drawn

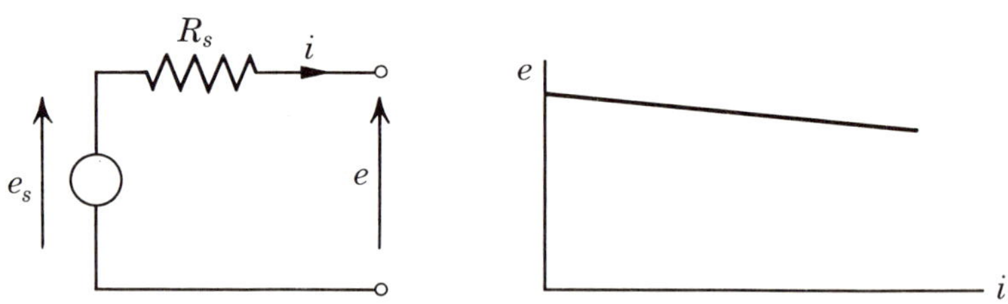

Fig. 6–6. Practical voltage source.

from this source will cause a voltage drop across this resistance and the terminal voltage will decrease with an increase in load current.

A voltage source is said to be idle when it is open-circuited. In this condition the source supplies no current and, therefore, the power supplied is zero. On the other hand, if it is desired to indicate that the terminal voltage is zero, the voltage source is shown in a short-circuited condition. If the voltage source as an energy source is removed from the circuit but circuit continuity must exist, a short circuit must be placed between the connection points.

6–7. Ideal and Practical Current Sources. An ideal current source is one in which the terminal current is independent of the voltage at that pair of terminals. Figure 6–7 shows a representation for an ideal current source and a volt-ampere plot indicating that the current supplied is independent of the terminal voltage. As in the case of the voltage source, the arrow beside the current source of Fig. 6–7 indicates that when the current is in the direction of the arrow it is to be considered a positive quantity.

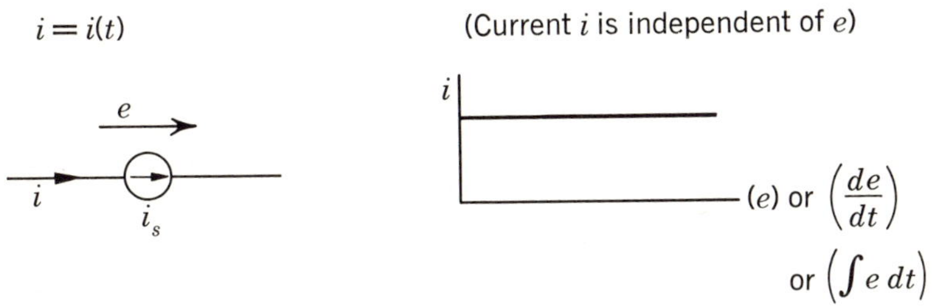

Fig. 6–7. Ideal current source.

Again, in practice, the current source can never exist without a current path, indicated by the internal conductance connected across the current source shown in Fig. 6–8. The volt-ampere plot for the non-ideal current source shown in Fig. 6–8 indicates that the current i_G is proportional to the voltage, and the terminal current of the current generator is then as shown by the curve of Fig. 6–8. The usual schematic representation for the current generator is as shown in Fig. 6–8, where the leakage impedance (here represented by a conductance G) is in parallel with the current source.

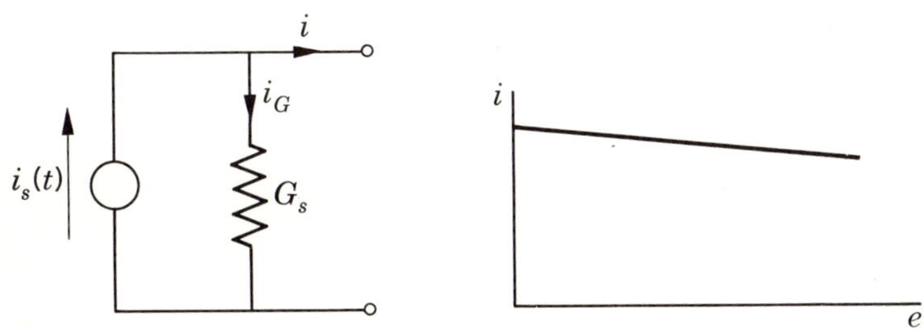

Fig. 6–8. Practical current source.

A current source is said to be idle when short-circuited at its terminals. For this situation all of the current passes through a short circuit and the product of current and voltage at the terminals is zero and no power is delivered. Removal of the current source from the circuit leaves an open circuit between the points to which the source was connected.

It is important to remember that e_s and i_s may be functions of time. The terminal voltage of the ideal voltage source and the terminal current from the ideal current source may vary with time, but must be independent of the value of any element connected to the terminals.

6–8. Equivalence of Current and Voltage Sources. A voltage source and a current source are said to be equivalent if:

1. The open-circuit voltages at their terminals are equal.
2. The short-circuit currents at their terminals are equal.

These statements are special forms of the general equivalence relationship that describes sources as equivalent if their volt-ampere characteristics are the same for all terminal conditions.

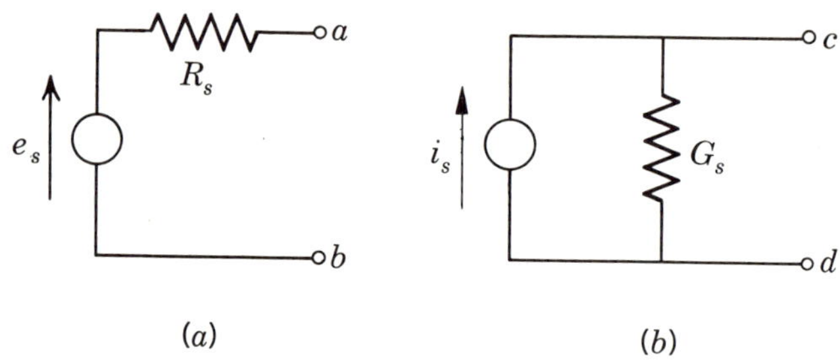

Fig. 6–9. Resistive voltage and current sources for comparison.

Figure 6–9*a* represents a non-ideal voltage source with terminals *ab*, and Fig. 6–9*b* represents a non-ideal current source with terminals *cd*. The open-circuit voltage across terminals *ab* of the voltage source is

$$e_{oc} = e_s \tag{6–16}$$

The open-circuit voltage at terminals *cd* of the current source is

$$e_{oc} = \frac{i_s}{G_s} \tag{6–17}$$

Hence for equivalence, Eqs. 6–16 and 6–17 must be equal, and

$$e_s = \frac{i_s}{G_s} \tag{6–18}$$

or

$$i_s = e_s G_s \tag{6–19}$$

The short-circuit current of the voltage source with terminals *ab* short-circuited is

$$i_{sc} = \frac{e_s}{R_s} \tag{6–20}$$

The short-circuited current of the current source with terminals *cd* short-circuited is

$$i_{sc} = i_s \tag{6–21}$$

For equivalence Eqs. 6–20 and 6–21 must be equal or

$$i_s = \frac{e_s}{R_s} \tag{6–22}$$

From Eqs. 6–19 and 6–22 we have

$$R_s = \frac{1}{G_s} \tag{6–23}$$

Therefore a current source equivalent to the voltage source of Fig. 6–9*a* is as

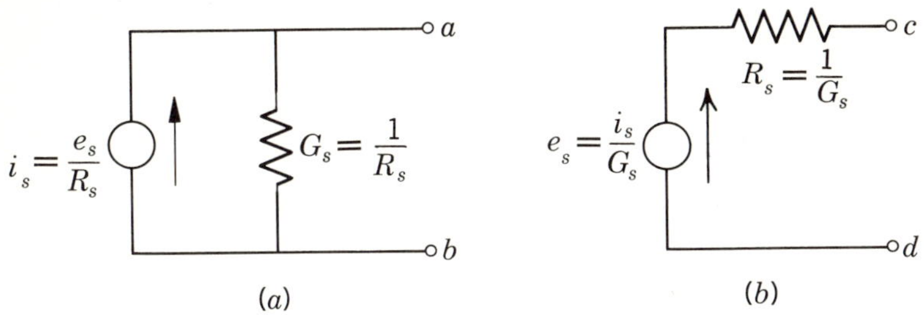

Fig. 6–10. Current and voltage sources equivalent to Fig. 6–9.

shown in Fig. 6–10*a* and a voltage source equivalent to the current source of Fig. 6–9*b* is as shown in Fig. 6–10*b*.

If a voltage source has an inductance L_s between the ideal source and its terminals as shown in Fig. 6–11*a*, an equivalent current source will be of the

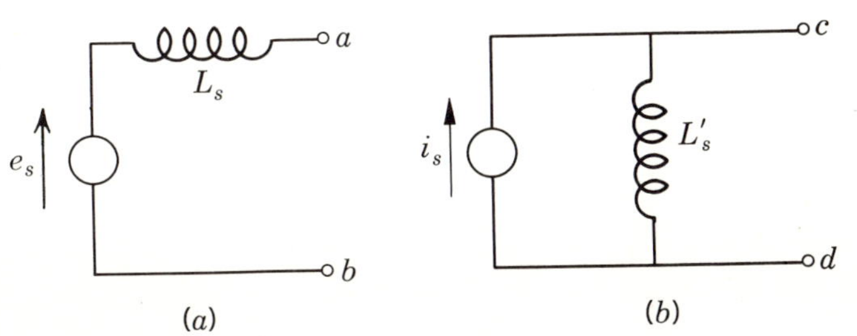

Fig. 6–11. Equivalent inductive current and voltage sources.

form shown in Fig. 6–11*b*. The open-circuit voltage at *ab* must equal the open-circuit voltage at *cd*, or

$$e_s = L'_s \frac{di_s}{dt} \tag{6–24}$$

The short-circuit current of the voltage source must be equal to the short-circuit current of the current source. With *ab* short-circuited

$$L_s \frac{di}{dt} = e_s \tag{6–25}$$

or

$$i_{sc} = \frac{1}{L_s}\int e_s\,dt \tag{6–26}$$

With cd short-circuited the current

$$i_{sc} = i_s \tag{6–27}$$

Therefore from Eqs. 6–27 and 6–26

$$i_s = \frac{1}{L_s}\int e_s\,dt \tag{6–28}$$

But from Eq. 6–24

$$i_s = \frac{1}{L'_s}\int e_s\,dt \tag{6–29}$$

or

$$L_s = L'_s \tag{6–30}$$

Therefore

$$i_s = \frac{1}{L_s}\int e_s\,dt \tag{6–31}$$

and

$$e_s = L_s\frac{di_s}{dt} \tag{6–32}$$

relate the values of e_s and i_s if the sources are to be equivalent. In other words, the current i_s in Fig. 6–11*b* must be $1/L_s$ times the integral of the voltage e_s of

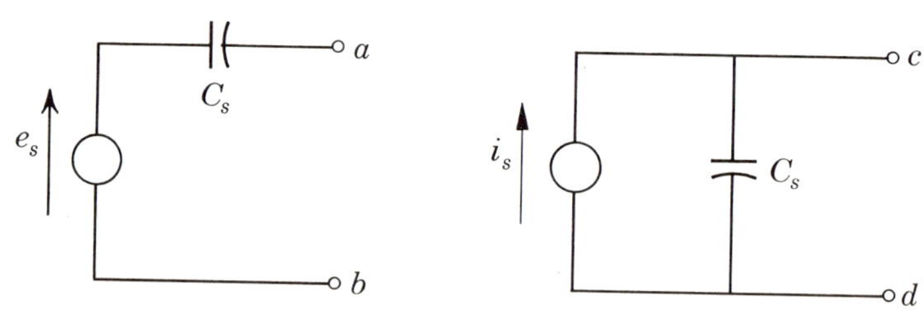

(*a*) (*b*)

Fig. 6–12. Equivalent capacitive current and voltage source.

Fig. 6–11*a*, or the voltage e_s of Fig. 6–11*a* must be L_s times the derivative of the current i_s of Fig. 6–11*b* if Fig. 6–11*a* and 6–11*b* are to be equivalent.

If a voltage source has a capacitance C_s between the ideal source and its terminals as shown in Fig. 6–12*a*, an equivalent current source will be of the form shown in Fig. 6–12*b*. Here the open-circuit voltage of the voltage source is

$$e_{oc} = e_s \tag{6–33}$$

and the open-circuit voltage of the current source is

$$e_{\text{oc}} = \frac{1}{C_s} \int i_s \, dt \tag{6–34}$$

Therefore for equivalence

$$e_s = \frac{1}{C_s} \int i_s \, dt \tag{6–35}$$

The short-circuit current for the voltage source is

$$i_{\text{sc}} = C_s \frac{de_s}{dt} \tag{6–36}$$

and for equivalence

$$i_s = C_s \frac{de_s}{dt} \tag{6–37}$$

For equivalence i_s must be C_s times the derivative of e_s, or e_s must be $1/C_s$ times the integral of i_s.

PROBLEMS

6–1. A conductance of value 3 mhos carries a current which is increasing linearly at the rate of 0.1 amp/sec. If $i = 0$ at $t = 0$ what is the instantaneous power at $t = 5$ sec? What is the total energy dissipated as heat during the first minute of conduction? Sketch the curve of power versus time for the first minute and comment on the physical significance of the area under the curve.

6–2. Sketch the waveforms of the voltage drop across a 15-ohm resistor for each of the current forms shown.

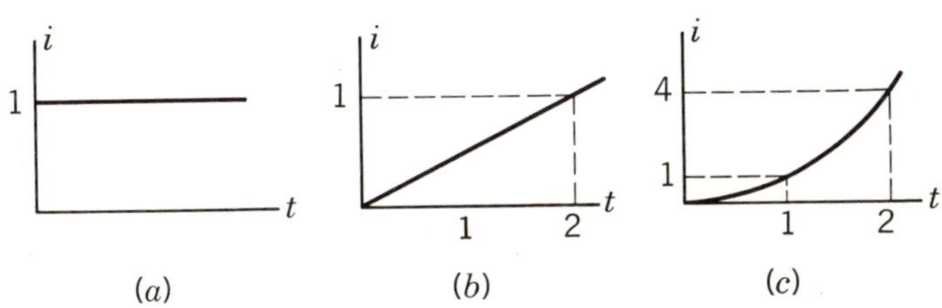

Prob. 6–2.

6–3. What is the voltage induced across a 10-mh inductor for each of the current waveforms of Prob. 6–2?

6–4. What is the potential across a capacitor if the displacement current assumes the waveform (*b*) of Prob. 6–2?

6–5. A coil of 1000 turns has a flux through it which changes linearly from 0 to 1 weber in 10 sec. Find the voltage induced in the coil.

6–6. The induced voltage in a certain coil is 200 v when the current is dropped linearly from 10 to 0 amp in 5 sec. Find the inductance of the coil.

6–7. The charge on a 0.4-μf capacitor is 20 μcoulombs. If the charge drops uniformly to zero in 5 sec, what is the displacement current in the capacitor?

6–8. The current flowing in an inductance has a form as plotted in the figure. Sketch to scale the induced voltage if the inductance has a value of 0.5 h.

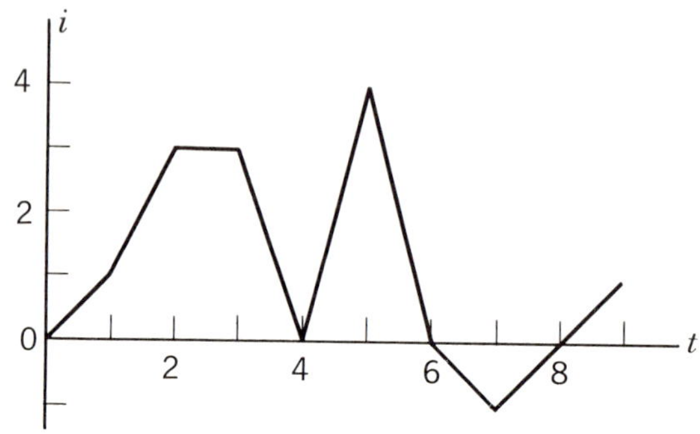

Prob. 6–8.

6–9. The voltage across a capacitor of 5 μf changes linearly from 0 to 10 v in 5 sec. Determine the current.

6–10. A 100-μf capacitor has an initial voltage of 100 v. At zero time the current through the capacitor starts to rise linearly at a rate of 10 amp/sec. Determine the value of the voltage across the capacitor at the end of 0.05 sec.

6–11. A parallel plate capacitor has plates of area 0.25 sq m and adjustable spacing. A fixed potential of 100 v is applied to the plates and the spacing varied from 10^{-3} to 10^{-2} m in 1 sec at such a rate that the capacitance change is linear with time. What current will flow in the external circuit?

6–12. A uniform electric field in air with an initial intensity of 1×10^6 newtons/coulomb collapses at a uniform rate in (a) 1 ms and (b) 1 μs. What is the displacement current density in each case?

6–13. It requires a force of 0.9 lb to move a 1-m conductor through a magnetic field of 0.5-weber/m^2 density at a constant velocity of 6.4 m/sec. Referring to Fig. 6–6, if R is replaced by a conductance of 0.25 mho what is the power dissipated by the conductance?

7

CIRCUIT EQUILIBRIUM EQUATIONS

In the preceding chapter the individual elements which make up electric circuits were discussed and their volt-ampere characteristics defined. Electric circuits consist of various combinations of these passive and active elements. To analyze combinations of elements it is necessary to establish a consistent and logical method of attack. The basic mathematical-electrical tools for this purpose are Kirchhoff's equations and a consistent notational system. The development of such a method is the topic of this chapter.

7–1. Kirchhoff's Current Law. At any instant and at any point in a continuous circuit, the number of electrons entering a chosen element of conductor volume is equal to the number of electrons leaving that conductor volume. Stated in another way, the number of free electrons per unit volume of a conductor in a circuit is constant.

If the point selected is the junction of several circuit elements (see Fig. 7–1), the sum of the currents toward the junction must be equal to the sum of the currents away from the junction, or

$$i_d = i_a + i_b + i_c \tag{7–1}$$

If the direction of the currents is taken into consideration, the sum of the currents at a junction is zero. This is expressed mathematically as

$$\sum i_n = 0 \tag{7–2}$$

and is known as *Kirchhoff's current law.*

EXAMPLE 7–1. KIRCHHOFF'S CURRENT LAW. A line representation for an electric circuit is as shown in Fig. 7–2, with i_1, i_2, and i_4 known. Express currents i_3, i_5, and i_6 in terms of the known currents.

$$\text{at } b \quad i_1 - i_2 - i_3 = 0$$
$$\text{or } i_3 = i_1 - i_2$$

$$\text{at } c \quad i_3 - i_4 - i_5 = 0$$
$$\text{or } i_5 = i_3 - i_4 = i_1 - i_2 - i_4$$

$$\text{at } d \quad i_2 + i_4 - i_6 = 0$$
$$\text{or } i_6 = i_2 + i_4$$

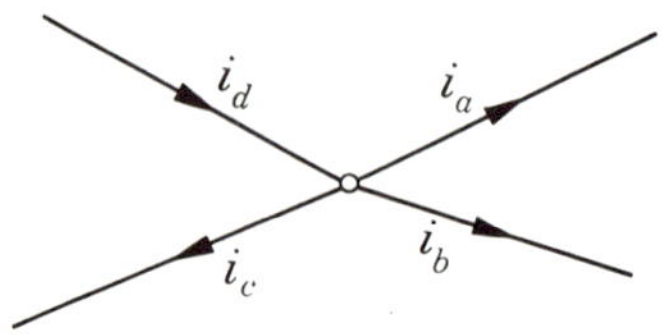

Fig. 7–1. Current junction illustrating Kirchhoff's current law.

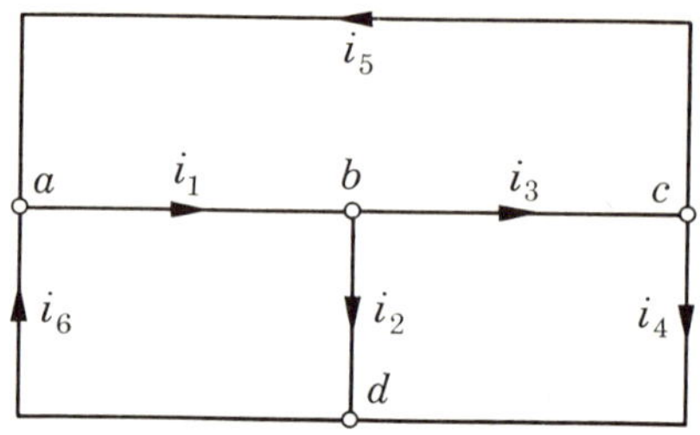

Fig. 7–2. Summations of currents at nodes.

7–2. Current-Voltage Notation. A convenient notational system relating current and voltage for any circuit element was established in Chapter 6 and is further discussed here.

Current and voltage sources are active circuit elements. The conventional current is assigned a direction such that it leaves the positive terminal of a

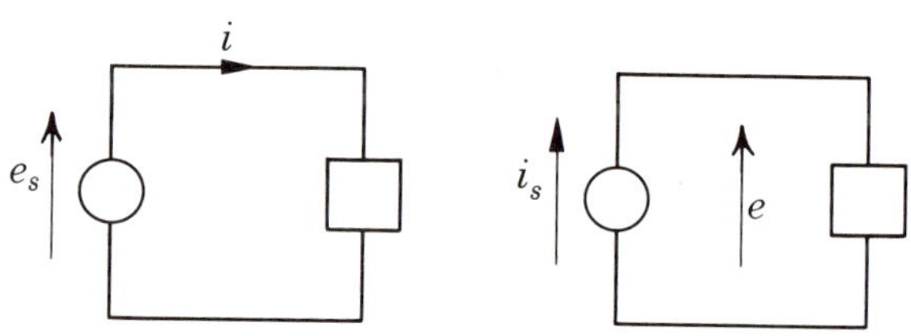

Fig. 7–3. Current-voltage notation for active elements.

source. Figure 7–3 shows such sources with their relative positive current and voltage senses indicated by arrows.

Resistance, inductance, and capacitance elements are passive and the conventional current is considered to enter the most positive terminal and leave the negative terminal of the element. A consistent voltage and current arrow sense designation is therefore as shown in Fig. 7–4.

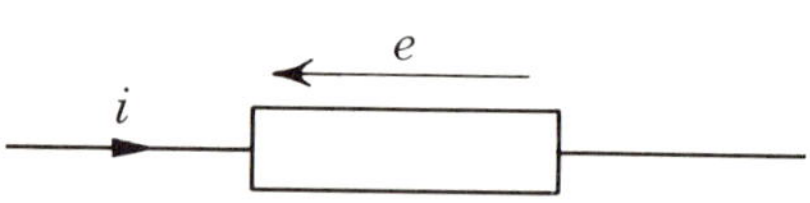

Fig. 7–4. Current-voltage notation for a passive element.

In summary, the most positive terminal of any element is indicated by the head of the voltage arrow. If a reference current direction is obtained from the sense of the source (or by arbitrary assignment in more complex circuits*), the sense arrow for potential is determined for all other elements. These concepts will be illustrated by the discussion of equilibrium equations following in this chapter and also later in a discussion of multimesh networks.

* Note particularly that the arrow direction may denote an arbitrarily assumed positive sense that is not necessarily the actual positive terminal of the physical network.

7-3. Kirchhoff's Voltage Law. In any closed circuit at a chosen instant the voltage between any two points is the same regardless of the path followed between those points. For example, in Fig. 7-5 a path may be traced

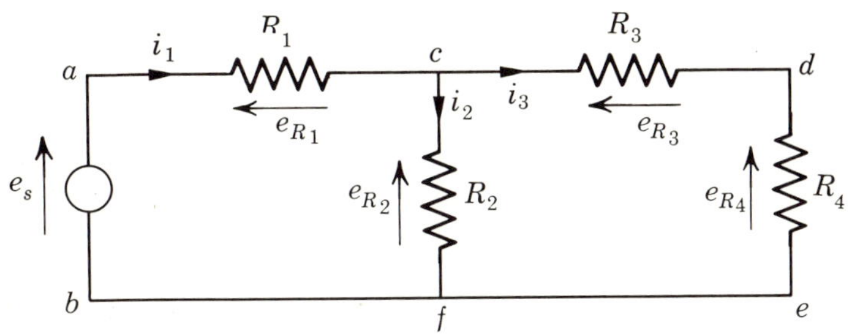

Fig. 7-5. Illustration of voltages along several paths.

between a and b through the source, through $acfb$, or through $acdefb$, and the voltages written*

$$e_s = e_{R_1} + e_{R_2} = e_{R_1} + e_{R_3} + e_{R_4} \tag{7-3}$$

This may also be stated: The sum of the voltages around any closed path is zero since

$$e_s - e_{R_1} - e_{R_2} = 0 \tag{7-4}$$

or

$$e_s - e_{R_1} - e_{R_3} - e_{R_4} = 0 \tag{7-5}$$

Also, around $cdef$

$$e_{R_2} - e_{R_3} - e_{R_4} = 0 \tag{7-6}$$

where a loop involving no active elements is considered.

A concise mathematical expression of the foregoing expressions is

$$\sum e_n = 0 \tag{7-7}$$

and is known as *Kirchhoff's voltage law.*

Equations 7-4 through 7-6 illustrate a use of the voltage sense arrow notation. Using this form of expression of Kirchhoff's laws, the sum of *all* element voltages is equated to zero. An element voltage is considered as positive if the arbitrarily chosen path direction passes through the element in the direction of the arrow and negative if against the arrow.

Another form commonly used is illustrated by Eq. 7-3, where the active element voltages are written to the left of the equality and the passive to the right. When using this form, an active element voltage is considered positive if passed through in the direction of the arrow, and a passive element voltage is positive if passed through against the arrow.

Actually, these two forms are the same and it is a matter of preference as to which is used.

* The form of e written here is without (t). In general, the use of a lower case symbol implies that the variable is a function of time and (t) will no longer be written.

7–4. Series Equilibrium Equations. The first step in the solution of a circuit problem is the schematic representation of the circuit elements and their manner of connection. With this schematic representation as a base, a current sense may be obtained from the source (or arbitrarily assigned for more complex combinations of sources) and the voltage sense for each element shown by arrows on the diagram as described in Art. 7–2. From this diagram, using Kirchhoff's laws, the equilibrium equations are written.

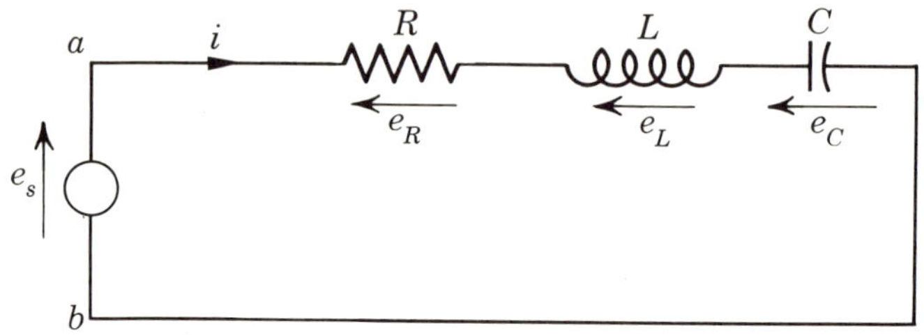

Fig. 7–6. Schematic diagram for a series circuit.

To illustrate this procedure, consider a circuit with resistance, inductance, and capacitance connected as shown in Fig. 7–6 with e_s the applied source voltage. From Kirchhoff's current law, it is known that the same current exists in every element. Let us denote this instantaneous current as i and choose its direction as the conventional current direction from the tip of the voltage source sense arrow. Across each element there is a voltage symbolized by e_R, e_L, and e_C with sense arrows determined from the current direction.

The voltage of the source is

$$e_{ba} = e_s \tag{7–8}$$

Kirchhoff's voltage law states that in a closed circuit the voltage between any two points must be the same. Therefore,

$$e_s = e_{ba} = e_R + e_L + e_C \tag{7–9}$$

or

$$e_s - e_R - e_L - e_C = 0 \tag{7–10}$$

which also shows that the sum of the voltages around this closed path is zero.

The voltages across each element may be obtained from the volt-ampere characteristics for that element, as

$$e_R = Ri \tag{7–11}$$

$$e_L = L\frac{di}{dt} \tag{7–12}$$

$$e_C = \frac{1}{C}\int i\,dt \tag{7–13}$$

and Eq. 7–9 rewritten as

$$e_{ba} = Ri + L\frac{di}{dt} + \frac{1}{C}\int i\,dt \tag{7–14}$$

In summary, for a series circuit the current in each circuit element is the same, and the voltage across each element may be secured from a knowledge of that element's volt-ampere characteristics.

The explicit solution of the equilibrium equation is dependent upon the source value as a function of time and will be discussed in later chapters.

7–5. Series Connection of Similar Circuit Elements. An equivalent element for a number of similar circuit elements connected in series may be

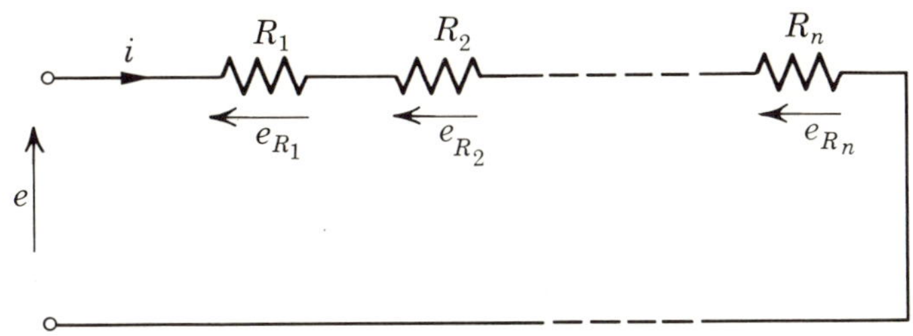

Fig. 7–7. Resistances in series.

obtained by the use of series circuit concepts. For example, if resistances are connected in series as shown in Fig. 7–7, the sum of the voltages across each of the resistances may be written as equal to the voltage at the terminals, or

$$e = e_{R_1} + e_{R_2} + \cdots + e_{R_n} \tag{7–15}$$

From the volt-ampere relation for the resistance, the voltage across each resistor is written as

$$e = Ri \tag{7–16}$$

and therefore Eq. 7–15 may be rewritten as

$$e = R_1 i + R_2 i + \cdots + R_n i \tag{7–17}$$

But, the current i is the same in each of the resistive elements and therefore

$$e = i(R_1 + R_2 + \cdots + R_n) \tag{7–18}$$

If the ratio of e to i is taken as defining an equivalent resistance R', Eq. 7–18 establishes

$$R' = \frac{e}{i} = R_1 + R_2 + \cdots + R_n \tag{7–19}$$

A resistance equivalent to a number of individual resistances in series may be obtained by the addition of the values of each of the individual resistances.

An inductance equivalent to inductances in series may be obtained by an analogous procedure. Figure 7–8 shows inductances in series together with the voltage across each inductive element and the current i through the elements. Equations similar to those written for the series resistance case, Eqs. 7–15 through 7–19, may be written for the inductance as follows:

$$e = e_{L_1} + e_{L_2} + \cdots + e_{L_n} \tag{7–20}$$

$$e = L\frac{di}{dt} \tag{7–21}$$

$$e = L_1\frac{di}{dt} + L_2\frac{di}{dt} + \cdots + L_n\frac{di}{dt} \tag{7–22}$$

$$e = \frac{di}{dt}(L_1 + L_2 + \cdots + L_n) \tag{7–23}$$

and finally

$$L' = \frac{e}{di/dt} = L_1 + L_2 + \cdots + L_n \tag{7–24}$$

An inductance equivalent to individual inductances in series may be obtained by taking the sum of the values of the individual inductance elements.

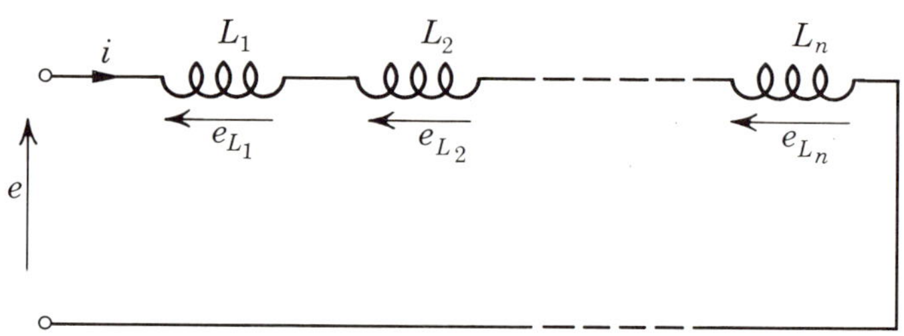

Fig. 7–8. Inductances in series.

For capacitances in series a similar approach may be made, and Fig. 7–9 indicates such a configuration having a common current i with the voltage sense across each element indicated by arrows. Again, equations similar to those written for the resistive case, Eqs. 7–15 through 7–19, may be written as

$$e = e_{C_1} + e_{C_2} + \cdots + e_{C_n} \tag{7–25}$$

$$e_{C_n} = \frac{1}{C_n}\int i\,dt \tag{7–26}$$

$$e = \frac{1}{C_1}\int i\,dt + \frac{1}{C_2}\int i\,dt + \cdots + \frac{1}{C_n}\int i\,dt \tag{7–27}$$

$$e = \int i\,dt\left(\frac{1}{C_1} + \frac{1}{C_2} + \cdots + \frac{1}{C_n}\right) \tag{7–28}$$

and finally

$$\frac{1}{C'} = \frac{e}{\int i\, dt} = \frac{1}{C_1} + \frac{1}{C_2} + \cdots + \frac{1}{C_n} \tag{7-29}$$

Equation 7-29 shows that the equivalent of a group of capacitors in series may be conveniently obtained.

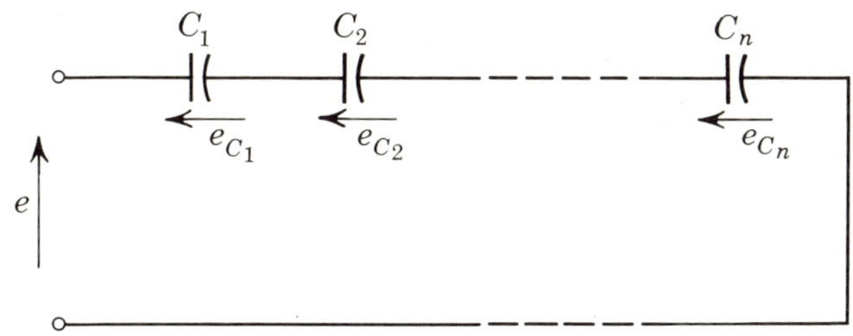

Fig. 7-9. Capacitances in series.

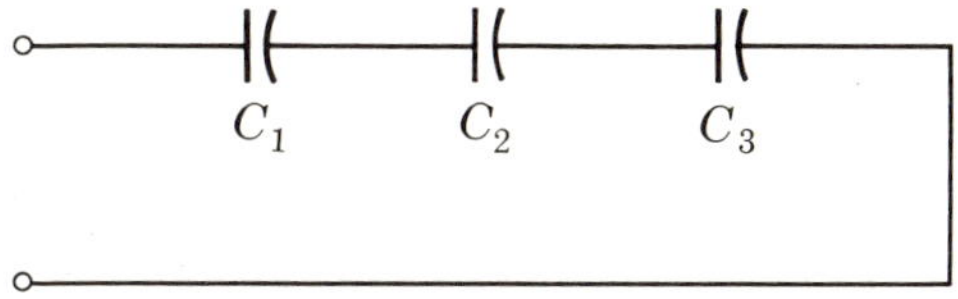

Fig. 7-10. Three capacitors in series.

Example 7-2. Capacitances in Series. For the circuit of Fig. 7-10, calculate the equivalent capacitance if the values of C_1, C_2, and C_3 are respectively 10, 15, and 20 μf.

$$\frac{1}{C'} = \left(\frac{1}{10} + \frac{1}{15} + \frac{1}{20}\right) \times 10^6$$

$$\frac{1}{C'} = \frac{6+4+3}{60} \times 10^6 = \frac{13}{60} \times 10^6$$

$$C' = \frac{60}{13} \times 10^{-6} = 4.61\ \mu\text{f}$$

7-6. Parallel Equilibrium Equations. A parallel combination of circuit elements is represented by Fig. 7-11. The current source i_s supplies a current which flows in the circuit elements in the direction shown. Since the three elements shown are in parallel, the voltage across each element must be the same and is designated as e with an arrow as shown on the figure used to indicate its sense.

From Kirchhoff's current law

$$i_s = i_R + i_C + i_L \tag{7-30}$$

and from the volt-ampere characteristics for the circuit elements,

$$\begin{aligned} i_R &= \frac{1}{R} e = Ge \\ i_C &= C \frac{de}{dt} \\ i_L &= \frac{1}{L} \int e \, dt \end{aligned} \tag{7-31}$$

If the currents for the individual elements of Eqs. 7–31 are substituted into Eq. 7–30, we have

$$i_s = Ge + C \frac{de}{dt} + \frac{1}{L} \int e \, dt \tag{7-32}$$

Equations 7–14 and 7–32 are termed *integrodifferential equations* because they contain both differential and integral terms. It should also be noted

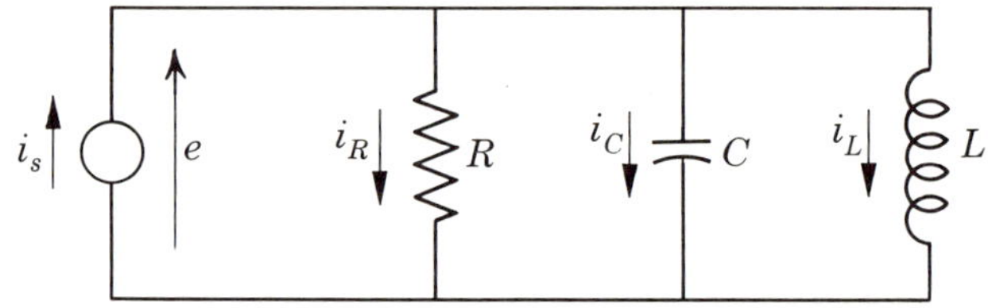

Fig. 7–11. Schematic representation of a parallel circuit.

that these two equations have the same basic mathematical form; as such, they are called *duals*. The properties of duals will be discussed in detail in Chapter 14.

Also, in comparing Eqs. 7–14 and 7–32, it may be seen that with a voltage source and series elements a common current was assumed in the circuit, and the equilibrium equation was written in terms of the assumed current; with a current source and parallel elements a voltage was assumed across the parallel combination and the equilibrium equation was written in terms of this assumed voltage. The procedure for a detailed solution of these equations will be considered in Chapter 8.

7–7. Parallel Combinations of Similar Elements. Analogously with the development of Art. 7–5 for series combinations of similar elements, an element equivalent to similar elements connected in parallel may be obtained as follows: If resistance elements are considered, as shown in

Fig. 7–12, and a voltage e appears across the circuit, then

$$i_1 = \frac{e}{R_1}$$

$$i_2 = \frac{e}{R_2} \tag{7–33}$$

$$i_n = \frac{e}{R_n}$$

where

$$i = i_1 + i_2 + \cdots + i_n \tag{7–34}$$

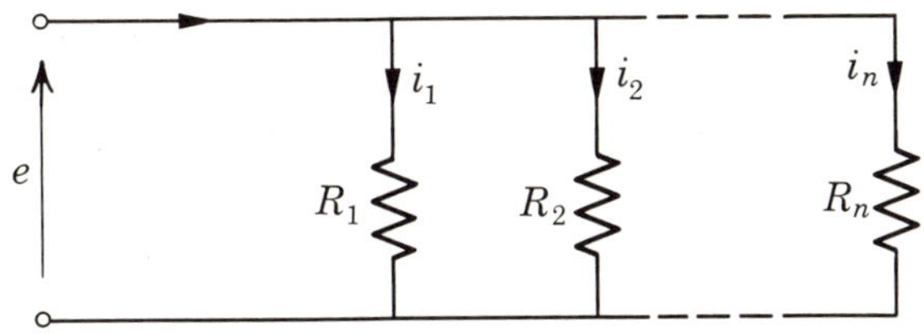

Fig. 7–12. Resistances in parallel.

Therefore

$$i = \frac{e}{R_1} + \frac{e}{R_2} + \cdots + \frac{e}{R_n} \tag{7–35}$$

and the equivalent resistance R' of the combination is

$$R' = \frac{e}{i} = \frac{1}{\dfrac{1}{R_1} + \dfrac{1}{R_2} + \cdots + \dfrac{1}{R_n}} \tag{7–36}$$

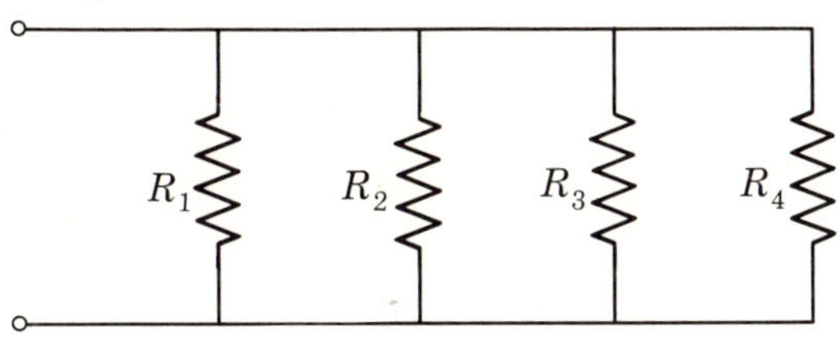

Fig. 7–13. Four resistors in parallel.

For two resistances in parallel Eq. 7–36 assumes a simple form

$$R' = \frac{1}{\dfrac{1}{R_1} + \dfrac{1}{R_2}} = \frac{R_1 R_2}{R_1 + R_2} \tag{7–37}$$

EXAMPLE 7–3. RESISTANCES IN PARALLEL. For the circuit shown in Fig. 7–13, calculate the equivalent resistance if R_1, R_2, R_3, and R_4 are respectively equal to 5, 10, 15, and 20 ohms.

$$\frac{1}{R'} = \frac{1}{5} + \frac{1}{10} + \frac{1}{15} + \frac{1}{20} = \frac{25}{60}$$

$$R' = \frac{60}{25} = \frac{12}{5} = 2.4 \text{ ohms}$$

If inductances are connected in parallel as shown in Fig. 7–14 and again a voltage e appears across the circuit, then

$$\frac{di_1}{dt} = \frac{e}{L_1}$$
$$\frac{di_2}{dt} = \frac{e}{L_2} \qquad (7\text{–}38)$$
$$\frac{di_n}{dt} = \frac{e}{L_n}$$

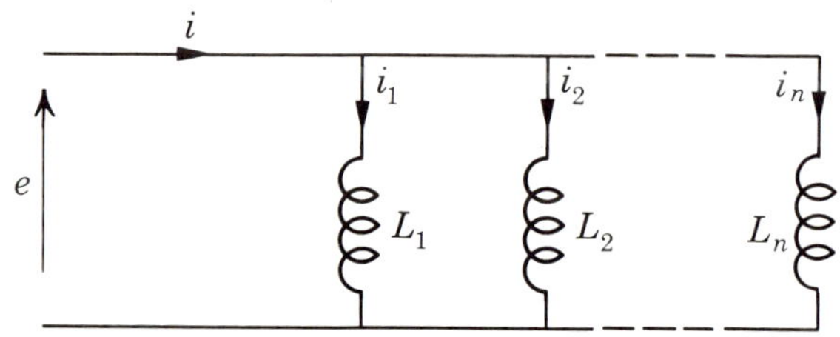

Fig. 7–14. Inductances in parallel.

where

$$i = i_1 + i_2 + \cdots + i_n \qquad (7\text{–}39)$$

Therefore

$$\frac{di}{dt} = \frac{e}{L_1} + \frac{e}{L_2} + \cdots + \frac{e}{L_n} \qquad (7\text{–}40)$$

but in general

$$L' \frac{di}{dt} = e \qquad (7\text{–}41)$$

and

$$L' = \frac{e}{\frac{di}{dt}} = \frac{1}{\frac{1}{L_1} + \frac{1}{L_2} + \cdots + \frac{1}{L_n}} \qquad (7\text{–}42)$$

Here it is assumed that there is no mutual inductance between the various inductance elements. When there are only two inductances in parallel Eq. 7–42 becomes

$$L' = \frac{1}{\frac{1}{L_1} + \frac{1}{L_2}} = \frac{L_1 L_2}{L_1 + L_2} \qquad (7\text{–}43)$$

EXAMPLE 7–4. INDUCTANCES IN PARALLEL. For the circuit of Fig. 7–15, calculate the equivalent inductance if L_1, L_2, and L_3 are respectively equal to 10 μh, 20 μh, and 0.1 mh.

$$\frac{1}{L'} = \left(\frac{1}{10} + \frac{1}{20} + \frac{1}{100}\right) \times 10^6$$

$$= \frac{10 + 5 + 1}{100} \times 10^6 = \frac{16}{100} \times 10^6$$

$$L' = \frac{100}{16} \times 10^{-6} = 6.25\ \mu\text{h}$$

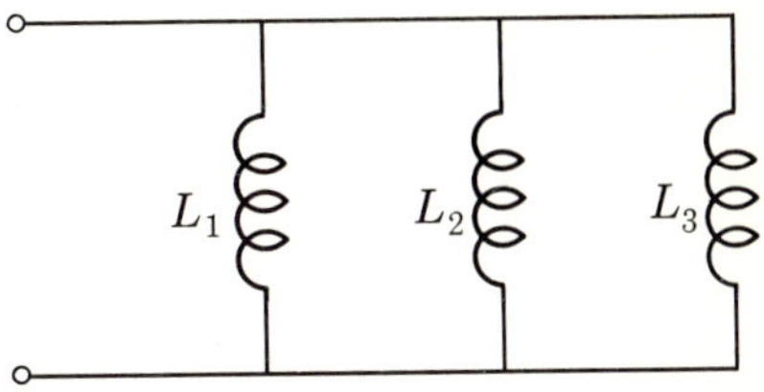

Fig. 7–15. Three inductances in parallel.

For capacitors connected in parallel as shown in Fig. 7–16 and again a potential across the circuit of e volts,

$$i_1 = C_1 \frac{de}{dt}$$

$$i_2 = C_2 \frac{de}{dt} \tag{7–44}$$

$$i_n = C_n \frac{de}{dt}$$

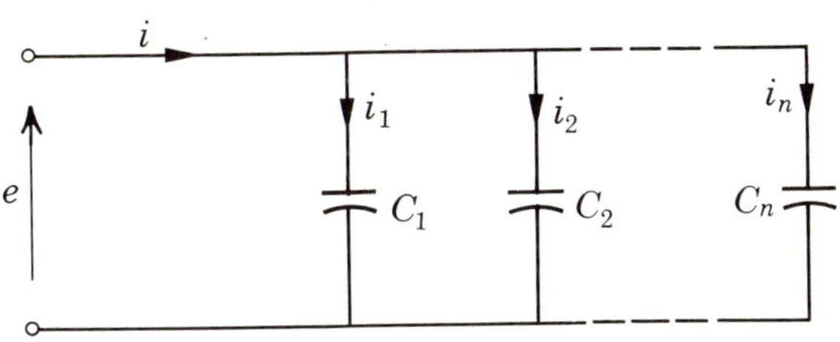

Fig. 7–16. Capacitors in parallel.

where

$$i = i_1 + i_2 + \cdots + i_n \tag{7–45}$$

Therefore

$$i = (C_1 + C_2 + \cdots + C_n) \frac{de}{dt} \tag{7–46}$$

and

$$i = C' \frac{de}{dt} \tag{7–47}$$

where

$$c' = C_1 + C_2 + \cdots + C_n \tag{7–48}$$

Capacitors in parallel have an equivalent capacitance obtained as the sum of their individual capacitances.

7–8. Superposition Theorem for Networks. Assume for simplicity a series circuit as shown in Fig. 7–17. If the source voltage is e'_s, then with current i' in the circuit the equilibrium equation is

$$e'_s = Ri' + L\frac{di'}{dt} + \frac{1}{C}\int i'\,dt \tag{7–49}$$

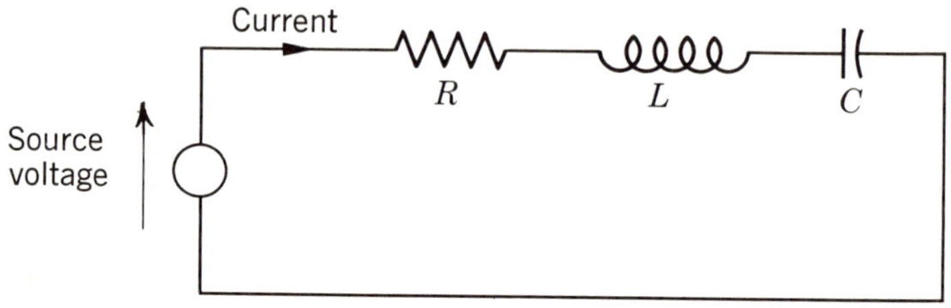

Fig. 7–17. *R-L-C* network.

Now, if instead of e'_s a source voltage e''_s is impressed causing a current i'', the equilibrium equation becomes

$$e''_s = Ri'' + L\frac{di''}{dt} + \frac{1}{C}\int i''\,dt \tag{7–50}$$

If R, L, and C are linear elements, then Eqs. 7–49 and 7–50 may be added and we have

$$e'_s + e''_s = R(i' + i'') + L\frac{d}{dt}(i' + i'') + \frac{1}{C}\int (i' + i'')\,dt \tag{7–51}$$

or

$$e_s = Ri + L\frac{di}{dt} + \frac{1}{C}\int i\,dt \tag{7–52}$$

where

$$e_s = e'_s + e''_s$$

and

$$i = i' + i'' \tag{7–53}$$

The principle of superposition is applicable in any situation in which the elements are linear and bilateral—that is, their value does not depend upon the current in or the voltage across the element either in magnitude or sense.

A statement of the superposition theorem is: If the elements in a network are linear and bilateral, the potential across or the current in any element caused by a combination of sources may be calculated by calculating the potentials and currents caused by each source acting alone* and adding the individual values algebraically with due regard for sign to obtain the total quantities.

* Although each source acts alone, all other sources are replaced by either open or short circuits for current or voltage sources, respectively.

EXAMPLE 7–5. SUPERPOSITION. Find the current in all branches of the circuit of Fig. 7–18 by superposition.

Note: An additional subscript added to those already present will designate the source to which the current or voltage component under inspection is

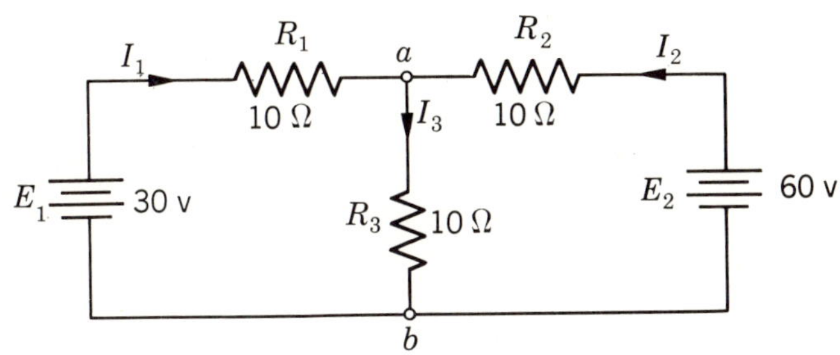

Fig. 7–18. Circuit with two sources of voltage.

due (that is, I_{12} is the component of I_1 caused by E_2). Current directions shown in Fig. 7–18 are *assumed* directions. A negative sign in an answer will indicate that the actual direction is opposite to that of the assumed one.

$$I_{11} = \frac{E_1}{R_1 + \dfrac{R_2R_3}{R_2 + R_3}} = \frac{30}{10 + 5} = 2 \text{ amp}$$

$$E_{ba1} = (I_{11})\left(\frac{R_2R_3}{R_2 + R_3}\right) = 2 \times 5 = 10 \text{ v}$$

$$I_{31} = \frac{E_{ba1}}{R_3} = \frac{10}{10} = 1 \text{ amp}$$

$$I_{21} = -\frac{E_{ba1}}{R_2} = -\frac{10}{10} = -1 \text{ amp}$$

$$I_{22} = \frac{E_2}{R_2 + \dfrac{R_1R_3}{R_1 + R_3}} = \frac{60}{10 + 5} = 4 \text{ amp}$$

$$E_{ba2} = (I_{22})\left(\frac{R_1R_3}{R_1 + R_3}\right) = 4 \times 5 = 20 \text{ v}$$

$$I_{32} = \frac{E_{ba2}}{R_3} = \frac{20}{10} = 2 \text{ amp}$$

$$I_{12} = -\frac{E_{ba2}}{R_1} = -\frac{20}{10} = -2 \text{ amp}$$

$$I_1 = I_{11} + I_{12} = 2 - 2 = 0$$

$$I_2 = I_{21} + I_{22} = -1 + 4 = 3 \text{ amp}$$

$$I_3 = I_{31} + I_{32} = 1 + 2 = 3 \text{ amp}$$

PROBLEMS

7–1. What is the magnitude and direction of the current through each of the resistances in the circuit shown?

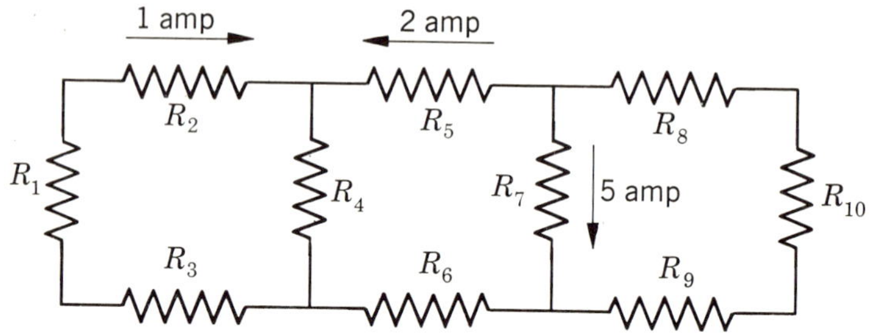

Prob. 7–1.

7–2. What is the magnitude and sense of all the unknown voltages in the adjacent circuit?

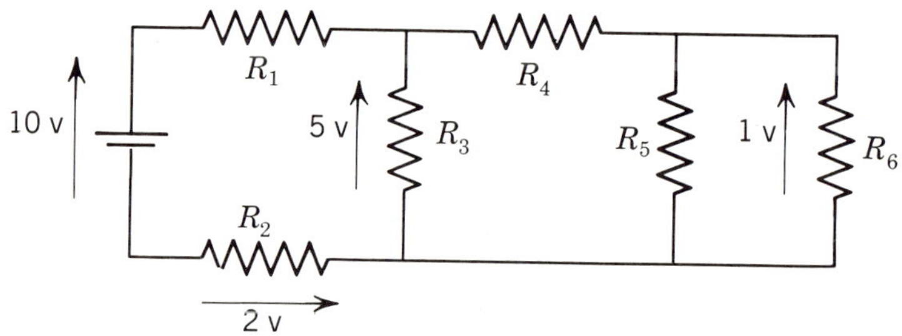

Prob. 7–2.

7–3. Two resistors of 5 ohms each are connected in series. What is the resistance of the combination?

7–4. Two resistors of 5 ohms each are connected in parallel. What is the resistance of the combination?

7–5. Two resistors of 10 ohms each are connected in parallel. The combination is then connected in series with a resistance of 15 ohms. What is the resistance of the combination?

7–6. If the total voltage applied across the circuit of Prob. 7–5 is 120 v, what is the current in each of the 10-ohm resistors?

7–7. If the entire combination of Prob. 7–5 is connected in parallel with a 25-ohm resistor, what is the resistance of this new circuit?

7–8. What is the equivalent resistance between a and b of the circuit shown?

7–9. What current will flow in each of the resistors of Prob. 7–8 if a 10-v source is connected to terminals a and b?

7–10. For the circuit illustrated (a) compute the equivalent resistance; (b) compute the current through each resistor; (c) compute the power dissipated by each; (d) compute the power for the entire circuit using the values of R_T and I_T

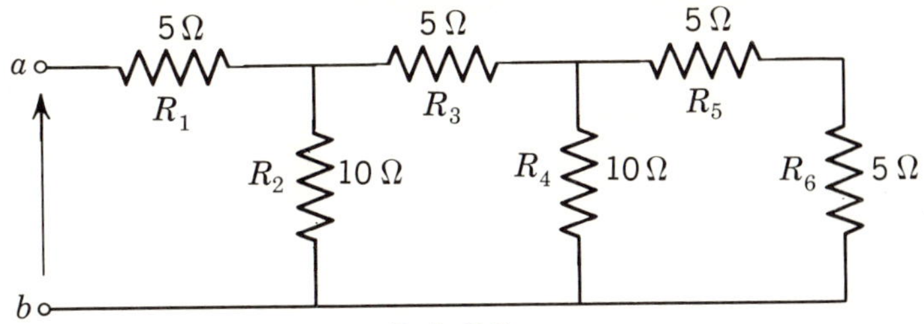

Prob. 7–8.

and compare this answer with that obtained by summing the individual powers obtained in (c).

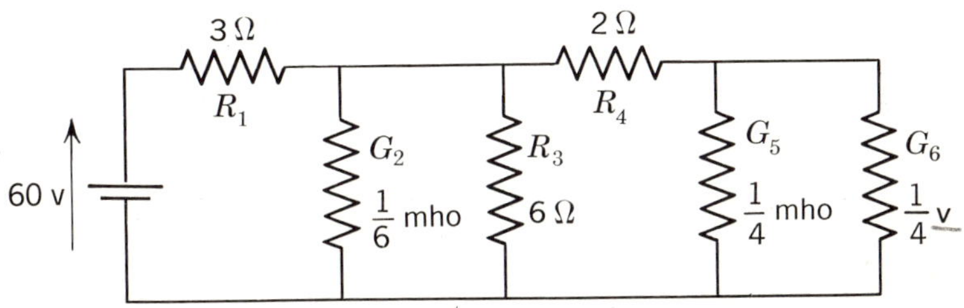

Prob. 7–10.

7–11. What is the internal resistance of the source shown in the figure?

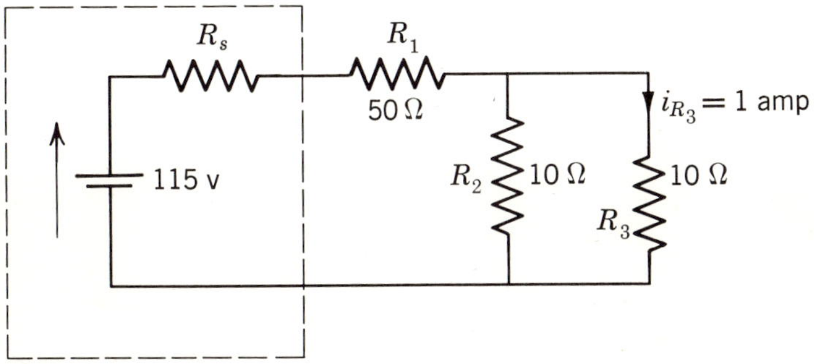

Prob. 7–11.

7–12. Two capacitors of 6×10^{-6} farad are connected in series. What is the capacitance of the combination? What would the capacitance be if they were connected in parallel?

7–13. Two capacitors of 0.5 μf are connected in series across a 100-v source. What is the voltage and charge across each?

7–14. If one of the capacitors in Prob. 7–13 is changed to 1.5 μf, what will be the new voltage and charge on each?

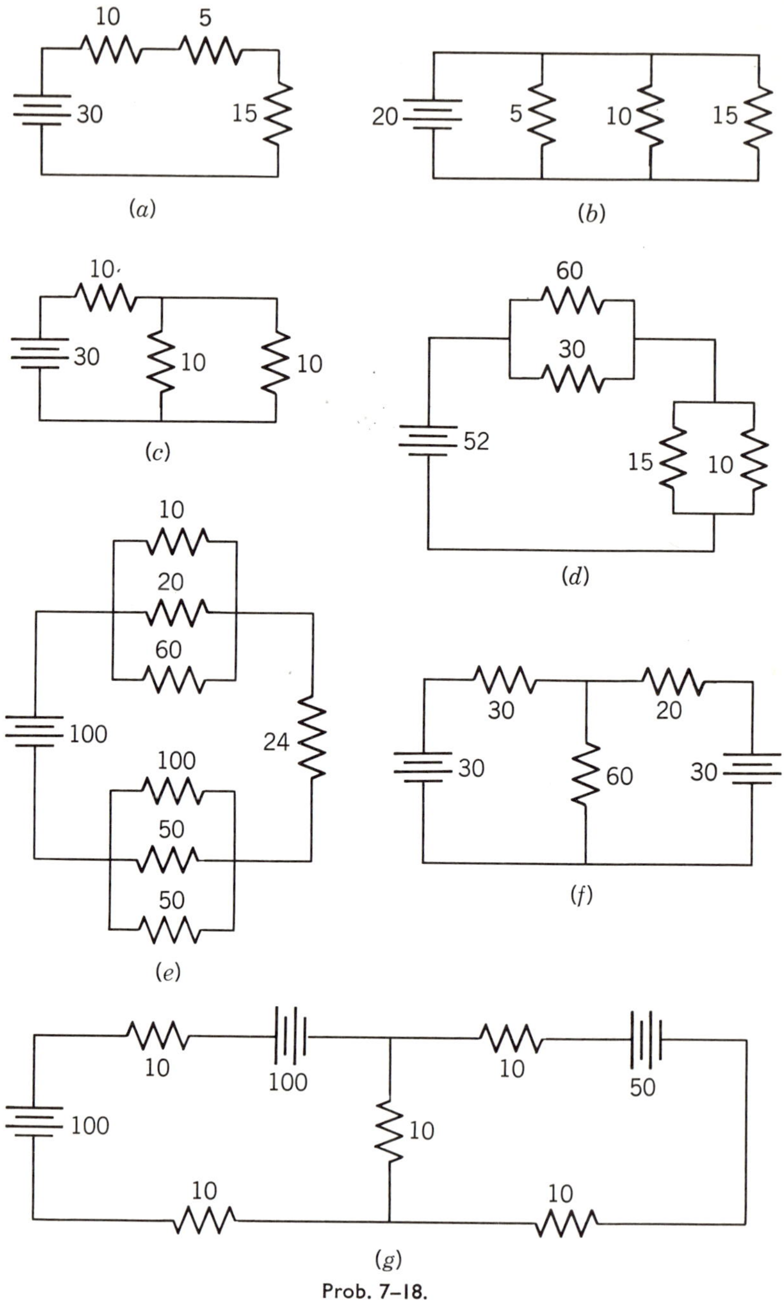
10
5
30
15
(a)
20
5
10
15
(b)
10
30
10
10
(c)
60
30
52
15
10
(d)
10
20
60
100
24
100
50
50
(e)
30
20
30
60
30
(f)
10
100
10
50
100
10
10
10
(g)

Prob. 7–18.

7–15. If one of the capacitors in Prob. 7–13 is changed to 1.5 μf and the two connected in parallel across the 100-v source, what will be the voltage and charge on each?

7–16. Two inductors of 10 mh are connected in parallel and the combination is connected in series with a 15-mh inductor. If the current of Prob. 6–2b is applied to this combination, what will be the voltage across each of the inductors?

7–17. Write the equilibrium equation in its simplest form for the circuit shown.

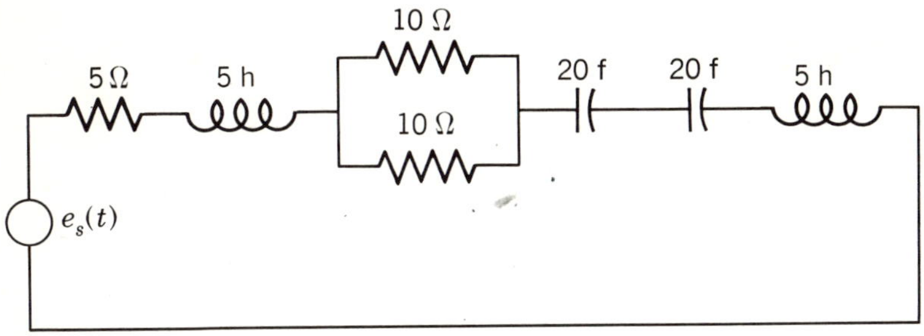

Prob. 7–17.

7–18. Find the potential across and the current through all resistors in each of the circuits shown.

8

SINGULAR FUNCTIONS AND SYSTEM RESPONSE

The previous two chapters introduced the concept of circuit elements and sources of energy for those circuit elements, where the energy source was either a current source or a voltage source. Furthermore, equilibrium equations were written that related the voltage or current applied to the circuit to the currents and voltages in the circuit and circuit elements. In these relationships no indication was made as to the form of these voltages or currents except to indicate that in general they are considered as functions of time, that is, their value at any instant of time depends upon the time of examination.

The great majority of commercial power available today provides a voltage which is a sinusoidal function of time. A second common energy source supplies voltages or currents that are constant with respect to time, such as might be obtained from batteries or so-called d-c generators or converters. Other energy sources provide waveforms that are arbitrary functions of time, such as step functions, impulses, and linear ramps.

The initial study of the response of passive network elements will be in terms of applied voltage and current waveforms that are arbitrary functions of time. In later chapters the response of elements to sinusoidal voltages and currents will be considered. It is important to remember that the use of any mathematical form for the source element function is only an approximation to the actual function which may exist in the physical circuit. The degree of refinement of approximation may be as great as desired, however, and the fact that the approximation exists should not be of concern to the user, if its limitations are understood.

This chapter discusses some of these arbitrary source functions, termed *singular functions*, and the response of series and parallel element combinations to them.

8–1. The Step Function. If a d-c source such as a battery could be connected instantaneously to a circuit, the instantaneous voltage, which is the same as the steady-state voltage, of the source would be applied to the circuit in zero time. Because it is physically necessary to close a switch to connect a voltage source to a network, it is impossible to obtain the instantaneous application of the voltage E across the element or the circuit in zero time. Instead, the actual applied voltage may appear somewhat as shown in Fig. 8–1. The voltage across the circuit is zero at $t = 0$ and increases in some fashion, reaching its final value E in δ seconds. If δ is very small compared with the amount of time it takes the circuit to achieve a steady-state response, then for simplicity the function described in this section may be assumed. This is a common engineering approximation.

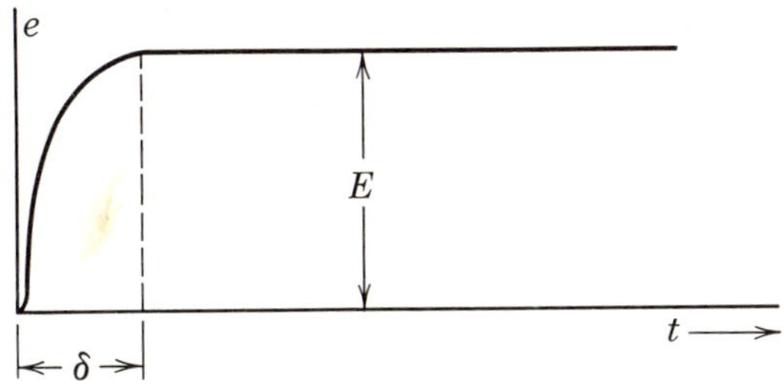

Fig. 8–1. A step function.

Figure 8–2 shows a graphical representation of the voltage assumed to be applied to the network in zero time. This applied voltage is termed a *step function* and is expressed mathematically as

$$\begin{aligned} e &= e(t) = 0 \qquad -\infty < t < 0 \\ e &= e(t) = E \qquad 0 \leqslant t < \infty \end{aligned} \tag{8–1}$$

If

$$E = 1 \tag{8–2}$$

the function is termed a *unit step function.*

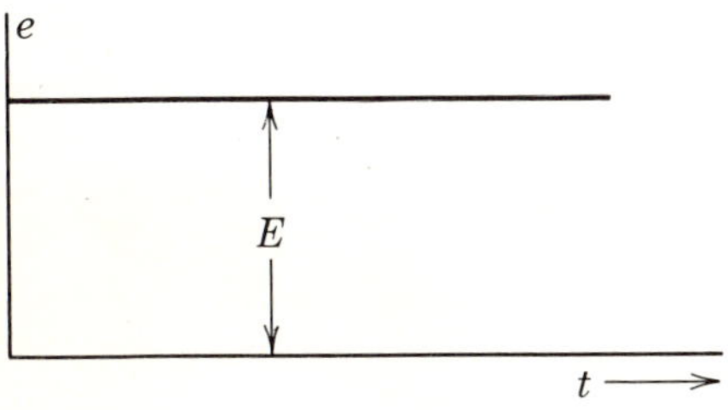

Fig. 8–2. A step function of voltage.

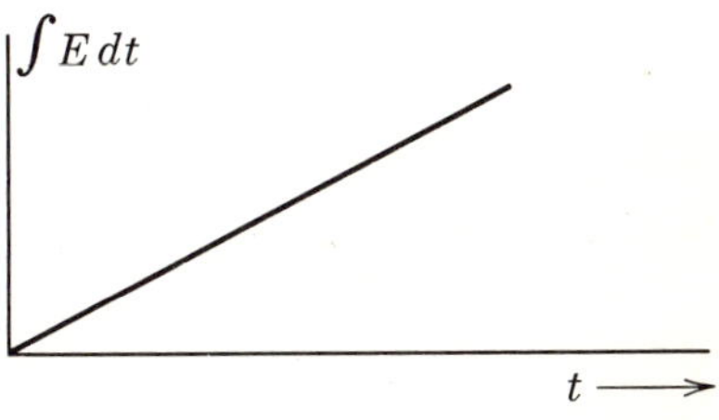

Fig. 8–3. A ramp function of voltage.

8–2. The Linear Ramp. If the step function is integrated as

$$\int_0^t e\,dt = \int_0^t E\,dt = Et \tag{8–3}$$

the resulting function is termed a *linear ramp*. Figure 8–3 is a plot of this integral against time and is the graphical representation of the linear ramp function. It will be seen that this curve is proportional to the area under the plot of e versus t as shown in Fig. 8–2.

A further integration of the ramp function will give a parabolic curve as

$$\int_0^t Et\,dt = \tfrac{1}{2}Et^2 \qquad (8\text{–}4)$$

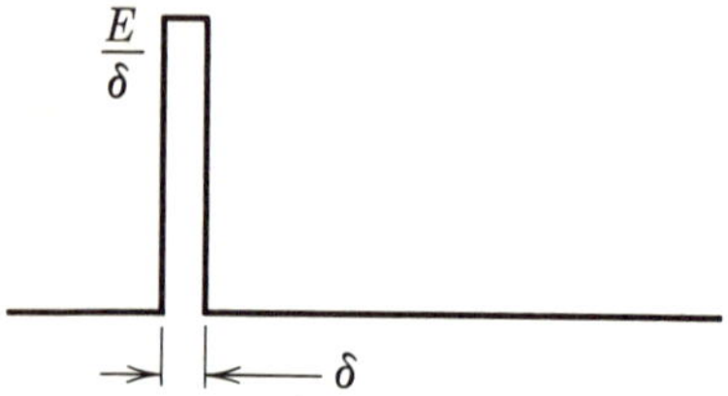

Fig. 8–4. Introduction to the impulse.

which is the area beneath the plot of Et versus t.

8–3. The Impulse. Assume a pulse of finite height and width as illustrated in Fig. 8–4. The height of the finite pulse is denoted as E/δ and the width of the pulse as δ. δ may be assumed as small as desired, that is, the height of the pulse allowed to approach infinity and the width to approach zero, as long as the product of the height and width remains finite and is equal to E. If $E = 1$, the function is defined as a *unit impulse*.

The mathematical representation for the unit impulse is

$$f(t) = 0 \qquad t \neq 0$$
$$\int_{-\infty}^{\infty} f(t)\,dt = 1 \qquad (8\text{–}5)$$

8–4. The Doublet. If an impulse is differentiated with respect to time, two infinite spikes will result, one at $t = 0-$ and the other at $t = 0+$, or, in other words, at the beginning and at the end of the time period δ. If $\delta \to 0$ these two spikes of opposite directions will occur consecutively in zero time. To permit a physical explanation, the vertical rise and vertical drop of the unit impulse sides must be avoided. Assume instead that the rise of the unit impulse occurs in $\delta/2$ seconds and the fall in another $\delta/2$ seconds. The unit impulse will then assume the form pictured in Fig. 8–5a. The area remains one unit. The derivative of a special unit impulse of this shape is then a doublet, which is pictured as Fig. 8–5b.

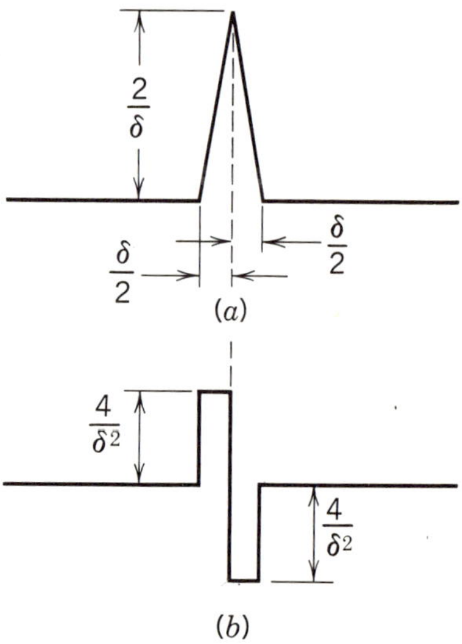

Fig. 8–5. Another unit impulse form and the doublet.

8–5. Nomenclature and Relationships of Singular Functions. The foregoing forms of time-varying functions have been termed singular functions, and the network response to voltages or currents having these forms called the response to singular functions. To conveniently distinguish among the various singular functions, a shorthand notational system has been established in which the unit impulse is designated as $u_0(t)$. As an example of the use of this designation, an impulse having an area of five units would be designated as $5u_0(t)$.

The unit step function has been designated as the integral of the unit impulse and is expressed by the symbol $u_{-1}(t)$. A linear ramp which is the integral of the unit step becomes $u_{-2}(t)$. The expression $Au_{-1}(t)$ means a step function of magnitude A.

A pictorial and mathematical representation of the singular functions in finite form is presented as Fig. 8–6, where the functions and their integrals and derivatives are shown. Figure 8–7 presents a pictorial representation of the unit singular functions together with the appropriate nomenclature derived from the symbolic system already described. It should be noted that if the singular function occurs at a time other than $t = 0$ a quantity within parentheses follows the u and possesses the form $(t - a)$, where a is the time at which the singular function begins. By definition, the quantity associated with $(t - a)$ is equal to zero when the quantity within parentheses is negative.

It should again be emphasized that physical phenomena cannot have finite changes in zero time. The step function, impulse, and so on, never appear in practical systems exactly as represented by the mathematical symbology, but it is convenient to utilize these mathematical approximations for the actual physical response phenomena.

In the following articles a study will be made of the effects of impressing various singular functions on combinations of circuit elements. The reasons for such studies are numerous and include the following:

1. Many physical systems are excited by functions which may be closely approximated by singular functions.
2. Any function of time may be considered as a summation of step functions, or
3. Any function of time can also be considered as a summation of impulses.
4. By the use of superposition in linear systems, the response to any forcing function of time can be obtained from the response to step functions or impulses.
5. The response to singular functions is the easiest mathematical procedure for the treatment of networks.

8–6. The Nature of System Response to Forcing Functions. The response of a linear circuit to any forcing function depends upon the characteristics of the circuit and upon the nature of the forcing function. In general, the response may be said to consist of two components. The first component depends upon and has the same form with respect to time as the forcing function *and* its derivatives, but its magnitude depends upon the characteristics of the circuit. The second component is a function of time which depends entirely upon the characteristics of the circuit; however, its magnitude depends upon the value of the forcing function and the initial state of excitation of the network.

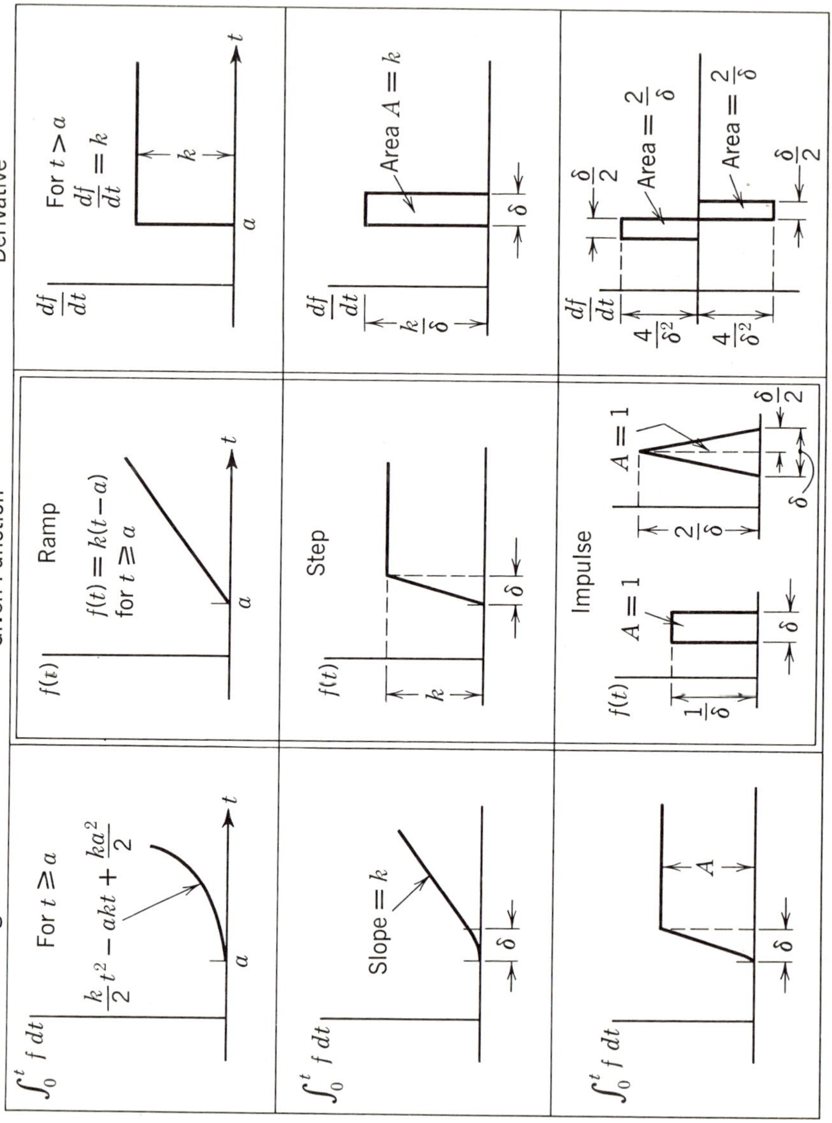

Fig. 8–6. Finite time basis for singular functions.

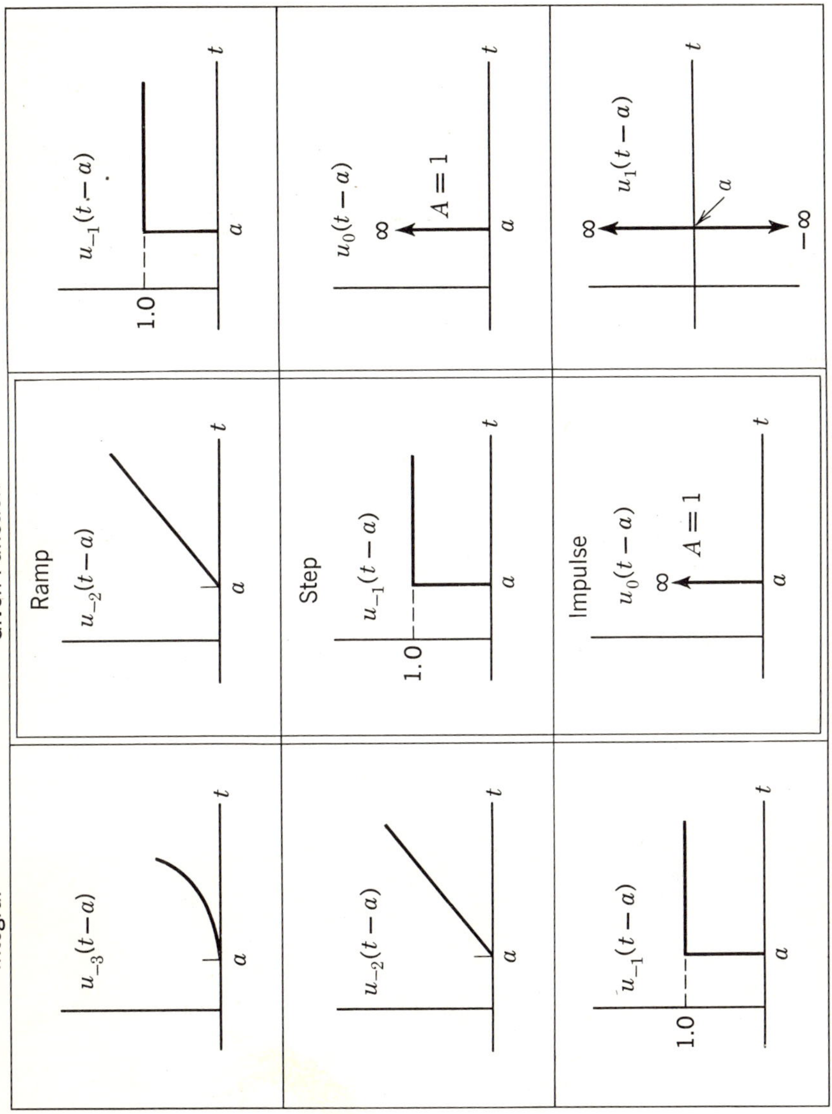

Fig. 8–7. Unit singular functions ($\delta \to 0$).

For example, consider a voltage forcing function of the form

$$e_p = E\epsilon^{-at} + k_1 \tag{8–6}$$

The first component of the response is a current of the form

$$i_p = I_p\epsilon^{-at} + k_2 \tag{8–7}$$

If the circuit contains only one element capable of storing energy, the second response component will be of the form

$$i_0 = I_0\epsilon^{s_1 t} \tag{8–8}$$

where s_1 is a function of the circuit elements and independent of e_p or a. I_0, however, depends upon the magnitude of the forcing function, the circuit elements, and the initial state of excitation of the network.

The first component is called the *particular* value of the response, and if a steady-state value of current exists the particular value will be that steady-state value. The second component of system response is the response of the system in the absence of any forcing function. It is introduced by the action of the energy storing elements of the circuit (inductance, capacitance, or their equivalent) and is called the *complementary* solution. The complementary solution is also referred to by engineers as the switching transient or the transient component of the solution.

The remaining sections of this chapter will illustrate this procedure by consideration of various combinations of circuit excitations and circuit configurations.

8–7. *R-L* Series Circuit—Step Voltage Source. To illustrate the response to a voltage step function of a network consisting of resistance and inductance in series (see Fig. 8–8), the following analysis is presented. Note

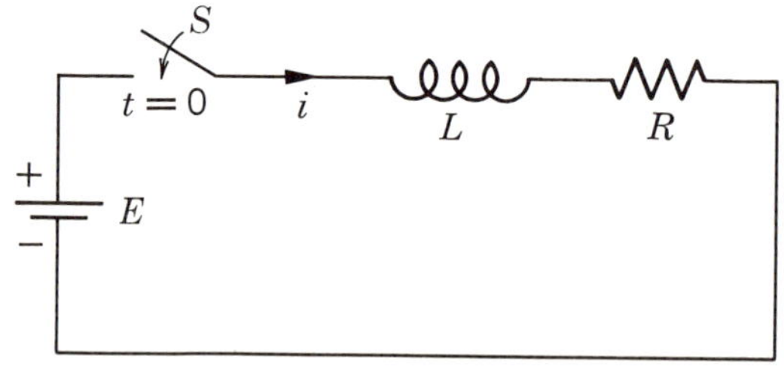

Fig. 8–8. *R-L* series circuit.

that the step function of voltage is illustrated in Fig. 8–8 as a battery of constant voltage E in series with a switch designated S.

Before making a quantitative analysis of the network, it is of interest to examine its behavior qualitatively. When switch S is closed the source voltage is impressed across the circuit elements. The battery and the switch may be considered as equivalent to a step voltage of magnitude E. Figure 8–8 is redrawn to show this as Fig. 8–9. A sudden application of voltage to a circuit containing an inductance cannot cause a current to be instantaneously established, since the energy stored in the inductance is proportional to the

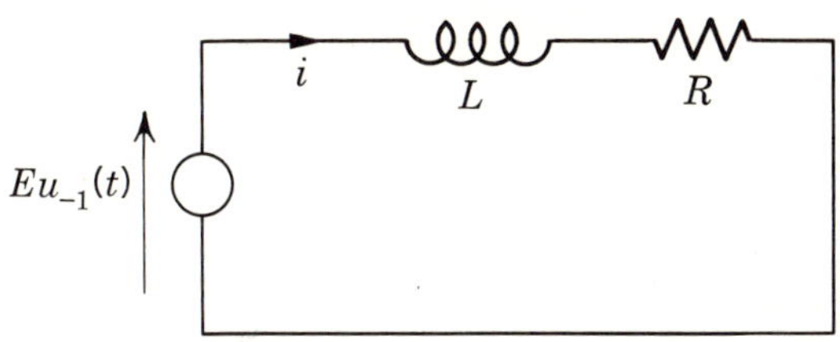

Fig. 8–9. *R-L* series circuit with step applied voltage.

square of the instantaneous value of current through the inductance and the establishment of a current in zero time would imply an infinite rate of energy flow or infinite power, which is physically unrealizable. Therefore, at the moment that the switch is closed (or the unit step function is applied)

$$i = 0 \tag{8–9}$$

and therefore

$$Eu_{-1}(t) = L\frac{di}{dt} \tag{8–10}$$

The initial rate of increase of the current in the circuit is

$$\frac{di}{dt} = \frac{E}{L} \tag{8–11}$$

Note that, although the current at this moment is zero, a finite rate of change of current exists. If the resistance were zero this rate of current increase would be constant and the current for any finite time would be given by

$$i = \frac{E}{L}t \tag{8–12}$$

which represents a linear ramp reaching infinity at infinite time.

Because of the existence of the resistance R in the circuit, however, a current i is established that causes a voltage drop iR. The voltage across the inductance then becomes $E - iR$ and the rate of increase of current is

$$\frac{di}{dt} = \frac{E - iR}{L} \tag{8–13}$$

The rate of increase of the current therefore decreases since the current increases until it acquires a value such that

$$E = iR \tag{8–14}$$

at which time di/dt becomes zero and the current has a so-called constant or steady-state value

$$i = I = \frac{E}{R} \tag{8–15}$$

The quantitative analysis of this network begins with the writing of the equilibrium equation as

$$E = Ri + L\frac{di}{dt} \tag{8–16}$$

in which i is the current at any time t after the switch is closed. Let the current i be written

$$i = i_p + i_0 \tag{8–17}$$

where i_p is the particular current and i_0 is the complementary current. Substitution of Eq. 8–17 into 8–16 yields

$$E = Ri_p + Ri_0 + L\frac{di_p}{dt} + L\frac{di_0}{dt} \tag{8–18}$$

The particular current has the same form with respect to time as the driving function E, hence is a constant which may be written as

$$i_p = I_p \tag{8–19}$$

and

$$\frac{di_p}{dt} = 0 \tag{8–20}$$

If Eqs. 8–19 and 8–20 are substituted into Eq. 8–18 the resultant expression is

$$E = RI_p + Ri_0 + L\frac{di_0}{dt} \tag{8–21}$$

But

$$I_p = \frac{E}{R} \tag{8–22}$$

or

$$E = RI_p \tag{8–23}$$

Substitution of Eq. 8–23 into 8–21 yields

$$0 = Ri_0 + L\frac{di_0}{dt} \tag{8–24}$$

The solution of Eq. 8–24 for the complementary current is of the form

$$i_0 = I_0\epsilon^{s_1t} \tag{8–25}$$

The validity of this solution can be checked by separating the variables and integrating Eq. 8–24.

Differentiation of Eq. 8–25 yields

$$\frac{di_0}{dt} = s_1I_0\epsilon^{s_1t} \tag{8–26}$$

Equation 8–24 becomes after the substitution of Eqs. 8–25 and 8–26

$$0 = I_0\epsilon^{s_1 t}(R + s_1 L) \tag{8–27}$$

For a non-trivial solution $I_0 \neq 0$, therefore

$$R + s_1 L = 0 \tag{8–28}$$

or

$$s_1 = -\frac{R}{L} \tag{8–29}$$

Therefore, Eq. 8–25 may be written as

$$i_0 = I_0\epsilon^{-\frac{R}{L}t} \tag{8–30}$$

and since

$$i = i_p + i_0$$

then

$$i = \frac{E}{R} + I_0\epsilon^{-\frac{R}{L}t} \tag{8–31}$$

The value of I_0 may be obtained if the value of the current at $t = 0$ is substituted into Eq. 8–31 as

$$0 = \frac{E}{R} + I_0 \tag{8–32}$$

which yields

$$I_0 = -\frac{E}{R} \tag{8–33}$$

Substitution of Eq. 8–33 into Eq. 8–31 yields

$$i = \frac{E}{R} - \frac{E}{R}\epsilon^{-\frac{R}{L}t} \tag{8–34}$$

or written in factored form

$$i = \frac{E}{R}\left(1 - \epsilon^{-\frac{R}{L}t}\right) \tag{8–35}$$

which is the complete solution for the current through the circuit.

Equation 8–35 gives the value of the current in the circuit at any time t after the switch is closed. From this equation it is seen that the total current is the sum of the steady-state current (the current that the forcing function will require the circuit to ultimately have) and the transient current which decays exponentially with time. The maximum value of the current I_0 is the difference between what the current in the circuit is at the time the switch is closed and what the forcing function requires it to be.

The term R/L in Eqs. 8–34 and 8–35 is an important constant. Dimensionally,

$$R = \text{volts per ampere}$$

and

$$L = \text{volt-seconds per ampere}$$

or

$$R/L = 1 \text{ per second}$$

The dimension of L/R is seconds and is given the name *time constant.*

The time constant is defined as the time in seconds necessary to make the numerical value to which ϵ is raised equal to unity. This corresponds to the time required for the current defined by Eq. 8–35 to reach 63.2% of its final value. This may be expressed mathematically as

$$\frac{R}{L}t = \frac{t}{T} = 1$$

or

$$\epsilon^{-\frac{t}{T}} = \epsilon^{-1} = 0.368$$

and

$$1 - 0.368 = 0.632$$

The time constant may also be defined as the time required for the current to reach its final value if it were to increase at the same rate which existed at the time $t = 0$.

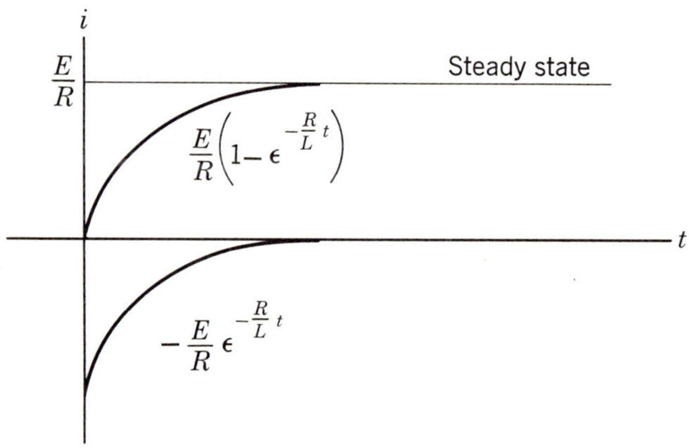

Fig. 8–10. Current components in R–L circuit.

A plot of Eq. 8–35 showing the current in the R-L circuit versus time is conveniently obtained by plotting the separate terms of that equation as shown in Fig. 8–10. The steady-state current E/R is constant and is represented by the continuous horizontal line above the axis. The transient current component is the negative of the steady state at $t = 0$ and increases exponentially to zero theoretically at $t = \infty$; in a practical sense this exponential decay usually approaches zero within a few seconds or a fractional part of a second, depending upon the time constant of the network (it is sometimes said that the exponential component of the current is effectively zero after 10 time constants of time have elapsed). The total current is the sum (with proper sign) of the steady-state and transient current components.

8–8. *R*-*L* Series Circuit—Impulse Voltage Source. To illustrate the use of the impulse, it is interesting to analyze the series R-L network of the preceding section using this form of excitation. Figure 8–11 shows the schematic representation of the network and the excitation. By definition of the voltage source and of the impulse, it will be seen that the impulse is applied to the circuit at $t = 0$ and the source after the impulse has passed becomes a short circuit.

A qualitative analysis of the circuit behavior may be made as follows. By definition, a unit voltage impulse has an area of 1 v-sec. Although physically the impulse cannot be produced in zero time, it is here mathematically assumed to have occurred in that manner. In the absence of the resistance R, the voltage impulse would cause a flux linkage of 1 v-sec or 1 weber-turn in the inductance and therefore establish a current of $1/L$ amperes. Furthermore, if R is zero this current will flow continuously in the inductance. If the resistance R exists in the circuit, the current established in the inductance by the voltage impulse must pass through the resistance, causing a voltage

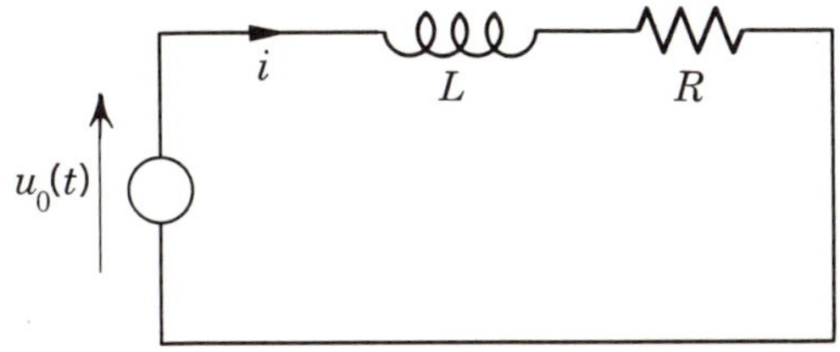

Fig. 8–11. *R-L* series circuit—impulse voltage applied.

drop of R/L. After the source discharges the impulse it becomes a short circuit. Therefore the current in the inductance must change at such a rate that

$$-L\frac{di}{dt} = Ri$$

The rate of change of current is negative, which implies that the current is decreasing with time. As the current decreases, the voltage across the resistance will also decrease and the rate of change of current must decrease. This process is cumulative and continues until the current finally becomes zero.

A quantitative analysis will begin with the writing of the equilibrium equation

$$iR + L\frac{di}{dt} = 0 \tag{8–36}$$

In this equation the forcing function is zero, for all times other than $t = 0$, and therefore the particular or steady-state current is zero. The complementary current is the same as that obtained as Eq. 8–30 and is repeated here:

$$i_0 = I_0\epsilon^{-\frac{R}{L}t} \tag{8–37}$$

The total current i is equal to i_0, or

$$i = i_0 = I_0\epsilon^{-\frac{R}{L}t} \tag{8–38}$$

At $t = 0$, the current i is $1/L$. Substitution of this value into Eq. 8–38 yields

$$I_0 = \frac{1}{L} \tag{8–39}$$

Therefore, Eq. 8–38 becomes

$$i = \frac{1}{L} \epsilon^{-\frac{R}{L}t} \tag{8–40}$$

It is interesting to note that Eq. 8–40 can be obtained from Eq. 8–35 by differentiation. As the impulse is the derivative of a step function, the response to a unit impulse is the derivative of the response to the unit step function, if the circuit is linear.

8–9. *R-C* Series Circuit—Step Voltage Source. Another simple and common network configuration involves resistance and capacitance in series. Schematically, this is shown in Fig. 8–12 with the excitation function designated as a step of value E.

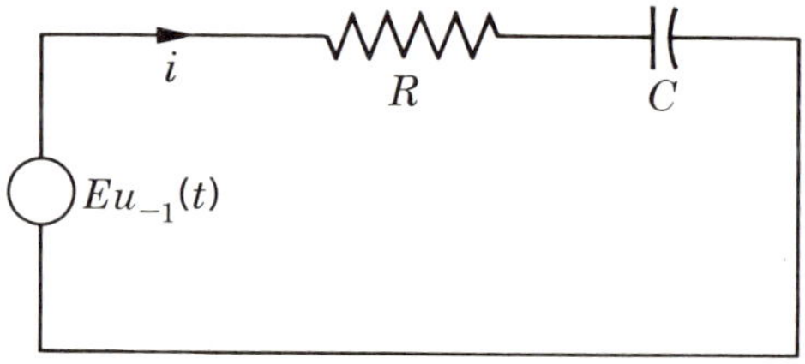

Fig. 8–12. *R-C* series circuit—step voltage source.

The equilibrium equation for this situation is written as

$$Eu_{-1}(t) = Ri + \frac{1}{C}\int i\,dt \tag{8–41}$$

When a voltage is applied a current flows instantaneously in the circuit since there is no inductance present. The capacitor is assumed to be initially uncharged and therefore has no voltage across it at $t = 0$. The initial current is then E/R. As the current begins, charge is deposited on the capacitor plates and a voltage appears across the capacitor. This capacitor voltage opposes the applied potential and decreases the current. The capacitor charge then increases at a slower rate and the voltage across the capacitor also increases, but more slowly. This process is cumulative until the current finally becomes zero when the voltage across the capacitance is equal to the applied voltage.

A procedure for the solution of Eq. 8–41 is as follows: Differentiate both sides of Eq. 8–41 to obtain

$$0 = \frac{i}{C} + R\frac{di}{dt} \qquad t \neq 0 \tag{8–42}$$

which is seen to be of the same mathematical form as Eq. 8–24, which had a solution of the form indicated by Eq. 8–25, rewritten here as

$$i_0 = I_0 \epsilon^{-\frac{t}{RC}} \tag{8–43}$$

At $t = 0$,

$$i_0 = I_0 = \frac{E}{R} \tag{8–44}$$

Substitution of Eq. 8–44 into Eq. 8–43 yields

$$i = i_0 = \frac{E}{R} \epsilon^{-\frac{t}{RC}} \tag{8–45}$$

As described quantitatively in the foregoing, it is seen that the current for this situation has the initial value of E/R and decreases exponentially from this value to zero.

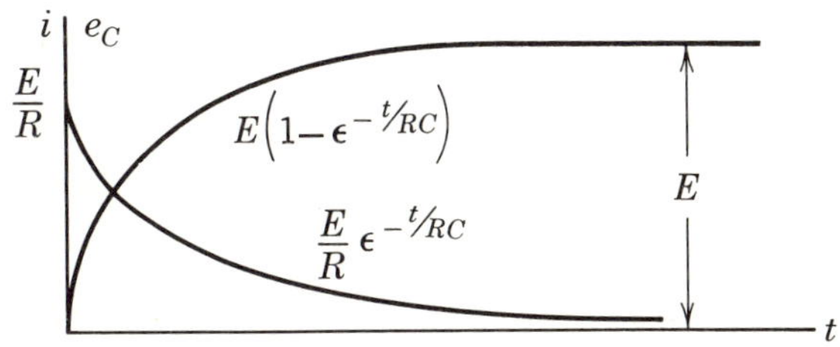

Fig. 8–13. Capacitor voltage and current for circuit of Fig. 8–12.

The voltage across the capacitor may be obtained by rewriting Eq. 8–41 as

$$e_C = E - e_R \tag{8–46}$$

where

$$e_R = Ri = E\epsilon^{-\frac{t}{RC}} \tag{8–47}$$

and by substituting Eq. 8–47 into Eq. 8–46 it is seen that the capacitor voltage is

$$e_C = E\left(1 - \epsilon^{-\frac{t}{RC}}\right) \tag{8–48}$$

This equation also verifies the qualitative analysis, which indicated that the initial voltage across the capacitor would be zero and increase toward a final value equal to the applied voltage.

Figure 8–13 is a plot of the current and also the capacitor voltage for the series R–C circuit with a step function of voltage applied.

8–10. *R-C* Series Circuit—Impulse Voltage Source. An analysis of the series resistance-capacitance circuit with an impulse voltage applied may be performed as follows. Figure 8–14 schematically represents the circuit configuration and the excitation function. A qualitative analysis indicates that the source delivers an impulse at $t = 0$ and becomes dead (that is, a short circuit). The capacitance at $t = 0$ has no voltage across it and therefore offers no opposition to any charge flow as a result of the application

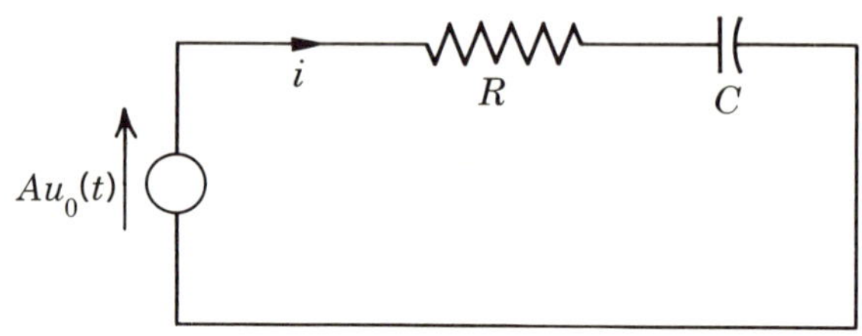

Fig. 8–14. *R-C* series circuit—impulse voltage source.

of the impulse. The only opposition to the current, therefore, is the resistance R. Mathematically, in zero time, the impulse places a charge on the capacitance of value

$$q_C = \frac{Au_0(t)}{R}$$

Effectively, the circuit now consists of a capacitor with an initial charge in series with a resistance. The capacitance will discharge through the resistance and the differential equation representing the equilibrium equations for the network may be written as

$$Ri + \frac{1}{C}\int i\,dt = 0 \tag{8–49}$$

or in terms of charge

$$R\frac{dq}{dt} + \frac{q}{C} = 0 \tag{8–50}$$

Again, this equation is of the form described by Eq. 8–24 and the solution may be written directly as

$$q_0 = Q_0\epsilon^{-\frac{t}{RC}} \tag{8–51}$$

At $t = 0$, however, $q_C = A/R$, and therefore

$$Q_0 = \frac{A}{R} \tag{8–52}$$

and Eq. 8–51 becomes

$$q = \frac{A}{R}\epsilon^{-\frac{t}{RC}} \tag{8–53}$$

If an expression for the instantaneous current is desired, it may be obtained by the differentiation of Eq. 8–53 and is given as

$$i = -\frac{A}{CR^2}\epsilon^{-\frac{t}{RC}} + \frac{A}{R}u_0(t) \tag{8–54}$$

Note the sign of the current. Compare this with the sign of the impulse which was initially taken as a reference.

8–11. *R-C* Parallel Circuit—Step Current Source. The schematic representation for this network is shown in Fig. 8–15, where the source excitation is a unit step function of current. By definition of the step function, the current source supplies the circuit with a constant value of current after $t = 0$. The capacitor is assumed initially uncharged and therefore at $t = 0$, $e = 0$. The initial action of the uncharged capacitor is a short circuit,

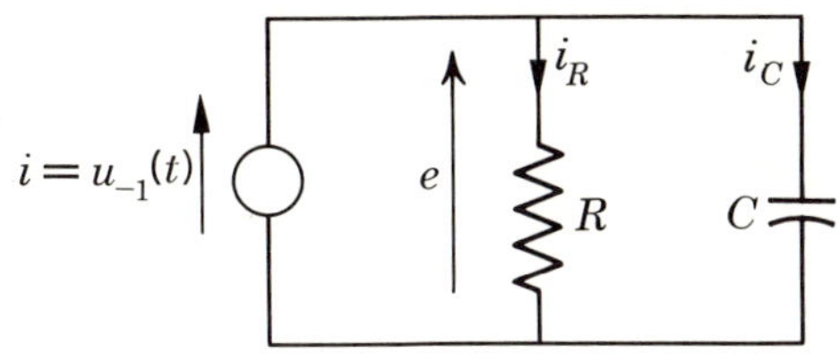

Fig. 8–15. *R-C* parallel circuit—step current source.

and therefore initially the circuit has no voltage across it. As charge flows into the capacitor, a voltage is established across it and less current then appears in the capacitive branch; simultaneously, more current appears in the resistive branch. This process continues until the capacitance is charged and the entire source current i passes through the resistive branch, or

$$e = Ri = Ru_{-1}(t) = R$$

No further current appears in the capacitive branch, and the voltage across the *R-C* parallel circuit has reached its steady state.

The foregoing qualitative analysis may be verified quantitatively by writing the equilibrium equations for the network of Fig. 8–15 as follows:

$$i_R + i_C = 1 \tag{8–55}$$

or

$$\frac{1}{R}e + C\frac{de}{dt} = 1 \tag{8–56}$$

At $t = 0$,

$$e = 0 \tag{8–57}$$

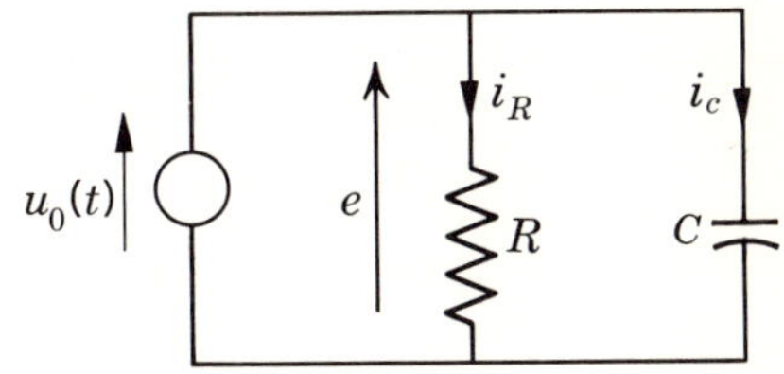

Fig. 8–16. *R-C* parallel circuit—impulse current source.

Equation 8–56 and the initial condition of Eq. 8–57 are identical in mathematical form to Eqs. 8–16 and 8–9; therefore, the solution may be obtained by similarity to Eq. 8–35 as

$$e = R\left(1 - \epsilon^{-\frac{t}{RC}}\right) \tag{8–58}$$

The term $1/RC$ has the dimension of 1 per second, and by analogy to the reasoning employed for the R-L circuit, RC is a time constant. This may be verified from

$$R = \text{volts per ampere}$$
$$C = \text{ampere-seconds per volt}$$
$$RC = \text{seconds}$$

8–12. R-C Parallel Circuit—Impulse Current Source. Figure 8–16 shows schematically the representation for a unit current impulse source and a parallel R-C network. The current source at $t = 0$ supplies a current impulse and then becomes idle (an open circuit). By definition, the unit impulse of current when applied to a capacitance instantaneously places a charge of 1 coulomb upon that capacitance. (In a practical sense note that the charge can be supplied physically only in a very small time period, termed δ seconds. However, since δ is small it may be said that the charge is placed upon the capacitance instantaneously.)

A charge of 1 coulomb across the capacitor establishes a voltage $e_C = 1/C$ volts, and the capacitor will then discharge through the resistance. The equilibrium equation written for the parallel situation illustrated is

$$Ge + C\frac{de}{dt} = 0 \tag{8–59}$$

or

$$Ge = -C\frac{de}{dt} \tag{8–60}$$

where

$$G = \frac{1}{R} \tag{8–61}$$

Equation 8–59 is mathematically identical in form to Eq. 8–36 and therefore the solution may be written directly for the new variables here involved as

$$e = \frac{1}{C}\epsilon^{-\frac{G}{C}t} \tag{8–62}$$

8–13. R-L-C Series Circuit—Impulse Voltage Source. The R-L-C series circuit as shown in Fig. 8–17 may be conveniently analyzed using the applied voltage of the impulse form. The current in the circuit is the unknown of initial interest.

A qualitative analysis of the behavior of the circuit of Fig. 8–17 may be based upon an extension of the reasoning employed for the R-L and R-C circuits. With the R-L elements only in the circuit, the unit voltage impulse was absorbed by the inductance and established an instantaneous current of $i = 1/L$ in that inductance and in the rest of the circuit. The resistance in series with the inductance was without effect since the voltage drop across the resistance was finite, and, when subtracted from the infinite pulse magnitude, still permitted an infinite voltage to appear across the inductance. The application of a unit impulse to the series R-C, on the other hand, caused a charge of value $1/R$ coulombs to be established on the capacitor. The capacitor voltage is therefore $1/RC$, a finite value. Therefore, in the *R-L-C* circuit the voltage across the resistance and capacitance can be only finite and effectively the infinite voltage appears across the inductance. The current at $t = 0$ is $1/L$ amperes. (Since the current here is finite, the initial charge on C is zero, not $1/R$.)

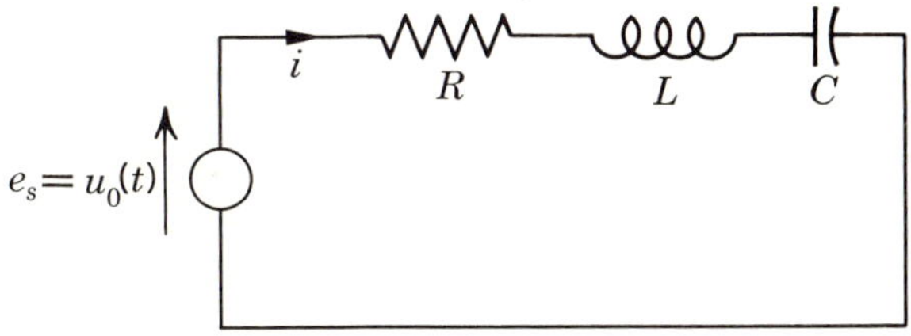

Fig. 8–17. *R-L-C* circuit with unit voltage impulse source.

Since the impulse is mathematically assumed to occur in zero time, an energy of value $Li^2/2$ is injected instantaneously into the inductance. With the disappearance of the impulse there is no restraint tending to maintain that energy in the inductance and therefore it is propagated in the circuit. Several possibilities as to the mode of this propagation exist that are dependent upon the circuit constants:

1. In the absence of R ($R = 0$) the energy stored in the magnetic field of the inductance is transferred to the electric field of the capacitance and then returns to the magnetic field. This process will continue indefinitely with no loss of energy and is termed *oscillation*.
2. In the presence of a small value of resistance, while the energy is oscillating between the magnetic and electric fields, a small amount of energy will be dissipated during each cycle as heat in the resistor. In time all of the energy will be dissipated in the resistance and the oscillation will die away.
3. In the presence of a large resistance the energy cannot oscillate; that is, during the initial propagation of energy from the inductance to the capacitance all of it is dissipated in the resistance.

These three conditions will be clearly recognized from the mathematical analysis which follows.

The quantitative analysis of the series *R-L-C* circuit may initially appear to be somewhat complicated; however, it will be seen that the complication consists primarily of a large number of steps, each of which is straightforward in itself. A study of the following material in sequence should yield a reasonable understanding of the behavior of the network. As in the previous cases, the initial procedure is the writing of the equilibrium equation for the network of Fig. 8–17 as

$$Ri + L\frac{di}{dt} + \frac{1}{C}\int i\,dt = 0 \tag{8–63}$$

where the right-hand side of the equation is zero, since the driving function, a unit impulse, existed and disappeared at $t = 0$ only.

The solution of this integrodifferential equation is of the standard form

$$i = A\epsilon^{st} \tag{8-64}$$

Differentiation and integration of Eq. 8–64 yield

$$\frac{di}{dt} = sA\epsilon^{st}$$

and (8–65)

$$\int i\,dt = \frac{A\epsilon^{st}}{s}$$

The quantities of Eqs. 8–65 may be substituted in Eq. 8–63 to yield

$$A\epsilon^{st}\left(R + sL + \frac{1}{sC}\right) = 0 \tag{8-66}$$

or

$$A\epsilon^{st}\left(\frac{sRC + s^2LC + 1}{sC}\right) = 0 \tag{8-67}$$

To establish Eq. 8–64 as a solution form, the following must be an identity:

$$LC\left(s^2 + \frac{R}{L}s + \frac{1}{LC}\right) = 0 \tag{8-68}$$

The following values of s from Eq. 8–68 satisfy this requirement:

$$\begin{aligned} s_1 &= -\frac{R}{2L} + \sqrt{\left(\frac{R}{2L}\right)^2 - \frac{1}{LC}} \\ s_2 &= -\frac{R}{2L} - \sqrt{\left(\frac{R}{2L}\right)^2 - \frac{1}{LC}} \end{aligned} \tag{8-69}$$

Three possible combinations of values for s_1 and s_2 exist:

1. s_1 and s_2 may be real and different if

$$\left(\frac{R}{2L}\right)^2 > \frac{1}{LC}$$

2. s_1 and s_2 may be real and equal if

$$\left(\frac{R}{2L}\right)^2 = \frac{1}{LC}$$

3. s_1 and s_2 may be complex and conjugate if

$$\left(\frac{R}{2L}\right)^2 < \frac{1}{LC}$$

Each case may be solved individually, but a solution for the third case, which is the most interesting one, may also be extended to cover the first two cases. Case 1 is called *overdamped*, case 2 is termed *critically damped*, and case 3 is termed *underdamped* or *oscillatory*.

Continuing the development for case 3, for which $(R/2L)^2 < 1/LC$, the values of s_1 and s_2 become

$$s_1 = -\frac{R}{2L} + j\sqrt{\frac{1}{LC} - \left(\frac{R}{2L}\right)^2} \qquad (8\text{–}70)^*$$
$$s_2 = -\frac{R}{2L} - j\sqrt{\frac{1}{LC} - \left(\frac{R}{2L}\right)^2}$$

For convenience the following new symbols are defined

$$\alpha = \frac{R}{2L}$$
$$\omega_0^{\,2} = \frac{1}{LC} \qquad (8\text{–}71)$$
$$\omega_d^{\,2} = \omega_0^{\,2} - \alpha^2$$

and the substitution of these new quantities from Eqs. 8–71 into Eqs. 8–70 yields

$$s_1 = -\alpha + j\omega_d \qquad (8\text{–}72)$$
$$s_2 = -\alpha - j\omega_d$$

* The symbol j in the expression for these roots is defined as $\sqrt{-1}$ (more commonly written as i in mathematics courses). This concept will be covered in detail in Chapter 11; however, certain identities are stated here without proof for convenient reference.

1. A complex number in rectangular form is written in the form $\boldsymbol{A} = a + jb$ and illustrated below.

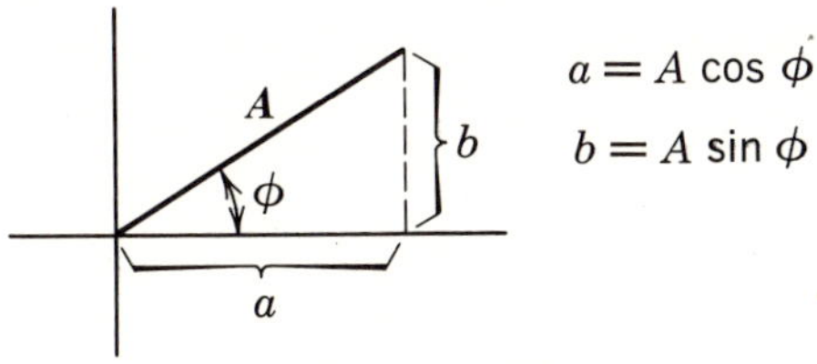

Complex number in rectangular coordinates.

2. $\epsilon^{j\phi} = \cos \phi + j \sin \phi$

   or

   $A\epsilon^{j\phi} = A \cos \phi + jA \sin \phi = a + jb = \boldsymbol{A}$

3. $\sin \phi = \dfrac{\epsilon^{j\phi} - \epsilon^{-j\phi}}{2j}$ and $\cos \phi = \dfrac{\epsilon^{j\phi} + \epsilon^{-j\phi}}{2}$

Equation 8–64 requires two constants of integration since two values for s have been obtained. It therefore will be rewritten as

$$i = A\epsilon^{s_1 t} + B\epsilon^{s_2 t} \tag{8–73}$$

The evaluation of constants A and B may be accomplished in terms of the physical characteristics of the system. At $t = 0$, the current from the unit impulse of voltage is $i = 1/L$, which when substituted into Eq. 8–73 gives

$$\frac{1}{L} = A + B \tag{8–74}$$

Again at $t = 0$, the charge on the capacitor is zero and the integral of Eq. 8–73 may be equated to zero as

$$\begin{aligned} \int i\,dt &= \frac{A}{s_1}\epsilon^{s_1 t} + \frac{B}{s_2}\epsilon^{s_2 t} \\ 0 &= \frac{A}{s_1} + \frac{B}{s_2} \end{aligned} \tag{8–75}$$

Equations 8–74 and 8–75 are two equations in two unknowns, A and B; therefore

$$A = \frac{s_1}{(s_1 - s_2)L} \tag{8–76}$$

$$B = -\frac{s_2}{(s_1 - s_2)L} \tag{8–77}$$

Substitution of these values into Eq. 8–73 yields an expression for i in terms of s_1 and s_2 as

$$= \frac{s_1}{(s_1 - s_2)L}\epsilon^{s_1 t} - \frac{s_2}{(s_1 - s_2)L}\epsilon^{s_2 t} \tag{8–78}$$

From Eqs. 8–72

$$s_1 - s_2 = 2j\omega_d \tag{8–79}$$

Substitution of Eqs. 8–79 and 8–72 into 8–78 yields the following several forms for the current in the series network

$$\begin{aligned} i &= \frac{\epsilon^{-\alpha t}}{L(2j\omega_d)}[(-\alpha + j\omega_d)\epsilon^{j\omega_d t} - (-\alpha - j\omega_d)\epsilon^{-j\omega_d t}] \\ &= \frac{\epsilon^{-\alpha t}}{L(2j\omega_d)}[-\alpha(\epsilon^{j\omega_d t} - \epsilon^{-j\omega_d t}) + j\omega_d(\epsilon^{j\omega_d t} + \epsilon^{-j\omega_d t})] \\ &= \frac{\epsilon^{-\alpha t}}{\omega_d L}(-\alpha \sin \omega_d t + \omega_d \cos \omega_d t) \\ &= \sqrt{\frac{C}{L}}\frac{\omega_0}{\omega_d}\epsilon^{-\alpha t}(\omega_d \cos \omega_d t - \alpha \sin \omega_d t) \end{aligned} \tag{8–80}$$

where α is termed a *damping factor*, which may also be written as

$$\alpha = \sqrt{\omega_0^2 - \omega_d^2} \tag{8–81}$$

Figure 8–18 shows the triangle formed by α, ω_d, and ω_0. (It will be remembered that quantities multiplied by j appear at right angles with respect to real quantities when geometrically represented.) By inspection of this figure, it may be seen that the last of Eqs. 8–80 may also be written as

$$i = -\frac{\omega_0}{\omega_d L}\,\epsilon^{-\alpha t}\sin(\omega_d t - \theta) \tag{8–82}$$

where

$$\theta = \tan^{-1}\frac{\omega_d}{\alpha} \tag{8–83}$$

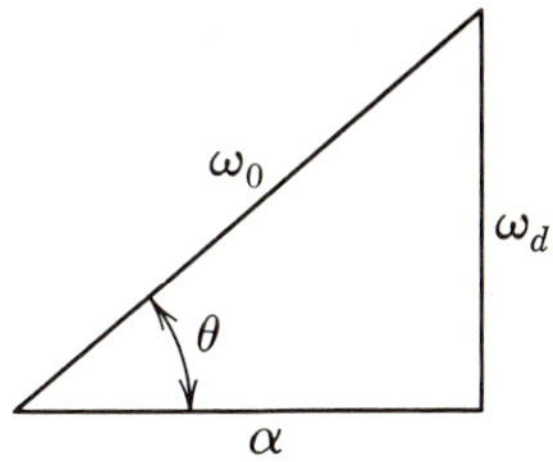

Fig. 8–18. Complex plane plot of α, ω_d, and ω_0.

From this more compact representation, Eq. 8–82, it may be seen that the current is oscillatory with a basic angular velocity ω_d. The oscillations, however, are damped and $\epsilon^{-\alpha t}$ forms an envelope for these oscillations. Figure 8–19 shows this current plotted against time. In $1/\alpha$ seconds the envelope of the oscillations decays to 36% of its initial value (see the previous discussions of time constant). The decay per cycle may be found as follows:

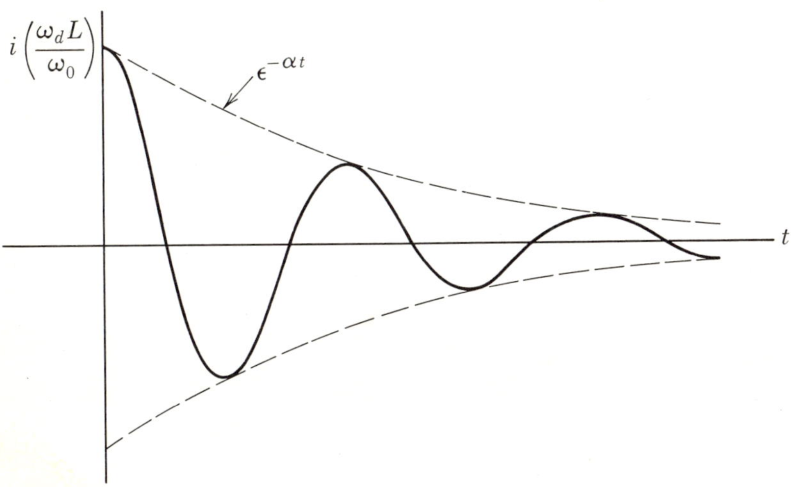

Fig. 8–19. Current form in *R-L-C* circuit—damped oscillatory condition (unit voltage impulse applied).

The time for one cycle is

$$\omega_d t = 2\pi \tag{8–84}$$

or

$$T = \frac{2\pi}{\omega_d} \tag{8–85}$$

where T is the time in seconds for one complete cycle. Using this value of T, the value of decay per cycle is

$$\epsilon^{\alpha T} = \epsilon^{\frac{2\pi\alpha}{\omega_d}} \tag{8–86}$$

If the natural logarithm of this factor is taken, it will be seen to be of the form

$$\Delta = \frac{2\pi\alpha}{\omega_d} = \frac{2\pi\alpha}{\sqrt{\omega_0^2 - \alpha^2}} \tag{8–87}$$

where Δ is termed the *logarithmic decrement*. Expressed in another way, $1/\Delta$ is the number of cycles occurring during the time in which the oscillation decays by the factor $1/\epsilon = 0.368$.

If α is very small compared with ω_0, Eq. 8–87 gives the logarithmic decrement approximately as

$$\begin{aligned} \Delta &\simeq \frac{2\pi\alpha}{\omega_0} = \frac{2\pi R}{2L\omega_0} \\ &= \pi \frac{R}{\omega_0 L} \end{aligned} \tag{8–88}$$

The term

$$\frac{\omega_0 L}{R} = Q \tag{8–89}$$

is an important quantity in electrical engineering. The Q of a system in a later discussion of tuned circuits will be shown to determine the selectivity of such a circuit. The logarithmic decrement may be written in terms of the defined Q as

$$\Delta = \frac{\pi}{Q} \tag{8–90}$$

The last of Eqs. 8–80 or Eq. 8–82 may be investigated for the specific case in which $R = 0$; then $\alpha = 0$, $\omega_d = \omega_0$, and the current equation becomes

$$i = \frac{1}{L} \cos \omega_0 t \tag{8–91}$$

where the current oscillates continuously with no damping as shown in Fig. 8–20.

An examination of this simplified system permits an understanding of the nature of the energy exchange in the system. At $t = 0$,

$$i = \frac{1}{L}$$

and the stored energy in the magnetic field of the inductance is

$$W_L = \frac{Li^2}{2} = \frac{1}{2L} \tag{8–92}$$

since the voltage across the capacitance at this time is zero. One-quarter cycle later when $\omega_0 t = \pi/2$ or $t = \pi/2\omega_0$, the voltage across the capacitor is

$$\begin{aligned} e_C &= \frac{1}{C}\int i\,dt \\ &= \frac{1}{CL}\int_0^{\frac{\pi}{2\omega_0}} \cos \omega_0 t\,dt \\ &= \frac{1}{CL}\frac{1}{\omega_0} \\ &= \frac{1}{\sqrt{LC}} \end{aligned}$$

The value of the current is zero at this time and the energy in the magnetic field is also zero. The energy in the dielectric field of the capacitor, however, is

$$\begin{aligned} W_C &= \frac{Ce_C^2}{2} \\ &= \frac{C}{2}\frac{1}{LC} = \frac{1}{2L} \end{aligned} \tag{8–93}$$

which is the same energy as was stored in the magnetic field one-quarter cycle earlier. As predicted in the qualitative analysis, the energy is transferred from the magnetic to the electric field. Another quarter-cycle will

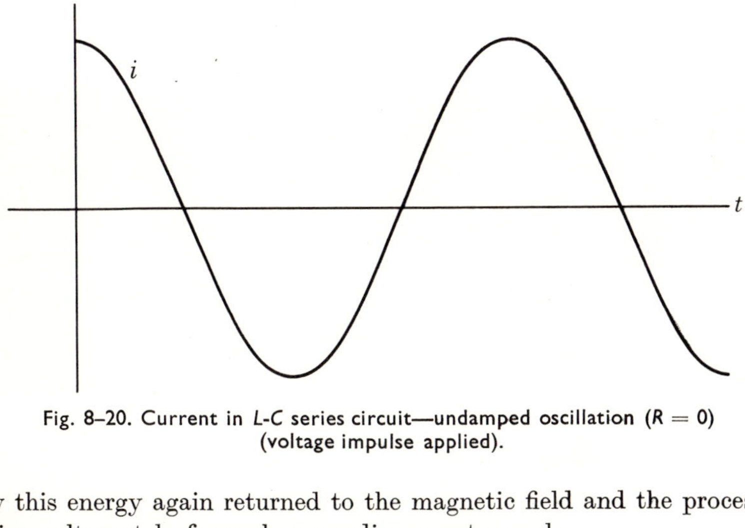

Fig. 8–20. Current in *L-C* series circuit—undamped oscillation ($R = 0$) (voltage impulse applied).

show this energy again returned to the magnetic field and the process will continue alternately for each succeeding quarter-cycle.

The critically damped case in which

$$\left(\frac{R}{2L}\right)^2 = \frac{1}{LC} \tag{8–94}$$

causes s_1 and s_2 to be equal in Eqs. 8–69. For this condition $\omega_d = 0$. The solution for the critically damped case may be obtained conveniently from the solution for the oscillatory case as expressed in Eqs. 8–80 by assuming that

$$\left(\frac{R}{2L}\right)^2 \to \frac{1}{LC} \tag{8–95}$$

The last form of Eqs. 8–80 rewritten for our convenience here is

$$i = \frac{1}{L}\,\epsilon^{-\alpha t}\left(\cos \omega_d t - \frac{\alpha}{\omega_d} \sin \omega_d t\right) \tag{8–96}$$

As $\omega_d \to 0$, $\sin \omega_d t \to \omega_d t$ and Eq. 8–96 becomes

$$i = \frac{1}{L}\,\epsilon^{-\alpha t}(1 - \alpha t) \tag{8–97}$$

The current form for this case is illustrated in Fig. 8–21.

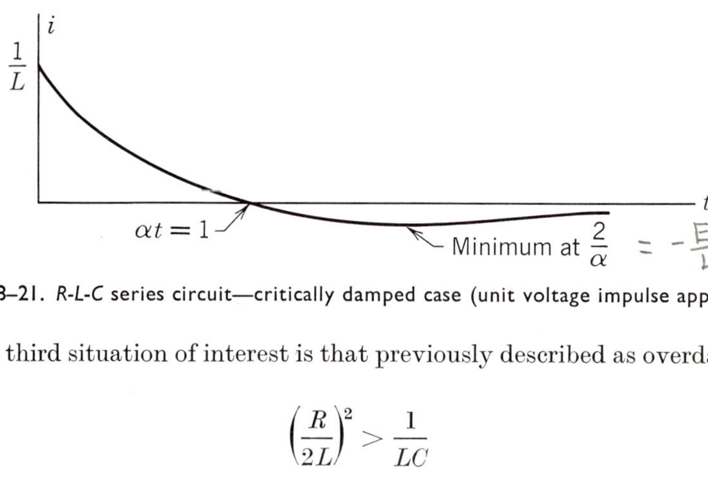

Fig. 8–21. *R-L-C* series circuit—critically damped case (unit voltage impulse applied).

The third situation of interest is that previously described as overdamped, where

$$\left(\frac{R}{2L}\right)^2 > \frac{1}{LC} \tag{8–98}$$

and Eqs. 8–69 may now be written for the roots s_1 and s_2 as

$$\begin{aligned} s_3 &= -\frac{R}{2L} + \sqrt{\left(\frac{R}{2L}\right)^2 - \frac{1}{LC}} = -\alpha + \beta \\ s_4 &= -\frac{R}{2L} - \sqrt{\left(\frac{R}{2L}\right)^2 - \frac{1}{LC}} = -\alpha - \beta \end{aligned} \tag{8–99}$$

The values of the previously described roots s_1 and s_2 are replaced here by s_3 and s_4, where the latter are real and different in value.

The constants A and B of Eq. 8–73 rewritten here become

$$A = \frac{s_3}{(s_3 - s_4)L}$$
$$B = -\frac{s_4}{(s_3 - s_4)L} \tag{8–100}$$

and the current in the circuit becomes

$$i = \frac{s_3}{(s_3 - s_4)L}\epsilon^{s_3 t} - \frac{s_4}{(s_3 - s_4)L}\epsilon^{s_4 t} \tag{8–101}$$

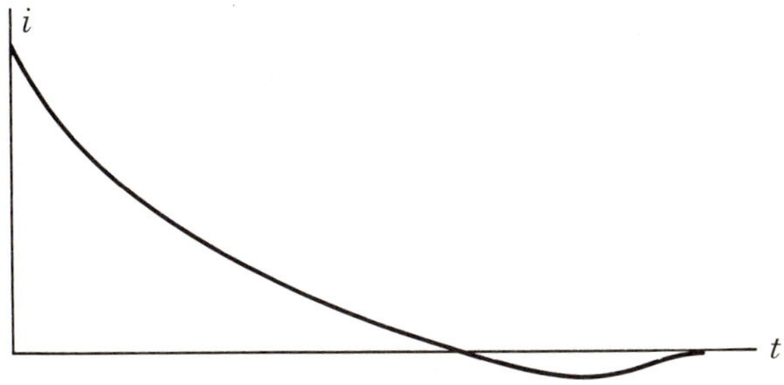

Fig. 8–22. *R-L-C* series circuit—overdamped case (unit impulse applied).

Equation 8–101 may be rewritten into several different forms by recognizing that

$$s_3 - s_4 = 2\beta \tag{8–102}$$

which yields the current expressions

$$i = \frac{(-\alpha + \beta)}{2\beta L}\epsilon^{(-\alpha+\beta)t} - \frac{(-\alpha - \beta)}{2\beta L}\epsilon^{(-\alpha-\beta)t}$$
$$= \frac{\epsilon^{-\alpha t}}{2\beta L}[\beta(\epsilon^{\beta t} + \epsilon^{-\beta t}) - \alpha(\epsilon^{\beta t} - \epsilon^{-\beta t})]$$
$$= \frac{\epsilon^{-\alpha t}}{\beta L}(\beta \cosh \beta t - \alpha \sinh \beta t)^* \tag{8–103}$$

Figure 8–22 is an illustration of the current of Eq. 8–103.

8–14. *R-L-C* Series Circuit—Step Voltage Source. The circuit of Fig. 8–17 is here redrawn as Fig. 8–23 with the voltage source changed to a step of value E.

* The hyperbolic sine and cosine are defined as

$$\sinh X = \frac{\epsilon^X - \epsilon^{-X}}{2}$$

and

$$\cosh X = \frac{\epsilon^X + \epsilon^{-X}}{2}$$

The equilibrium equation may be written for all values of time as

$$E = Ri + L\frac{di}{dt} + \frac{1}{C}\int i\,dt \tag{8–104}$$

A procedure for the solution of this integrodifferential equation employs the differentiation of both sides of the equation to yield

$$0 = R\frac{di}{dt} + L\frac{d^2i}{dt^2} + \frac{1}{C} \tag{8–105}$$

The solution, as in the previous situation, is of the form

$$i = A\epsilon^{st} \tag{8–106}$$

Differentiation of Eq. 8–106 and substitution into Eq. 8–105 yield the following values of s:

$$\begin{aligned} s_1 &= -\frac{R}{2L} + \sqrt{\left(\frac{R}{2L}\right)^2 - \frac{1}{LC}} \\ s_2 &= -\frac{R}{2L} - \sqrt{\left(\frac{R}{2L}\right)^2 - \frac{1}{LC}} \end{aligned} \tag{8–107}$$

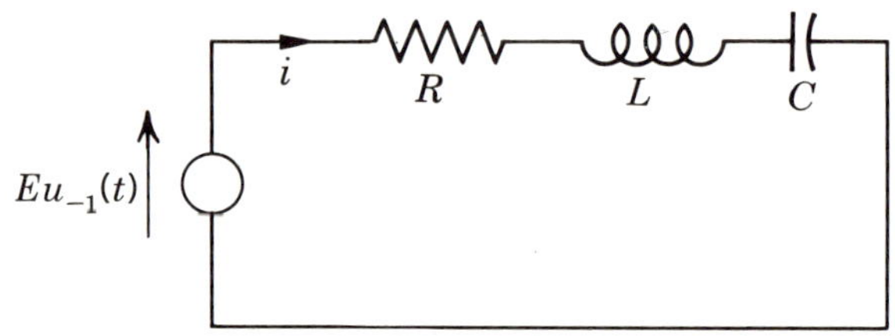

Fig. 8–23. *R-L-C* series circuit—step voltage applied.

Therefore, the current solution must again have two constants of integration and is written as

$$i = A\epsilon^{s_1t} + B\epsilon^{s_2t} \tag{8–108}$$

At $t = 0$, $i = 0$, and $di/dt = E/L$ from previous discussions of a circuit containing inductance. The constants A and B may be evaluated utilizing this information as

$$A + B = 0$$

and

$$L\frac{di}{dt} = L(s_1A + s_2B) = E$$

Therefore

$$A = \frac{E}{(s_1 - s_2)L} \quad \text{and} \quad B = -\frac{E}{(s_1 - s_2)L} \tag{8–109}$$

To illustrate the variation possible in notational form, this problem will be solved utilizing different equivalents from the previous R-L-C solution involving the impulse. Define

$$\frac{R}{2L} = \alpha = \rho\omega_0$$
$$\frac{1}{LC} = \omega_0^{\ 2} \tag{8–110}$$

from which

$$s_1 = -\rho\omega_0 + \omega_0\sqrt{\rho^2 - 1}$$
$$s_2 = -\rho\omega_0 - \omega_0\sqrt{\rho^2 - 1} \tag{8-111}$$

and

$$s_1 - s_2 = 2\omega_0\sqrt{\rho^2 - 1} \tag{8–112}$$

The expression for the current through the circuit may be written after substitution of the above defined quantities into Eq. 8–108 as

$$i = \frac{E}{2L\omega_0\sqrt{\rho^2 - 1}}\left(\epsilon^{-\rho\omega_0 t}\epsilon^{\omega_0\sqrt{\rho^2-1}t} - \epsilon^{-\rho\omega_0 t}\epsilon^{-\omega_0\sqrt{\rho^2-1}t}\right)$$
$$= \frac{E\epsilon^{-\rho\omega_0 t}}{2L\omega_0\sqrt{\rho^2 - 1}}\left(\epsilon^{\omega_0\sqrt{\rho^2-1}t} - \epsilon^{-\omega_0\sqrt{\rho^2-1}t}\right) \tag{8–113}$$

As before, three possible situations must be examined:

1. If $\rho > 1$, the square root term of Eq. 8–111 is positive, and s_1 and s_2 are real and not equal. This corresponds to the previously defined overdamped case and the expression for the current becomes

$$i = \frac{E\epsilon^{-\rho\omega_0 t}}{\omega_0 L\sqrt{\rho^2 - 1}}\sinh \omega_0\sqrt{\rho^2 - 1}t \tag{8–114}$$

2. If $\rho = 1$, then $s_1 = s_2$, and the critically damped situation exists. The current for this condition is obtained as

$$i = \frac{E}{L}t\epsilon^{-\omega_0 t} \tag{8–115}$$

3. If $\rho < 1$, s_1 and s_2 are conjugate complex quantities, and the previously defined underdamped oscillation exists in the current, which assumes the form

$$i = \frac{E\epsilon^{-\rho\omega_0 t}}{\omega_0 L\sqrt{1 - \rho^2}}\sin \omega_0\sqrt{1 - \rho^2}t$$
$$= E\sqrt{\frac{C}{L}}\frac{\omega_0}{\omega_d}\epsilon^{-\rho\omega_0 t}\sin \omega_d t \tag{8–116}$$

If $\rho = 0$, then

$$i = \frac{E}{\omega_0 L}\sin \omega_0 t \tag{8–117}$$

Many similarities exist between the form of the solution for the step voltage function and the form of the solution for the impulse voltage function. The basic procedures involved in the solutions are the same, but different notation has been employed. It is suggested that the student carry out a detailed comparison of the two solutions; and, further, analyze the step function case in terms of energies as was done for the unit impulse.

8–15. Initial Conditions. When solving the differential equations for the various combinations of elements, initial conditions of current or of charge, or both current and charge, were used to find the constants of integration. For example, in Eq. 8–32 the current at $t = 0$ was assumed zero, and in Eq. 8–39 the current at $t = 0$ was assumed $1/L$. In Eq. 8–57 the charge stored on the capacitance at $t = 0$ was assumed to be zero. All of these assumptions meant that the circuit at $t = 0$ (that is, the time of application of the excitation) was at rest. (The value $1/L$ of Eq. 8–39 came from the impulse and not from the previous history of the circuit.)

In general, the forcing function may be applied while there is a current in an inductance or a charge on a capacitance. The initial conditions at $t = 0$ must then include the value of this current and the value of this charge. To illustrate, reconsider the R-L-C series circuit. If a current i_0 exists in the inductance and a charge q_0 is on the capacitance at the instant of the voltage impulse, then at $t = 0$ the total initial current is $(i_0 + 1/L)$ and the initial charge is q_0. The equilibrium equations for the network do not change and Eq. 8–73 is still correct. At $t = 0$, however, the current now becomes

$$\frac{1}{L} + i_0 = A + B \tag{8–118}$$

and the charge across the capacitance becomes

$$q_0 = \frac{A}{s_1} + \frac{B}{s_2} \tag{8–119}$$

The solution of Eqs. 8–118 and 8–119 for A and B, remembering that

$$s_1 s_2 = \omega_0^2$$

and

$$s_1 - s_2 = j2\omega_d$$

yields

$$A = \frac{s_1\left(\frac{1}{L} + i_0\right) - \omega_0^2 q_0}{2j\omega_d} \tag{8–120}$$

and

$$B = -\frac{s_2\left(\frac{1}{L} + i_0\right) - \omega_0^2 q_0}{2j\omega_d} \tag{8–121}$$

The substitution of Eqs. 8–120 and 8–121 into Eq. 8–73 yields after reduction and simplification the following form

$$i = -\epsilon^{-\alpha t}\,\frac{\omega_0}{\omega_d}\left[\left(\frac{1}{L} + i_0\right)\sin\,(\omega_d t - \theta) + \omega_0 q_0 \sin\,\omega_d t\right] \tag{8–122}$$

where

$$\theta = \tan^{-1}\frac{\omega_d}{\alpha}$$

The foregoing procedure is a general one for the solution of circuit problems where initial values of current in an inductance or charge on a capacitance exist. It is important that the physical concepts involving energies in the circuit be understood so that the effect of initial conditions may be anticipated and the validity of the solution checked.

PROBLEMS

8–1. Express the voltages in the figure as a sum of singular functions.

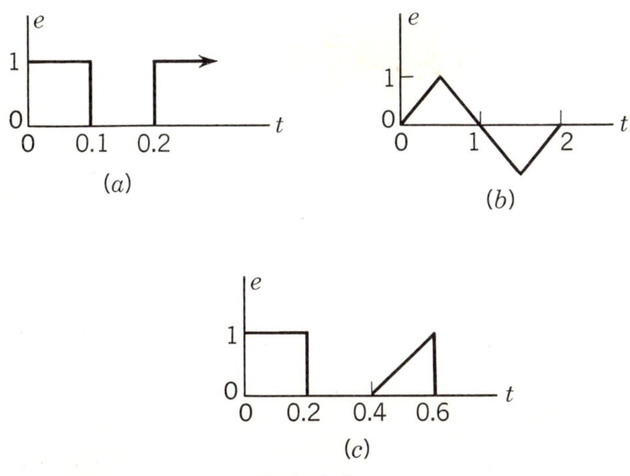

Prob. 8–1.

8–2. Write in notational form and sketch the waveforms for the derivatives and integrals of the voltages in Prob. 8–1.

8–3. A pure inductance (no resistance) has a value of 0.5 h. What will be the variation of voltage with respect to time if a current source is applied to it having the following form? Sketch the voltage waveform.

$$i_s(t) = 4u_{-1}(t) - 4u_{-1}(t - 0.1)$$

8–4. A unit impulse of voltage is applied to a pure inductor of value 0.4 h. (a) What is the value of current at $t = 10$ sec, at $t = \infty$? (b) What is the energy stored in the inductor at $t = 10$ sec? (c) What is the voltage across the inductor at $t = 10$ sec?

8–5. A unit impulse of current is applied to a 10-μf capacitor. What is the charge on the capacitor at $t = 5$ sec, at $t = \infty$?

8–6. What is the current-time variation in the inductance of Prob. 8–3 if the following voltage source is applied? Sketch the current waveform.

$$e_s(t) = 5u_{-1}(t) - 5u_{-1}(t - 0.1)$$

8–7. A current having a value $i_s(t) = 4u_{-1}(t)$ is applied to a 5-ohm resistor. What is the voltage across the resistor?

8–8. A voltage source which produces voltage as shown is connected to a circuit element and it is found that the current which results varies as shown. What kind of circuit element is involved and what is its value?

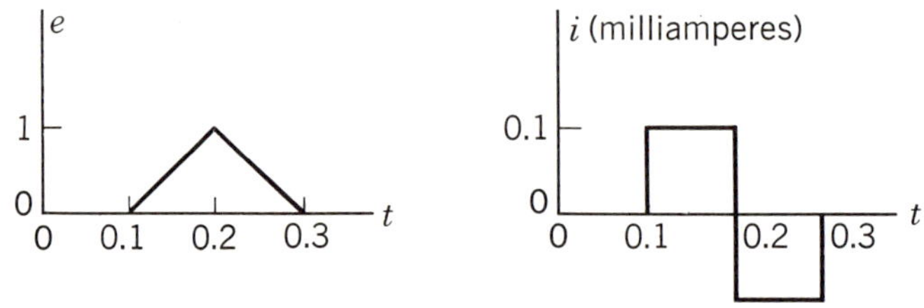

Prob. 8–8.

8–9. In the circuit illustrated find: (a) the equation for the circuit; (b) the time constant in seconds; (c) the value of current when $t = 4$ time constants (compare this value with the steady-state current); (d) the value of i at $t = 0$, at $t = \infty$; (e) the value of di/dt at $t = 0$, at $t = \infty$; (f) the plot of i versus t.

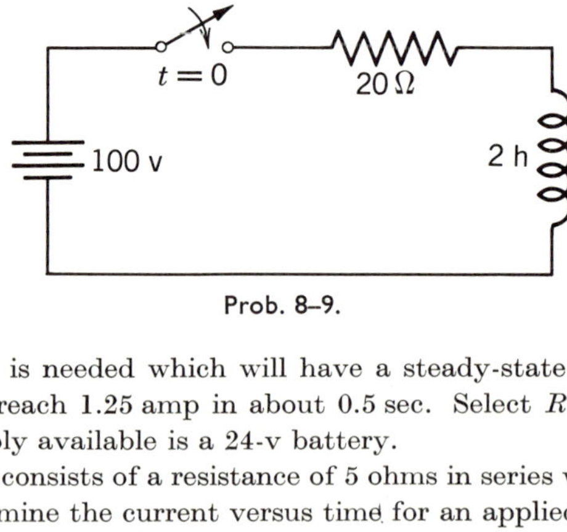

Prob. 8–9.

8–10. A circuit is needed which will have a steady-state current of 2 amp. The current must reach 1.25 amp in about 0.5 sec. Select R and L if the only direct-voltage supply available is a 24-v battery.

8–11. A circuit consists of a resistance of 5 ohms in series with an inductance of 0.5 h. (a) Determine the current versus time for an applied voltage source of value $5u_{-1}(t)$. (b) Determine the current versus time for an applied voltage source of value $5u_{-1}(t - 0.1)$. (c) Suppose the applied source voltage is

$$e_s(t) = 5u_{-1}(t) - 5u_{-1}(t - 0.1)$$

What is the current as a function of time? Draw a rough sketch of current versus time out to about $t = 0.3$ sec.

8–12. At $t = 0$ the switch S in the circuit shown is closed. (a) Write the equation for current as a function of time. (b) How long after closing the switch will it take for the current to reach 1 amp? (c) When the current is 2 amp, determine the (1) rate of change of current; (2) voltage across the inductor; (3) energy stored in the magnetic field; (4) power in the resistor; (5) power in the coil; (6) power output of the battery.

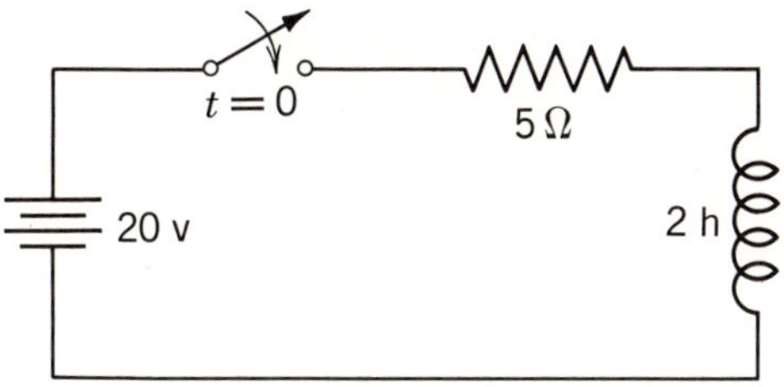

Prob. 8–12.

8–13. At the instant that the power dissipated in the resistor for the circuit illustrated is 20 w, what is the energy stored in the inductor?

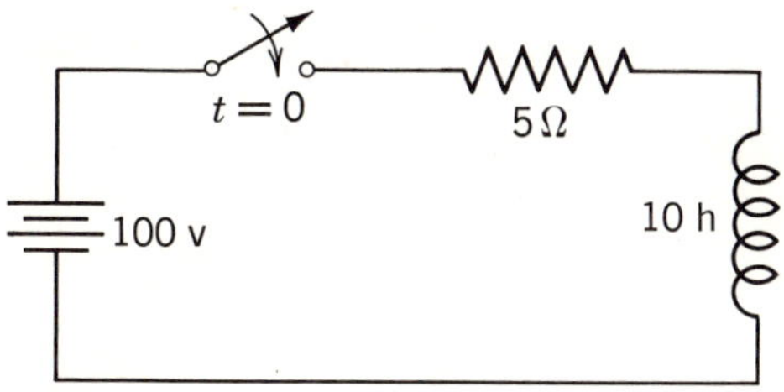

Prob. 8–13.

8–14. For the adjacent circuit find (a) the voltage across the inductor at $t = 5$ sec; (b) the energy being dissipated as heat in the resistor from $t = 0$ to $t = 1$ min.

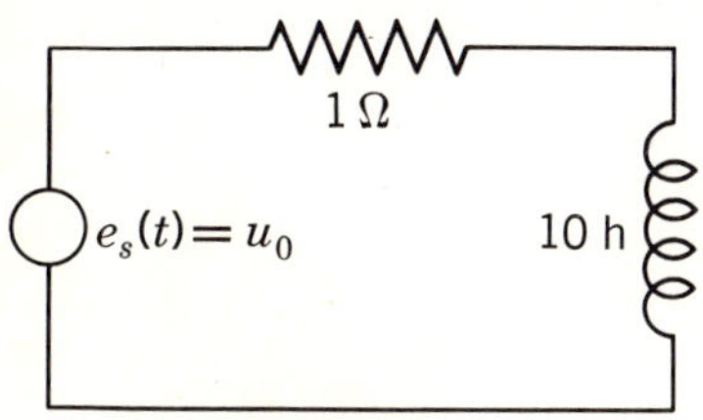

Prob. 8–14.

8–15. Calculate the energy stored in an inductance of value 0.1 h connected in series with a 10-ohm resistance if a unit voltage impulse is applied; further, calculate the energy dissipated during the time that a current flows in the circuit and compare to the amount of energy stored.

8–16. For the circuit shown if $i_s(t) = u_{-1}(t)$, determine the (a) equilibrium equation relating $i_s(t)$ and e; (b) expression for i_R and i_C in terms of e, employing Kirchhoff's current law; (c) time constant of the circuit; (d) charge on the capacitor at the end of 0.3 sec if $R = 1000$ ohms and $C = 100\ \mu$f.

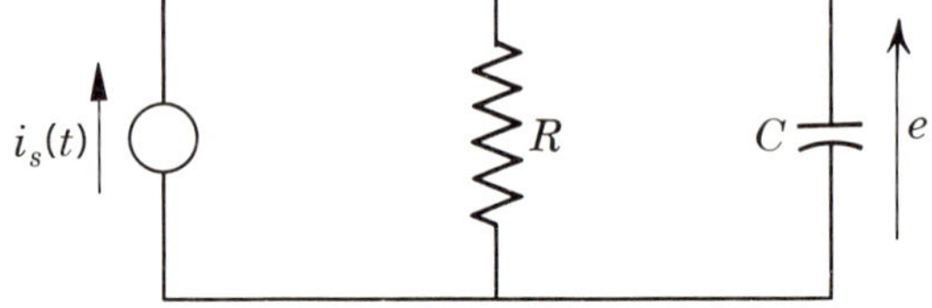

Prob. 8–16.

8–17. The current in an R-C series circuit with a battery voltage source is allowed to reach its steady-state value of E/R amperes. The battery is then instantaneously replaced by a short circuit. Develop the expression for the circuit current, the voltage across R, and the voltage across C as a function of time.

8–18. A unit impulse of current is applied to an R-L parallel network. Derive an expression for the voltage across the inductance as a function of time.

8–19. Show that the solution of any response problem solved for an impulse applied is the derivative of the solution when a step function is applied. Demonstrate this for the R-L series network.

8–20. Find the current equation for the circuit shown. The switch was initially open for a long time and is now suddenly closed.

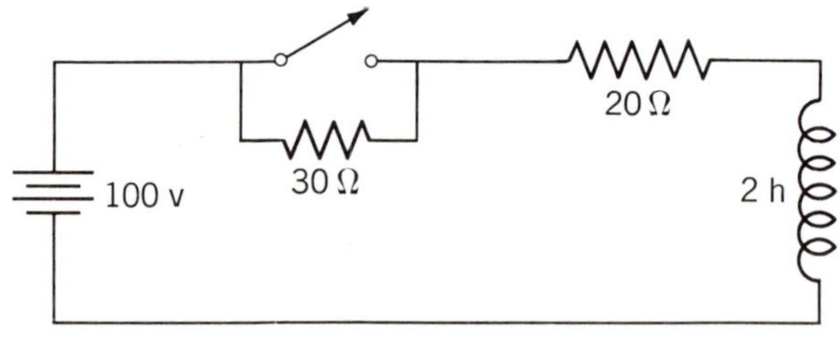

Prob. 8–20.

8–21. An R-L-C series circuit is excited by a unit step function. (a) If $R = 10$ ohms, $L = 10^{-2}$ h, and $C = 40 \times 10^{-6}$ farad, express i as a function of time and plot it. (b) For the parameters of (a) determine the voltage across L and C and plot on the same diagram. (c) From the plot of e_L and e_C, *sketch* the energy stored versus time in L and C. Comment.

9

NETWORK TRANSFORMATIONS

After a circuit is schematically represented and before its equilibrium equations are formulated it is oftentimes convenient to rearrange the elements in that circuit into a more convenient pattern for solution. This rearrangement of circuit elements may consist of establishing an equivalent for a combination of series or parallel elements, the conversion of a current source to a voltage source or vice versa, or the application of other simplifying theorems which may permit a major reduction in the complexity of the network configuration.

Also, it is often desirable to secure the current or voltage response of a certain element in the circuit or the response of a new element to be added between two points in a circuit. Although these results can certainly be obtained by a complete solution of the system, the labor involved may be appreciably shortened by the utilization of network transformation theorems. This chapter will discuss two of the most common, Thévenin's theorem and Norton's theorem.

9–1. Thévenin's Theorem. Thévenin's theorem is a general transformation which reduces any combination of active and passive circuit elements to a simple equivalent consisting of a voltage source in series with an equivalent passive element. This is shown in Fig. 9–1. Thévenin's theorem is a general

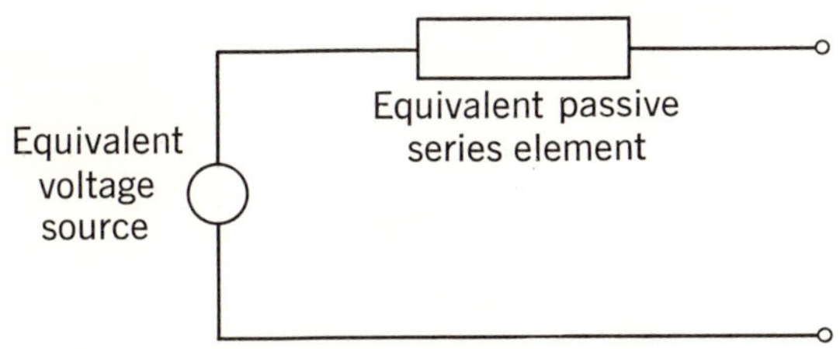

Fig. 9–1. Thévenin's equivalent circuit.

theorem applicable to all combinations of passive circuit elements, but the developments shown here will be in terms of combinations of single types of elements only, that is, only resistance, only inductance, or only capacitance.

The resistive case may be established by consideration of Fig. 9–2. Figure 9–2a represents a general situation in which a network consisting of resistive elements and active elements, such as current and voltage sources, is contained within the box with the exception of one loop containing *any* general circuit element shown connected externally between points a and b. The behavior of the network is such that a current i flows in the branch containing the generalized element. Figure 9–2b represents the same network with an

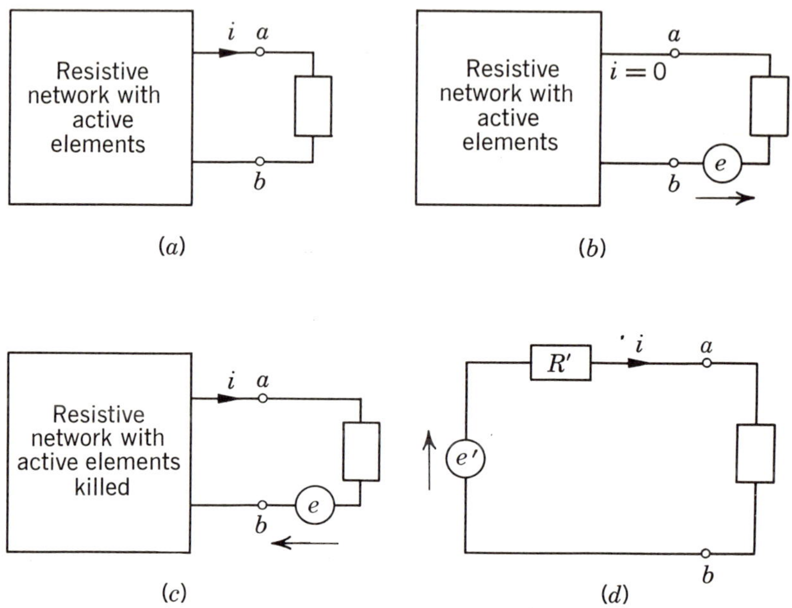

Fig. 9–2. Networks for the development of Thévenin's theorem.

external source of emf of value e connected in such a way that the current through the external branch is reduced to zero. By superposition it may be seen that source e is producing a current which is equal in magnitude and opposite in sense to that established by all of the active elements within the box. If all of the active elements within the box are rendered inactive, that is, voltage sources replaced by short circuits and current elements replaced by open circuits, then the voltage e will cause a current to exist in the external mesh equal in value to the original current but with the opposite direction. If the polarity of this external emf is reversed as shown in Fig. 9–2c, then exactly the same current flows through the external branch as appeared in Fig. 9–2a. In Fig. 9–2b the voltage existing between points a and b may be considered an open-circuit voltage since no current is flowing in the external branch; therefore, the voltage e is equivalent to the open-circuit voltage which would be measured across points a and b if the general circuit element were removed. From this reasoning, Fig. 9–2d is derived as the Thévenin equivalent of the network, where e' is the open-circuit voltage across points a and b of the original network and R' is the resistance which would be seen looking into terminals a and b (with the external network element removed) and all of the sources within the network "dead."

EXAMPLE 9–1. AN ILLUSTRATION OF THÉVENIN'S THEOREM. Simplify the resistive network of Fig. 9–3a by replacing the portions of the network to the left of the dashed line with a Thévenin equivalent circuit.

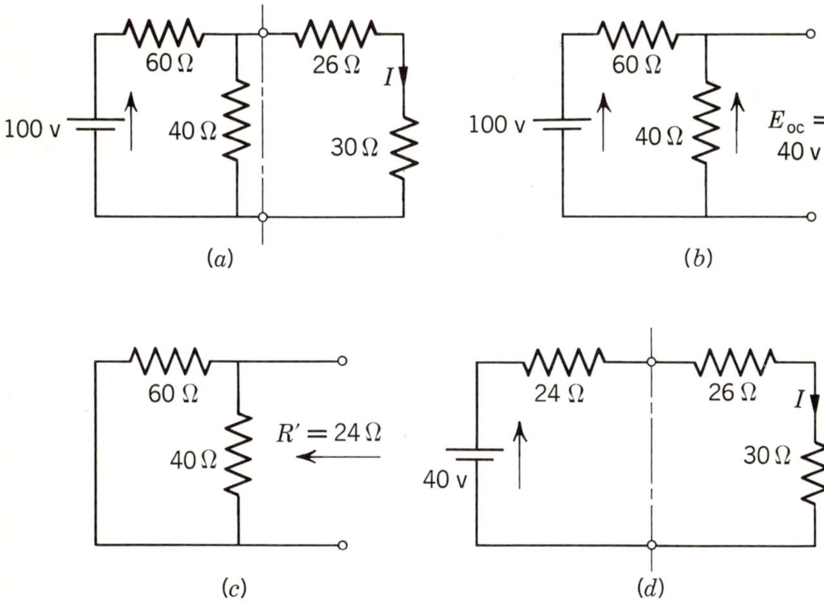

Fig. 9–3. Steps to obtain Thévenin's circuit.

The left-hand portion of the circuit is detached from the remainder as shown in (b) and the open-circuit voltage for this part computed as

$$E' = E_{oc} = \frac{40}{40 + 60} \times 100 = 40 \text{ v}$$

In (c) the 100-v source has been killed (that is, replaced by a short circuit) and the equivalent R' seen looking into the terminals calculated as the parallel combination of 60 ohms and 40 ohms, or 24 ohms. The final step, (d), is the reconnection of the circuit portion to the right to 40 v in series with 24 ohms. The current flowing through the portion of the circuit to the right of the dashed line may be computed as

$$I = \frac{40}{24 + 26 + 30} = 0.5 \text{ amp}$$

As a check, the original value of the current flowing through the portion of this circuit to the right of the dashed line may be computed by the use of series and parallel relationships as

$$I_s = \frac{100}{60 + \dfrac{40 \times 56}{40 + 56}} = \frac{100}{60 + \frac{70}{3}} = \frac{300}{250} = 1.2 \text{ amp}$$

and $$I = I_s \times \tfrac{40}{96} = \tfrac{6}{5} \times \tfrac{40}{96} = 0.5 \text{ amp}$$

As anticipated, the value of the current is the same for both situations. Although the example cited here is of course extremely simple, the principles may be applied to any combination of resistive elements within the original network and the element through which the current is desired may be a resistance, an inductance, or a capacitance.

The reasoning employed in conjunction with Fig. 9–2 may be applied to networks within the box that consist solely of active and inductive elements or active and capacitive elements. For these cases, the equivalent network assumes the same form as that indicated in Fig. 9–2*d* except that R' is replaced by L' and C' as shown in Fig. 9–4*a* and 9–4*b*, respectively. The

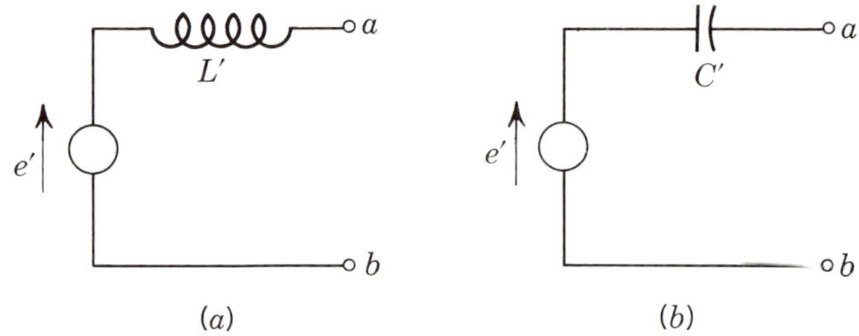

Fig. 9–4. Thévenin's network for inductive and capacitive elements.

values of L' and C' may be obtained by rendering all active elements inactive and calculating the equivalent element values existing across the terminals a and b with the external element removed. The voltage e' again is the open-circuit voltage existing between points a and b with the external circuit element removed.

EXAMPLE 9–2. THÉVENIN'S THEOREM FOR AN INDUCTIVE NETWORK. Assume the box to have a voltage source and two inductance elements L_1 and L_2 as shown in Fig. 9–5*a*. The Thévenin equivalent diagram is shown in Fig. 9–5*b*, where

$$e' = \frac{L_2}{L_1 + L_2} e$$

and $$L' = \frac{L_1 L_2}{L_1 + L_2}$$

The current i can now be found by solution of the equilibrium equation

$$e' = L' \frac{di}{dt} + Ri$$

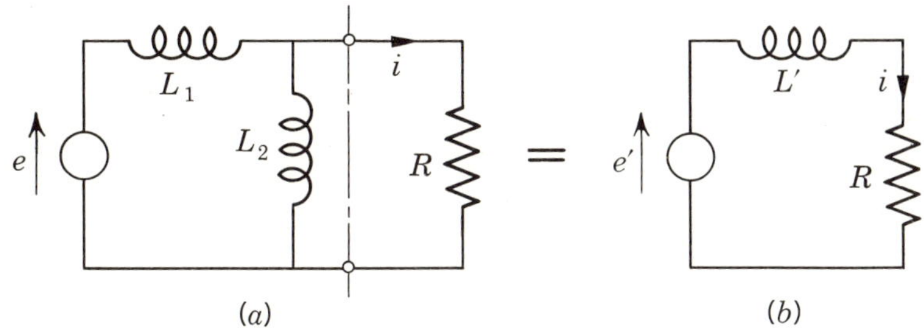

Fig. 9–5. Inductive circuit and its Thévenin equivalent.

EXAMPLE 9–3. THÉVENIN'S THEOREM FOR A CAPACITIVE NETWORK. If the box consists of a current source and two capacitances as shown in Fig. 9–6a, the Thévenin circuit is given by Fig. 9–6b, where

$$e' = \frac{1}{C_2} \int i_s \, dt$$

and

$$C' = C_2$$

The current i can now be found by solution of the equilibrium equation

$$e' = \frac{1}{C'} \int i \, dt + Ri$$

Again it must be emphasized that the application of Thévenin's theorem given here is a specialized one. In general, it is possible to treat combinations

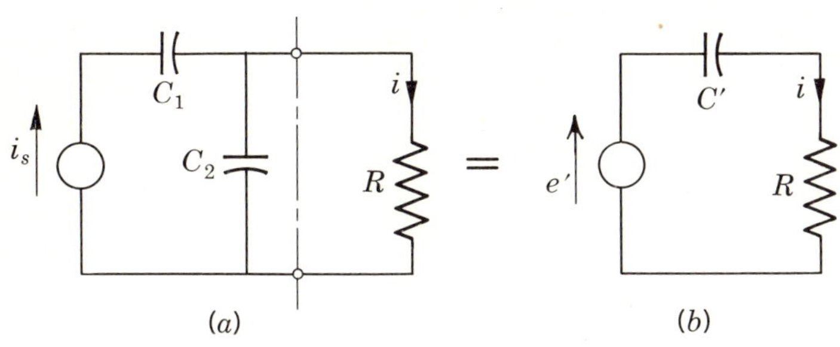

Fig. 9–6. Capacitive circuit and its Thévenin equivalent.

of elements within the box where resistance, inductance, and capacitance are combined. A more generalized proof for Thévenin's theorem in terms of generalized circuit impedances is presented in Chapters 16 and 22.

9–2. Norton's Theorem. Thévenin's theorem is often referred to as the constant-voltage equivalent circuit of a network. Chapter 6 indicated that there existed for every voltage source an equivalent current source; therefore, a constant-current equivalent of the same nature as the Thévenin equivalent may be derived directly by the use of source transformation procedures. The resultant network obtained is described as Norton's equivalent network and is shown in general form in Fig. 9–7. Figure 9–7 indicates that a network may be represented by an equivalent current source connected in parallel with an equivalent parallel element.

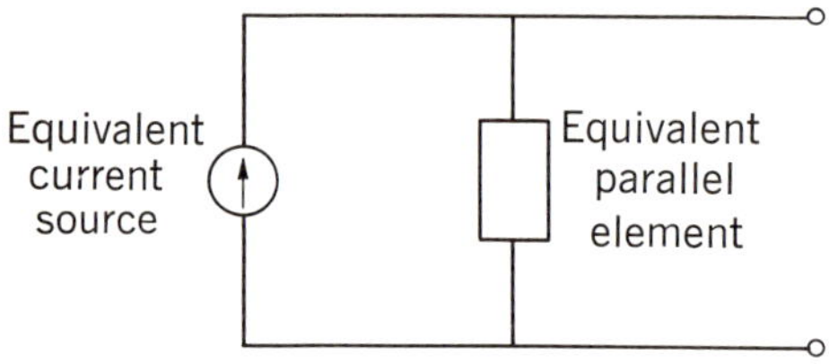

Fig. 9–7. Norton's equivalent circuit.

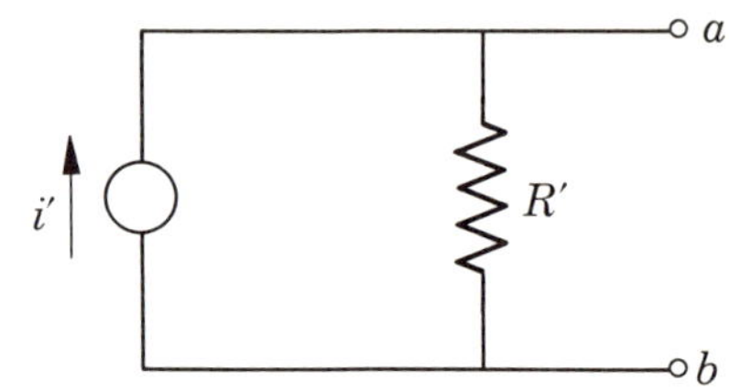

Fig. 9–8. Norton's equivalent circuit for a resistive network.

Equation 6–23 related the series resistance of a voltage source to the equivalent shunt resistance of a current source. The current source consisted of this equivalent series resistance connected between the two arms of the circuit and supplied by a current source whose current was obtained as the short-circuit current of the voltage source. The resistive form of Norton's theorem may be obtained now, as shown in Fig. 9–8. The current source for this figure is the short-circuit current obtained through a short circuit connected between terminals a and b of the general resistive networks. The shunt resistance is the same as the equivalent series resistance of the Thévenin form.

A similar equivalence exists between voltage sources having series inductive or capacitive elements and current sources having parallel inductive or capacitive elements. The equivalence relationships of Chapter 6 indicated that in each case the series element of the voltage source became the shunt element of the current source; therefore, Fig. 9–9a and 9–9b are the equivalents of the voltage sources shown in Fig. 9–4. The current sources indicated for each equivalent of Fig. 9–9 may be obtained from Thévenin's equivalent voltage sources by the transformations of Eqs. 6–31 and 6–37, or obtained directly from the chosen network as the short-circuit current existing between the terminals across which the general element of interest is normally connected.

As with Thévenin's theorem, it is important to note that Norton's theorem is a general transformation which may be applied to networks containing combinations of resistance, inductance, and capacitance, and that these derivations for networks containing only one variety of element represent a specialized case. The general development for Norton's theorem is discussed in detail in a later chapter.

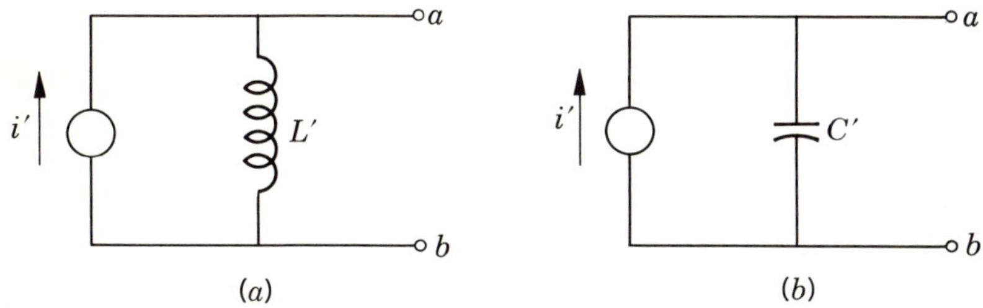

Fig. 9-9. Norton's equivalent circuit for inductive and capacitive networks.

Example 9-4. Application of Norton's Theorem to a Resistive Network. Reconsider the situation of Example 9-1 and apply Norton's theorem to the portion of the circuit to the left of the dashed line in Fig. 9-3*a*, as shown in Fig. 9-10*a*.

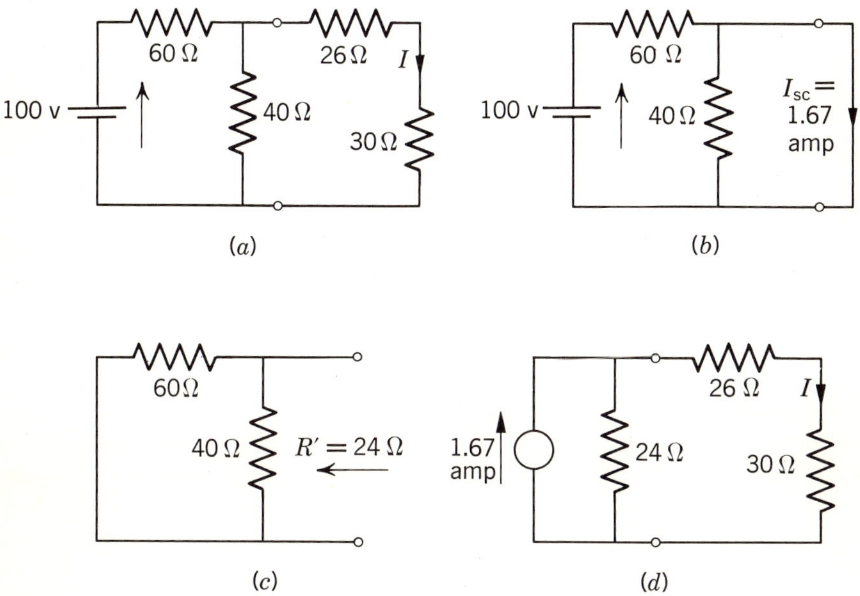

Fig. 9-10. Steps to obtain Norton's equivalent circuit.

The left-hand portion of the circuit is detached and short-circuited as shown in (*b*). The current flowing in the short circuit is

$$I_{sc} = \tfrac{100}{60} = 1.67 \text{ amp}$$

This may be easily seen by noting that the 40-ohm resistor is in parallel with zero ohm and therefore no current will exist in it. In (*c*), the 100-v source has been killed and the equivalent R' looking into ab calculated as

60 ohms in parallel with 40 ohms, or 24 ohms. The final step shown in (*d*) is the connection of the equivalent value of 24 ohms in parallel with the current source. The right-hand portion of the circuit is then reconnected and the current through the right-hand portion is computed as

$$E_{ab} = \frac{24(26 + 30)}{24 + 26 + 30} \times 1.667 = 28 \text{ v}$$

$$I = \tfrac{28}{56} = 0.5 \text{ amp}$$

EXAMPLE 9–5. NORTON'S THEOREM AND AN INDUCTIVE NETWORK. The Norton circuit equivalent to Fig. 9–11*a* is given by Fig. 9–11*b*, where

$$i' = \frac{1}{L_1}\int e_s\, dt$$

and

$$L' = \frac{L_1 L_2}{L_1 + L_2}$$

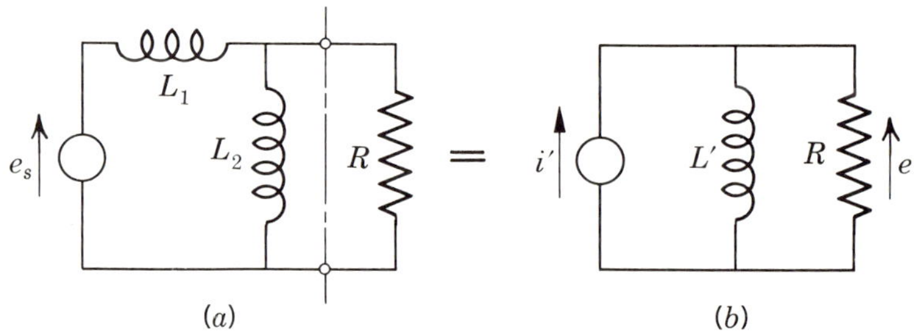

Fig. 9–11. Inductive circuit and its Norton equivalent.

The voltage e can now be found by solution of the equilibrium equation

$$i' = \frac{1}{L'}\int e\, dt + \frac{1}{R}e$$

EXAMPLE 9–6. NORTON'S THEOREM ON A CAPACITIVE NETWORK. The Norton circuit equivalent to Fig. 9–12*a* is given by Fig. 9–12*b*, where

$$i' = i$$

and

$$C' = C_2$$

The voltage e can now be found by the solution of the equilibrium equation

$$i' = C'\frac{de}{dt} + \frac{1}{R}e$$

The application of Thévenin's and Norton's theorems, even in the simplified form presented here, makes it possible in the process of circuit analysis to interchange voltage and current sources at will and to simplify portions of a circuit by introducing equivalent sources.

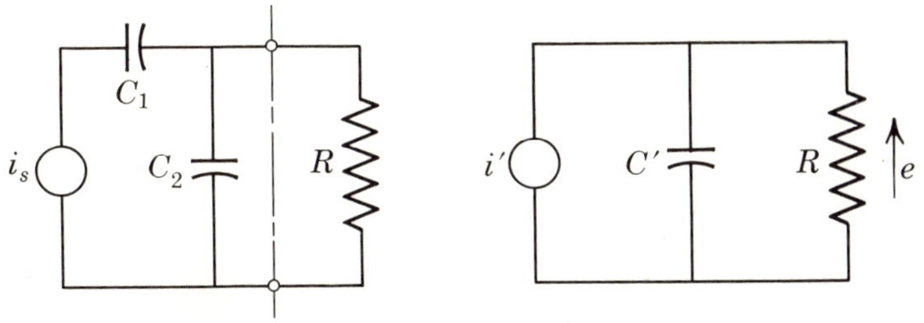

(a) (b)

Fig. 9–12. Capacitive circuit and its Norton equivalent.

PROBLEMS

9–1. Utilize Thévenin's or Norton's theorem to simplify the networks shown on pages 128–129 and to obtain the currents or voltages requested. Values are given in ohms, henrys, and farads unless otherwise specified.

9–2. An ideal voltage source of 100 v d-c is in series with a 5-ohm resistance. This may represent an actual voltage source. What will the terminal voltage of this actual source be when the current flowing to an external circuit is 2 amp?

9–3. An actual constant-current source is formed by a d-c ideal current source having a current of 1 amp and a parallel resistor of 100 ohms. (a) For a load connected to the terminals of 100 ohms, what is the terminal voltage and what is the load current? (b) When the load is changed, a voltage of 80 v is noted. What, then, is the load current and load resistance?

9–4. A d-c generator has a terminal voltage of 120 v at no load (open circuit), and a voltage of 112 v when the load current is 40 amp. Find the values of the constant-voltage equivalent circuit.

9–5. Convert the results of Prob. 9–4 to a constant-current equivalent circuit.

9–6. A vacuum-tube circuit has two terminals. With the terminals short-circuited, the current flowing is observed to be 15 ma. With a 100,000-ohm resistor connected in place of the short circuit, the current is 10 ma. Find the constant-current equivalent circuit of this tube circuit.

9–7. Convert the data of Prob. 9–6 to find the constant-voltage equivalent circuit.

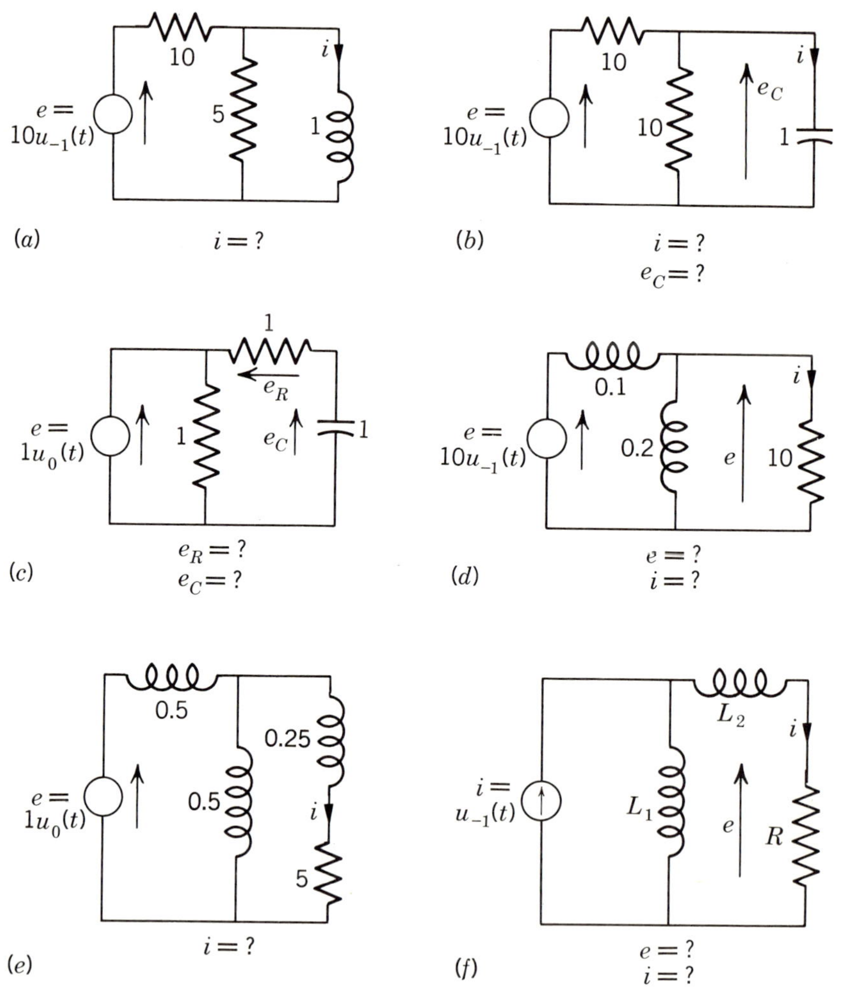
(a)
e = 10u_{-1}(t)
10
5
1
i
i = ?
(b)
e = 10u_{-1}(t)
10
10
e_C
1
i
i = ?
e_C = ?
(c)
e = 1u_0(t)
1
1
e_R
e_C
1
e_R = ?
e_C = ?
(d)
e = 10u_{-1}(t)
0.1
0.2
e
10
i
e = ?
i = ?
(e)
e = 1u_0(t)
0.5
0.5
0.25
i
5
i = ?
(f)
i = u_{-1}(t)
L_1
L_2
e
R
i
e = ?
i = ?

Prob. 9–1

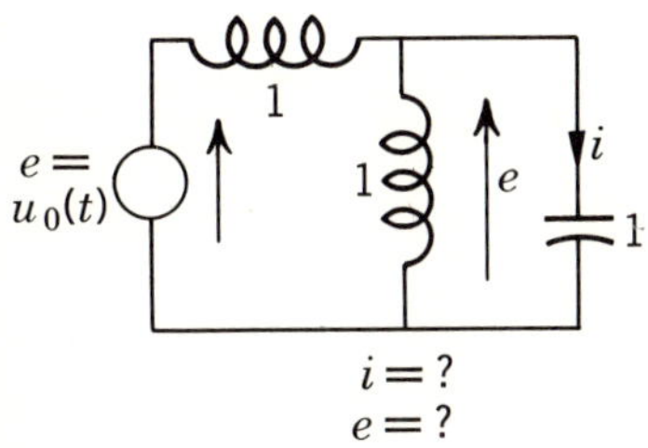

(g)

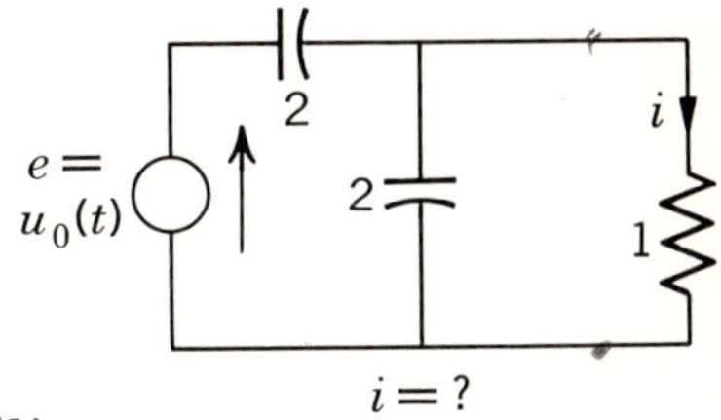

(h)

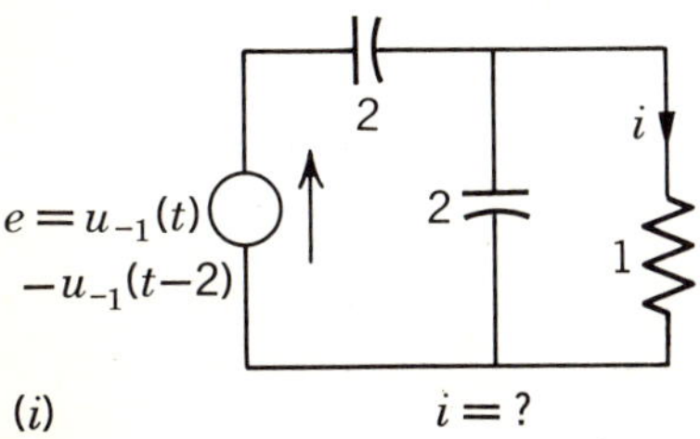

(i)

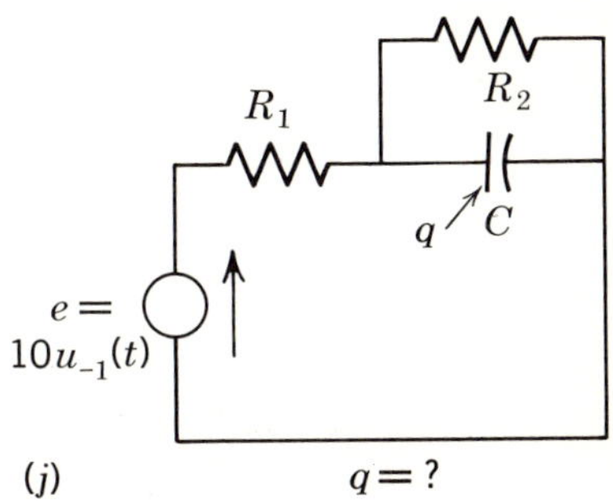

(j)

Prob. 9–1. (Cont.)

10

STEADY-STATE RESPONSE TO SINUSOIDAL FORCING FUNCTIONS

A knowledge of sinusoidal functions is basic to the study of electrical engineering, since in practice nearly all commercial current and voltage sources generate a waveform which is an approximation to a sinusoid. A second important reason for the study of sinusoids is, as will be shown later, that any periodic function of time may be considered as the sum of sinusoids, and the response of a system to this periodic function determined as the sum of the response to each of the sinusoidal components.

A sinusoid is mathematically a most amenable function. For example, the sum of two sinusoids of the same frequency will be a sinusoid. Multiplication of sinusoids yields a new function also of sinusoidal form, and the integral and derivative of a sinusoid are, in turn, sinusoids. The response of any linear passive bilateral system to a sinusoidal forcing function is also a sinusoid. This chapter will illustrate some of these properties and show how they are applied to electrical networks and elements.

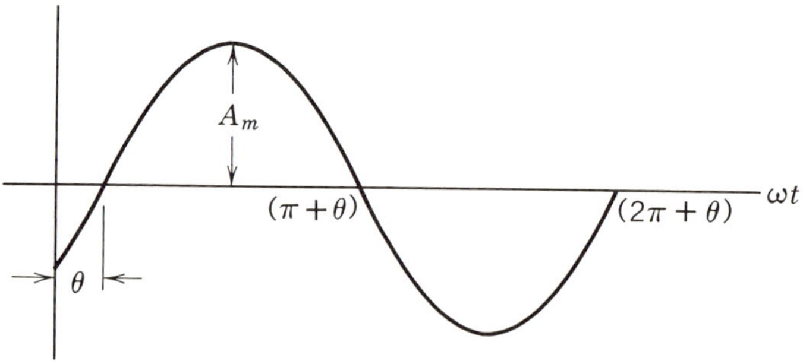

Fig. 10–1. Plot of a sine function.

10–1. Sinusoidal Notation and Nomenclature. The usual electrical engineering expression for a sinusoidal* function is

$$A_m \sin(\omega t - \theta) \tag{10–1}$$

* "Sinusoid" may refer either to the sine or the cosine function. Since $\cos x = \sin(x + \pi/2)$, only the sine will be discussed in this section.

which may be plotted as shown in Fig. 10–1. ω is defined as $2\pi f$, where f is the frequency of the function in cycles per second, or $f = 1/T$, where T is the time in seconds for one complete cycle. The phase angle θ is measured negatively to the right from $\omega t = 0$ as a reference. Note that the horizontal axis is calibrated in units of radians (ωt = angle in radians) for simplicity.* A_m is an amplitude factor which determines the maximum displacement of the wave above and below the horizontal axis.

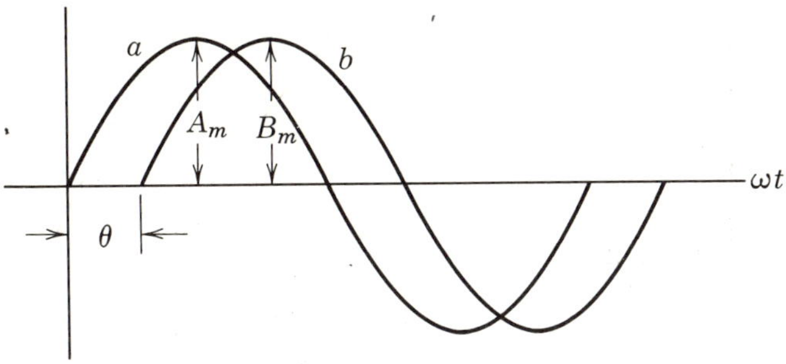

Fig. 10–2. Phase difference between sinusoids.

Two sine waves of the same frequency are plotted against ωt in Fig. 10–2. Curve a is said to lead curve b by the angle θ, or, conversely, curve b lags curve a by θ. Either one of the curves, a or b, may be taken as a reference. If curve a is mathematically expressed as

$$a = A_m \sin \omega t \tag{10–2}$$

then curve b is expressed as

$$b = B_m \sin (\omega t - \theta) \tag{10–3}$$

If the origin of time is changed so that b is expressed as

$$b = B_m \sin \omega t \tag{10–4}$$

* The sine function of Fig. 10–1 may also be plotted versus time, as shown below.

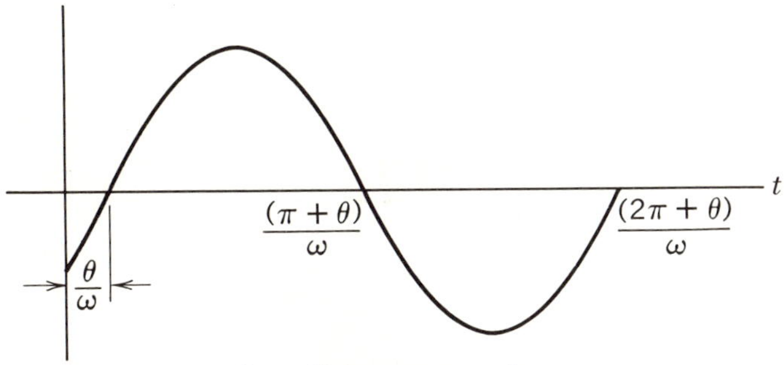

Sinusoidal wave versus time.

then a will become

$$a = A_m \sin(\omega t + \theta) \tag{10–5}$$

The origin or reference may be taken at some point other than when a or b is passing through zero. In Fig. 10–3, for example, curves a and b are expressed mathematically as

$$a = A_m \sin(\omega t - \alpha) \tag{10–6}$$

$$b = B_m \sin(\omega t - \alpha - \theta) \tag{10–7}$$

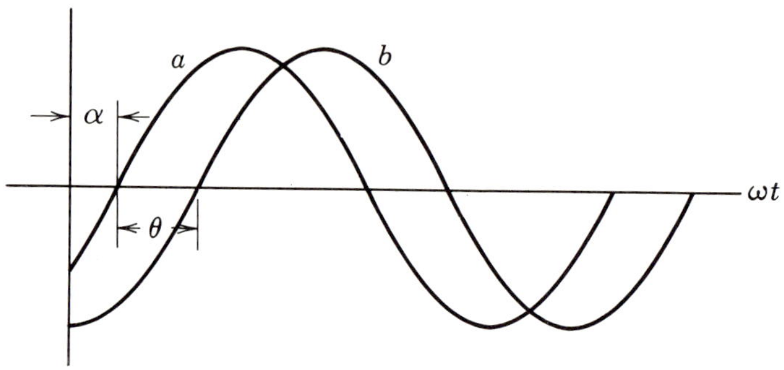

Fig. 10–3. Sinusoids with general phase relation.

10–2. Particular Integral or Steady-State Value. Assume a voltage source of the form

$$e = E_m \cos(\omega t + \alpha) \tag{10–8}$$

applied to an R-L series circuit as in Fig. 10–4. The equilibrium equation for the circuit is

$$Ri + L\frac{di}{dt} = E_m \cos(\omega t + \alpha) \tag{10–9}$$

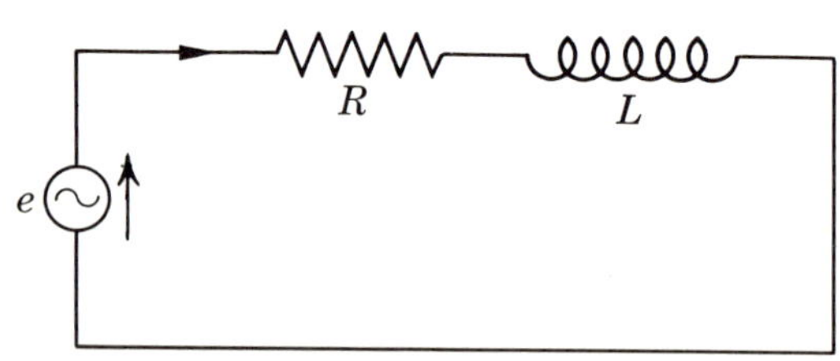

Fig. 10–4. *R-L* series circuit—sinusoidal excitation.

The particular or steady-state solution of the equilibrium equation has the same form as the *forcing function and all its derivatives*. The derivative of a cosine is a sine, and vice versa. Therefore, the form of the particular integral after all the factors of the cosine and the sine have been collected in I_1 and I_2 is:

$$i = I_1 \cos(\omega t + \alpha) + I_2 \sin(\omega t + \alpha) \tag{10–10}$$

If Eq. 10–10 is substituted into Eq. 10–9,

$$R[I_1 \cos(\omega t + \alpha) + I_2 \sin(\omega t + \alpha)] - \omega L[I_1 \sin(\omega t + \alpha) - I_2 \cos(\omega t + \alpha)] = E_m \cos(\omega t + \alpha) \quad (10\text{–}11)$$

The sine terms on the left of Eq. 10–11 are equated to the sine terms on the right, and the cosine terms on the left to the cosine terms on the right:

$$\sin(\omega t + \alpha)(RI_2 - \omega LI_1) = 0 \quad (10\text{–}12)$$

and

$$\cos(\omega t + \alpha)(RI_1 + \omega LI_2) = E_m \cos(\omega t + \alpha) \quad (10\text{–}13)$$

Equation 10–12 gives

$$I_2 = \frac{\omega L}{R} I_1 \quad (10\text{–}14)$$

and the substitution of Eq. 10–14 into Eq. 10–13 yields

$$I_1\left(R + \frac{\omega^2 L^2}{R}\right) = E_m \quad (10\text{–}15)$$

or

$$I_1 = \frac{E_m R}{R^2 + \omega^2 L^2} \quad (10\text{–}16)$$

From Eqs. 10–16 and 10–14, I_2 is obtained as

$$I_2 = \frac{E_m \omega L}{R^2 + \omega^2 L^2} \quad (10\text{–}17)$$

Therefore, the final form of Eq. 10–10 after Eqs. 10–16 and 10–17 are substituted is

$$i = \frac{E_m R}{R^2 + \omega^2 L^2} \cos(\omega t + \alpha) + \frac{E_m \omega L}{R^2 + \omega^2 L^2} \sin(\omega t + \alpha) \quad (10\text{–}18)$$

10–3. Impedance. The units of ωL, which is termed *inductive reactance*, are volts per ampere or ohms. If a right triangle is constructed with R as the base and ωL as the height as shown in Fig. 10–5, the hypotenuse also has units of ohms and is called *impedance*, denoted by Z. The value of the impedance is

$$Z = \sqrt{R^2 + \omega^2 L^2} \quad (10\text{–}19)$$

From Fig. 10–5 it is seen that

$$\cos\theta = \frac{R}{Z} \quad (10\text{–}20)$$

$$\sin\theta = \frac{\omega L}{Z} = \frac{X_L}{Z} \quad (10\text{–}21)$$

and

$$\tan\theta = \frac{X_L}{R} \quad (10\text{–}22)$$

where $X_L = \omega L$

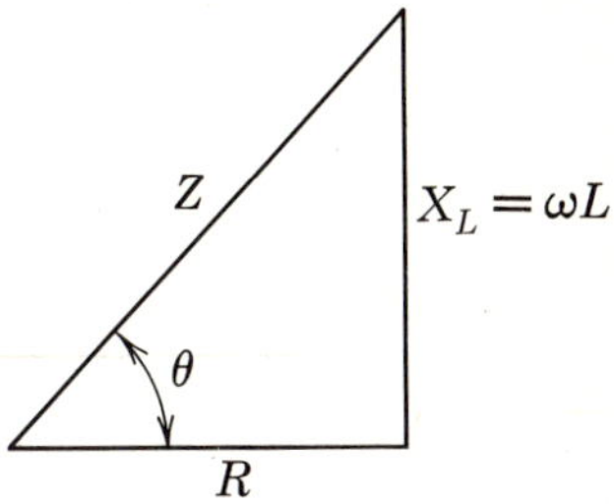

Fig. 10–5. The impedance triangle.

In general, for circuits containing resistance and reactance (either inductive or capacitive), an impedance triangle may be constructed. The reactance always appears at right angles to the resistance in such a representation. Capacitive reactance is negative, if inductive reactance is chosen positive, and will appear below the vector representing R. This will be seen in the following sections discussing sinusoidal volt-ampere characteristics.

10–4. Instantaneous Current in Terms of *Z*. If Eqs. 10–19 through 10–22 are substituted into Eq. 10–18, we have

$$i = \frac{E_m}{Z} \times \frac{R}{Z} \cos(\omega t + \alpha) + \frac{E_m}{Z} \times \frac{X_L}{Z} \sin(\omega t + \alpha)$$

$$= \frac{E_m}{Z} [\cos\theta \cos(\omega t + \alpha) + \sin\theta \sin(\omega t + \alpha)]$$

$$= \frac{E_m}{Z} \cos(\omega t + \alpha - \theta) \tag{10–23}$$

Letting

$$\frac{E_m}{Z} = I_m \tag{10–24}$$

Eq. 10–23 becomes

$$i = I_m \cos(\omega t + \alpha - \theta) \tag{10–25}$$

Equation 10–25 shows that the current in the R-L series circuit is also a cosine lagging the voltage by θ, the impedance angle.

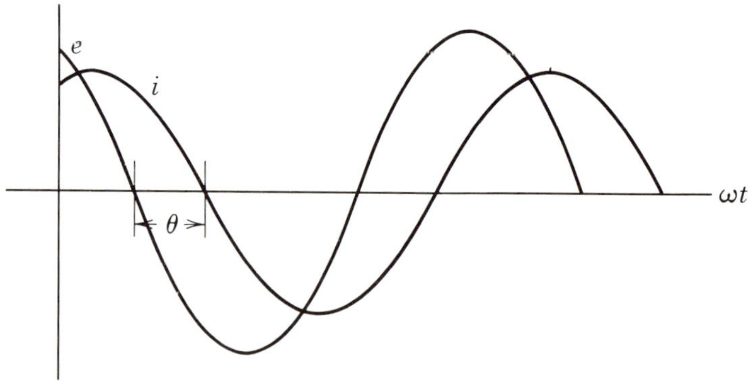

Fig. 10–6. Instantaneous current and voltage in the *R-L* series circuit.

From Eq. 10–24 it is seen that the maximum current is equal to the maximum voltage divided by the impedance. This is Ohm's law for the impedance circuit.

The current, however, is out of phase with and lags the voltage by the impedance angle θ, where

$$\theta = \tan^{-1} \frac{\omega L}{R} \tag{10–26}$$

Figure 10–6 is a plot of the instantaneous current and voltage described by Eqs. 10–8 and 10–25.

If the origin ($\omega t = 0$) is moved to the left by the angle $\omega t = \alpha$, the voltage may be re-expressed as

$$e = E_m \cos \omega t \tag{10–27}$$

and the current as

$$i = I_m \cos (\omega t - \theta) \tag{10–28}$$

where i is lagging the voltage by the angle θ.

10–5. Steady-State Sinusoidal Volt-Ampere Characteristics of Circuit Elements. RESISTANCE. Assume a resistance in which a current i exists, where

$$i = I_m \sin \omega t \tag{10–29}$$

The steady-state voltage across the resistance is

$$\begin{aligned} e_R &= iR \\ &= I_m R \sin \omega t \\ &= E_{m_R} \sin \omega t \end{aligned} \tag{10–30}$$

where

$$E_{m_R} = I_m R \tag{10–31}$$

Equation 10–30 indicates that the voltage across the resistance is also a sine wave in phase with the current. The maximum value E_m of the voltage is R times the maximum current.

Similarly, if a sine wave voltage is applied across a resistance R, the current in the resistance is

$$i = \frac{E_m}{R} \sin \omega t \tag{10–32}$$

$$= I_m \sin \omega t \tag{10–33}$$

The current too is a sine wave in phase with the voltage and of maximum value

$$I_m = \frac{E_m}{R} \tag{10–34}$$

Figure 10–7 shows a plot of the current in and the voltage across a resistance.

INDUCTANCE. Assume an inductance with a current in it of value

$$i = I_m \sin \omega t$$

The steady-state voltage across the inductance is then

$$e_L = L \frac{di}{dt} = \omega L I_m \cos \omega t \tag{10–35}$$

but

$$\cos \omega t = \sin \left(\omega t + \frac{\pi}{2} \right) \tag{10–36}$$

and therefore

$$\begin{aligned} e_L &= \omega L I_m \sin\left(\omega t + \frac{\pi}{2}\right) \\ &= E_{m_L} \sin\left(\omega t + \frac{\pi}{2}\right) \end{aligned} \tag{10–37}$$

where

$$E_{m_L} = \omega L I_m \tag{10–38}$$

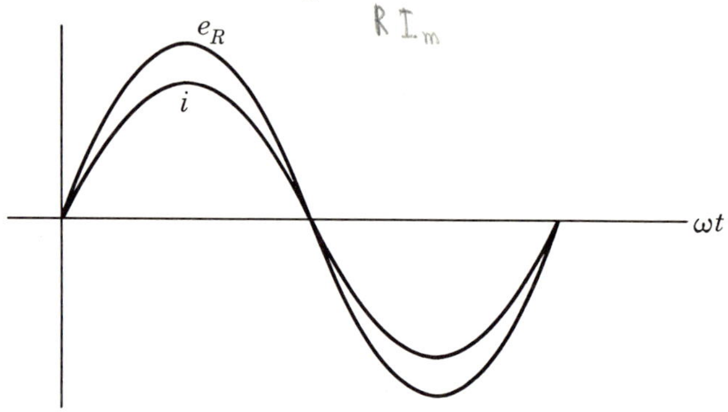

Fig. 10–7. Sinusoidal volt-ampere characteristics of a resistance.

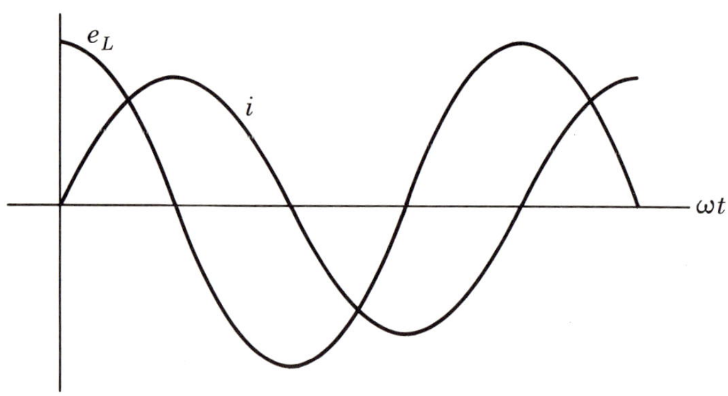

Fig. 10–8. Sinusoidal volt-ampere characteristics of an inductance.

The voltage is also a sine wave 90° ahead of the current in phase.

If a voltage

$$e = E_m \sin \omega t \tag{10–39}$$

is applied across an inductance, the steady-state current is

$$i = -\frac{E_m}{\omega L} \cos \omega t = \frac{E_m}{\omega L} \sin\left(\omega t - \frac{\pi}{2}\right) \tag{10–40}$$

$$= I_m \sin\left(\omega t - \frac{\pi}{2}\right) \tag{10–41}$$

where

$$I_m = \frac{E_m}{\omega L} \tag{10–42}$$

From Eq. 10–40 it is seen that the current is also a sine wave but lagging the voltage by a 90° phase angle.

Figure 10–8 is a plot of the current in and voltage across an inductance.

Capacitance. Assume a capacitor with a current in it of value

$$i = I_m \sin \omega t \tag{10–43}$$

The steady-state voltage across the capacitor is then

$$e_C = \frac{1}{C}\int i\,dt$$

$$= -\frac{I_m}{\omega C}\cos \omega t \tag{10–44}$$

$$= \frac{I_m}{\omega C}\sin\left(\omega t - \frac{\pi}{2}\right)$$

$$= E_{mC}\sin\left(\omega t - \frac{\pi}{2}\right) \tag{10–45}$$

where

$$E_{mC} = \frac{I_m}{\omega C} = I_m X_C \tag{10–46}$$

and

$$X_C = \frac{1}{\omega C} \tag{10–47}$$

From Eq. 10–45 it may be seen that the voltage across the capacitance is also a sine wave lagging the current by 90° in phase. X_C in Eqs. 10–46 and 10–47 has units of ohms and is called *capacitive reactance.*

If a voltage

$$e = E_m \sin \omega t \tag{10–48}$$

is across a capacitance element, the current is

$$i = C\frac{de}{dt}$$

$$= \omega C E_m \cos \omega t \tag{10–49}$$

$$= \omega C E_m \sin\left(\omega t + \frac{\pi}{2}\right) \tag{10–50}$$

$$= I_m \sin\left(\omega t + \frac{\pi}{2}\right) \tag{10–51}$$

where

$$I_m = \omega C E_m \tag{10–52}$$

and the current is a sine wave leading the voltage by 90° in phase.

Figure 10–9 is a plot of the current in and the voltage across a capacitance.

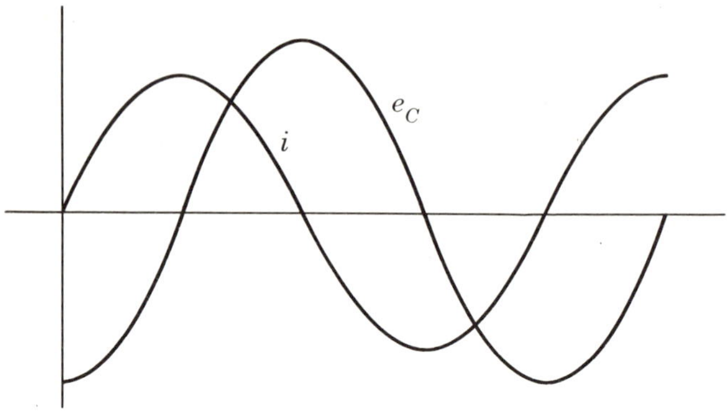

Fig. 10–9. Sinusoidal volt-ampere characteristics of a capacitance.

10–6. Addition of Sine Waves. Consider two sine waves

$$\begin{aligned} a_1 &= A_{m_1} \sin(\omega t + \alpha_1) \\ a_2 &= A_{m_2} \sin(\omega t + \alpha_2) \end{aligned} \tag{10–53}$$

We desire to find their sum.

The addition can be effected mathematically as follows:

$$\begin{aligned} a_1 &= A_{m_1} \sin \omega t \cos \alpha_1 + A_{m_1} \cos \omega t \sin \alpha_1 \\ a_2 &= A_{m_2} \sin \omega t \cos \alpha_2 + A_{m_2} \cos \omega t \sin \alpha_2 \end{aligned} \tag{10–54}$$

and

$$\begin{aligned} a_1 + a_2 = {} & (A_{m_1} \cos \alpha_1 + A_{m_2} \cos \alpha_2) \sin \omega t \\ & + (A_{m_1} \sin \alpha_1 + A_{m_2} \sin \alpha_2) \cos \omega t \end{aligned} \tag{10–55}$$

where the quantities A_{m_1}, A_{m_2}, α_1, and α_2 are constant.

Let

$$A_{m_1} \cos \alpha_1 + A_{m_2} \cos \alpha_2 = C_m \cos \delta \tag{10–56}$$

and

$$A_{m_1} \sin \alpha_1 + A_{m_2} \sin \alpha_2 = C_m \sin \delta \tag{10–57}$$

From Eqs. 10–56 and 10–57, δ must have a value given by

$$\tan \delta = \frac{A_{m_1} \sin \alpha_1 + A_{m_2} \sin \alpha_2}{A_{m_1} \cos \alpha_1 + A_{m_2} \cos \alpha_2} \tag{10–58}$$

If Eqs. 10–56 and 10–57 are squared and added

$$\begin{aligned} C_m^{\,2} &= A_{m_1}^{\,2} + A_{m_2}^{\,2} + 2A_{m_1}A_{m_2} \cos(\alpha_2 - \alpha_1) \\ &= A_{m_1}^{\,2} + A_{m_2}^{\,2} - 2A_{m_1}A_{m_2} \cos[\pi - (\alpha_2 - \alpha_1)] \end{aligned} \tag{10–59}$$

From Eqs. 10–55, 10–56, and 10–57

$$\begin{aligned} a_1 + a_2 &= C_m \sin \omega t \cos \delta + C_m \cos \omega t \sin \delta \\ &= C_m \sin(\omega t + \delta) \end{aligned} \tag{10–60}$$

and the sum of two sine waves is also a sine wave.

Equation 10–59 suggests a geometric method of finding the maximum value of the resultant sine wave as follows: If the maximum values of the original sine waves are assumed to be the legs of a triangle separated by an angle $\pi - (\alpha_2 - \alpha_1)$, then the third side of the triangle facing this angle is the maximum value C_m of the resultant sine wave. This is shown in Fig. 10–10.

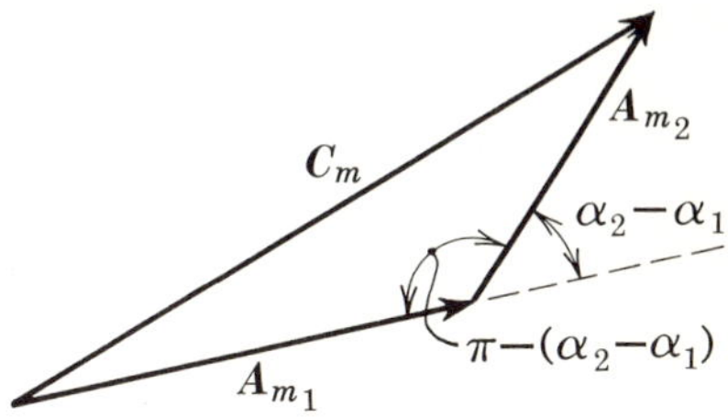

Fig. 10–10. The geometric addition of sinusoidal functions.

10–7. Phasors. The geometric addition of sine waves in Art. 10–6 suggests that the maximum values of a sine wave may be considered as a phasor.* Consider again the functions of Eq. 10–53:

$$a_1 = A_{m_1} \sin (\omega t + \alpha_1)$$

$$a_2 = A_{m_2} \sin (\omega t + \alpha_2)$$

On a rectangular XY coordinate system draw a line ON, as shown in Fig. 10–11, where ON rotates about O in the XY plane with angular velocity ω. The angle between ON and OX is ωt (Fig. 10–11a). Draw line $OA = A_{m_1}$ at angle α_1 with respect to ON, and line $OB = A_{m_2}$ at angle α_2 with respect to ON. Provided ω is the same for these phasors, the lines OA and OB will rotate with ON, keeping their angles constant with respect to ON and themselves.

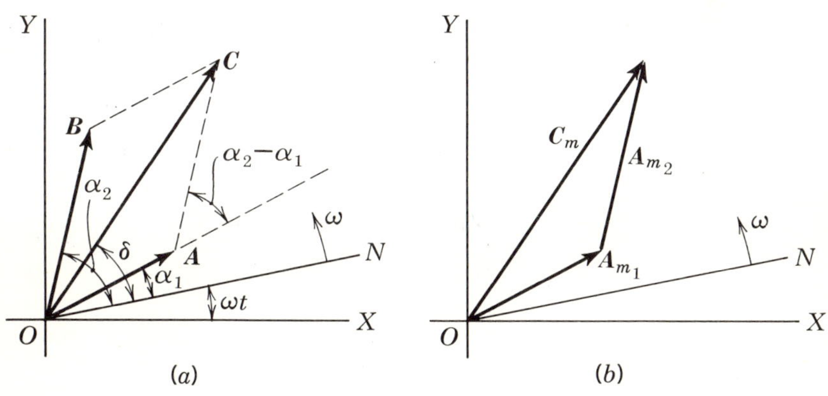

Fig. 10–11. Phasor representation of sinusoidal quantities.

The phasor sum of OA and OB is obtained by completing parallelogram $OACB$, and obtaining $OC = C_m$ at an angle δ with respect to ON, or

$$a_1 + a_2 = C_m \sin (\omega t + \delta)$$

* In the older literature this is termed a *vector*. To avoid confusion with three-dimensional directed lines, also called vectors, the AIEE has recommended use of the term *phasor* for two-dimensional directed lines representing electrical quantities.

Note that a_1 is the projection of OA on the Y axis, a_2 is the projection of OB on the Y axis, and $C_m \sin(\omega t + \delta)$ is the projection of OC on the Y axis. The projection of OA, OB, and OC on the X axis gives $A_{m_1} \cos(\omega t + \alpha_1)$, $A_{m_2} \cos(\omega t + \alpha_2)$, and $C_m \cos(\omega t + \delta)$ respectively.

When a sine or cosine wave is represented by a rotating phasor set as shown in Fig. 10–11, the phasor may be called a *sinor* or *cosinor*, but the usual term for it is phasor.

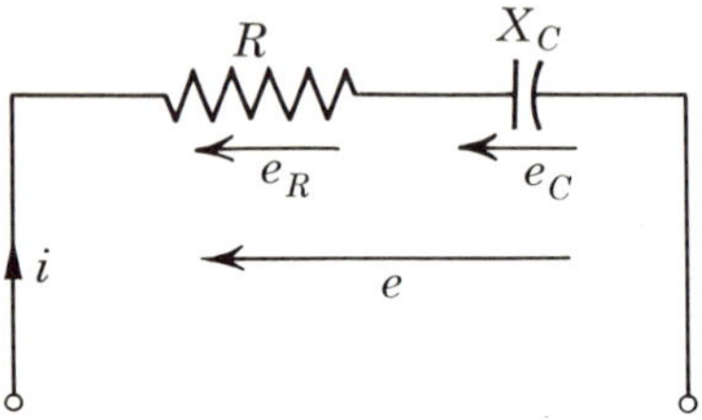

Fig. 10–12. Series *R-C* circuit with sinusoidal current.

Figure 10–11*b* presents a simplified form of Fig. 10–11*a*, showing A_{m_1} and A_{m_2} directly. The representation may be further simplified by placing one phasor and ON on the X axis, thus making α_1 and ωt zero.

10–8. Application of Phasor Method. Consider an R-C series circuit with a current $i = I_m \sin(\omega t + \alpha)$. The phasor method will be used to find the voltage e across the elements, as shown in Fig. 10–12.

Plot, as shown on Fig. 10–13, the phasor I_m making an angle $(\omega t + \alpha)$ with the X axis. From the volt-ampere characteristics for a resistance the

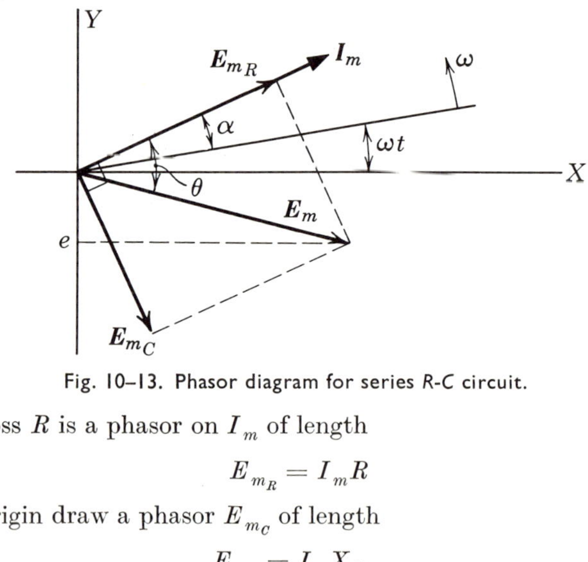

Fig. 10–13. Phasor diagram for series *R-C* circuit.

voltage across R is a phasor on I_m of length

$$E_{m_R} = I_m R \tag{10–61}$$

From the origin draw a phasor E_{m_C} of length

$$E_{m_C} = I_m X_C \tag{10–62}$$

having an angle of $-\pi/2$ with respect to I_m. The instantaneous value of the voltage e is

$$e = e_R + e_C \tag{10–63}$$

$$= E_{m_R} \sin(\omega t + \alpha) + E_{m_C} \sin\left(\omega t + \alpha - \frac{\pi}{2}\right) \tag{10–64}$$

and the maximum value of e is E_m, where

$$E_m^2 = E_{m_C}^2 + E_{m_C}^2$$
$$= (I_m R)^2 + (I_m X_C)^2 \tag{10–65}$$

or

$$E_m = I_m \sqrt{R^2 + X_C^2} \tag{10–66}$$

The angle between E_m and I_m is $-\theta$, where

$$\theta = \tan^{-1} \frac{X_C}{R} \tag{10–67}$$

The projections of the phasors on the Y axis give the instantaneous value

$$e = I_m \sqrt{R^2 + X_C^2} \sin (\omega t + \alpha - \theta) \tag{10–68}$$

and the term $\sqrt{R^2 + X_C^2}$ is the circuit impedance.

If the voltage applied to a series R-C circuit is known, and it is desired to find the current, a similar procedure may be employed. Let the voltage be

$$e = E_m \sin (\omega t + \alpha) \tag{10–69}$$

Then the value of the maximum current is

$$I_m = \frac{E_m}{\sqrt{R^2 + X_C^2}} \tag{10–70}$$

and the angle by which current leads the voltage is given by

$$\theta = \tan^{-1} \frac{X_C}{R} \tag{10–71}$$

The current is then

$$i = I_m \sin (\omega t + \alpha + \theta) \tag{10–72}$$

and is the projection of the phasor I_m on the Y axis as shown in Fig. 10–14.

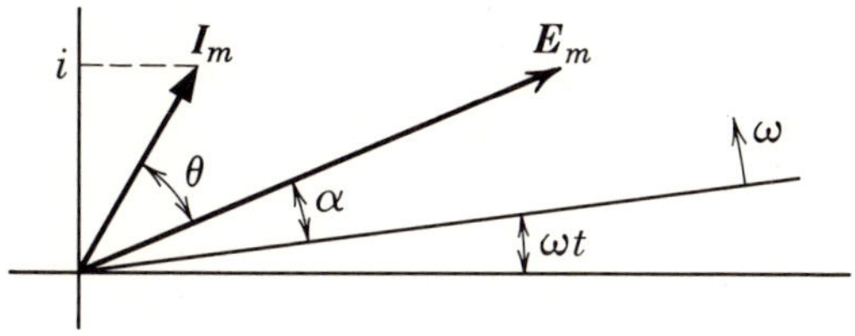

Fig. 10–14. Phasor diagram for series R-C circuit.

10–9. Phasors in Terms of Effective Values. In the preceding sections the phasors were assumed to be rotating with an angular velocity ω. Relative to each other, however, the current and voltage phasors are stationary.

Usually it is preferred to plot the phasor in terms of effective values, where "effective value" is defined as the value of current or voltage as read from a conventionally calibrated a-c ammeter or an a-c voltmeter. This effective value of a sine wave will be shown later to be equal to the maximum value divided by $\sqrt{2}$.

If the phasors are plotted in terms of their effective values, the phasor relationship remains the same as if plotted for maximum values. A scale factor has been applied to the diagram but no other change has occurred. As usual, either the voltage or the current may be chosen as reference, and if one is selected the other is plotted relative to it. The phasors are then treated as if at a standstill. Since the phase relationship is constant for phasors of the same frequency, the time of investigation is unimportant and may be arbitrarily chosen.

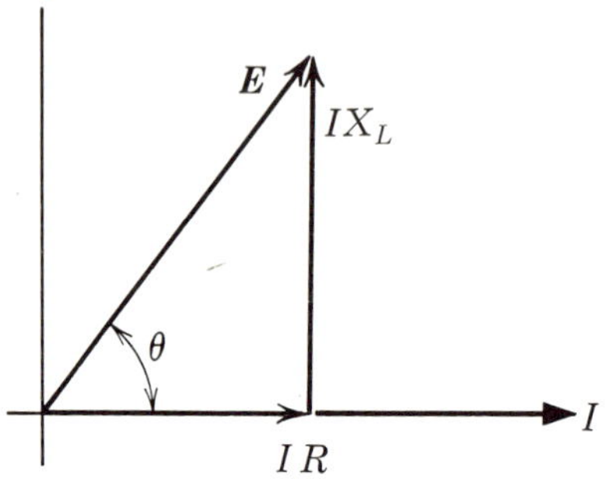

Fig. 10–15. Phasor diagram using effective values for an *R-L* series circuit.

The effective value of voltage and current is written in capital letters without subscripts: E and I. A phasor diagram, drawn on this basis, for an *R-L* series circuit is shown as Fig. 10–15.

EXAMPLE 10–1. USE OF PHASORS IN THE SOLUTION OF A STEADY-STATE A-C PROBLEM. Given a voltage

$$e = E_m \sin \omega t$$

applied to the circuit of Fig. 10–16, find the current in the circuit.

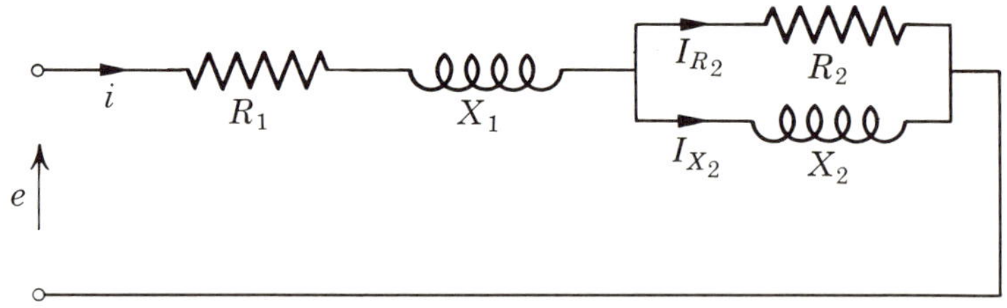

Fig. 10–16. Phasors in complex circuit.

The first step is to find the impedance of the parallel circuit and find its equivalent resistive and reactive components. Secondly, find the total resistive component of the circuit and its total reactive component. From this, the total impedance and its angle may be found. From the magnitude of the total impedance the maximum current is calculated, and from the total impedance angle the phase angle of the current is available and the current equation can then be written.

Following this procedure, assume 1 v across R_2 and X_2 in parallel and take that voltage as reference (Fig. 10-17). The current in phase with the 1 v is

$$I_{R_2} = \frac{1}{R_2}$$

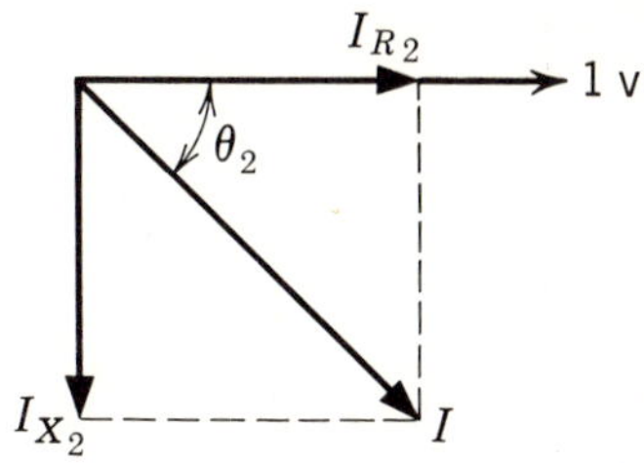

Fig. 10-17. The in-phase and out-phase current.

The current in X_2 (90° behind the reference) is

$$I_{X_2} = \frac{1}{X_2}$$

The total current in the parallel combination caused by the 1 v is

$$I = \sqrt{\left(\frac{1}{R_2}\right)^2 + \left(\frac{1}{X_2}\right)^2} = \frac{\sqrt{R_2^2 + X_2^2}}{R_2X_2}$$

The impedance of the R_2X_2 parallel circuit is

$$Z_2 = \frac{E}{I} = \frac{1}{I} = \frac{R_2X_2}{\sqrt{R_2^2 + X_2^2}}$$

This impedance has a resistive component given by

$$R'_2 = Z_2 \cos \theta_2 = \frac{R_2X_2}{\sqrt{R_2^2 + X_2^2}}\left(\frac{X_2}{\sqrt{R_2^2 + X_2^2}}\right) = \frac{R_2X_2^2}{R_2^2 + X_2^2}$$

and a reactive component of

$$X'_2 = Z_2 \sin \theta_2 = \frac{R_2X_2}{\sqrt{R_2^2 + X_2^2}}\left(\frac{R_2}{\sqrt{R_2^2 + X_2^2}}\right) = \frac{R_2^2X_2}{R_2^2 + X_2^2}$$

The total resistive component of the circuit is

$$R = R_1 + R'_2 = R_1 + \frac{R_2X_2^2}{R_2^2 + X_2^2}$$

and the total reactive component of the circuit is

$$X = X_1 + X'_2 = X_1 + \frac{R_2^2X_2}{R_2^2 + X_2^2}$$

and the total impedance of the circuit is

$$Z = \sqrt{R^2 + X^2}$$

The maximum value of current is

$$I_m = \frac{E_m}{Z}$$

and the angle by which I_m lags E_m is

$$\theta = \tan^{-1}\frac{X}{R} = \tan^{-1}\frac{X_1 + \dfrac{R_2{}^2X_2}{R_2{}^2 + X_2{}^2}}{R_1 + \dfrac{R_2X_2{}^2}{R_2{}^2 + X_2{}^2}}$$

Finally, the steady-state equation of the current is

$$i = I_m \sin(\omega t - \theta)$$

PROBLEMS

10–1. A phasor of length 1 unit is rotating at a uniform angular velocity $\omega = 2\pi$ radians/sec. Assume the phasor to be in the position shown for $t = 0$. Consider the phasor rotated through one complete revolution, and plot accurately against time its projection on the X axis and the Y axis for this travel. Describe the resulting waveforms and discuss their relation.

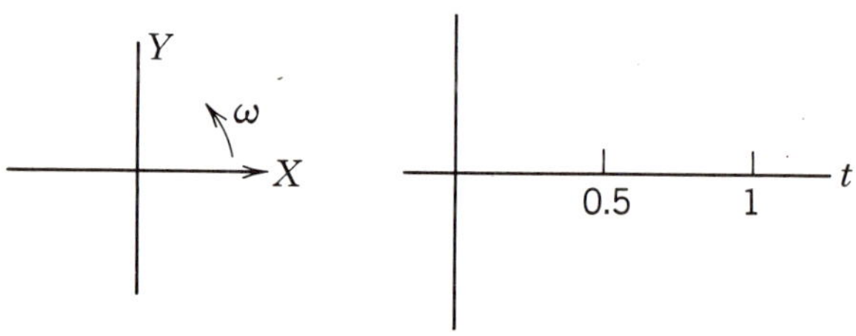

Prob. 10–1.

10–2. Determine the steady-state sinusoidal current in the network illustrated if $e = E_m \sin \omega t$. Express the answer in a form similar to Eq. 10–18.

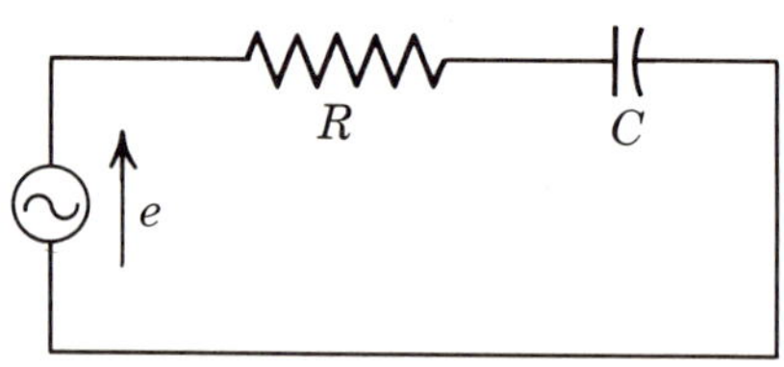

Prob. 10–2.

10–3. Draw the impedance triangle for the R-C series circuit of Prob. 10–2. Discuss similarities and differences compared to the impedance triangle for the series R-L circuit.

10–4. A series circuit consists of a 10-ohm resistance and a 10^{-4}-farad capacitance. If a potential of

$$e = 14.14 \sin 1000t$$

is applied, what are the instantaneous expressions for the current and voltages across R and C? Plot against time for one cycle of the applied voltage the values of i, e_R, and e_C.

10–5. A series circuit consists of a coil having an inductance of 4×10^{-4} h and 3 ohms resistance connected in series with a 2-ohm resistance. For a voltage of

$$e = 20 \cos 10^4 t$$

determine the expression for the circuit current, the voltage across the 2-ohm resistance, and the voltage across the terminals of the *coil*. Plot these values against time for one cycle of the applied voltage.

10–6. A voltage of value

$$e = 200 \sin (377t - 80°)$$

causes a current of 20 sin $(377t - 30°)$ in a circuit. What is the resistance and reactance of the circuit?

10–7. An R-C series circuit has a 60-cps sinusoidal potential of magnitude 100 v effective applied. The voltage across the resistance is 70.7 v. What is the voltage across the capacitance, and what is the size of C if $R = 35.35$ ohms? Draw the phasor diagram and the impedance triangle for this circuit, and also express in instantaneous form all voltages and the circuit current.

10–8. A series circuit has a 100-ohm resistor, a 10^{-6}-farad capacitor, and an 0.5-h inductance excited by a sinusoidal source having $\omega = 1000$ radians/sec. Draw the phasor diagram and the impedance triangle for the circuit and express the current and the voltages across R, L, and C in instantaneous form.

10–9. Repeat Prob. 10–8 if $L = 1$ h. Discuss the results.

11

COMPLEX ALGEBRA

The previous chapter demonstrated that a phasor representation could be employed to assist the steady-state analysis of sinusoidally excited networks and elements. Phasor notation and procedure are appreciably simpler than direct trigonometric manipulation, but an even greater ease of analysis is possible through the introduction of a systematic notation and manipulation procedure for the phasor. This occurs if the concept of "complex numbers" is employed to represent the phasors.

11-1. The *j* Operator. In algebra it is shown that there are n distinct nth roots for any number and that the sum of these roots is zero. For example, to find the four 4th roots of unity the procedure is as follows:

Let x be the 4th root of unity, written

$$x^4 - 1 = 0 \tag{11-1}$$

Factoring yields

$$(x^2 - 1)(x^2 + 1) = 0 \tag{11-2}$$

or

$$(x - 1)(x + 1)(x - j)(x + j) = 0 \tag{11-3}$$

where j is the square root of (-1).* Therefore the four roots of unity are

$$\begin{aligned} x_1 &= 1 \\ x_2 &= j \\ x_3 &= -1 \\ x_4 &= -j \end{aligned} \tag{11-4}$$

If these roots are plotted on the complex plane, Fig. 11-1 results. j is seen to be a directed line (vector) having a magnitude of unity and making an angle of $+90°$ with the real axis, or

$$j = 1\underline{/90°} \tag{11-5}$$

But j was obtained from

$$x^2 + 1 = 0$$

and therefore

$$j^2 = -1 \tag{11-6}$$

where -1 is the negative of $+1$ and from trigonometry is equal to 1 rotated through 180°, or

$$j^2 = -1 = 1\underline{/180°} \tag{11-7}$$

* Mathematics texts usually employ the symbol i for $\sqrt{-1}$ but j is the common symbol in electrical engineering.

Similarly,
$$j^3 = (1\angle 90°)^3 = -j = 1\angle 270° = 1\angle -90° \tag{11-8}$$
and
$$j^4 = (1\angle 90°)^4 = 1\angle 360° = 1\angle 0° = 1 \tag{11-9}$$

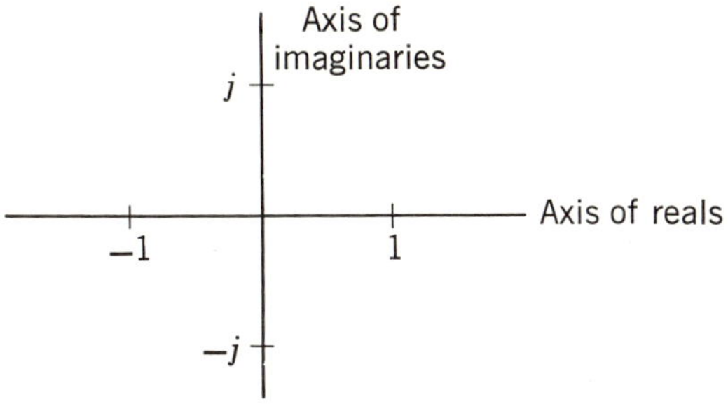

Fig. 11-1. Four roots of unity plotted on the complex plane.

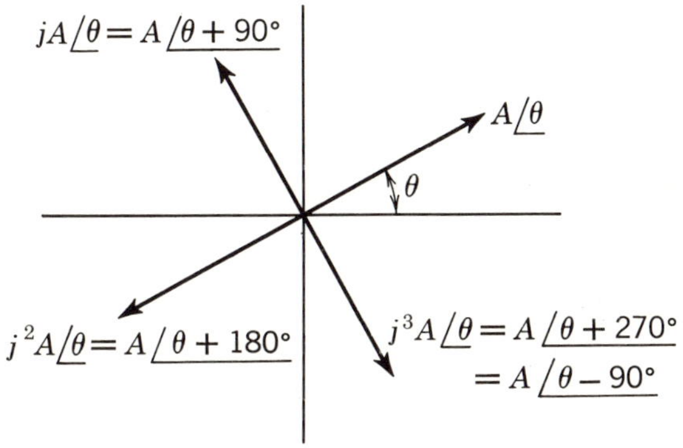

Fig. 11-2. Rotation of a complexor by j operators.

From Eqs. 11-5 and 11-7 through 11-9 the following conclusions may be drawn:

1. j before any number (real or complex) represented by a vector on the complex plane operates on that vector by rotating it through 90°.
2. j^2 before any number (real or complex) represented by a vector on the complex plane operates on that vector by rotating it through 180°.
3. j^3 before any number (real or complex) represented by a vector on the complex plane operates on that vector by rotating it through 270° (or −90°).
4. j^4 before any number (real or complex) represented by a vector on the complex plane operates on that vector by rotating it through 360° (or 0°).

This is demonstrated for a complexor* of magnitude A and angle θ in Fig. 11-2.

* *Complexor* is used here to represent a complex quantity plotted on the complex plane. When the complexor represents an electrical quantity, the term *phasor* is commonly used.

11–2. Complexor Forms. RECTANGULAR FORM. Figure 11–3 shows a complexor, denoted $\boldsymbol{A}$, having a magnitude A and an angle θ with the real axis. The magnitude of $\boldsymbol{A}$, written $|\boldsymbol{A}|$ or A, is given by geometric relations as

$$A = \sqrt{a^2 + b^2} \tag{11–10}$$

where

$$a = A \cos \theta$$

and

$$b = A \sin \theta \tag{11–11}$$

The phasor sum of a and b is the complexor $\boldsymbol{A}$, and therefore the rectangular form of $\boldsymbol{A}$ is

$$\boldsymbol{A} = a + jb \tag{11–12}$$

or

$$\boldsymbol{A} = A \cos \theta + jA \sin \theta = A(\cos \theta + j \sin \theta) \tag{11–13}$$

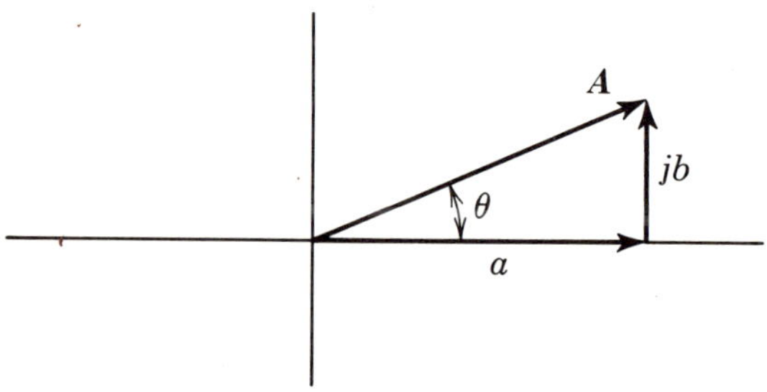

Fig. 11–3. Rectangular components of a complexor.

EXPONENTIAL FORM. $(\cos \theta + j \sin \theta)$ always has a magnitude of unity. Therefore this complexor may be written as

$$\boldsymbol{u} = \cos \theta + j \sin \theta \tag{11–14}$$

Now if the derivative of $\boldsymbol{u}$ is taken with respect to θ,

$$\begin{aligned} \frac{d\boldsymbol{u}}{d\theta} &= -\sin \theta + j \cos \theta \\ &= j(\cos \theta + j \sin \theta) \\ &= j\boldsymbol{u} \end{aligned} \tag{11–15}$$

or

$$\frac{d\boldsymbol{u}}{\boldsymbol{u}} = j\, d\theta \tag{11–16}$$

Integration of Eq. 11–16 yields

$$\ln \boldsymbol{u} = j\theta \tag{11–17}$$

or

$$\boldsymbol{u} = \epsilon^{j\theta} \tag{11–18}$$

But from Eq. 11–14

$$\boldsymbol{u} = \cos \theta + j \sin \theta$$

and therefore

$$\epsilon^{j\theta} = \cos\theta + j\sin\theta \tag{11-19}$$

The complexor previously described by Eq. 11–12 as

$$\boldsymbol{A} = a + jb$$

may then be written as

$$\boldsymbol{A} = A\epsilon^{j\theta} \tag{11-20}$$

where the notation of Eq. 11–20 is termed the *exponential form* of the complexor.

Also, from Eq. 11–19 $\epsilon^{-j\theta}$ may be written as

$$\epsilon^{-j\theta} = \cos\theta - j\sin\theta \tag{11-21}$$

By addition and subtraction, Eqs. 11–19 and 11–21 may be combined to obtain the following complex exponential forms for the sine and cosine:

$$\sin\theta = \frac{\epsilon^{j\theta} - \epsilon^{-j\theta}}{2j} \tag{11-22}$$

$$\cos\theta = \frac{\epsilon^{j\theta} + \epsilon^{-j\theta}}{2} \tag{11-23}$$

Polar Form. Since $\epsilon^{j\theta}$ has a magnitude of unity and makes an angle θ with the real axis, a shorter notation may be employed, representing the same quantity as

$$\epsilon^{j\theta} = 1\underline{/\theta} \tag{11-24}$$

or the complexor $\boldsymbol{A}$ may be written in *polar form* as

$$\boldsymbol{A} = A\underline{/\theta} \tag{11-25}$$

11–3. Complexor Manipulation. The addition, multiplication, raising to a power, and so on, of real numbers are common mathematical procedures. Similar manipulation of complex quantities may be performed, as shown in the following material.

Addition and Subtraction. The procedure for addition of complex quantities requires that they first be expressed in rectangular form. Real parts are added to real parts and imaginary to imaginary to obtain an answer in rectangular form.

To illustrate this, add two phasors of the following form:

$$\boldsymbol{A} = A\epsilon^{j\alpha} \qquad \text{and} \qquad \boldsymbol{B} = B\epsilon^{j\beta}$$

$\boldsymbol{A}$ and $\boldsymbol{B}$ may be expressed in rectangular form as

$$\begin{aligned}\boldsymbol{A} &= A\cos\alpha + jA\sin\alpha \\ \boldsymbol{B} &= B\cos\beta + jB\sin\beta\end{aligned} \tag{11-26}$$

Adding reals to reals and imaginaries to imaginaries yields

$$\mathbf{A} + \mathbf{B} = (A \cos \alpha + B \cos \beta) + j(A \sin \alpha + B \sin \beta) \tag{11–27}$$

$$= M + jN \tag{11–28}$$

$$= P\epsilon^{j\phi} \tag{11–29}$$

where

$$\begin{aligned} M &= A \cos \alpha + B \cos \beta \\ N &= A \sin \alpha + B \sin \beta \end{aligned} \tag{11–30}$$

$$P = \sqrt{M^2 + N^2} \tag{11–31}$$

and

$$\phi = \tan^{-1} \frac{N}{M} \tag{11–32}$$

The subtraction of complexors proceeds in the same manner as their addition. The complex numbers are first expressed in rectangular form and then reals are subtracted from reals and imaginaries from imaginaries.

To illustrate this, subtract

$$\mathbf{B} = B\epsilon^{j\beta} \text{ from } \mathbf{A} = A\epsilon^{j\alpha}$$

By similarity to Eq. 11–27,

$$\mathbf{A} - \mathbf{B} = (A \cos \alpha - B \cos \beta) + j(A \sin \alpha - B \sin \beta) \tag{11–33}$$

$$= R + jQ \tag{11–34}$$

$$= S\epsilon^{j\gamma} \tag{11–35}$$

where

$$\begin{aligned} R &= A \cos \alpha - B \cos \beta \\ Q &= A \sin \alpha - B \sin \beta \end{aligned} \tag{11–36}$$

$$S = \sqrt{R^2 + Q^2} \tag{11–37}$$

and

$$\gamma = \tan^{-1} \frac{Q}{R} \tag{11–38}$$

EXAMPLE 11–1. ADDITION OF COMPLEX NUMBERS. Three complex quantities are given as

$$\mathbf{A} = a + jb$$

$$\mathbf{B} = c + jd$$

$$\mathbf{C} = e + jf$$

Find their sum.

$$\mathbf{A} + \mathbf{B} + \mathbf{C} = (a + c + e) + j(b + d + f)$$

MULTIPLICATION AND DIVISION. Multiplication of complex quantities may be performed using any form for the complexor. This is now illustrated.

Multiply $\boldsymbol{A}$ times $\boldsymbol{B}$, where

$$\boldsymbol{A} = A\epsilon^{j\alpha} \qquad \text{and} \qquad \boldsymbol{B} = B\epsilon^{j\beta}$$

$$\boldsymbol{AB} = A(\cos\alpha + j\sin\alpha)B(\cos\beta + j\sin\beta) \tag{11–39}$$

$$= AB(\cos\alpha\cos\beta - \sin\alpha\sin\beta)$$

$$+ jAB(\sin\alpha\cos\beta + \cos\alpha\sin\beta) \tag{11–40}$$

$$= AB\cos(\alpha+\beta) + jAB\sin(\alpha+\beta) \tag{11–41}$$

$$= AB\epsilon^{j(\alpha+\beta)} \tag{11–42}$$

$$= A\epsilon^{j\alpha}B\epsilon^{j\beta} \tag{11–43}$$

or in polar form

$$\boldsymbol{AB} = A\underline{/\alpha}B\underline{/\beta} = AB\underline{/\alpha+\beta} \tag{11–44}$$

If the polar form of the complex quantities is used, multiplication is accomplished by multiplying magnitudes and adding angles.

Division is accomplished similarly. Again using $\boldsymbol{A}$ and $\boldsymbol{B}$,

$$\frac{\boldsymbol{A}}{\boldsymbol{B}} = \frac{A(\cos\alpha + j\sin\alpha)}{B(\cos\beta + j\sin\beta)} \tag{11–45}$$

Rationalization of a complex fraction is accomplished by multiplying numerator and denominator by the conjugate* of the denominator. Therefore

$$\frac{\boldsymbol{A}}{\boldsymbol{B}} = \frac{A\,(\cos\alpha + j\sin\alpha)}{B\,(\cos\beta + j\sin\beta)} \times \frac{(\cos\beta - j\sin\beta)}{(\cos\beta - j\sin\beta)} \tag{11–46}$$

or

$$\frac{\boldsymbol{A}}{\boldsymbol{B}} = \frac{A}{B}\frac{[(\cos\alpha\cos\beta + \sin\alpha\sin\beta) + j(\sin\alpha\cos\beta - \cos\alpha\sin\beta)]}{(\cos^2\beta + \sin^2\beta)} \tag{11–47}$$

and

$$\frac{\boldsymbol{A}}{\boldsymbol{B}} = \frac{A}{B}\frac{\cos(\alpha-\beta) + j\sin(\alpha-\beta)}{1} \tag{11–48}$$

$$= \frac{A}{B}\epsilon^{j(\alpha-\beta)} \tag{11–49}$$

Using the polar form of A and B,

$$\frac{\boldsymbol{A}}{\boldsymbol{B}} = \frac{A\underline{/\alpha}}{B\underline{/\beta}} = \frac{\boldsymbol{A}}{\boldsymbol{B}}\underline{/\alpha-\beta} \tag{11–50}$$

and division is accomplished by dividing magnitudes and subtracting angles.

* The conjugate of a complex quantity has the same magnitude but the negative of the original angle, that is, the conjugate of $\boldsymbol{B} = B\underline{/\beta}$ is $B\underline{/-\beta}$ or, if $\boldsymbol{B}$ is written as $B(\cos\beta + j\sin\beta)$, the conjugate form is $B(\cos\beta - j\sin\beta)$.

Example 11–2. Multiplication and Division of Complexors. Evaluate

$$\frac{A_1A_2}{A_3}$$

where

$$A_1 = 3 + j4$$

$$A_2 = 4 + j4$$

and

$$A_3 = 8 - j6$$

In polar form

$$A_1 = \sqrt{3^2 + 4^2}\,\underline{/\tan^{-1}\tfrac{4}{3}} = 5\,\underline{/53.2^\circ}$$

$$A_2 = \sqrt{4^2 + 4^2}\,\underline{/\tan^{-1}\tfrac{4}{4}} = 4\sqrt{2}\,\underline{/45^\circ}$$

$$A_3 = \sqrt{8^2 + 6^2}\,\underline{/-\tan^{-1}\tfrac{6}{8}} = 10\,\underline{/-36.8^\circ}$$

therefore

$$\frac{A_1A_2}{A_3} = \frac{(5\,\underline{/53.2^\circ})(4\sqrt{2}\,\underline{/45^\circ})}{10\,\underline{/-36.8^\circ}}$$

$$= 2\sqrt{2}\,\underline{/135^\circ}$$

or using rectangular form

$$\frac{A_1A_2}{A_3} = \frac{(3 + j4)(4 + j4)}{(8 - j6)}$$

$$= \frac{(12 - 16) + j(16 + 12)}{(8 - j6)} = \frac{-4 + j28}{8 - j6}$$

$$= \frac{(-4 + j28)(8 + j6)}{(8 - j6)(8 + j6)} = \frac{(-32 - 168) + j(224 - 24)}{(64 + 36)}$$

$$= \frac{-200 + j200}{100} = -2 + j2$$

$$= 2\sqrt{2}\,\underline{/135^\circ}$$

Raising to a Power. The nth power of a complexor may be obtained as follows: Let $\boldsymbol{A} = A\epsilon^{j\beta}$ be raised to the nth power by

$$(\boldsymbol{A})^n = (A\epsilon^{j\beta})^n \tag{11–51}$$

$$= A^n\epsilon^{jn\beta} \tag{11–52}$$

$$= A^n\,\underline{/n\beta} \tag{11–53}$$

The magnitude of the complexor is raised to the nth power and the angle multiplied by n.

EXAMPLE 11–3. RAISING A COMPLEXOR TO A POWER. Raise $\mathbf{N} = 2\underline{/20^\circ}$ to the 5th power.

$$(\mathbf{N})^5 = 2^5\underline{/5.20^\circ}$$

$$= 32\underline{/100^\circ}$$

ROOTS. A complexor $\mathbf{A}$ may be written as before as

$$\mathbf{A} = A\epsilon^{j\alpha}$$

However, if an angle of $2\pi m$ radians is added to the angle α, the complexor does not change in magnitude or direction if m is an integer. Thus, in general, $\mathbf{A}$ should be expressed as

$$\mathbf{A} = A\epsilon^{j(\alpha+2m\pi)} \tag{11–54}$$

where m is an integer 0, 1, 2, 3, $\cdots$.

If it is desired to find the nth root of a complexor $\mathbf{A}$ it must be remembered that there are n distinct nth roots. These are found as

$$\sqrt[n]{\mathbf{A}} = \sqrt[n]{A}\exp\left[j\frac{(\alpha + 2m\pi)}{n}\right] \tag{11–55}$$

$$= \sqrt[n]{A}\exp\left(j\frac{\alpha}{n}\right), \sqrt[n]{A}\exp\left[j\frac{(\alpha + 2\pi)}{n}\right], \cdots, \sqrt[n]{A}\exp\left[j\frac{[\alpha + (n-1)2\pi]}{n}\right] \tag{11–56}$$

If $n = m$ the roots begin to repeat, since

$$\sqrt[n]{A}\exp\left[j\left(\frac{\alpha}{n} + 2\pi\right)\right] \text{ is the same as } \sqrt[n]{A}\exp\left(j\frac{\alpha}{n}\right)$$

Unless otherwise specified, the desired root is

$$\sqrt[n]{A}\exp\left(j\frac{\alpha}{n}\right) \tag{11–57}$$

which is obtained by taking the desired root of the magnitude and dividing the angle by the order of the root.

EXAMPLE 11–4. ROOT OF A COMPLEX QUANTITY. Determine the 4th root of $10\underline{/60^\circ}$.

$$(10\underline{/60^\circ})^{1/4} = 10^{1/4}\underline{/15^\circ}$$

$$= 1.78\underline{/15^\circ}$$

if only the usual single root is calculated.

LOGARITHMS. If it is desired to find ln $\boldsymbol{A}$, where

$$\boldsymbol{A} = A\epsilon^{j(\alpha+2m\pi)}$$

then

$$\ln \boldsymbol{A} = \ln A + j(\alpha + 2m\pi) \tag{11-58}$$

The natural logarithm of a complexor is also a complexor with a single-valued real part and a multivalued imaginary component. Usually the component desired is for $m = 0$, or

$$\ln \boldsymbol{A} = \ln A + j\alpha \tag{11-59}$$

The logarithm of negative numbers may also be obtained by assuming the negative number as a complexor with an angle of π radians.

EXAMPLE 11–5. LOGARITHM OF A COMPLEXOR. Determine the natural logarithm of -9.

$$\begin{aligned}\ln(-9) &= \ln(9\underline{/180^\circ}) \\ &= \ln 9\epsilon^{j\pi} \\ &= \ln 9 + j\pi \\ &= 2.2 + j\pi = 2.2 + j3.14\end{aligned}$$

PROBLEMS

11–1. Complex quantities are normally expressed in (i) rectangular, (ii) polar, or (iii) exponential form. Write the quantities below in the other two forms, and sketch the complexor on the complex plane.

(a) $3 + j4$ (b) $1 + j10$

(c) $0.1 - j10$ (d) $a + jb$

(e) $a - jb$ (f) $-5 + j10$

(g) $-5 - j5$ (h) ϵ^{j60°

(i) ϵ^{-j25° (j) ϵ^{j135°

(k) ϵ^{j270° (l) $10\underline{/-10^\circ}$

(m) $-10\underline{/170^\circ}$ (n) $5\underline{/90^\circ}$

(o) $100\underline{/-135^\circ}$ (p) $1\underline{/180^\circ}$

11–2. Evaluate j^4, j^7, j^{15}, j^{220}.

11–3. Three complex quantities are given as

$$\begin{aligned}\boldsymbol{A} &= 10 + j10 \\ \boldsymbol{B} &= 10 - j10 \\ \boldsymbol{C} &= 6 + j8\end{aligned}$$

Perform the indicated operation and express the answer in polar *and* rectangular form.

(a) $\boldsymbol{AB}$

(b) $\boldsymbol{A} + \boldsymbol{B}$

(c) $\boldsymbol{A} - \boldsymbol{B}$

(d) $\dfrac{\boldsymbol{A}}{\boldsymbol{B}}$

(e) $\boldsymbol{A} + \boldsymbol{B} + \boldsymbol{C}$

(f) $\boldsymbol{A} + \boldsymbol{B} - \boldsymbol{C}$

(g) $\boldsymbol{A} - \boldsymbol{B} - \boldsymbol{C}$

(h) $\boldsymbol{BC}$

(i) $\boldsymbol{CA}$

(j) $\dfrac{1}{\boldsymbol{A}} - \dfrac{1}{\boldsymbol{B}}$

(k) $\dfrac{1}{\boldsymbol{A}} + \dfrac{1}{\boldsymbol{C}}$

(l) $\dfrac{\boldsymbol{BC}}{\boldsymbol{A}}$

(m) $\dfrac{\boldsymbol{A} + \boldsymbol{C}}{\boldsymbol{B} + \boldsymbol{C}}$

(n) $(\boldsymbol{A})^2$

(o) $(\boldsymbol{B})^{1/4}$

(p) $\ln \boldsymbol{C}$

(q) $\dfrac{1}{\boldsymbol{ABC}}$

(r) $\dfrac{1}{\boldsymbol{C} - \boldsymbol{A}} - \dfrac{1}{\boldsymbol{B} - \boldsymbol{C}}$

(s) $\dfrac{\boldsymbol{A}}{\boldsymbol{B}} + \boldsymbol{C}^2$

(t) $\boldsymbol{A}\underline{/90^\circ} + \boldsymbol{B}\underline{/-60^\circ} + \boldsymbol{C}\underline{/20^\circ}$

11–4. Rationalize the following quantities and express the answer in polar *and* rectangular form.

(a) $\dfrac{1}{3 + j4}$

(b) $\dfrac{3 + j4}{2 + j3}$

(c) $\dfrac{10 - j10}{1 + j1}$

(d) $\dfrac{25 + j30}{10 - j15}$

(e) $\dfrac{4 + j3}{(8 + j8)10\underline{/40^\circ}}$

12

STEADY-STATE SINUSOIDAL ANALYSIS BY COMPLEX ALGEBRA

The representation of the phasors of Chapter 10 by means of the complexor notation of Chapter 11 is the basis for most steady-state sinusoidal circuit analysis. Complex notation permits the establishment of a rapid and convenient procedure for the solution of this type of circuit.

12–1. Complex Representation of Phasors. Figure 10–15 is reproduced here as Fig. 12–1 with the X axis now designated as the real axis and the Y axis as the imaginary axis.

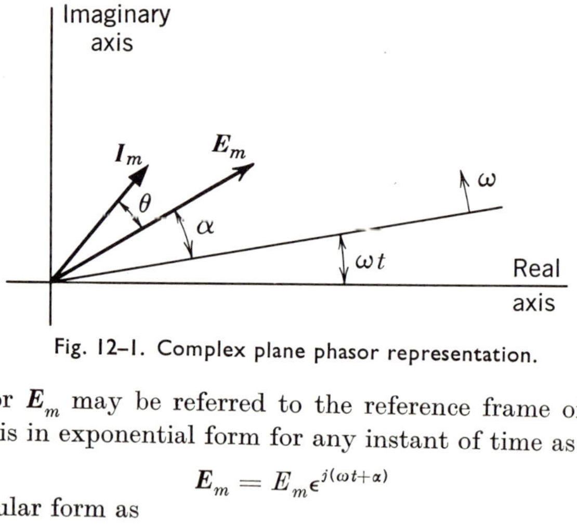

Fig. 12–1. Complex plane phasor representation.

The phasor $\boldsymbol{E}_m$ may be referred to the reference frame of the real and imaginary axis in exponential form for any instant of time as

$$\boldsymbol{E}_m = E_m \epsilon^{j(\omega t + \alpha)} \tag{12–1}$$

or in rectangular form as

$$\boldsymbol{E}_m = E_m \cos(\omega t + \alpha) + jE_m \sin(\omega t + \alpha) \tag{12–2}$$

$\boldsymbol{E}_m$ is a quantity whose angular position on the complex plane varies with time, but whose magnitude does not. Note, however, that the real part of $\boldsymbol{E}_m$ is

$$e_1 = E_m \cos(\omega t + \alpha) \tag{12–3}$$

and the imaginary part of $\boldsymbol{E}_m$ is

$$e_2 = E_m \sin(\omega t + \alpha) \tag{12–4}$$

Therefore $\boldsymbol{E}_m$ may also be considered as representing two stationary phasors whose directions are along the real and imaginary axes and whose magnitudes vary with time.

If $\boldsymbol{E}_m$ is applied to an R-L series circuit, the steady-state response may be obtained from the equilibrium equation

$$Ri + L\frac{di}{dt} = \boldsymbol{E}_m = E_m\epsilon^{j(\omega t+\alpha)} \tag{12-5}$$

by assuming a current of the same form as $\boldsymbol{E}_m$ but differing in phase by the angle θ as

$$\boldsymbol{I}_m = I_m\epsilon^{j(\omega t+\alpha-\theta)} \tag{12-6}$$

and substituting this into Eq. 12–5. From this, $\boldsymbol{I}_m$ becomes

$$\boldsymbol{I}_m = \frac{\boldsymbol{E}_m}{R + jX_L} \tag{12-7}$$

where $(R + jX_L)$ is the complex form of the impedance $\boldsymbol{Z}$ previously introduced, or

$$\frac{\boldsymbol{E}_m}{\boldsymbol{I}_m} = \boldsymbol{Z} = (R + jX_L) \tag{12-8}$$

$$= Z\epsilon^{j\theta} \tag{12-9}$$

where
$$Z = \sqrt{R^2 + X_L{}^2} \tag{12-10}$$

and
$$\theta = \tan^{-1}\frac{X_L}{R} \tag{12-11}$$

It must be emphasized that the complex form of $\boldsymbol{Z}$ may be represented on a complex plane diagram, such as Fig. 12–1, but it does not rotate with time as do the phasors representing current and voltage. The impedance of the circuit is a constant whose magnitude and angle do not change with time. $\boldsymbol{Z}$ may be described as a complex proportionality factor relating the phasors $\boldsymbol{E}_m$ and $\boldsymbol{I}_m$.

$\boldsymbol{I}_m$ from Eq. 12–7 may be written as

$$\boldsymbol{I}_m = \frac{\boldsymbol{E}_m}{\sqrt{R^2 + X_L{}^2}\epsilon^{j\theta}} = \frac{\boldsymbol{E}_m}{Z}\,\epsilon^{-j\theta} \tag{12-12}$$

and using Eq. 12–6

$$\boldsymbol{I}_m = \frac{E_m}{\boldsymbol{Z}}\,\epsilon^{j(\omega t+\alpha)} \tag{12-13}$$

or
$$\boldsymbol{I}_m = \frac{E_m}{Z}\,\epsilon^{j(\omega t+\alpha-\theta)} \tag{12-14}$$

from which $\boldsymbol{I}_m$ may be written as

$$\begin{aligned}\boldsymbol{I}_m &= i_1 + ji_2\\ &= \frac{E_m}{Z}\cos(\omega t + \alpha - \theta) + j\,\frac{E_m}{Z}\sin(\omega t + \alpha - \theta)\end{aligned} \tag{12-15}$$

where
$$i_1 = \frac{E_m}{Z} \cos (\omega t + \alpha - \theta) \tag{12-16}$$

and
$$i_2 = \frac{E_m}{Z} \sin (\omega t + \alpha - \theta) \tag{12-17}$$

From the superposition concept, i_1 may be considered the component of current corresponding to e_1, and i_2 the component of current corresponding to e_2, where e_1 and e_2 are defined by Eqs. 12–3 and 12–4. Thus the current $\boldsymbol{I}_m$ in the network in response to the phasor $\boldsymbol{E}_m$ is actually a response to both a sine and cosine voltage excitation, and the instantaneous response to either may be secured from $\boldsymbol{I}_m$. The response to a sine function alone is then the response to the imaginary part of $\boldsymbol{E}_m$, written as Im $(\boldsymbol{E}_m)$ or Im $[E_m\epsilon^{j(\omega t+\alpha)}]$, and the response to a cosine function is the response to the real part of $\boldsymbol{E}_m$, written Re $(\boldsymbol{E}_m)$ or Re $[E_m\epsilon^{j(\omega t+\alpha)}]$. The currents corresponding to these excitations are respectively Im $(\boldsymbol{I}_m)$ or Im $[I_m\epsilon^{j(\omega t+\alpha-\theta)}]$ and Re $(\boldsymbol{I}_m)$ or Re $[\boldsymbol{I}_m\epsilon^{j(\omega t+\alpha-\theta)}]$.

The phasor representation of a sinusoidal function includes both the sine and cosine terms, but usual electrical engineering phasor notation omits the Re or Im prefix since the instantaneous value is not usually sought. *The phasor notation relates the maximum value of the current and voltage via the impedance constant and this notation may be used interchangeably for either sine or cosine excitation.* Only if the instantaneous current or voltage response is required is it necessary to consider the real or imaginary parts.

A further notational simplification may be made if the ωt axis is superimposed on the real axis.* For this condition, $\omega t = 0$ and Eq. 12–1 becomes

$$\boldsymbol{E}_m = E_m\epsilon^{j\alpha} \tag{12-18}$$

and Eq. 12–14 becomes

$$\boldsymbol{I}_m = \frac{E_m}{Z}\epsilon^{j(\alpha-\theta)} \tag{12-19}$$

If the time of investigation is always chosen at $\omega t = 0$ (ωt axis on the real axis) the time term is eliminated from the phasor expression. *This is the usual way in which the phasor form is used in electrical engineering.* Equations 12–18 and 12–19 written in polar form then appear as

$$\boldsymbol{E}_m = E_m\underline{/\alpha} \tag{12-20}$$

and
$$\boldsymbol{I}_m = I_m\underline{/\alpha - \theta} \tag{12-21}$$

Figure 12–2 illustrates this.

It would be logical to further ease the computational efforts involved by rotating the phasor system so that either the current or voltage phasor is also superimposed on the real axis (that is, either α or $\alpha - \theta = 0$ in Eq. 12–20 or

* The phase relation between $\boldsymbol{E}_m$ and $\boldsymbol{I}_m$ is constant for all time; therefore, any chosen instant of time may be chosen to take a "snapshot" of the phasors. Logic dictates choosing the most convenient time, usually $\omega t = 0$.

12–21). If the current is chosen as the reference quantity in a system, its phasor is commonly placed on the real axis; similarly, if voltage is the reference quantity, its phasor is placed on the real axis. These conditions correspond to those usually assumed for series and parallel arrangements of elements respectively.

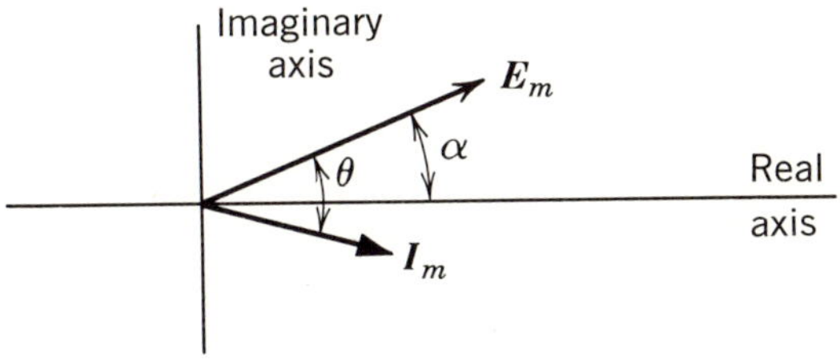

Fig. 12–2. Phasors represented for $\omega t = 0$.

As noted in Chapter 10, effective instead of maximum values are commonly plotted on phasor diagrams. This is equivalent to a change in scale for the voltage and current phasors, but the impedance, which is their ratio, is the same for either representation.

12–2. Complex Volt-Ampere Relations for Circuit Elements.

In complex notation the volt-ampere relationships for the passive circuit elements are as follows:

Resistance Element

$$I_m = \frac{E_m}{R} = GE_m \qquad (12\text{–}22)$$

or

$$E_m = RI_m \qquad (12\text{–}23)$$

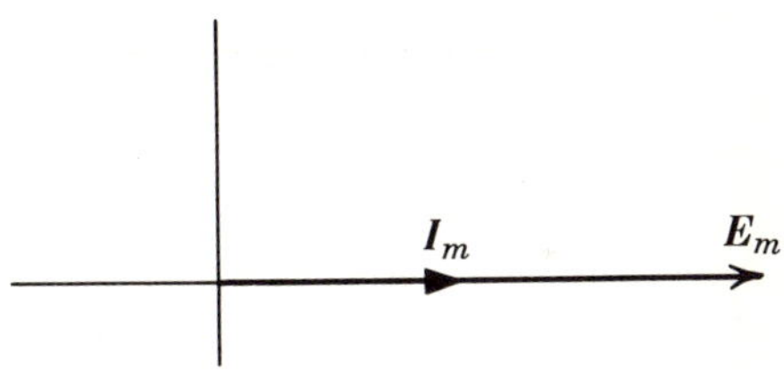

Fig. 12–3. Phasor relations for a resistance element.

The current and voltage phasors are always in phase with each other, as shown in Fig. 12–3.

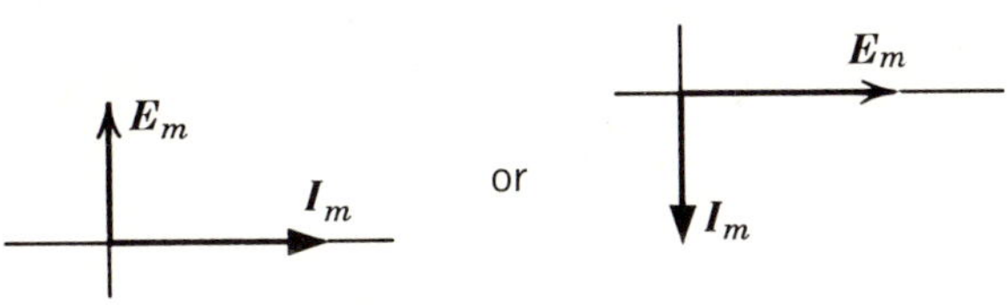

Fig. 12–4. Phasor relations for an inductance element.

Inductance Element

$$I_m = -j\frac{E_m}{X_L} \qquad (12\text{–}24)$$

or

$$E_m = jX_L I_m \qquad (12\text{–}25)$$

The current phasor lags the voltage phasor by 90°, or the voltage phasor leads the current phasor by 90°. The phasor diagrams of Fig. 12–4 illustrate this.

CAPACITANCE ELEMENT

$$\boldsymbol{I}_m = j\frac{\boldsymbol{E}_m}{X_C} \tag{12–26}$$

or

$$\boldsymbol{E}_m = -jX_C\boldsymbol{I}_m \tag{12–27}$$

The current phasor leads the voltage phasor by 90°, or the voltage phasor lags the current phasor by 90°. Figure 12–5 illustrates this.

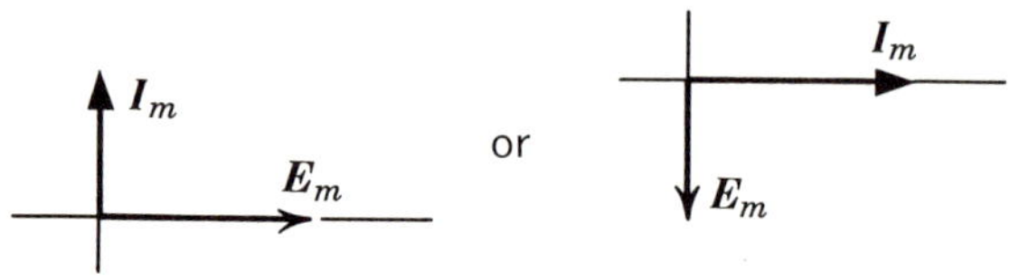

Fig. 12–5. Phasor relations for a capacitance element.

In general, the voltage across any element may be written

$$\boldsymbol{E}_m = \boldsymbol{I}_m\boldsymbol{Z} \tag{12–28}$$

where $\boldsymbol{Z}$ is the complex impedance of the element.*

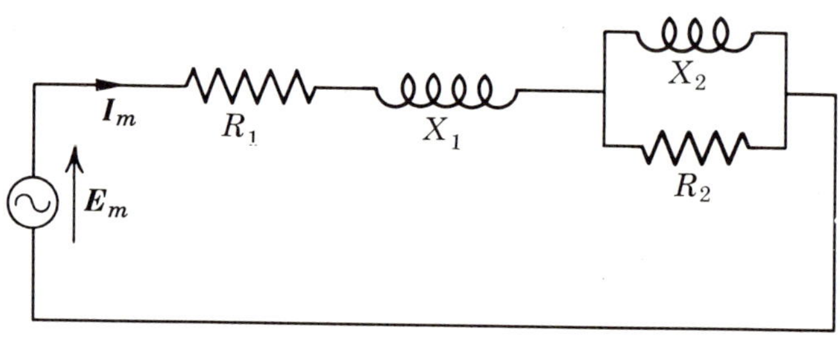

Fig. 12–6. Use of complex notation for circuit solution.

EXAMPLE 12–1. CIRCUIT SOLUTION USING COMPLEX NOTATION. Repeat the solution of Example 10–1, using complex notation. (See Fig. 12–6.)

The impedance of the parallel circuit is

$$\boldsymbol{Z}_2 = \frac{jX_2R_2}{R_2 + jX_2} = \frac{R_2X_2^2}{R_2^2 + X_2^2} + j\frac{X_2R_2^2}{R_2^2 + X_2^2}$$

and the impedance of the circuit as a whole is

$$\boldsymbol{Z} = \boldsymbol{Z}_1 + \boldsymbol{Z}_2$$

* Element as used here may include the results of more than one electrical effect. For example, a physical inductor has both the properties of resistance and inductance.

where

$$Z_1 = R_1 + jX_1$$

or

$$Z = \left(R_1 + \frac{R_2 X_2^{\,2}}{R_2^{\,2} + X_2^{\,2}}\right) + j\left(X_1 + \frac{X_2 R_2^{\,2}}{R_2^{\,2} + X_2^{\,2}}\right)$$
$$= R + jX$$

The phasor current is then

$$I_m = \frac{E_m}{Z} = \frac{E_m}{R + jX}$$
$$= E_m\left(\frac{R}{R^2 + X^2} - j\,\frac{X}{R^2 + X^2}\right)$$
$$= \frac{E_m}{Z}\,\epsilon^{-j\theta}$$

where

$$\theta = \tan^{-1}\frac{X}{R}$$

If the phasor E_m represents

$$e = E_m \sin \omega t$$

then

$$i = I_m \sin(\omega t - \theta)$$

This example was introduced at this point to show the simplicity of the complex phasor representation. A detailed justification of the several steps will be found in the remaining portions of this chapter.

12-3. Complex Notation for the Series Circuit. Kirchhoff's laws, as previously written, were considered valid for instantaneous voltages and currents. They are also correct for voltages and currents expressed in complex notation providing that all summations are performed "vectorially."

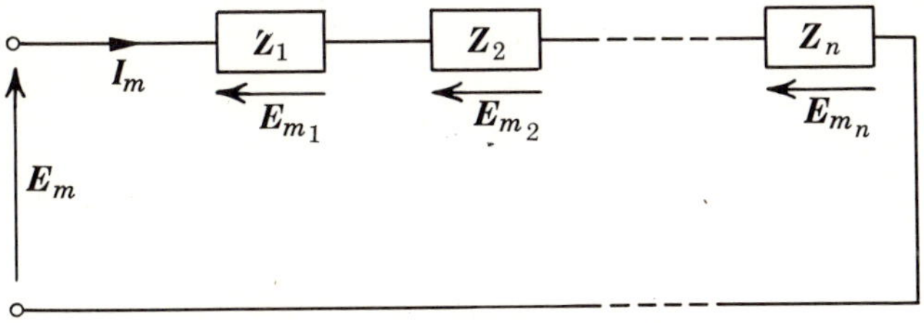

Fig. 12-7. Impedances in series.

For example, if several impedances (which may be made up of single elements or combinations of single elements) are connected in series as shown in Fig. 12-7, a common current I_m will exist in each and the phasor voltage E_m must equal the complex sum of the phasor voltages across each impedance, or

$$E_m = E_{m_1} + E_{m_2} + \cdots + E_{m_n} \tag{12-29}$$

and from the complex volt-ampere relations

$$\boldsymbol{E}_m = \boldsymbol{I}_m\boldsymbol{Z}_1 + \boldsymbol{I}_m\boldsymbol{Z}_2 + \cdots + \boldsymbol{I}_m\boldsymbol{Z}_n \tag{12–30}$$

where $\boldsymbol{I}_m\boldsymbol{Z}_1$, $\boldsymbol{I}_m\boldsymbol{Z}_2, \cdots$, are phasors and must be added as such. $\boldsymbol{I}_m$ is common to all elements and therefore Eq. 12–13 becomes

$$\boldsymbol{E}_m = \boldsymbol{I}_m(\boldsymbol{Z}_1 + \boldsymbol{Z}_2 + \cdots + \boldsymbol{Z}_n) \tag{12–31}$$

Since the ratio of phasor voltage to phasor current is impedance, as noted in Eq. 12–26, an impedance $\boldsymbol{Z}'$ which is equivalent to a number of impedances in series is obtained as

$$\boldsymbol{Z}' = \frac{\boldsymbol{E}}{\boldsymbol{I}} = \boldsymbol{Z}_1 + \boldsymbol{Z}_2 + \cdots + \boldsymbol{Z}_n \tag{12–32}$$

where the impedances must be added as complexors. To do this, each impedance is expressed as the sum of its real and imaginary parts or

$$\boldsymbol{Z}' = (R_1 + R_2 + \cdots + R_n) + j(X_1 + X_2 + \cdots + X_n) \tag{12–33}$$

$\boldsymbol{I}_m$ of Eq. 12–31 is related to $\boldsymbol{E}_m$ by $\boldsymbol{Z}'$ as

$$\boldsymbol{I}_m = \frac{\boldsymbol{E}_m}{\boldsymbol{Z}'} = \frac{\boldsymbol{E}_m}{Z'\epsilon^{j\theta}} \tag{12–34}$$

where

$$\theta = \tan^{-1}\frac{\Sigma X}{\Sigma R} \tag{12–35}$$

and

$$Z' = \sqrt{(\Sigma R)^2 + (\Sigma X)^2} \tag{12–36}$$

To illustrate these concepts, several series configurations are analyzed in the remainder of this section.

Series R-L Circuit. A resistance R and an inductance L are connected in series and excited by a sinusoidal voltage $e = E_m \sin \omega t$ as shown in Fig. 12–8. The complex sum of the $\boldsymbol{E}_{R_m}$ and $\boldsymbol{E}_{L_m}$ must be the applied voltage $\boldsymbol{E}_m$, or

$$\boldsymbol{E}_m = \boldsymbol{E}_{R_m} + \boldsymbol{E}_{L_m} \tag{12–37}$$

But

and

$$\boldsymbol{E}_{R_m} = \boldsymbol{I}_m R \qquad \boldsymbol{E}_{L_m} = j\boldsymbol{I}_m X_L \tag{12–38}$$

and therefore

$$\boldsymbol{E}_m = \boldsymbol{I}_m(R + jX_L) \tag{12–39}$$

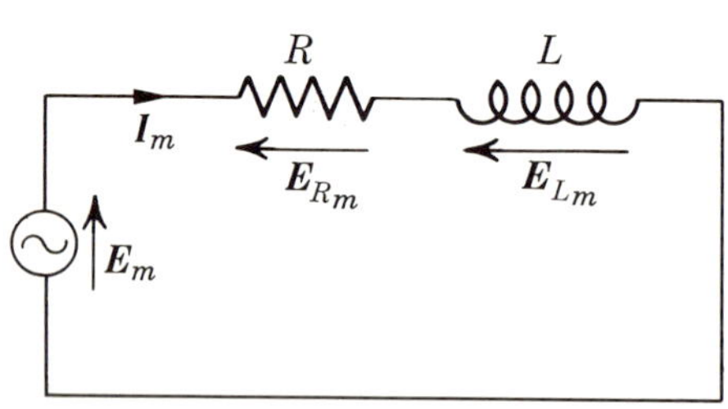

Fig. 12–8. *R-L* series circuit—sinusoidal excitation.

The circuit current is

$$\boldsymbol{I}_m = \frac{\boldsymbol{E}_m}{R + jX_L} = \frac{\boldsymbol{E}_m}{\boldsymbol{Z}} \tag{12–40}$$

or
$$\boldsymbol{I}_m = \frac{\boldsymbol{E}_m}{Z\epsilon^{j\theta}} \tag{12-41}$$

where
$$Z = \sqrt{R^2 + X_L^2} \tag{12-42}$$

and
$$\theta = \tan^{-1}\frac{X_L}{R} \tag{12-43}$$

The phasor for $\boldsymbol{E}_m$ and the phasor for $\boldsymbol{I}_m$ are related in magnitude by the magnitude of the impedance complexor $\boldsymbol{Z}$, and the phase relation between the current and voltage phasor is determined by θ, as shown in Fig. 12-9.

(a) Current as reference

(b) Voltage as reference

Fig. 12-9. Phasor diagrams for R-L series circuit.

SERIES R-C CIRCUIT. In a similar manner for the R-C series circuit shown in Fig. 12-10, the voltage equations may be written (using effective values) as

$$\boldsymbol{E} = \boldsymbol{I}(R - jX_C) \tag{12-44}$$

and
$$\boldsymbol{I} = \frac{\boldsymbol{E}}{R - jX_C} = \frac{\boldsymbol{E}}{\boldsymbol{Z}} = \frac{\boldsymbol{E}}{Z\epsilon^{-j\theta}} \tag{12-45}$$

where
$$\theta = \tan^{-1}\frac{X_C}{R}$$

The phasor diagram is as shown in Fig. 12-11.

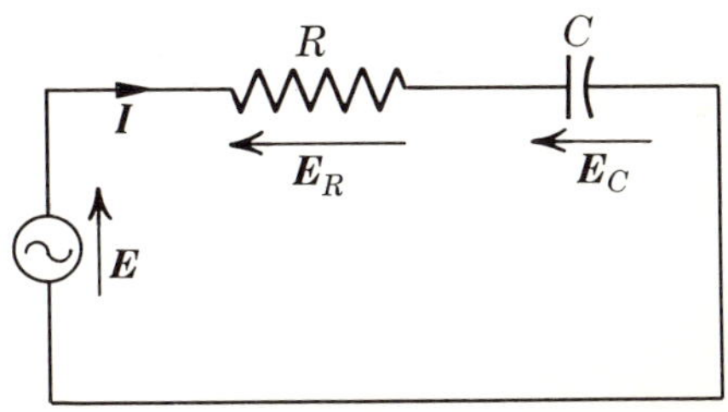

Fig. 12-10. R-C series circuit—sinusoidal excitation.

SERIES R-L-C CIRCUIT. This network configuration is illustrated as Fig. 12-12. The series impedance is

$$\begin{aligned}\boldsymbol{Z} &= R + jX_L - jX_C \\ &= R + j(X_L - X_C)\end{aligned} \tag{12-46}$$

and the phasor current for $e = E_m \sin \omega t$ applied is

$$\boldsymbol{I}_m = \frac{\boldsymbol{E}_m}{R + j(X_L - X_C)} \tag{12-47}$$

Equation 12-47 will be discussed in some detail for varying values of R, X_L, and X_C.

Fig. 12–11. Phasor diagrams for *R-C* series circuit.

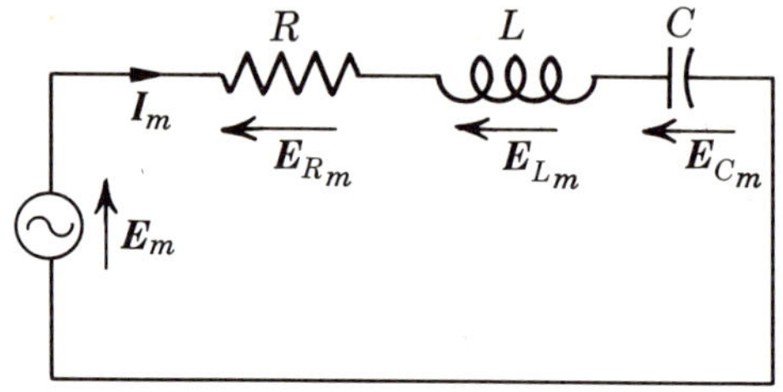

Fig. 12–12. *R-L-C* series circuit—sinusoidal excitation.

1. If $R = 0$ and $X_L > X_C$, the current becomes

$$\boldsymbol{I}_m = \frac{\boldsymbol{E}_m}{j(X_L - X_C)} = \frac{\boldsymbol{E}_m}{X_L - X_C}\,\epsilon^{-j(\pi/2)} \tag{12–48}$$

and the current lags the voltage by 90°.

2. If $R = 0$ and $X_L < X_C$, the current becomes

$$\boldsymbol{I}_m = \frac{\boldsymbol{E}_m}{-j(X_C - X_L)} = \frac{\boldsymbol{E}_m}{X_C - X_L}\,\epsilon^{j(\pi/2)} \tag{12–49}$$

and the current leads the voltage by 90°.

3. If $R = 0$ and $X_L \to X_C$, the current approaches an infinite value and the phase angle approaches zero.

4. In general, writing ωL for X_L and $1/\omega C$ for X_C, the current expression of Eq. 12–47 becomes

$$\boldsymbol{I}_m = \frac{\boldsymbol{E}_m}{R + j\left(\omega L - \dfrac{1}{\omega C}\right)} \tag{12–50}$$

If $\omega \to 0$, $1/\omega C \to \infty$, and the magnitude of the current approaches zero while the leading phase angle approaches 90°. If ω increases from zero, the current becomes finite and initially leads the voltage in phase. ωL increases and $1/\omega C$ decreases until $X_L = X_C$ and the current becomes

$$\boldsymbol{I}_m = \frac{\boldsymbol{E}_m}{R} \tag{12–51}$$

and is in phase with the voltage. A further increase in frequency results in a decrease of current from the value of Eq. 12–51 while the phase angle between current and voltage becomes lagging. As $\omega \to \infty$ the current again approaches zero magnitude and the lagging phase angle approaches 90°.

5. The behavior of the network for frequencies in the vicinity of $X_L = X_C$ is of major importance. If

$$\omega L = \frac{1}{\omega C} \tag{12-52}$$

the value of ω obtained is

$$\omega = \omega_R = \frac{1}{\sqrt{LC}} \tag{12-53}$$

where ω_R is termed the *resonant angular velocity* of the system and

$$f_R = \frac{\omega_R}{2\pi} \tag{12-54}$$

is the resonant frequency, which also may be written

$$f_R = \frac{1}{2\pi\sqrt{LC}} \tag{12-55}$$

The magnitude of the voltage across X_L and X_C at resonance is

$$E_{L_m} = E_{C_m} = \frac{E_m}{R} X_L = \frac{E_m}{R} X_C \tag{12-56}$$

Note that these voltages may be much higher than E_m. Equation 12–56 may be rewritten as

$$\frac{E_m}{R} \omega_R L = \frac{E_m}{R} \frac{1}{\omega_R C} \tag{12-57}$$

or

$$E_m \frac{\omega_R L}{R} = E_m \frac{1}{\omega_R C R} \tag{12-58}$$

If

$$Q_R = \frac{\omega_R L}{R} = \frac{1}{\omega_R C R} \tag{12-59}$$

is defined as a figure of merit of the circuit at a specified frequency, this definition may be used to rewrite Eq. 12–56, the voltage across L or C at resonance, as

$$E_{L_m} = E_{C_m} = Q_R E_m \tag{12-60}$$

In the vicinity of resonance the current in the series circuit is high and is a function of frequency and Q. This may be demonstrated by rewriting Eq. 12–50, repeated here,

$$I_m = \frac{E_m}{R + j\left(\omega L - \dfrac{1}{\omega C}\right)}$$

as

$$\boldsymbol{I}_m = \frac{\boldsymbol{E}_m}{R + jL\left(\omega - \dfrac{1}{\omega LC}\right)} \tag{12–61}$$

or

$$\boldsymbol{I}_m = \frac{\boldsymbol{E}_m}{R}\left[\frac{1}{1 + j\dfrac{L}{R}\left(\omega - \dfrac{1}{\omega LC}\right)}\right] \tag{12–62}$$

But

$$\omega_R^{\ 2} = \frac{1}{LC}$$

and therefore

$$\boldsymbol{I}_m = \frac{\boldsymbol{E}_m}{R}\left[\frac{1}{1 + j\dfrac{L}{R}\left(\omega - \dfrac{\omega_R^{\ 2}}{\omega}\right)}\right] \tag{12–63}$$

$$= \frac{\boldsymbol{E}_m}{R}\left[\frac{1}{1 + j\dfrac{\omega_R L}{R}\left(\dfrac{\omega}{\omega_R} - \dfrac{\omega_R}{\omega}\right)}\right] \tag{12–64}$$

$$= \frac{\boldsymbol{E}_m}{R}\left[\frac{1}{1 + jQ_R\left(\dfrac{\omega}{\omega_R} - \dfrac{\omega_R}{\omega}\right)}\right] \tag{12–65}$$

The magnitude of $\boldsymbol{I}_m$ becomes

$$I_m = \frac{E_m}{R}\left[\frac{1}{\sqrt{1 + Q_R^{\ 2}\left(\dfrac{\omega}{\omega_R} - \dfrac{\omega_R}{\omega}\right)^2}}\right] \tag{12–66}$$

Note that

$$\left(\frac{\omega}{\omega_R} - \frac{\omega_R}{\omega}\right)^2 = \left(\frac{\omega_R}{\omega} - \frac{\omega}{\omega_R}\right)^2 \tag{12–67}$$

which means that on either side of ω_R (at which the current is a maximum) there exists an angular velocity (or frequency) at which the value of I_m is the same; the angle of the current referred to a voltage reference associated with the value of ω higher than ω_R is a lagging one, and for the value of ω lower than ω_R it is a leading one.

The sharpness of resonance in the series R-L-C circuit is determined from the rate at which the current decreases as the frequency of the applied voltage departs from ω_R. By definition, the sharpness is measured by a ratio of the frequency difference existing between the two frequencies on either side of resonance at which the current is 0.707 of the value at resonance to the frequency at resonance. These quantities are illustrated in Fig. 12–13.

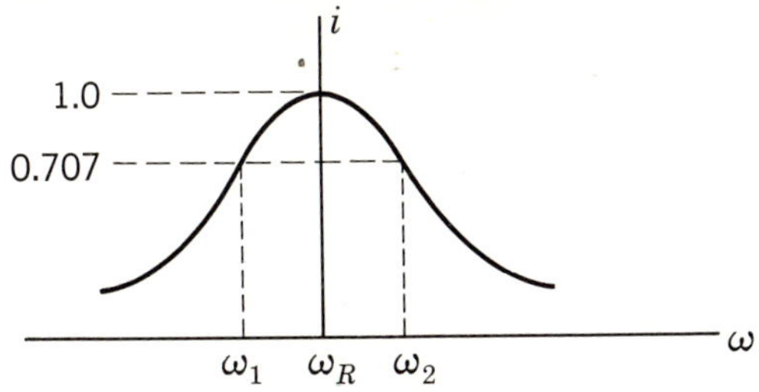

Fig. 12-13. Series circuit current plot about resonance.

From an expression for the magnitude of the current

$$I_m = \frac{E_m}{\sqrt{R^2 + (X_L - X_C)^2}} \tag{12-68}$$

it may be seen that if

$$X_L - X_C = R \tag{12-69}$$

or

$$X_C - X_L = R \tag{12-70}$$

the current will be $1/\sqrt{2} = 0.707$ of that at resonance.

Equation 12-69 is rewritten as

$$\omega L - \frac{1}{\omega C} = R \tag{12-71}$$

or

$$\omega^2 LC - \omega RC - 1 = 0 \tag{12-72}$$

or

$$\omega^2 - \frac{R}{L}\omega - \frac{1}{LC} = 0 \tag{12-73}$$

which yields

$$\omega = \frac{R}{2L} \pm \sqrt{\left(\frac{R}{2L}\right)^2 + \frac{1}{LC}} \tag{12-74}$$

as the angular velocity at which the current is 0.707 of that at resonance. If

$$\frac{1}{LC} \gg \left(\frac{R}{2L}\right)^2 \tag{12-75}$$

then

$$\omega = \frac{R}{2L} + \frac{1}{\sqrt{LC}} = \frac{R}{2L} + \omega_R \tag{12-76}$$

where the negative sign is neglected since at the moment negative frequencies are not of interest.

Therefore, in Fig. 12-13,

$$\omega_2 = \omega_R + \frac{R}{2L} \tag{12-77}$$

Similarly, from Eq. 12–70,

$$\omega_1 = \omega_R - \frac{R}{2L} \tag{12–78}$$

If Eq. 12–78 is subtracted from Eq. 12–77,

$$\omega_2 - \omega_1 = \frac{R}{L} \tag{12–79}$$

and

$$\frac{\omega_2 - \omega_1}{\omega_R} = \frac{R}{\omega_R L} = \frac{1}{Q_R} \tag{12–80}$$

The larger the value of Q, the sharper will be the resonance peak and the more selective the circuit.

12–4. Complex Notation for the Parallel Circuit. If several impedances are connected in parallel as shown in Fig. 12–14, a procedure similar to that employed in Art. 12–3 for series-connected elements may be used to establish an equivalent impedance.

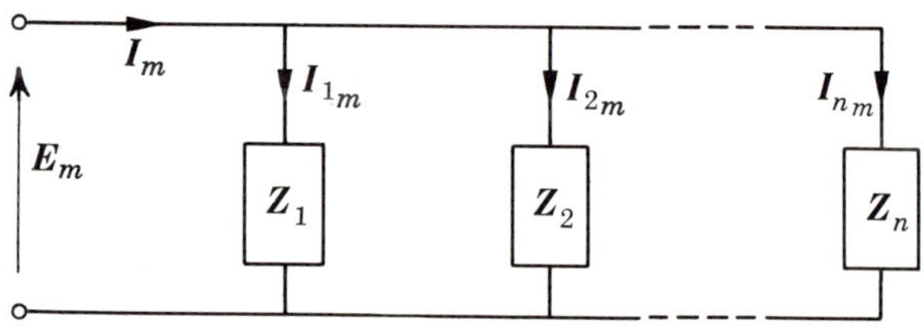

Fig. 12–14. Parallel impedances.

Kirchhoff's current law in vector form may be written as

$$\mathbf{I}_m = \mathbf{I}_{1_m} + \mathbf{I}_{2_m} + \cdots + \mathbf{I}_{n_m} \tag{12–81}$$

and using the volt-ampere relation for an impedance from Eq. 12–28

$$\mathbf{I}_m = \frac{\mathbf{E}_m}{\mathbf{Z}_1} + \frac{\mathbf{E}_m}{\mathbf{Z}_2} + \cdots + \frac{\mathbf{E}_m}{\mathbf{Z}_n} \tag{12–82}$$

or

$$\mathbf{I}_m = \mathbf{E}_m\left(\frac{1}{\mathbf{Z}_1} + \frac{1}{\mathbf{Z}_2} + \cdots + \frac{1}{\mathbf{Z}_n}\right) \tag{12–83}$$

The expression for the equivalent complex impedance $\mathbf{Z}'$ is then

$$\frac{1}{\mathbf{Z}'} = \frac{\mathbf{I}_m}{\mathbf{E}_m} = \frac{1}{\mathbf{Z}_1} + \frac{1}{\mathbf{Z}_2} + \cdots + \frac{1}{\mathbf{Z}_n} \tag{12–84}$$

or

$$\mathbf{Z}' = \frac{1}{\dfrac{1}{\mathbf{Z}_1} + \dfrac{1}{\mathbf{Z}_2} + \cdots + \dfrac{1}{\mathbf{Z}_n}} \tag{12–85}$$

The reciprocal of impedance is termed *admittance,* denoted Y, or in complex form

$$\boldsymbol{Y} = \frac{1}{\boldsymbol{Z}} \tag{12–86}$$

with units of mhos. Equation 12–84 may then be conveniently written

$$\boldsymbol{Y}' = \boldsymbol{Y}_1 + \boldsymbol{Y}_2 + \cdots + \boldsymbol{Y}_n \tag{12–87}$$

For example, if

$$\boldsymbol{Z}_1 = R_1 + jX_1 \tag{12–88}$$

then

$$\boldsymbol{Y}_1 = \frac{1}{R_1 + jX_1} \tag{12–89}$$

$$= \frac{R_1}{R_1^2 + X_1^2} - j\frac{X_1}{R_1^2 + X_1^2} \tag{12–90}$$

$$= G_1 - jB_1 \tag{12–91}$$

where

$$G_1 = \frac{R_1}{R_1^2 + X_1^2} \tag{12–92}$$

and

$$B_1 = \frac{X_1}{R_1^2 + X_1^2} \tag{12–93}$$

G_1 is called the *conductive component* of $\boldsymbol{Y}_1$, and B_1 the *susceptive component* of $\boldsymbol{Y}_1$. Note that if $\boldsymbol{Z}_1$ had been *only* R or *only* X, then we would have $G = 1/R$ or $B = 1/X$. Where a single element only is involved, G is a conductance and B a susceptance. For the combination of elements as exemplified by the R-L series combination chosen in the foregoing, it is better to term G and B conductive and susceptive components since the value of each depends upon both R and L.

Equation 12–87 shows that the equivalent admittance of several admittances in parallel is the complex sum of the several admittances. To illustrate this concept, some parallel combinations of elements are considered in the remainder of this section.

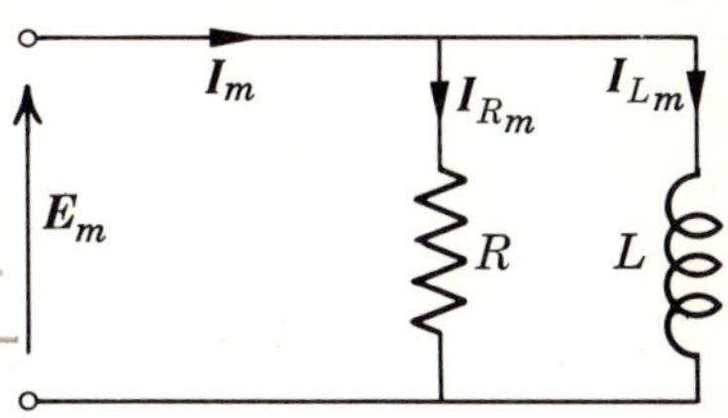

Fig. 12–15. *R-L* parallel circuit.

PARALLEL R-L ELEMENTS. Figure 12–15 shows an assumed pure resistance in parallel with a purely inductive branch. The current $\boldsymbol{I}_m$ is related to $\boldsymbol{E}_m$ by

$$\boldsymbol{I}_m = \boldsymbol{E}_m \boldsymbol{Y} \tag{12–94}$$

where

$$Y = \frac{1}{Z_1} + \frac{1}{Z_2} = \frac{1}{R} + \frac{1}{jX_L} \tag{12–95}$$

or

$$Y = G - jB_L \tag{12–96}$$

A phasor diagram of currents and voltage for this configuration is shown in Fig. 12–16.

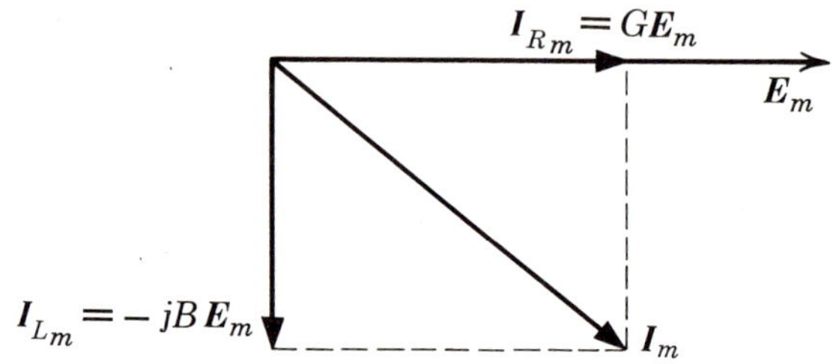

Fig. 12–16. Phasor diagram for a parallel *R-L* circuit.

PARALLEL *R-C* ELEMENTS. A parallel combination of R and C is shown in Fig. 12–17. Again, the current I_m is obtained as

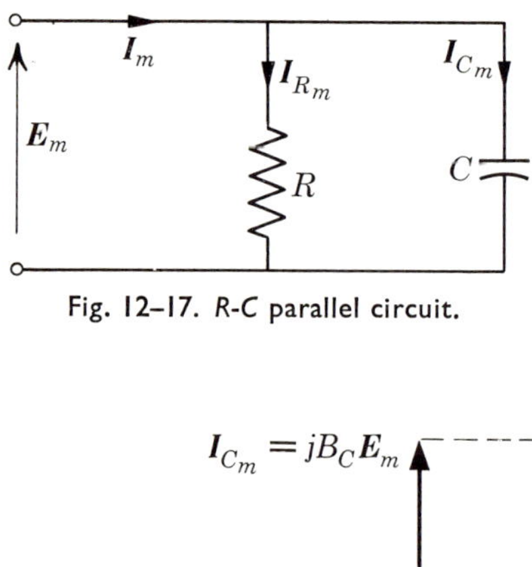

Fig. 12–17. *R-C* parallel circuit.

$$I_m = E_m Y \tag{12–97}$$

$$= E_m\left(\frac{1}{R} + \frac{1}{-jX_C}\right) \tag{12–98}$$

or

$$I_m = E_m(G + jB_C) \tag{12–99}$$

and the phasor diagram is as shown in Fig. 12–18.

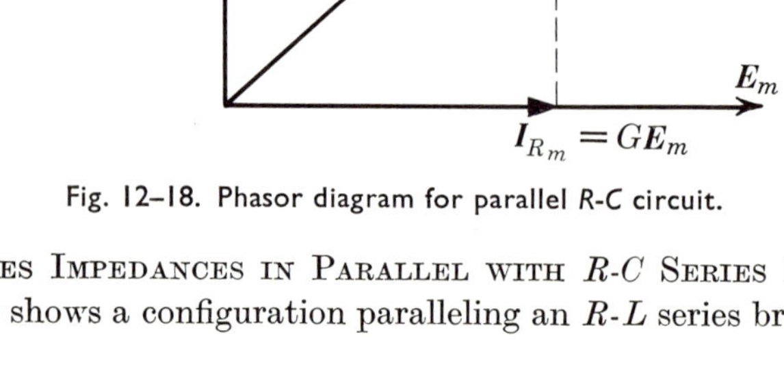

Fig. 12–18. Phasor diagram for parallel *R-C* circuit.

R-L SERIES IMPEDANCES IN PARALLEL WITH *R-C* SERIES IMPEDANCES. Figure 12–19 shows a configuration paralleling an *R-L* series branch with an

R-C series branch. The total admittance of the combination is the sum of the admittances of the two branches, or

$$\boldsymbol{Y} = \frac{1}{R_L + jX_L} + \frac{1}{R_C - jX_C} \qquad (12\text{–}100)$$

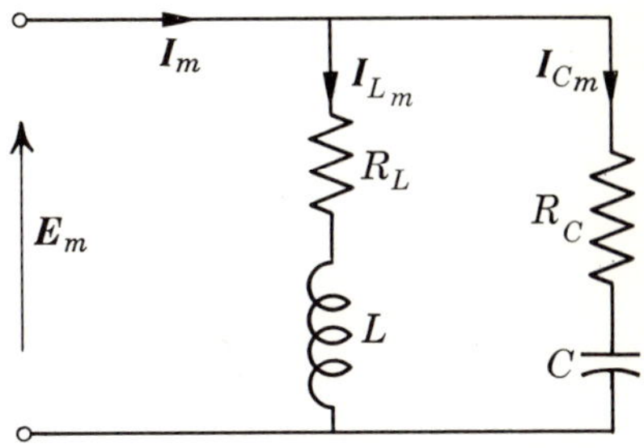

Fig. 12–19. *R-L* series–*R-C* series in parallel.

After rationalization, Eq. 12–100 becomes

$$\boldsymbol{Y} = \frac{R_L}{R_L^2 + X_L^2} - j\frac{X_L}{R_L^2 + X_L^2} + \frac{R_C}{R_C^2 + X_C^2} + j\frac{X_C}{R_C^2 + X_C^2} \qquad (12\text{–}101)$$

and

$$\boldsymbol{I}_m = \boldsymbol{E}_m\left[\left(\frac{R_L}{R_L^2 + X_L^2} + \frac{R_C}{R_C^2 + X_C^2}\right) - j\left(\frac{X_L}{R_L^2 + X_L^2} - \frac{X_C}{R_C^2 + X_C^2}\right)\right] \qquad (12\text{–}102)$$

$$= \boldsymbol{E}_m(G' - jB') \qquad (12\text{–}103)$$

If $\boldsymbol{I}_m$ and $\boldsymbol{E}_m$ are in phase, the parallel R-L–R-C combination is said to be in *parallel resonance.* For this to occur, we must have

$$B' = 0 \qquad (12\text{–}104)$$

or

$$\frac{X_L}{R_L^2 + X_L^2} = \frac{X_C}{R_C^2 + X_C^2} \qquad (12\text{–}105)$$

This may be rewritten as

$$\frac{\omega L}{R_L^2 + \omega^2 L^2} = \frac{\dfrac{1}{\omega C}}{R_C^2 + \dfrac{1}{\omega^2 C^2}} \qquad (12\text{–}106)$$

or

$$\frac{\omega L}{R_L^2 + \omega^2 L^2} = \frac{\omega C}{R_C^2 \omega^2 C^2 + 1} \qquad (12\text{–}107)$$

from which

$$\omega = \omega_R = \frac{1}{\sqrt{LC}}\sqrt{\frac{L - CR_L^2}{L - CR_C^2}} \qquad (12\text{–}108)$$

where ω_R is the parallel resonant frequency. $(L - CR_L^2)$ and $(L - CR_C^2)$ must both be either positive or negative if ω_R is to be real and possible.

If $R_L = R_C$ in Eq. 12–108, there are two situations which may exist:

(1) $$R_L = R_C \neq \sqrt{L/C}$$

For this case

$$\sqrt{\frac{L - CR_L^2}{L - CR_C^2}} = 1 \tag{12–109}$$

and

$$\omega_R = \frac{1}{\sqrt{LC}} \tag{12–110}$$

which is the same form as Eq. 12–53, which defined resonance in the series circuit.

Most engineering applications involve parameter sizes such that CR_L^2 and CR_C^2 are very small in comparison with L. If this is assumed, again

$$\omega_R = \frac{1}{\sqrt{LC}}$$

(2)

$$R_L = R_C = \sqrt{L/C}$$

For this particular case

$$\sqrt{\frac{L - CR_L^2}{L - CR_C^2}} = \sqrt{\frac{0}{0}} \tag{12–111}$$

which is indeterminate. A return to Eq. 12–107, however, permits an investigation as follows:

$$\frac{\omega L}{R_L^2 + \omega^2 L^2} = \frac{\omega C}{R_C^2 \omega^2 C^2 + 1} \tag{12–107}$$

Substitution of L/C for R_L^2 and R_C^2 yields

$$\frac{L}{\dfrac{L}{C} + \omega^2 L^2} = \frac{C}{\dfrac{L}{C}\,\omega^2 C^2 + 1} \tag{12–112}$$

or

$$\frac{LC}{L + \omega^2 L^2 C} = \frac{C^2}{C + L\omega^2 C^2} \tag{12–113}$$

from which it may be seen that

$$\frac{1}{1 + \omega^2 LC} = \frac{1}{1 + \omega^2 LC} \tag{12–114}$$

Equation 12–114 shows that, for $R_C = R_L = \sqrt{L/C}$, Eq. 12–107 is satisfied for all frequencies. That is, for this condition the admittance has no susceptive component at any frequency. The current is then in phase with the voltage at all frequencies, or the circuit is described as being in resonance at all frequencies. For this state, the current is

$$\boldsymbol{I}_m = \boldsymbol{E}_m\left(\frac{R_L}{R_L^2 + X_L^2} + \frac{R_C}{R_C^2 + X_C^2}\right) \tag{12–115}$$

$$= \boldsymbol{E}_m\sqrt{\frac{C}{L}} = \frac{\boldsymbol{E}_m}{R_C} = \frac{\boldsymbol{E}_m}{R_L} \tag{12–116}$$

and the magnitude of $\boldsymbol{I}_m$ is independent of frequency.

If $R_C = 0$ and $R_L \neq 0$, the usual engineering situation is encountered. From Eq. 12–107, then,

$$\frac{\omega L}{R_L^2 + \omega^2 L^2} = \omega C \tag{12–117}$$

or

$$\frac{L}{C} = R_L^2 + \omega^2 L^2 \tag{12–118}$$

from which

$$\omega = \omega_R = \sqrt{\frac{1}{LC} - \left(\frac{R_L}{L}\right)^2} \tag{12–119}$$

At this frequency the magnitude of $\boldsymbol{I}_m$ is given by

$$I_m = E_m \frac{R_L}{R_L^2 + \omega^2 L^2} \tag{12–120}$$

or, substituting Eq. 12–119, by

$$I_m = E_m \frac{R_L}{R_L^2 + L^2\left[\dfrac{1}{LC} - \left(\dfrac{R_L}{L}\right)^2\right]} \tag{12–121}$$

and

$$I_m = E_m \frac{R_L}{L/C} = \frac{E_m}{R} \tag{12–122}$$

in which

$$R = \frac{L}{CR_L} \tag{12–123}$$

$$= \frac{\omega_R L}{\omega_R R_L C} = Q_R X_C = Q_R X_L \tag{12–124}$$

which indicates that the resistance R_L has been transformed into an equivalent resistance R.

The condition for parallel resonance has been defined as zero phase angle between the applied voltage and the current in the parallel branches. Another definition is: the frequency at which the admittance of the parallel branches is a minimum. For R-L branch Q's larger than approximately 5, these two conditions are the same and $\omega_R = 1/\sqrt{LC}$. If a lower Q exists, separate frequencies for the two definitions will be noted. A discussion of this second definition will not be undertaken here, but a knowledge of its existence is desirable.

12–5. The Parallel Equivalent of Series Impedances. In this study the usual procedure has been the reduction of element combinations to a series equivalent. Sometimes, however, it is desired that a series impedance be replaced by a parallel equivalent.

For example, consider the networks of Fig. 12–20 in which a parallel $R_p - L_p$ combination is desired equivalent to a series $R_s - L_s$ combination.

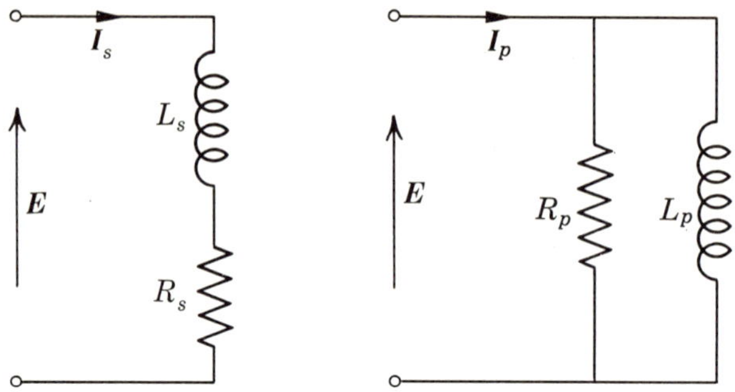

Fig. 12–20. Parallel equivalent of a series circuit.

If the two are to be equivalent, the admittance of both circuits must be the same at all frequencies, or

$$\frac{1}{Z_s} = \frac{1}{R_p} + \frac{1}{jX_p} = \frac{1}{R_s + jX_s} \tag{12–125}$$

From this

$$\frac{1}{R_p} + \frac{1}{jX_p} = \frac{R_s}{R_s^2 + X_s^2} - j\frac{X_s}{R_s^2 + X_s^2} \tag{12–126}$$

or

$$\frac{1}{R_p} = \frac{R_s}{R_s^2 + X_s^2} \tag{12–127}$$

and

$$\frac{1}{X_p} = \frac{X_s}{R_s^2 + X_s^2} \tag{12–128}$$

Equations 12–127 and 12–128 may be rewritten as

$$R_p = R_s + \frac{X_s^2}{R_s} \tag{12–129}$$

and

$$X_p = X_s + \frac{R_s^2}{X_s} \tag{12–130}$$

or also as

$$R_p = R_s + Q\omega L_s \tag{12–131}$$

and

$$X_p = X_s + \frac{R_s}{Q} \tag{12–132}$$

Further approximations may be made if Q is larger (above 5).

PROBLEMS

12–1. A voltage source of value.

$$e = 1.414 \sin (t + 30°)$$

is applied to a series combination of $R = 1$ ohm and $L = 1$ h. (a) Represent the applied voltage as a phasor expressed in rectangular, polar, and exponential form. (b) Using phasor operations, calculate the phasor expression for the current in the circuit. (c) Determine the phasor expressions for the voltage across R and L. (d) Draw a phasor diagram showing all currents and voltages. (e) From the results of (b) and (c), write the instantaneous expressions for the current and element voltages.

12–2. A current source of value

$$i = 20 \cos (1000t + 45°)$$

is applied to a parallel combination of $R = 1000$ ohms and $C = 0.5\ \mu$f. (a) Represent the applied current as a phasor expressed in rectangular, polar, and exponential form. (b) Using phasor notation, calculate the voltage across each element and the current in each element. (c) Express the parallel combination of R and C in terms of an equivalent series impedance. (d) Draw a phasor diagram showing all voltages and currents. (e) From the answers of (b), write the instantaneous expressions for the current in each element and the voltages across them.

12–3. A series R-L-C circuit is shown. (a) Calculate its resonant frequency. (b) Determine Q_R and the frequencies at which the power dissipated by the circuit is one-half of that at resonance. (c) Plot against frequency the current and the voltage across R, L, and C. Discuss. (d) Repeat (a) and (b) for R changed to 10 ohms, 500 ohms, 1000 ohms, and 10,000 ohms.

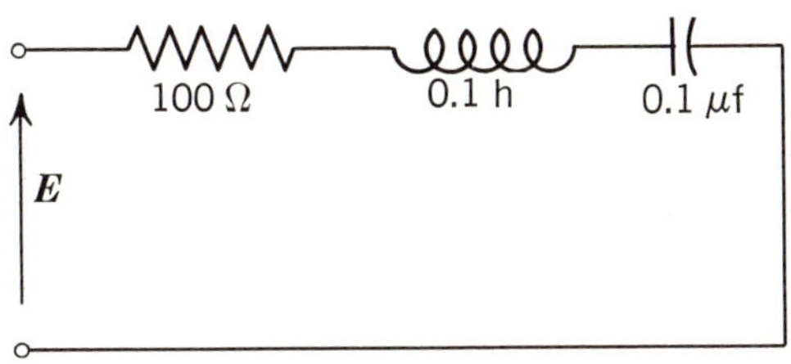

Prob. 12–3.

12–4. A parallel combination of elements is shown. (a) Calculate the parallel resonant frequency, approximately and exactly. Discuss the difference. (b) Determine Q_R for the network. (c) Plot against frequency the total current, the current in the branches, and the impedance of the combination. (d) Determine and sketch an equivalent of this network, showing resistance, inductance, and capacitance all in parallel. (e) Repeat (d) for R increased to 1000 ohms.

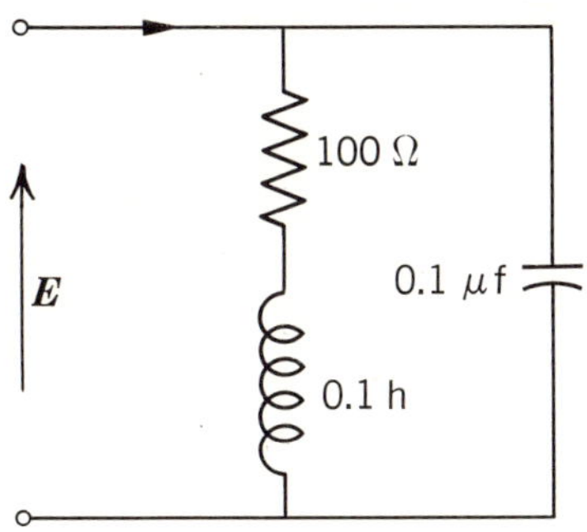

Prob. 12–4.

12–5. The circuit of Prob. 12–4 is supplied from a non-ideal voltage source having a resistive internal impedance as shown. (a) Develop an expression for the exact parallel resonant frequency of this configuration and compare to that of Prob. 12–4a. (b) Determine Q_R for the network. (c) Plot against frequency the total current and the ratio of load voltage to total current. Is this later curve a plot of Z, and how does it compare to the plot of Z of Prob. 12–4c?

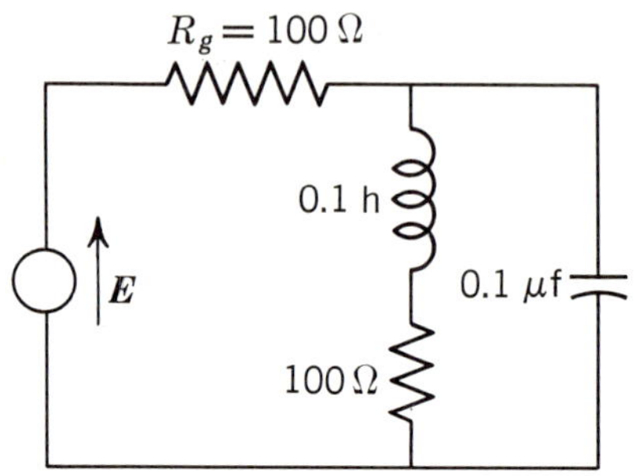

Prob. 12–5.

12–6. The network configuration illustrated is excited by a voltage source of variable frequency and constant amplitude. (a) Use phasor notation to derive a general expression for the current as a function of ω. (b) Determine the complex expressions for Z and Y and plot the magnitude and phase angle of Z versus ω. Explain the resultant plots.

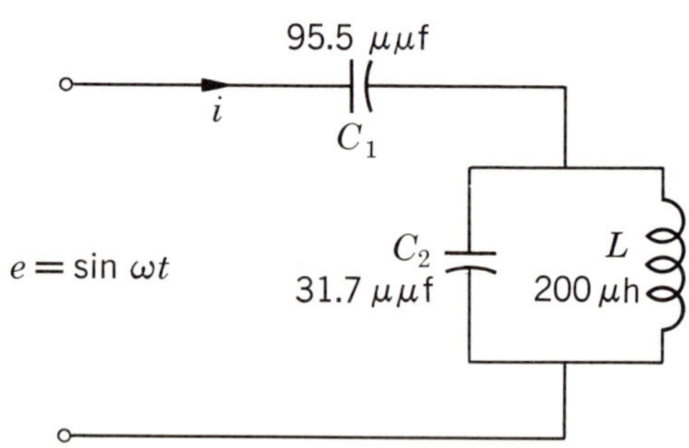

Prob. 12–6.

12–7. Repeat Prob. 12–6 for the situation in which the inductance L has a resistance of $R = 250$ ohms in series with it. Compare the results with those of Prob. 12–6 and discuss the effect of R.

12–8. Express in complex form the admittance seen between points a and b in the network shown.

12–9. Use phasor notation and procedure to obtain the value of $\boldsymbol{E}_2$ in the circuit shown, where the instantaneous form of $\boldsymbol{E}_1$ is

$$e_1 = 10 \sin 100t$$

Also write the instantaneous form of $\boldsymbol{E}_2$.

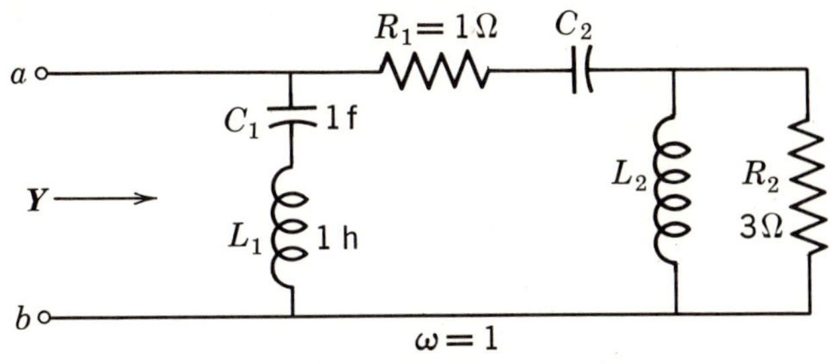

Prob. 12–8.

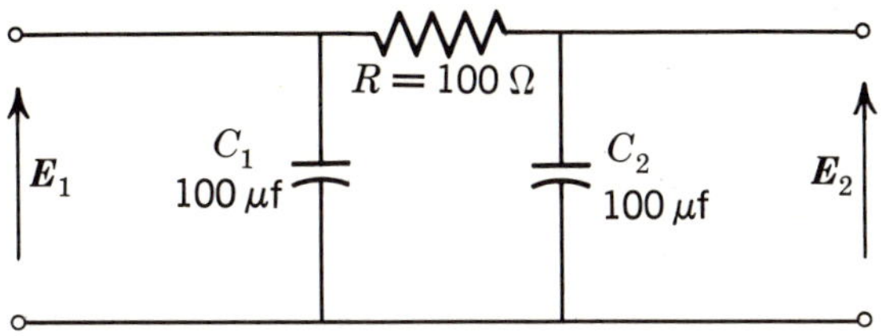

Prob. 12–9.

12–10. Determine the phasors $\boldsymbol{I}$ and $\boldsymbol{E}$ for the network illustrated.

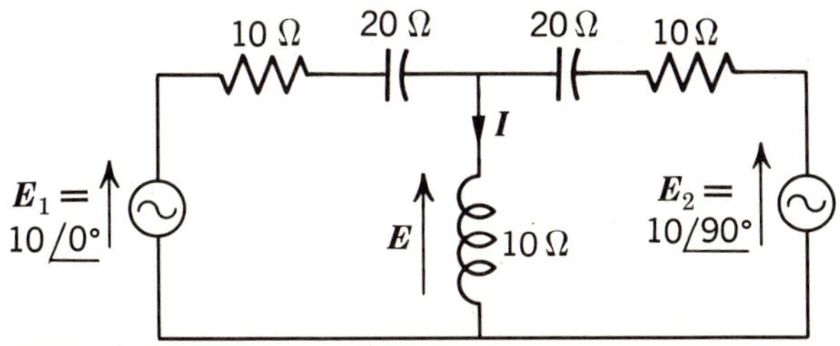

Prob. 12–10.

13

EFFECTIVE VALUES AND POWER

The preceding chapters on sinusoidal analysis made reference to the effective value of a sinusoidal waveform. The concept of effective values is also applicable to any waveform, and this chapter develops this in detail.

Closely tied to the effective value of current and voltage is the concept of instantaneous and average power associated with time-varying currents and voltages. This is also considered here.

13–1. Effective or Root Mean Square Value. The power dissipated in a resistance carrying a direct current is

$$P = I_{dc}^2 R \tag{13–1}$$

If the current in a resistance varies with time, the instantaneous power will also vary with time. If the current is also periodic with respect to time, the average power per period may be computed or measured.

If a *constant current I* is chosen such that I^2R gives the same average power as the periodic current, then I is said to be the *effective value* of the *time-varying periodic current.*

Mathematically, the instantaneous power dissipated in a resistance R is i^2R and the average power per period is computed as

$$P_{av} = \frac{1}{2\pi} \int_0^{2\pi} i^2 R \, d(\omega t) \tag{13–2}$$

where 2π is the value of ωt for one period, and

$$P = P_{av} = \text{average power}$$

From the definition of effective value, then,

$$I^2 R = \frac{1}{2\pi} \int_0^{2\pi} i^2 R \, d(\omega t) \tag{13–3}$$

Since R is a constant,

$$I^2 = \frac{1}{2\pi} \int_0^{2\pi} i^2 \, d(\omega t) \tag{13–4}$$

where

$$\frac{1}{2\pi} \int_0^{2\pi} i^2 \, d(\omega t) \tag{13–5}$$

is the mean (or average) of the square of the instantaneous current. The effective value I is then

$$I = \sqrt{\frac{1}{2\pi}\int_0^{2\pi} i^2\, d(\omega t)} \tag{13-6}$$

and the effective value is "the root of the mean of the square of the instantaneous current over a period." Another name for effective value is *root mean square*, abbreviated rms.

EXAMPLE 13-1. RMS VALUE OF A TRIANGULAR WAVE. Find the effective value of the triangular periodic current shown in Fig. 13-1.

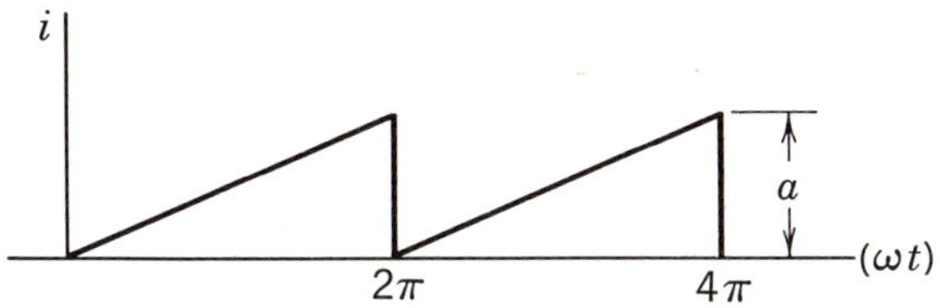

Fig. 13-1. Triangular wave of current.

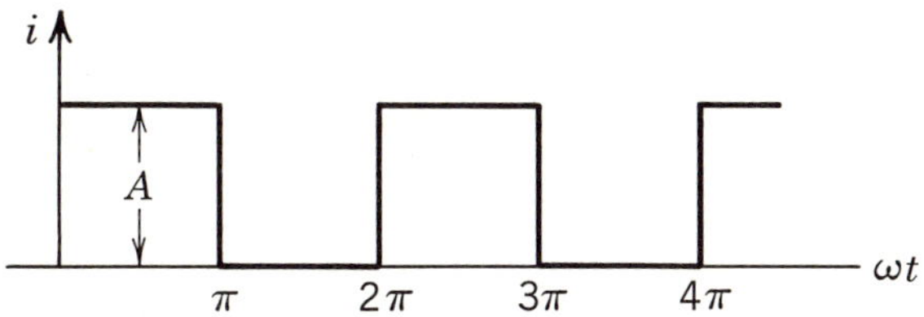

Fig. 13-2. Periodic current pulses.

The expression from analytic geometry for the instantaneous current for one period is

$$i = \frac{a}{2\pi}\,\omega t, \qquad 0 < \omega t < 2\pi$$

and the effective value is

$$I = \sqrt{\frac{1}{2\pi}\int_0^{2\pi} i^2\, d(\omega t)}$$

$$= \sqrt{\frac{1}{2\pi}\int_0^{2\pi}\left(\frac{a}{2\pi}\,\omega t\right)^2 d(\omega t)}$$

or

$$I = \sqrt{\frac{a^2}{8\pi^3}\int_0^{2\pi}(\omega t)^2\, d(\omega t)} = \sqrt{\frac{a^2}{3}} = \frac{a}{\sqrt{3}}$$

EXAMPLE 13-2. RMS VALUE OF A SERIES OF RECTANGULAR PULSES. Find the effective value of the periodic current shown in Fig. 13-2.

The instantaneous expression for the function may be written for one cycle as

$$f(\omega t) = A, \qquad 0 < (\omega t) < \pi$$
$$f(\omega t) = 0, \qquad \pi < (\omega t) < 2\pi$$

and the effective current is

$$I = \sqrt{\frac{1}{2\pi}\int_0^{2\pi} i^2\, d(\omega t)}$$
$$= \sqrt{\frac{1}{2\pi}\left[\int_0^{\pi} A^2\, d(\omega t) + \int_{\pi}^{2\pi} 0\, d(\omega t)\right]}$$
$$= \sqrt{\frac{A^2}{2}} = \frac{A}{\sqrt{2}}$$

13–2. Effective Value of a Sinusoidal Function. The definition of the effective value may be applied to sinusoidal functions of time to find the effective value of a sine wave of current or voltage. Again, from Eq. 13–6,

$$I = \sqrt{\frac{1}{2\pi}\int_0^{2\pi} i^2\, d(\omega t)}$$

and if

$$i = I_m \sin \omega t \tag{13–7}$$

then

$$I = \sqrt{\frac{1}{2\pi}\int_0^{2\pi} I_m{}^2 \sin^2 \omega t\, d(\omega t)} \tag{13–8}$$

$\sin^2 \omega t\, d(\omega t)$ may be integrated by using the trigonometric identity

$$\sin^2 \omega t = \frac{1 - \cos 2\omega t}{2} \tag{13–9}$$

Substituting Eq. 13–9 into 13–8,

$$I = \sqrt{\frac{I_m{}^2}{4\pi}\int_0^{2\pi} (1 - \cos 2\omega t)\, d(\omega t)} \tag{13–10}$$

$$= \sqrt{\frac{I_m{}^2}{4\pi} 2\pi} \tag{13–11}$$

$$= \frac{I_m}{\sqrt{2}} = 0.707 I_m \tag{13–12}$$

For a sine or cosine wave the effective value of the current is equal to its maximum value divided by $\sqrt{2}$. This is the value that a conventional a-c ammeter will read.

13–3. Average Value. The average value of a periodic function is the area under the periodic curve divided by the base of the period.

From this definition the average value of a sinusoidal function of time is zero since it has as much area under the positive half as under the negative half of the cycle. However, from an arbitrary definition *applied to sinusoidal waves,* the average value for this wave is defined as the average value over the positive half-cycle.

Therefore, if

$$i = I_m \sin \omega t$$

then

$$I_{\text{av}} = \frac{1}{\pi} \int_0^{\pi} I_m \sin \omega t \, d(\omega t) \tag{13-13}$$

$$= \frac{2}{\pi} I_m = 0.637 I_m \tag{13-14}$$

Example 13-3. The Average Value of a Triangular Wave. Find the average value of the current function of Example 13-1.

The average value may be obtained from

$$I_{\text{av}} = \frac{1}{2\pi} \int_0^{2\pi} \frac{a\omega t}{2\pi} \, d(\omega t)$$

$$= \frac{a}{2}$$

Example 13-4. The Average Value of a Series of Rectangular Pulses. Find the average value of the function shown in Example 13-2.

$$I_{\text{av}} = \frac{\pi A}{2\pi} = \frac{A}{2}$$

13-4. Power and Energy in Sinusoidally Excited Circuit Elements. The sinusoidal form of the volt-ampere characteristics of Art. 10-5 may be used to develop expressions for the instantaneous and average power associated with the element.

Resistance Element. The voltage across a resistance element and the current in the element are in phase with each other. If the current is written

$$i = I_m \sin \omega t \tag{13-15}$$

the voltage is

$$e = E_m \sin \omega t \tag{13-16}$$

and the instantaneous power in the element is

$$p = ei = (E_m \sin \omega t)(I_m \sin \omega t) \tag{13-17}$$

$$= E_m I_m \sin^2 \omega t \tag{13-18}$$

$$= \frac{E_m I_m}{2} (1 - \cos 2\omega t) \tag{13-19}$$

$$= EI(1 - \cos 2\omega t) \tag{13-20}$$

where E and I are the effective values of the voltage across and the current in the resistance.

The power is a double frequency quantity superimposed on a constant value. Graphically this is shown in Fig. 13–3.

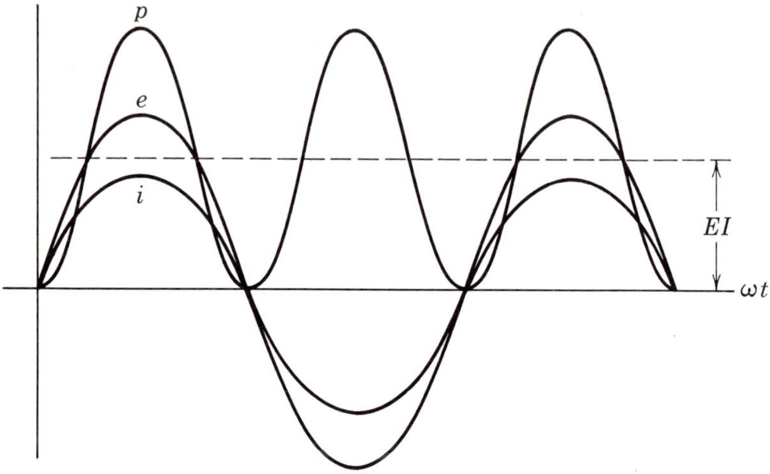

Fig. 13–3. Instantaneous power in a resistance.

The average power in a resistance may be obtained by integrating Eq. 13–19 over one cycle. The result is

$$P = P_{\text{av}} = \frac{E_m I_m}{4\pi} \int_0^{2\pi} (1 - \cos 2\omega t)\, d(\omega t)$$

$$= \frac{E_m I_m}{2} = EI \tag{13–21}$$

For a resistance, however,

$$I_m = \frac{E_m}{R}$$

and therefore

$$P = \frac{I_m^{\,2} R}{2} = \frac{E_m^{\,2}/R}{2} \tag{13–22}$$

$$= I^2 R = \frac{E^2}{R} \tag{13–23}$$

$$= E^2 G \tag{13–24}$$

Inductance Element. The voltage across an inductance is 90° ahead of the current in it. If the instantaneous current is

$$i = I_m \sin \omega t \tag{13–25}$$

the instantaneous voltage is

$$e = E_m \sin (\omega t + 90°) = E_m \cos \omega t \tag{13–26}$$

and the instantaneous power is

$$p = ei = (E_m \cos \omega t)(I_m \sin \omega t) \tag{13-27}$$

$$= \frac{E_m I_m}{2} \sin 2\omega t \tag{13-28}$$

$$= EI \sin 2\omega t \tag{13-29}$$

This is represented by Fig. 13-4.

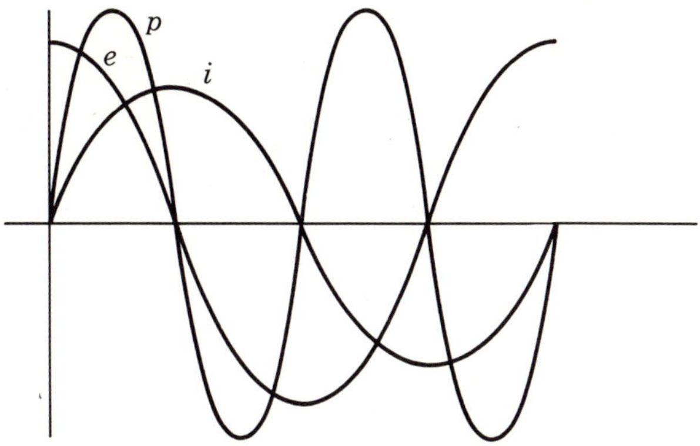

Fig. 13-4. Instantaneous power in an inductance.

The instantaneous power is a double frequency quantity. During some portions of the cycle the power is positive and the inductance is absorbing power from the source. During other portions of the cycle the power is negative and the inductance is returning power to the source. *The average power per cycle, however, is zero.*

Capacitance Element. The voltage across the capacitance lags the current by 90° in phase. Choosing the current as reference as

$$i = I_m \sin \omega t \tag{13-30}$$

the voltage is

$$e = E_m \sin (\omega t - 90°) = -E_m \cos \omega t \tag{13-31}$$

and the instantaneous power is

$$p = ei = -E_m I_m \sin \omega t \cos \omega t \tag{13-32}$$

$$= -\frac{E_m I_m}{2} \sin 2\omega t \tag{13-33}$$

$$= -EI \sin 2\omega t \tag{13-34}$$

This is shown in Fig. 13-5. Here, too, power is a double frequency quantity whose average value is zero. During portions of the cycle power is absorbed by the capacitance and during other portions power is returned to the circuit.

R-L-C Series Circuit. A comparison of Eqs. 13–29 and 13–34 shows that in a series circuit when power is delivered by the capacitance it is absorbed by the inductance and vice versa. This is true also in parallel *L-C* circuits.

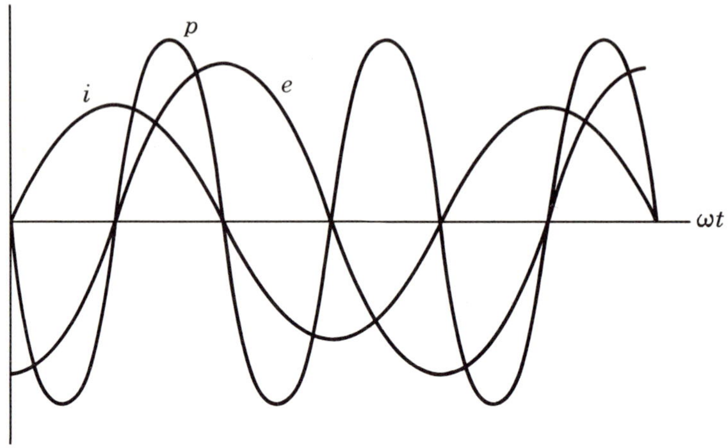

Fig. 13–5. Instantaneous power in a capacitance.

Consider an *R-L-C* series circuit with a current expressed as

$$i = I_m \sin \omega t \tag{13–35}$$

The voltage may then be expressed as

$$e = E_m \sin (\omega t + \theta) \tag{13–36}$$

where

$$\theta = \tan^{-1} \frac{X_L - X_C}{R} \tag{13–37}$$

The instantaneous power is then given by

$$p = ei = E_m I_m \sin \omega t \sin (\omega t + \theta) \tag{13–38}$$

$$= \frac{E_m I_m}{2} \cos \theta - \frac{E_m I_m}{2} \cos (2\omega t + \theta) \tag{13–39}$$

$$= \frac{E_m I_m}{2} [\cos \theta (1 - \cos 2\omega t) + \sin \theta \sin 2\omega t] \tag{13–40}$$

$$= EI[\cos \theta (1 - \cos 2\omega t) + \sin \theta \sin 2\omega t] \tag{13–41}$$

The first term is called the *instantaneous active* (or real) *power* and the second term is called the *instantaneous reactive power*. The value of $(E_m I_m/2) \cos \theta(1 - \cos 2\omega t)$ varies between zero and $E_m I_m \cos \theta$ and is always positive since θ for a passive circuit lies between $+90°$ and $-90°$. It is the power which is consumed by the resistive part of the circuit and transformed into heat. The instantaneous reactive power $(E_m I_m/2) \sin \theta \sin 2\omega t$ is both positive and negative during the cycle.

The average active power is

$$P_{\text{av}} = \frac{E_m I_m}{2} \cos \theta \tag{13–42}$$

$$= EI \cos \theta \tag{13–43}$$

Note that the active power is the product of the in-phase components of the phasors $\boldsymbol{E}$ and $\boldsymbol{I}$ or one-half the product of the in-phase components of $\boldsymbol{E}_m$ and $\boldsymbol{I}_m$.

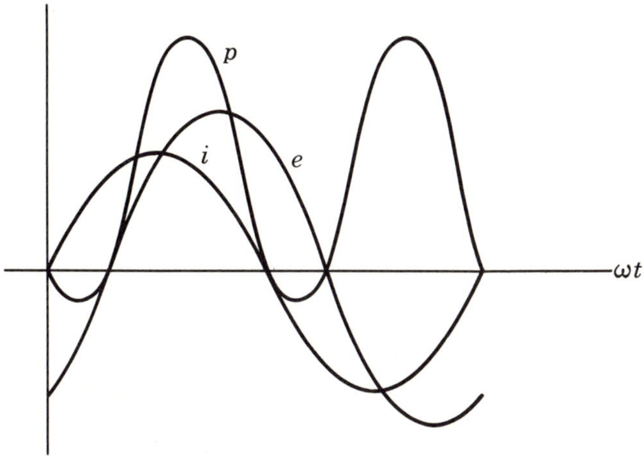

Fig. 13–6. Instantaneous power in the *R-L-C* circuit.

By definition, the out-of-phase product $(E_M I_M/2) \sin \theta$ is called *reactive power* and expressed in var units (var denotes volt-ampere reactive). The usual symbol for reactive power is Q_{av},* where

$$Q_{\text{av}} = \frac{E_m I_m}{2} \sin \theta \tag{13–44}$$

$$= EI \sin \theta \tag{13–45}$$

Notice that Q_{av} may be positive or negative, depending upon whether θ is positive or negative. A physical meaning for Q_{av} will be given in a later section.

Equation 13–41 is shown graphically in Fig. 13–6 for a positive value of θ. Equation 13–41 shows that instantaneous power is a double frequency function. From Fig. 13–6 it is seen that the instantaneous power is positive during parts of the cycle and negative during other parts. The area under the positive loops is larger than that under the negative loops, however, and the net area is positive.

* A number of symbols have two or more meanings. Q here is *not* the same as the symbol used to define the properties of a resonant circuit.

13–5. Vector Power. Active or real power is $EI \cos \theta$ and reactive power is $EI \sin \theta$. Vectorially the two powers are 90° apart with respect to each other. If the average active power, denoted P_{av}, is represented by a vector and the average reactive power, denoted Q_{av}, also is represented by a vector, the vectorial sum of the two is called *vector power,* or *apparent power*, and termed volt-amperes (va).

The average active power is always positive in a passive circuit and is taken as a reference. Reactive power is considered negative if the current lags the voltage and positive if the current leads the voltage.

To demonstrate, assume phasors for $\boldsymbol{E}_M$ and $\boldsymbol{I}_M$ as represented by Fig. 13–7, where

$$\boldsymbol{E}_m = E_m \epsilon^{j\alpha} \tag{13–46}$$

and

$$\boldsymbol{I}_m = I_m \epsilon^{j\beta} \tag{13–47}$$

Since α is larger than β, the current lags the voltage and the reactive power should have a negative sign. The vector power is

$$P_{av} + jQ_{av} = \frac{E_m I_m}{2} \cos \theta + j \frac{E_m I_m}{2} \sin \theta \tag{13–48}$$

But

$$\cos \theta = \cos (\alpha - \beta) = \cos \alpha \cos \beta + \sin \alpha \sin \beta \tag{13–49}$$

and

$$\sin \theta = \sin (\alpha - \beta) = \sin \alpha \cos \beta - \cos \alpha \sin \beta \tag{13–50}$$

Therefore

$$\begin{aligned} P_{av} + jQ_{av} &= \frac{E_m I_m}{2} (\cos \alpha \cos \beta + \sin \alpha \sin \beta) \\ &\quad - j \frac{E_m I_m}{2} (\sin \alpha \cos \beta - \cos \alpha \sin \beta) &(13–51) \\ &= \tfrac{1}{2}[(E_m \cos \alpha\, I_m \cos \beta + E_m \sin \alpha\, I_m \sin \beta) \\ &\quad - j(E_m \sin \alpha\, I_m \cos \beta - E_m \cos \alpha\, I_m \sin \beta)] &(13–52) \\ &= \tfrac{1}{2}[I_m \cos \beta\, (E_m \cos \alpha - jE_m \sin \alpha) \\ &\quad + jI_m \sin \beta (E_m \cos \alpha - jE_m \sin \alpha)] &(13–53) \\ &= \tfrac{1}{2}(E_m \cos \alpha - jE_m \sin \alpha)(I_m \cos \beta + jI_m \sin \beta) &(13–54) \\ &= \tfrac{1}{2}(E_m \epsilon^{-j\alpha}\, I_m \epsilon^{j\beta}) &(13–55) \\ &= \tfrac{1}{2}\boldsymbol{E}_m{}^* \boldsymbol{I}_m = \boldsymbol{E}^* \boldsymbol{I} &(13–56) \end{aligned}$$

where $\boldsymbol{E}_M{}^*$ is the conjugate of $\boldsymbol{E}_M$. The vector power is given by the product of $\boldsymbol{E}^*$, the conjugate of $\boldsymbol{E}$, and the phasor $\boldsymbol{I}$.

Equations 12–3, 12–4, and 12–15 demonstrated that the phasors $\boldsymbol{E}_m$ and $\boldsymbol{I}_m$ could be written as the complex sum of sine and cosine components:

$$\boldsymbol{E}_m = e_1 + je_2$$

and

$$\boldsymbol{I}_m = i_1 + ji_2$$

$\boldsymbol{E}_m$ and $\boldsymbol{I}_m$ are shown in Fig. 13–8.

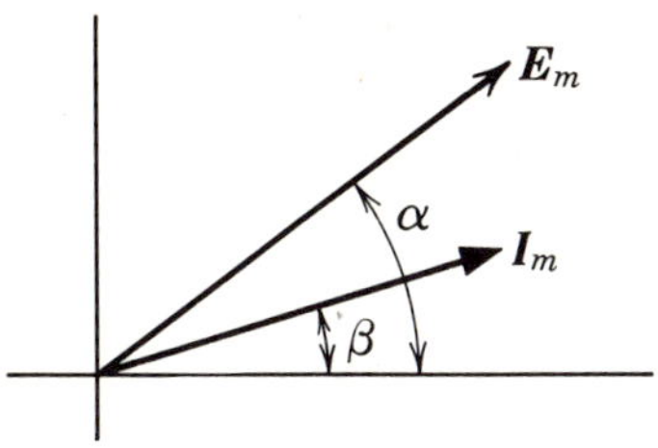

Fig. 13–7. Voltage and current phasors.

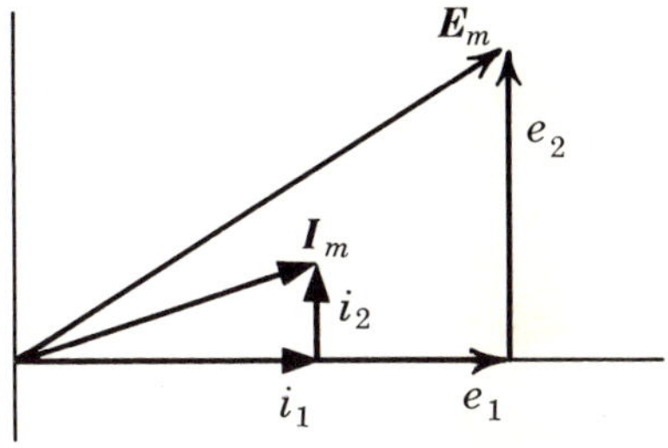

Fig. 13–8. Components of $\boldsymbol{E}_m$ and $\boldsymbol{I}_m$.

The conjugate of $\boldsymbol{E}_m$ in this form is

$$\boldsymbol{E}_m^* = e_1 - je_2 \tag{13–57}$$

and in component form the vector power expressed by Eq. 13–56 becomes

$$P_{\text{av}} - jQ_{\text{av}} = \tfrac{1}{2}(e_1 - je_2)(i_1 + ji_2) \tag{13–58}$$

$$= \tfrac{1}{2}[(e_1 i_1 + e_2 i_2) - j(e_2 i_1 - e_1 i_2)] \tag{13–59}$$

The active power is seen to be the sum of the products of the in-phase components of the current and voltage phasors, and the reactive power the difference of the products of the quadrature (or 90° out of phase) components.

If the phasor volt-ampere relation for an impedance, Eq. 12–28,

$$\boldsymbol{E} = \boldsymbol{IZ}$$

is written in conjugate form,

$$\boldsymbol{E}^* = \boldsymbol{I}^*\boldsymbol{Z}^* \tag{13–60}$$

and substituted into Eq. 13–56, the vector power may be obtained as

$$P_{\text{av}} - jQ_{\text{av}} = \boldsymbol{E}^*\boldsymbol{I} = I^2\boldsymbol{Z}^* \tag{13–61}$$

since

$$\boldsymbol{I}^*\boldsymbol{I} = I^2 \tag{13–62}$$

Equation 13–61 may also be written as

$$\frac{P_{\text{av}} + jQ_{\text{av}}}{I^2} = \boldsymbol{Z}^* \tag{13–63}$$

The vector power may also be obtained by combining the volt-ampere relation for an admittance,

$$I = EY$$

with Eq. 13–56 to obtain

$$P_{av} + jQ_{av} = E^*EY^* = E^2Y^* \tag{13–64}$$

Equation 13–64 may also be written

$$\frac{P_{av} - jQ_{av}}{E^2} = Y \tag{13–65}$$

13–6. Energy Storage in Circuit Elements. INDUCTANCE. The energy stored in the magnetic field of an inductance element is

$$w_L = \frac{Li^2}{2} \tag{13–66}$$

If the current is a sine wave the energy is

$$w_L = \frac{LI_m^{\ 2}}{2} \sin^2 \omega t \tag{13–67}$$

But

$$\sin^2 \omega t = \frac{1 - \cos 2\omega t}{2} \tag{13–68}$$

and therefore

$$w_L = \frac{LI_m^{\ 2}}{4} (1 - \cos 2\omega t) \tag{13–69}$$

The maximum energy storage occurs when $\cos 2\omega t = -1$ (where the current is maximum) and the amount stored is

$$W_{L_m} = \frac{LI_m^{\ 2}}{2} \tag{13–70}$$

The average energy is obtained from Eq. 13–69 as

$$W_{L_{av}} = \frac{LI_m^{\ 2}}{4} \tag{13–71}$$

and is equal to half the maximum stored energy. Differentiating Eq. 13–69 yields the power absorbed by the inductance as

$$p_L = \frac{\omega LI_m^{\ 2}}{2} \sin 2\omega t \tag{13–72}$$

$$= \frac{E_mI_m}{2} \sin 2\omega t \tag{13–73}$$

which is the same result obtained in Eq. 13–28.

CAPACITANCE. The energy stored in the electric field of a capacitance element is

$$w_C = \frac{Ce^2}{2} \tag{13–74}$$

For a sine wave of current

$$e^2 = \left(\frac{I_m}{\omega C} \cos \omega t\right)^2 \tag{13–75}$$

and therefore

$$w_C = \frac{C}{4}\left(\frac{I_m^{\ 2}}{\omega^2 C^2}\right)(1 + \cos 2\omega t) \tag{13–76}$$

The maximum energy is stored when $\cos 2\omega t = 1$. Its value is

$$W_{C_m} = \frac{I_m^{\ 2}}{2\omega^2 C} \tag{13–77}$$

For this value of ωt the current is zero.

The average stored energy from Eq. 13–76 is

$$W_{C_{\mathrm{av}}} = \frac{I_m^{\ 2}}{4\omega^2 C} \tag{13–78}$$

and is half the maximum stored energy. Differentiating Eq. 13–76 yields the power absorbed by the capacitance as

$$p_C = -\frac{I_m^{\ 2}}{2\omega C} \sin 2\omega t \tag{13–79}$$

$$= -\frac{E_m I_m}{2} \sin 2\omega t \tag{13–80}$$

Equations 13–80 and 13–73 show that when the inductance is absorbing power the capacitance is releasing power that was previously absorbed.

Equation 13–78 subtracted from Eq. 13–71 yields

$$W_{L_{\mathrm{av}}} - W_{C_{\mathrm{av}}} = \frac{LI_m^{\ 2}}{4} - \frac{I_m^{\ 2}}{4\omega^2 C} \tag{13–81}$$

$$= \frac{I_m^{\ 2}}{4\omega}\left(\omega L - \frac{1}{\omega C}\right) \tag{13–82}$$

$$= \frac{I_m^{\ 2}}{4\omega}(X_L - X_C) \tag{13–83}$$

$$= \frac{I_m^{\ 2}}{4\omega}(X) \tag{13–84}$$

where X is the net reactance, the difference between X_L and X_C. The magnitude of $\boldsymbol{I}_m$ may be written in terms of E_m and Z as

$$I_m = \frac{E_m}{Z}$$

and Eq. 13–84 then becomes

$$W_{L_{\text{av}}} - W_{C_{\text{av}}} = \left(\frac{I_m}{4\omega}\right)\left(\frac{E_m}{Z}\right)X \tag{13–85}$$

From the impedance triangle,

$$\frac{X}{Z} = \sin\theta$$

and therefore

$$W_{L_{\text{av}}} - W_{C_{\text{av}}} = \frac{I_m}{4\omega} E_m \sin\theta \tag{13–86}$$

or

$$\frac{I_m E_m \sin\theta}{2} = 2\omega(W_{L_{\text{av}}} - W_{C_{\text{av}}}) \tag{13–87}$$

But from Eq. 13–44

$$Q_{\text{av}} = \frac{E_m I_m}{2} \sin\theta$$

and therefore

$$Q_{\text{av}} = 2\omega(W_{L_{\text{av}}} - W_{C_{\text{av}}}) \tag{13–88}$$

The reactive power is seen to be the difference between the average energy stored in the inductance and capacitance elements. If the stored energies are equal, the reactive power required from the circuit during steady-state operation is zero.

13–7. Circuit *Q*. The average active or real power delivered to the circuit is consumed by the resistive component of the circuit. The energy per cycle dissipated by this equivalent resistance may be expressed as

$$P_{\text{av}}T = I^2RT \tag{13–89}$$

where $T = 2\pi/\omega$, the period of (or time for) one cycle, and R is an equivalent resistive component of circuit impedance. Substituting $2\pi/\omega$ for T in Eq. 13–89 yields

$$P_{\text{av}}T = \frac{2\pi I^2 R}{\omega} \tag{13–90}$$

$$= \frac{\pi I_m{}^2 R}{\omega} \tag{13–91}$$

At resonance the maximum energy stored in the magnetic field portions of the circuit is

$$W_{L_m} = L\frac{I_m{}^2}{2}$$

The ratio of this stored energy to the energy dissipated in the resistance is then for the resonant angular velocity ω_R

$$\frac{W_{L_m}}{P_{\text{av}}T} = \frac{\omega_R L I_m^{\,2}}{2\pi I_m^{\,2} R} \tag{13–92}$$

$$= \frac{\omega_R L}{2\pi R} \tag{13–93}$$

$$= \frac{Q_R}{2\pi} \tag{13–94}$$

where Q_R is the same factor defined in Eqs. 12–59 and 12–80. Therefore

$$Q_R = 2\pi\left(\frac{\text{maximum energy stored during a cycle}}{\text{energy dissipated per cycle}}\right)$$

This is the basic definition of Q_R.

13–8. Power Factor. It has been shown that when an inductance is absorbing power a capacitance in the circuit will be returning power to the circuit, and that the reactive power in a circuit is the difference of the average stored energy in the inductance and capacitance times twice the angular velocity of the source. If the average energy in the capacitance is equal to the average energy in the inductance, the reactive power supplied by the source becomes zero. This corresponds to a resonant condition for the system.

The average active power in watts in a sinusoidally excited system is given by

$$P_{\text{av}} = EI\cos\theta \tag{13–95}$$

where E is the rms magnitude of the applied voltage in volts
I is the rms magnitude of the current in amperes
θ is the angle between the phasors $\boldsymbol{E}$ and $\boldsymbol{I}$, known as the power factor angle

From this, the *power factor* of the system is defined as

$$\text{pf} = \cos\theta \tag{13–96}$$

In a system with d-c excitation the steady-state product of E and I always yields power in watts. For sinusoidal excitation, however, the product of current, voltage, *and* power factor is required to yield the power in watts. The power factor may vary between zero and unity since θ may be only between $\pm 90°$ for passive circuit elements.

Consider an inductive load as shown in Fig. 13–9*a*. The current will lag the voltage, as represented in the phasor diagram of Fig. 13–9*b*.

The current in the load referred to the *load* voltage is

$$\boldsymbol{I} = I_1 - jI_2 = I(\cos\theta - j\sin\theta) \tag{13–97}$$

I_1 is the component of current in phase with $\mathbf{E}$. The product of E and I_1 is real power since

$$P_{\text{av}} = EI \cos \theta = EI_1 \tag{13–98}$$

I_2 is the reactive or quadrature component of current giving rise to the reactive power since

$$Q_{\text{av}} = EI \sin \theta = EI_2 \tag{13–99}$$

The load of Fig 13–9a is supplied through a distribution system simulated by R. The power loss in the supply system is proportional to the square of I

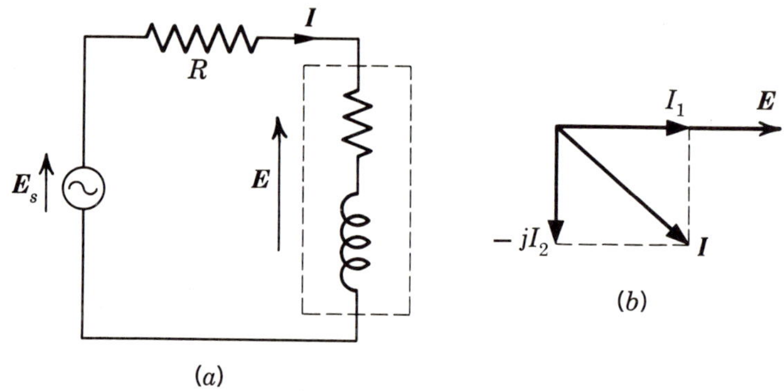

Fig. 13–9. Inductive load and phasor diagram.

since the power dissipated in R does no useful work at the load. Maximum efficiency for the system as a whole will be secured for a minimum value of I. If E is fixed, this corresponds to $\cos \theta = 1$ or unity power factor.

Since I_2 is a component of current existing by virtue of the magnetic field (inductive) properties of the load, it is logical to consider the addition of a capacitance at the load to partially or completely cancel I_2. This is possible since it has been shown that the phase relation between capacitive and inductive current components is 180° (cancellation).

If complete or partial cancellation of I_2 is desired, a capacitor may be connected across the load as shown in Fig. 13–10a. The capacitor current will appear as the phasor $\mathbf{I}_C$ in Fig. 13–10b.

For complete power factor correction (to pf = 1)

$$I_C = I_2 \tag{13–100}$$

But

$$I_C = \frac{E}{X_C} \tag{13–101}$$

or

$$I \sin \theta = I_C = E\omega C \tag{13–102}$$

Therefore, the value of C in farads required will be

$$C = \frac{I \sin \theta}{\omega E} \tag{13–103}$$

The new line current I_L then becomes

$$I_L = I_1 - jI_2 + jI_C = I_1 \tag{13–104}$$

Power factor correction is possible since the reactive power required by the inductive load may be supplied by reactive power from the capacitor. After a steady-state condition is reached, reactive power is no longer obtained from the source, but merely interchanged between electric and magnetic fields at the load.

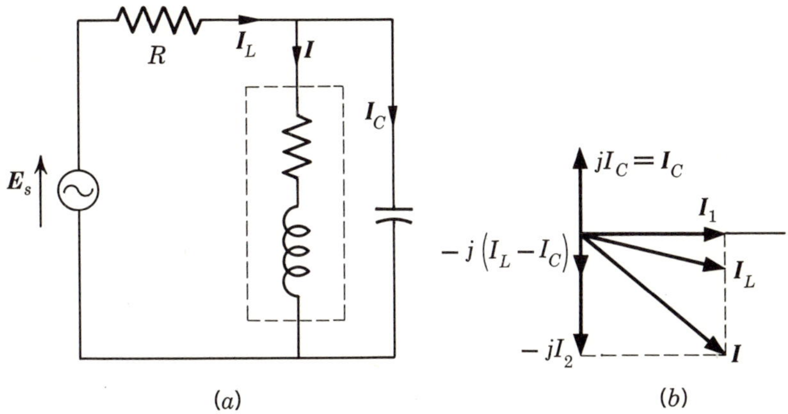

Fig. 13–10. Power factor correction.

Power factor correction may be complete as illustrated, or partial if only a portion of the needed reactive power is supplied by a capacitor. The amount of correction is usually determined by economic factors; that is, the balancing of capacitor cost against the cost of energy lost in the distribution system.

13–9. Maximum Power Transfer. The preceding section introduced the concept of a load whose energy was supplied through a distribution system. This is a general situation and it is of interest to investigate the conditions under which a load will receive maximum power.

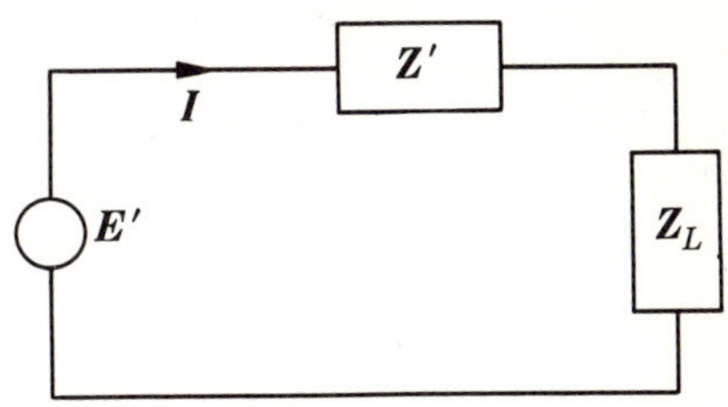

Fig. 13–11. Equivalent circuit for maximum power transfer study.

Any network containing passive impedances and energy sources may be represented by a Thévenin equivalent circuit of the form shown in Fig. 13–11. In this figure, Z' is an equivalent source and system impedance, Z_L is the impedance of the load, E' is an equivalent source voltage, and I is the current in the loop.

The most common situation involves a fixed value of $\mathbf{Z}'$ and a variable $\mathbf{Z}_L$. Three cases are of interest for this situation.

Case 1. $\mathbf{Z}'$ and $\mathbf{Z}_L$ are pure resistances, or

$$\mathbf{Z}' = R' \tag{13–105}$$

and

$$\mathbf{Z}_L = R_L \tag{13–106}$$

The current then becomes

$$\mathbf{I} = \frac{\mathbf{E}'}{R_L + R'} \tag{13–107}$$

and the average power supplied to R_L is

$$P = I^2 R_L = \frac{(E')^2 R_L}{(R_L + R')^2} \tag{13–108}$$

The value of R_L for maximum power delivered may be obtained by differentiating Eq. 13–108 with respect to R_L and setting the result equal to zero, or

$$\frac{dP}{dR_L} = (E')^2 \frac{[(R' + R_L)^2 - 2R_L(R' + R_L)]}{(R' + R_L)^4} = 0 \tag{13–109}$$

From this,

$$(R' + R_L)^2 = 2R_L(R' + R_L) \tag{13–110}$$

and

$$R_L = R' \tag{13–111}$$

for maximum power transfer.

If the load on a circuit and the circuit elements are pure resistances, the maximum power will be delivered to the load when the load resistance is equal to the equivalent resistance of the system.

Case 2. $\mathbf{Z}'$ and $\mathbf{Z}_L$ are general impedances containing resistance and reactance, or

$$\mathbf{Z}' = R' + jX' \tag{13–112}$$

and

$$\mathbf{Z}_L = R_L + jX_L \tag{13–113}$$

The magnitude of the current now becomes

$$I = \frac{E'}{\sqrt{(R' + R_L)^2 + (X' + X_L)^2}} \tag{13–114}$$

The average active power in the load is

$$P = I^2 R_L \tag{13–115}$$

and it will be a maximum if I is a maximum. From Eq. 13–114, I will be a maximum if the denominator is a minimum, or

$$(X' + X_L) = 0 \tag{13–116}$$

or

$$X_L = -X' \tag{13–117}$$

If Eq. 13–117 is satisfied, the power in the load is again expressed by Eq. 13–108 and the maximum power there was obtained for $R_L = R'$.

Maximum power is obtained in the load if the general load impedance is the conjugate of the equivalent impedance of the system.

Case 3. $\mathbf{Z}'$ and $\mathbf{Z}_L$ are again general impedances. The magnitude and angle of $\mathbf{Z}'$ are fixed, as is the phase angle θ_L of the load. Only the magnitude of $\mathbf{Z}_L$ may be varied. The expression for load power then becomes

$$P = \frac{(E')^2 Z_L \cos\theta_L}{(R' + Z_L \cos\theta_L)^2 + (X' + Z_L \sin\theta_L)^2} \tag{13–118}$$

since

$$X_L = Z_L \sin\theta_L \tag{13–119}$$

and

$$R_L = Z_L \cos\theta_L \tag{13–120}$$

Differentiating Eq. 13–118 with respect to Z_L and equating to zero as follows

$$\frac{dP}{dR_L} = \frac{[(R' + Z_L\cos\theta_L)^2 + (X' + Z_L\sin\theta_L)^2](E')^2\cos\theta_L}{[(R' + Z_L\cos\theta_L)^2 + (X' + Z_L\sin\theta_L)^2]^2}$$
$$- \frac{(E')^2 Z_L\cos\theta_L[2(R' + Z_L\cos\theta_L)\cos\theta_L + 2(X' + Z_L\sin\theta_L)\sin\theta_L]}{[(R' + Z_L\cos\theta_L)^2 + (X' + Z_L\sin\theta_L)^2]^2} = 0 \tag{13–121}$$

yields

$$(R' + R_L)^2 + (X' + X_L)^2 = 2(R'R_L + R_L{}^2 + X'X_L + X_L{}^2) \tag{13–122}$$

or

$$R_L{}^2 + X_L{}^2 = (R')^2 + (X')^2 \tag{13–123}$$

which shows that

$$Z_L = Z' \tag{13–124}$$

for maximum power transfer. If only the magnitude of $\mathbf{Z}_L$ can be varied, the magnitude of $\mathbf{Z}_L$ should equal the magnitude of $\mathbf{Z}'$ for maximum power transfer.

PROBLEMS

13–1. Calculate the effective and average values of the waveforms illustrated on p. 196.

13–2. Consider the R-C parallel network shown in the figure on p. 197. (a) Determine an expression for the instantaneous power in the network and plot against time. (b) What is the average active power and the average reactive power? (c) Draw the vector diagram relating the active, reactive, and vector power.

13–3. Consider the R-L-C network shown on p. 197. (a) Determine $\mathbf{I}$. (b) Calculate the power values using $EI\cos\theta$ and $EI\sin\theta$ and compare with $\mathbf{E}^*\mathbf{I}$. (c) Draw the vector power diagram.

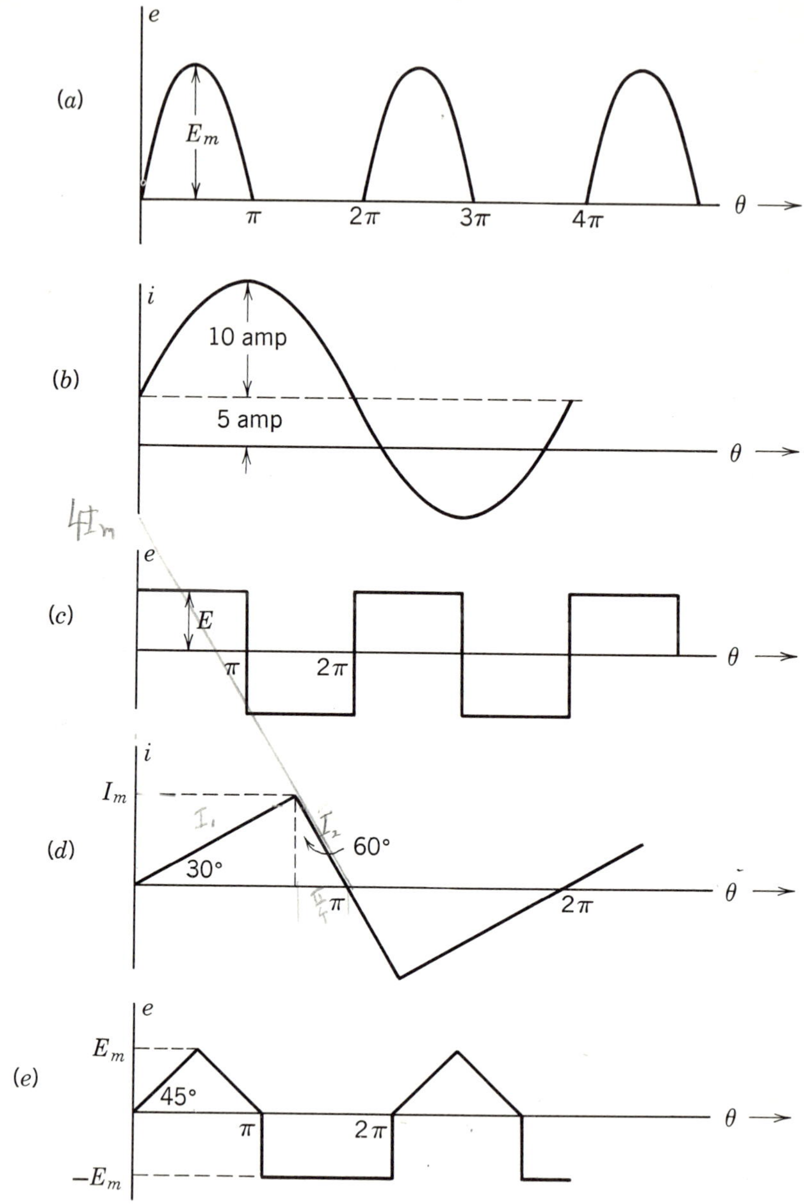
(a)
e
E_m
π
2π
3π
4π
θ →
(b)
i
10 amp
5 amp
θ →
(c)
e
E
π
2π
θ →
(d)
i
I_m
30°
60°
π
2π
θ →
(e)
e
E_m
45°
π
2π
−E_m
θ →

Prob. 13-1.

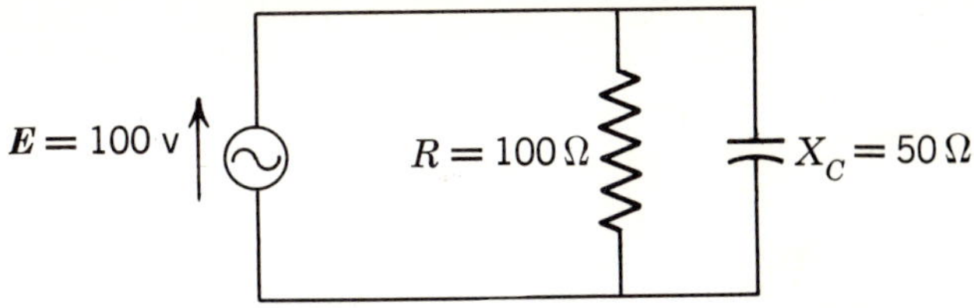

Prob. 13–2.

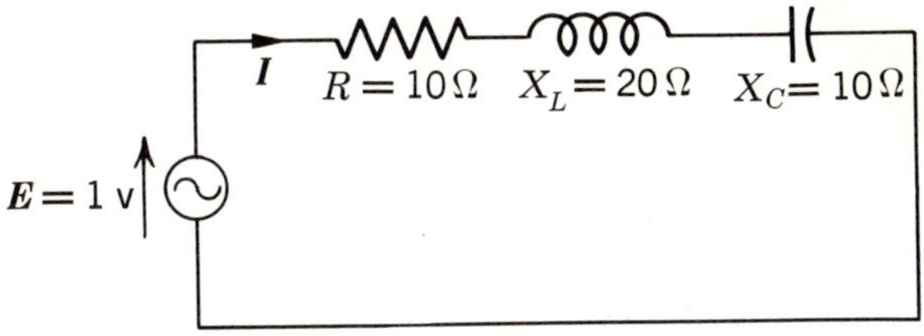

Prob. 13–3.

13–4. For a given network, $\boldsymbol{E} = 10\angle 0°$ and $\boldsymbol{I} = 1\angle 60°$. Determine **Z**, **Y**, and the active and reactive power in the network.

13–5. A network consumes 100 w and −60 vars of power when connected to 100 v at 60 cps. Determine **Z**, **Y**, and the component values for an equivalent series and equivalent parallel circuit representing the network.

13–6. A 2300-v electrical load draws 5000 kw and −3000 kvars. Compute the vars of capacitive reactance needed and the actual capacitor sizes if the power factor is to be corrected to (a) 1.0, (b) 0.9, (c) 0.8, and (d) 0.7.

13–7. An inductive load of 25 kw and 0.8 power factor is supplied at 440 v from two sources in parallel. If the source currents are equal and one source has unity power factor, what is the power factor of the other source and what is the real power supplied by each source?

13–8. Two loads in parallel draw respectively 1000 w at −0.8 power factor and 2000 w at +0.9 power factor. What is their combined power factor and how much reactive power must be added to secure unity power factor?

13–9. A battery having an internal impedance of 2.0 ohms and an emf of 48 v is connected to a resistive load by 20 miles of No. 14 AWG copper wire. Calculate and plot the value of the power delivered versus the value of load resistance. Show that the maximum power at the load is obtained for conditions satisfying the maximum power transfer conditions.

13–10. A fixed load of value $100\angle 30°$ ohms is supplied through a source impedance of variable magnitude and angle from a voltage source of value 200 v. Determine the value of the source impedance for maximum power in the load and calculate the load power for this condition.

14

REVIEW OF THE IMPEDANCE CONCEPT

In the preceding study of the response of circuits to steady-state sinusoids it was found that an appreciable simplification could be achieved by the introduction of complex notation to *represent* sinusoidal time functions. As a result of this simplification it became possible to utilize the concept of *impedance* as a means of describing the ratio of voltage to current. In this way a concise and complete technique for the computation of the steady-state sinusoidal response of any circuit was arrived at.

It is now appropriate to examine ways of extending these ideas to make them even more useful in the interpretation of both steady-state response and transient response. To ensure an adequate background for this extension, this chapter will review the impedance concept and express it in a slightly different form.

14–1. Complex Representation of Sinusoids. The basis for the complex representation of sinusoids is the identities which relate sines and cosines to exponential functions. These identities are

$$\cos \omega t = \tfrac{1}{2}(\epsilon^{j\omega t} + \epsilon^{-j\omega t}) \tag{14–1}$$

$$\sin \omega t = \frac{1}{2j}(\epsilon^{j\omega t} - \epsilon^{-j\omega t}) \tag{14–2}$$

If a voltage is given as

$$e(t) = E_m \cos (\omega t + \psi) \tag{14–3}$$

it may be written as

$$\begin{aligned} e(t) &= \tfrac{1}{2}E_m[\epsilon^{j(\omega t+\psi)} + \epsilon^{-j(\omega t+\psi)}] \\ &= \tfrac{1}{2}E_m(\epsilon^{j\psi}\epsilon^{j\omega t} + \epsilon^{-j\psi}\epsilon^{-j\omega t}) \end{aligned} \tag{14–4}$$

The *complex voltage amplitude* may be defined as

$$\mathbf{E}_m = E_m\epsilon^{j\psi} \tag{14–5}$$

and its conjugate as

$$\mathbf{E}_m{}^* = E_m\epsilon^{-j\psi} \tag{14–6}$$

This complex voltage amplitude is simply a designation of the magnitude and phase (*with respect to* $\cos \omega t$) of the actual sinusoidal time function. Because it is *not* a function of time it may be represented graphically (as a phasor) and may be used as a compact representation for the actual time function for purposes of computation. It is important, however, not to lose sight of the fact that $\boldsymbol{E}_m$ is not the actual voltage but only a convenient *representation* of it. This application has already been discussed in detail.

The actual voltage, as a time function, may now be written in terms of $\boldsymbol{E}_m$ and its conjugate as

$$e(t) = \tfrac{1}{2}(\boldsymbol{E}_m \epsilon^{j\omega t} + \boldsymbol{E}_m{}^* \epsilon^{-j\omega t}) \tag{14–7}$$

The significance of returning to this exact representation for the time function will become apparent in the succeeding work.

14–2. Introduction of the Impedance Concept. A simple example will serve to review the manner in which the concept of impedance became associated with the complex representation of sinusoids. Consider the series R-L circuit shown in Fig. 14–1, to which a sinusoidal steady-state voltage is applied. The voltage equilibrium equation which applies in this case is

$$Ri + L\frac{di}{dt} = e(t) \tag{14–8}$$

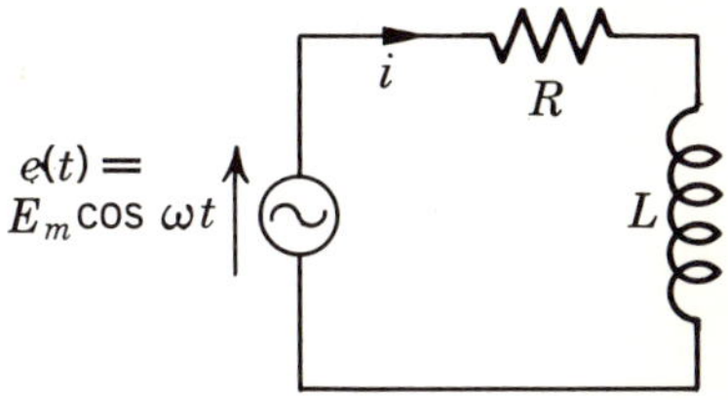

Fig. 14–1. *R-L* series circuit—sinusoidal voltage applied.

When this same problem was previously considered, the solution to the differential equation was obtained by substituting a sinusoidal function into Eq. 14–8 and determining the necessary coefficients and phase angles. This time, however, the exponential form of the voltage will be used. The applied voltage is now represented by

$$e(t) = \tfrac{1}{2}(\boldsymbol{E}_m \epsilon^{j\omega t} + \boldsymbol{E}^* \epsilon^{-j\omega t}) \tag{14–9}$$

where if a phase angle of zero is assumed

$$\boldsymbol{E}_m = E_m \epsilon^{j0} = E_m$$

and

$$\boldsymbol{E}_m{}^* = E_m \epsilon^{-j0} = E_m$$

The principle of superposition may be applied here by determining the solution to each part of the voltage of Eq. 14–9 and then combining results to obtain the total solution.

Let us consider the $+j\omega$ term first (and temporarily ignore the $\frac{1}{2}$). Equation 14–8 may then be rewritten as

$$Ri_a + L\frac{di_a}{dt} = \boldsymbol{E}_m \epsilon^{j\omega t} \tag{14–10}$$

where i_a is the component of current due to the first voltage term only. By analogy to previous results it may be expected that a function of the form

$$i_a = \boldsymbol{I}_m \epsilon^{j\omega t} \tag{14–11}$$

will be a solution to the differential equation, and this may be verified by direct substitution as follows:

$$R\boldsymbol{I}_m \epsilon^{j\omega t} + j\omega L\boldsymbol{I}_m \epsilon^{j\omega t} = \boldsymbol{E}_m \epsilon^{j\omega t} \tag{14–12}$$

From this

$$I_m = \frac{E_m}{R + j\omega L} \tag{14-13}$$

is a non-trivial solution. The complex quantity $(R + j\omega L)$ was designated as the *impedance* of the circuit and led to the form

$$I_m = \frac{E_m}{Z} \tag{14-14}$$

where

$$Z = R + j\omega L$$

This result is in complete agreement with the one previously obtained.

Now we consider the $-j\omega$ term, and rewrite Eq. 14–8 as

$$Ri_b + L\frac{di_b}{dt} = E_m{}^*\epsilon^{-j\omega t} \tag{14-15}$$

where i_b is the component of current due to the second voltage term only. A function of the form

$$i_b = I_m{}^*\epsilon^{-j\omega t} \tag{14-16}$$

may be expected to be a solution. Substitution of Eq. 14–16 into Eq. 14–15 yields

$$RI_m{}^*\epsilon^{-j\omega t} - j\omega LI_m{}^*\epsilon^{-j\omega t} = E_m{}^*\epsilon^{-j\omega t} \tag{14-17}$$

from which

$$I_m{}^* = \frac{E_m{}^*}{R - j\omega L} \tag{14-18}$$

Examination of Eq. 14–18 reveals that $I_m{}^*$ is the conjugate of I_m and may be written as

$$I_m{}^* = \frac{E_m{}^*}{Z^*} \tag{14-19}$$

where

$$Z^* = R - j\omega L$$

The total current (as a function of time) may now be written as

$$\begin{aligned} i &= \tfrac{1}{2}(i_a + i_b) \\ &= \tfrac{1}{2}(I_m\epsilon^{j\omega t} + I_m{}^*\epsilon^{-j\omega t}) \\ &= I_m \cos(\omega t + \phi) \end{aligned} \tag{14-20}$$

where

$$I_m = I_m\epsilon^{j\phi} = \frac{E_m}{Z} = \frac{E_m}{R + j\omega L}$$

and

$$I_m{}^* = I_m\epsilon^{-j\phi} = \frac{E_m{}^*}{Z^*} = \frac{E_m{}^*}{R - j\omega L}$$

If $\boldsymbol{Z}$ is written

$$\boldsymbol{Z} = R + j\omega L = Z\epsilon^{j\theta} = \sqrt{R^2 + \omega^2 L^2} \exp\left(j \tan^{-1} \frac{\omega L}{R}\right) \qquad (14\text{–}21)$$

it is apparent that

$$I_m = \frac{E_m}{\sqrt{R^2 + \omega^2 L^2}} \qquad (14\text{–}22)$$

and

$$\phi = -\tan^{-1} \frac{\omega L}{R} \qquad (14\text{–}23)$$

Again, these results are seen to be entirely consistent with those previously obtained.

14-3. Impedance in Terms of Time Function. An additional interpretation of impedance may be made by considering the two components of the voltage and current expressions. From Eq. 14-9 the first component may be written as $\frac{1}{2}\boldsymbol{E}_m\epsilon^{j\omega t}$ and designated the *positive-frequency* term since the exponential contains $(+j\omega)$. Likewise, the second term is $\frac{1}{2}\boldsymbol{E}_m{}^*\epsilon^{-j\omega t}$ and may be designated as the *negative-frequency* term since the exponential contains $(-j\omega)$. The positive- and negative-frequency terms of the current may be designated similarly. Now, considering the ratio of the positive-frequency voltage to the positive-frequency current, we find that this ratio is

$$\frac{\frac{1}{2}\boldsymbol{E}_m\epsilon^{j\omega t}}{\frac{1}{2}\boldsymbol{I}_m\epsilon^{j\omega t}} = \frac{\boldsymbol{E}_m}{\boldsymbol{I}_m} = \boldsymbol{Z} \qquad (14\text{–}24)$$

Likewise, the ratio of negative-frequency components is

$$\frac{\frac{1}{2}\boldsymbol{E}_m{}^*\epsilon^{-j\omega t}}{\frac{1}{2}\boldsymbol{I}_m{}^*\epsilon^{-j\omega t}} = \frac{\boldsymbol{E}_m{}^*}{\boldsymbol{I}_m{}^*} = \boldsymbol{Z}^* \qquad (14\text{–}25)$$

Thus, it may be seen that the impedance $\boldsymbol{Z}$ is simply the ratio of positive-frequency components of voltage and current at *every instant of time*. The conjugate impedance $\boldsymbol{Z}^*$ is the ratio of negative-frequency components at every instant of time.

Although the discussion so far has been restricted to an impedance consisting of resistance and inductance only, it is not difficult to extend this to include circuits with capacitance also. It will be recalled that the volt-ampere characteristic for capacitance is

$$i = C \frac{dv}{dt} \qquad (14\text{–}26)$$

If the positive-frequency component of a sinusoidal voltage is considered, then this may be written as

$$v = e = \boldsymbol{E}_m\epsilon^{j\omega t} \qquad (14\text{–}27)$$

and

$$i = C \frac{d(\boldsymbol{E}_m\epsilon^{j\omega t})}{dt} = j\omega C\boldsymbol{E}_m\epsilon^{j\omega t} \qquad (14\text{–}28)$$

Since the impedance has been shown to be the instantaneous ratio of positive-frequency components, the impedance of a single capacitor becomes

$$\boldsymbol{Z} = \frac{e}{i} = \frac{\boldsymbol{E}_m\epsilon^{j\omega t}}{j\omega C\boldsymbol{E}_m\epsilon^{j\omega t}} = \frac{1}{j\omega C} \tag{14–29}$$

which agrees with previous results.

The preceding discussion may be summarized by reiterating the *two* ways in which earlier concepts have been extended:

1. The complex representation of sinusoids has been shown to be more than just a convenient representation of the magnitude and phase. By combining it with its conjugate, it becomes an exact mathematical representation at every instant of time.
2. The concept of impedance has been extended to become more than a complex relationship between the magnitudes and phases of the voltage and current. It has further been shown to be the *instantaneous* ratio of positive-frequency components of voltage and current. Similarly, the conjugate impedance is the instantaneous ratio of the negative-frequency components.

14–4. Dual Impedances. The concept of duality was briefly mentioned in Chapter 7 but no extensive discussion of it has yet been presented. In this chapter the concept will again be introduced briefly and then further elaborated on in succeeding chapters.

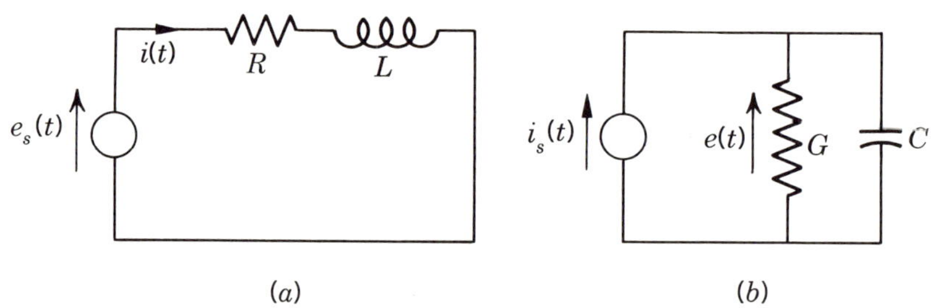

Fig. 14–2. An example of dual circuits.

Duality is both a geometrical and a mathematical relationship between two circuits. It is perhaps best illustrated by means of an example of dual circuits. For this purpose consider the two circuits shown in Fig. 14–2.

The voltage equilibrium equation for the circuit of Fig. 14–2*a* is

$$Ri + L\frac{di}{dt} = e_s(t) \tag{14–30}$$

Likewise, the current equilibrium equation for the circuit of Fig. 14–2*b* is

$$Ge + C\frac{de}{dt} = i_s(t) \tag{14–31}$$

Examination of these two equations reveals that they have the same mathematical form (hence, can be expected to have the same form of solution). However, there has been an interchange of elements and variables. In particular, the identifications in Table 14–1 may be made.

TABLE 14–1
Interchange of Elements and Variables for Fig. 14–2

Circuit a	Circuit b
Resistance R	Conductance G
Inductance L	Capacitance C
Voltage source $e_s(t)$	Current source $i_s(t)$
Current response $i(t)$	Voltage response $e(t)$
Series connection	Parallel connection
Voltage equilibrium	Current equilibrium

Two circuits which can be related in this way are designated as duals. Duality is a mutual and unique relationship. Given either circuit, the other one can be obtained uniquely. Note that it is necessary for the numerical values to be uniquely related as well as the form of circuit and sources. Hence, in the example just discussed, it is necessary that the conductance in mhos be numerically equal to the resistance in ohms and that the capacitance in farads be numerically equal to the inductance in henrys.

Duality may also be applied to impedances and admittances apart from any source. In Fig. 14–2a the *impedance* of the circuit can be shown to be

$$Z = R + j\omega L \tag{14–21}$$

In the circuit of Fig. 14–2b the *admittance* is

$$Y = G + j\omega C \tag{14–32}$$

and this is seen to be the dual of the impedance.

The practical application of duality lies in our being able to apply the form of solution to any circuit problem to *two* different circuits and thereby avoid the necessity of carrying out the solution for both. However, the main reason for emphasizing the concept of duality is not its practical application but the added understanding and insight to circuit behavior that is gained by constant awareness of this unifying concept.

PROBLEMS

14–1. Express each of the following trigonometric functions in exponential form, that is, functions of the form of Eq. 14–7. Determine the complex voltage (or current) amplitude for each case.

(a) $e = 100 \cos 1000t$ (b) $e = 10 \sin 100t$
(c) $e = 50 \cos (10t + 60°)$ (d) $i = 0.1 \sin (1000t + 30°)$
(e) $i = 2 \sin (100t + 120°)$ (f) $e = 5 \cos (100t - 135°)$

14–2. Verify the following expressions by writing the functions in exponential form and then performing the indicated operations.

(a) $\frac{d}{dt}\cos\omega t = -\omega \sin \omega t$ (b) $\frac{d}{dt}\sin\omega t = \omega \cos\omega t$

(c) $\int \sin \omega t = -\frac{1}{\omega}\cos\omega t$ (d) $\int \cos\omega t = \frac{1}{\omega}\sin\omega t$

(e) $A\cos\omega t + B\sin\omega t = \sqrt{A^2 + B^2}\cos\left(\omega t - \tan^{-1}\frac{B}{A}\right)$

(f) $(A\cos\omega t)(B\sin\omega t) = \frac{AB}{2}\sin 2\omega t$

(g) $(A\cos\omega t)[B\cos(\omega t - \theta)]^2 = \frac{AB}{2}\cos\theta + \frac{AB}{2}\cos(2\omega t - \theta)$

14–3. Determine the current of the figure as a function of time (in trigonometric form) by expressing the voltage as a sum of exponentials, determining the current for each exponential term considered separately, and combining these by use of the principle of superposition. Sketch a phasor diagram showing the complex amplitudes of voltage and current.

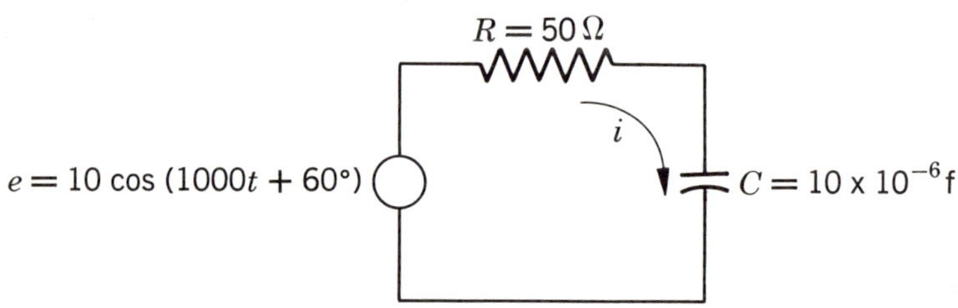

Prob. 14–3.

14–4. Determine e, i_1, i_2, and i_3 of the figure (in trigonometric form) by the use of exponential representation of voltages and currents. Draw a phasor diagram showing the complex amplitudes of $\boldsymbol{I}_s$, $\boldsymbol{E}$, $\boldsymbol{I}_1$, $\boldsymbol{I}_2$, and $\boldsymbol{I}_3$.

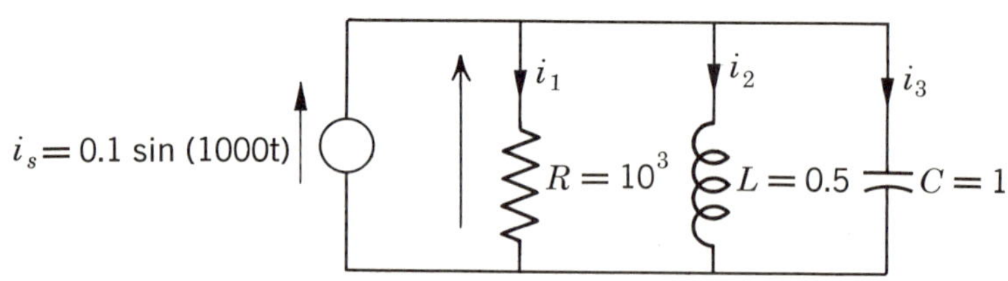

Prob. 14–4.

14–5. Use the results of Probs. 14–3 and 14–4 to demonstrate that it is only necessary to know the positive-frequency component of voltage or current in order to write the equation for the quantity in trigonometric form.

14–6. Determine the impedance or admittance (as indicated) for each of the circuits illustrated. The value of ω to be used is given for each circuit.

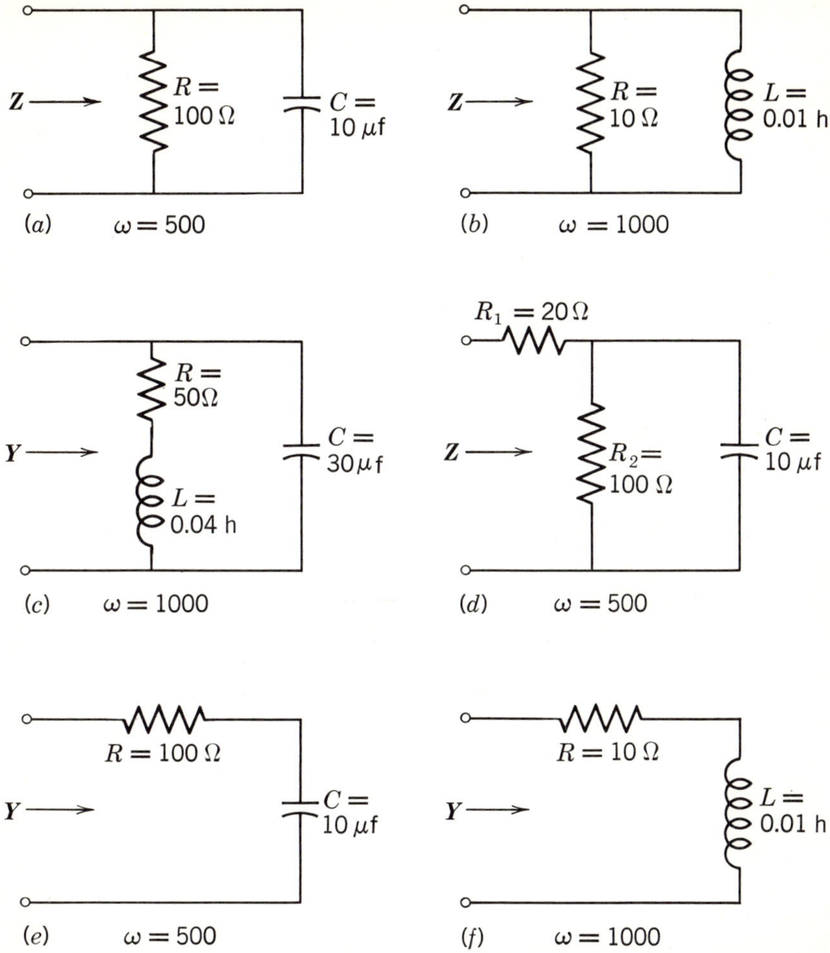

Prob. 14–6.

14–7. Determine the duals of the circuits of parts (a), (c), and (d) of Prob. 14–6.

14–8. Determine $i_1(t)$, $i_2(t)$, $i_3(t)$, $e_1(t)$, and $e_2(t)$ of the figure. Draw a phasor diagram showing the complex amplitudes of these quantities.

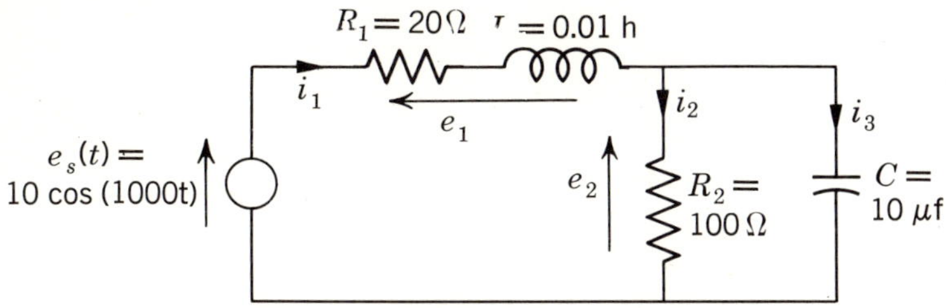

Prob. 14–8.

14–9. Determine $i_1(t)$, $i_2(t)$, $i_3(t)$, $e_1(t)$, and $e_2(t)$ of the figure. Draw a phasor diagram showing the complex amplitude of these quantities.

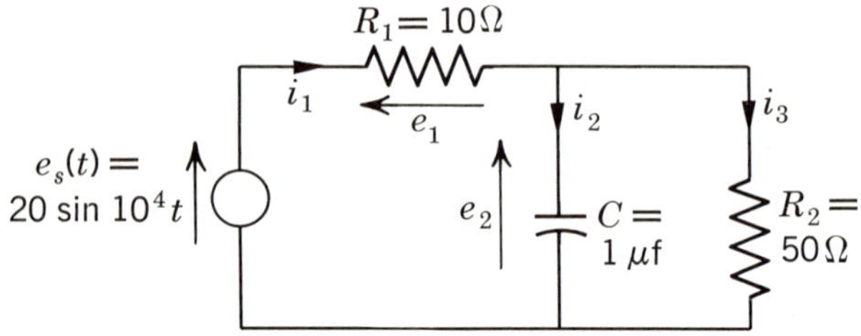

Prob. 14–9.

15

THE COMPLEX FREQUENCY CONCEPT

In the study of sinusoidal functions the term "frequency" was considered to mean the number of cycles that the sine wave completed in 1 sec. When used in this sense, the frequency will, of course, be a *positive, real number.* However, if these ideas are to be extended to make them useful in applications other than the sinusoidal steady state, then this concept is too restricted and it becomes necessary to think of frequency in more general terms. Fortunately, it is not necessary to sacrifice anything in the way of physical interpretation in order to make this generalization. Instead, the more general concept unifies a number of previously isolated ideas and thus leads to a more satisfying physical picture than before.

15-1. An Extended Definition of Frequency. Any concept of frequency which is to be really useful in general circuit analysis should be applicable to practically any time function rather than just to sinusoids. Since most time functions do not oscillate, as the sinusoid does, it is clear that any definition of frequency which is to be general cannot be phrased in terms of "cycles per second" but must be based upon some other characteristic of the time function which is more general. One such characteristic might be the slope or rate of change since any physically realizable time function must have a unique time derivative. However, the rate of change cannot be directly associated with frequency because it does not have the proper dimensions. Sinusoidal frequency has dimensions of 1 per second (*not* cycles per second) and any other concept of frequency which is to be consistent with this must have the same dimensions. If the time function being considered is a voltage, the rate of change of this time function has dimensions of volts per second, which is not satisfactory for frequency.

On the basis of the preceding argument it appears that it might be reasonable to consider the *ratio* of the rate of change of a function to the value of the function itself as a definition of frequency since this ratio does have the dimensions of 1 per second. Thus, a generalized frequency may be defined for any time function $e(t)$ as

$$s = \frac{de/dt}{e} \tag{15-1}$$

and this definition compared with previous ideas.

Although it might appear logical to first examine the sinusoidal function in terms of the definition of Eq. 15–1, this is not advisable for reasons that will shortly become apparent. Instead, the time function for which s is a constant, independent of time, will be determined. It is certainly true that the sinusoidal frequency is such a constant when defined as shown by Eqs. 14–1 and 14–2 and it is also true that not all time functions, when substituted into Eq. 15–1, will yield a constant. As a matter of fact, there is only one class of time function which will yield a constant value of s and this function may be obtained by means of a straightforward solution of Eq. 15–1. Rewriting this as

$$\frac{de}{e} = s\,dt \tag{15–2}$$

integration immediately yields

$$\ln e = st \tag{15–3}$$

from which

$$e = \epsilon^{st} \tag{15–4}$$

The exponential time function has been previously encountered in the study of electric circuits. For example, if a unit impulse of voltage is applied to a resistance R in series with an inductance L, the resulting current is found to be

$$i = \frac{1}{L}\,\epsilon^{-\frac{R}{L}t} \tag{15–5}$$

On the basis of the definition of Eqs. 15–1 and 15–4, this current would be said to have a frequency of

$$s = -\frac{R}{L} \tag{15–6}$$

which has dimensions of 1 per second. Similarly, if a unit step of voltage is applied to a resistance R in series with a capacitance C, the resulting current is

$$i = \frac{1}{R}\,\epsilon^{-\frac{1}{RC}t} \tag{15–7}$$

in which the frequency of the current is

$$s = -\frac{1}{RC} \tag{15–8}$$

which again has the dimensions of 1 per second.

It is not necessary, however, that the frequency s be a real number; it may be imaginary or complex. In order to consider this possibility, let us return to the sinusoidal function and relate it to the new definition. Assume a voltage of the form

$$e = E_m \cos \omega t \tag{15–9}$$

from which

$$\frac{de}{dt} = -\omega E_m \sin \omega t \tag{15–10}$$

Substitution of these values into Eq. 15–1 yields

$$s = -\frac{\omega E_m \sin \omega t}{E_m \cos \omega t} = -\omega \tan \omega t \tag{15–11}$$

which is *not* independent of time. In order to resolve this difficulty, we must recall that in Chapter 14 it was found convenient to express the cosine function as

$$E_m \cos \omega t = \tfrac{1}{2}E_m(\epsilon^{j\omega t} + \epsilon^{-j\omega t}) = e_a(t) + e_b(t) \tag{15–12}$$

in which the first term was arbitrarily designated as the positive-frequency term and the second as the negative-frequency term. If these terms are considered individually the *two* frequencies may be defined as

$$s_a = \frac{de_a/dt}{e_a} = \frac{j\omega(\frac{1}{2}E_m)\epsilon^{j\omega t}}{\frac{1}{2}E_m\epsilon^{j\omega t}} = j\omega \tag{15–13}$$

and

$$s_b = \frac{de_b/dt}{e_b} = \frac{-j\omega(\frac{1}{2}E_m)\epsilon^{-j\omega t}}{\frac{1}{2}E_m\epsilon^{-j\omega t}} = -j\omega \tag{15–14}$$

It may be seen that in terms of our new definition the cosine function contains not one but *two* frequency components, each independent of time, and related as conjugates. A similar result may be obtained for the sine function and this may be demonstrated by the student.

15–2. Complex Frequency Terminology. Before proceeding further with the discussion of complex frequencies, a possible source of confusion regarding terminology should be clarified. Originally, the term "frequency" was interpreted to mean the number of cycles per second of a sine wave and was designated by the symbol f. Most circuit calculations, however, involve the rate of change of angle and so the angular velocity, $\omega = 2\pi f$ (radians per second), was introduced. The ω used in the preceding discussion is exactly this quantity, hence, should properly be designated as an angular velocity rather than a frequency. Likewise, the real values of s obtained in Eqs. 15–6 and 15–8 should properly be designated as linear velocities measured in *nepers per second*. It is convenient, however, to have a single term which can be applied to the quantity s regardless of whether s is real, imaginary, or complex. The term which is most commonly used is "frequency," in spite of the fact that it is not accurate. It has the advantages of being simpler to use and more compact than any strictly correct designation and for these reasons its use has become so widespread as to be almost universal.

In accordance with this custom, the following policy will be observed here. *The quantity* s, *or any specific value of that quantity, will be designated as a frequency and carry dimensions of* 1 *per second regardless of whether* s *is real, imaginary, or complex.*

The term "frequency" will also be used to designate the cycles per second of a sinusoidal function but when used in this sense will be symbolized by f and carry units of cycles per seconds (cps).

15–3. Complex Values of s. To extend the application of s still further, consider a time function of the form

$$i = I_m \epsilon^{-\alpha t} \cos \omega t \tag{15–15}$$

Time functions very similar to this were obtained in previous discussions of the response of R-L-C circuits to singular functions. Writing this in exponential form yields

$$\begin{aligned} i &= I_m \epsilon^{-\alpha t}(\tfrac{1}{2})(\epsilon^{j\omega t} + \epsilon^{-j\omega t}) \\ &= \tfrac{1}{2} I_m (\epsilon^{-\alpha t}\epsilon^{j\omega t} + \epsilon^{-\alpha t}\epsilon^{-j\omega t}) \\ &= \tfrac{1}{2} I_m [\epsilon^{-(\alpha - j\omega)t} + \epsilon^{-(\alpha + j\omega)t}] \end{aligned} \tag{15–16}$$

This result contains two exponential terms from which the two frequency components are seen to be

$$\begin{aligned} s_1 &= -\alpha + j\omega \\ s_2 &= -\alpha - j\omega \end{aligned} \tag{15–17}$$

It may be observed that frequencies representing this time function are complex and are related as conjugates.

A significant point in connection with both complex and imaginary values of s is that they *must* appear in conjugate pairs in order to represent a physically realizable time function. A single complex frequency of the form

$$s = -\alpha + j\omega \tag{15–18}$$

would pertain to a time function of the form

$$\begin{aligned} f(t) &= \epsilon^{(-\alpha + j\omega)t} = \epsilon^{-\alpha t}\epsilon^{j\omega t} \\ &= \epsilon^{-\alpha t} \cos \omega t + j\epsilon^{-\alpha t} \sin \omega t \end{aligned} \tag{15–19}$$

Such a time function is complex and has *no* physical significance. *Time is always a real quantity and any function of time which can exist physically must likewise be real.* Therefore, whenever a complex frequency appears in a physical problem it must be accompanied by its conjugate in order that the imaginary parts of the time function representing each frequency may cancel and leave a purely real time function.

It should also be noted that a single frequency component may physically represent a time function if that frequency is purely real. For example, if

$$s = -\alpha \tag{15–20}$$

the corresponding time function is

$$f(t) = \epsilon^{-\alpha t} \tag{15–21}$$

which is a real time function and can exist physically. The frequency

$$s = +\alpha \tag{15–22}$$

which represents the time function

$$f(t) = \epsilon^{\alpha t} \tag{15–23}$$

is also a possibility but generally this time function does not appear in physical circuits because its value increases without limit as t increases. Hence, the limitation on this time function is a practical rather than a theoretical one.

15–4. Phasor Representation of Complex Frequencies. It was found convenient in the study of sinusoidal functions to represent such functions by means of phasors. A review of this procedure and its extension to complex frequencies is desirable as a further aid to understanding.

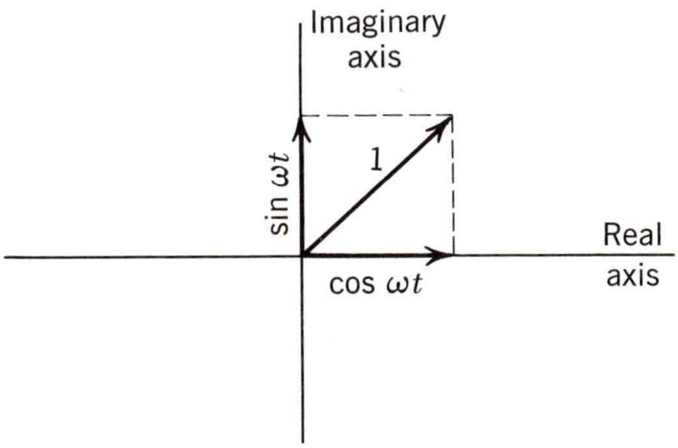

Fig. 15–1. Phasor representation of $\epsilon^{j\omega t}$.

First, consider the exponential

$$\epsilon^{j\omega t} = \cos \omega t + j \sin \omega t \tag{15–24}$$

and interpret it as a phasor set in a complex plane for any value of t. As shown in Fig. 15–1, the real component of this phasor is cos ωt and the imaginary, or quadrature, component is sin ωt. The length of this phasor is

$$\sqrt{\cos^2 \omega t + \sin^2 \omega t} = 1 \tag{15–25}$$

and the angle it makes with the positive real axis is

$$\tan^{-1}\left(\frac{\sin \omega t}{\cos \omega t}\right) = \tan^{-1}(\tan \omega t) = \omega t \tag{15–26}$$

Next consider a real, sinusoidal time function of the form

$$I_m \cos \omega t = \tfrac{1}{2}I_m(\epsilon^{j\omega t} + \epsilon^{-j\omega t}) \tag{15–27}$$

It may be seen that this time function may be interpreted as the *sum* of two phasors in the complex plane as shown in Fig. 15–2. As time increases the two phasors rotate in opposite directions with a constant angular velocity and with constant length. The resultant of these two phasors always lies on the real axis (as it must for real time functions) and has a length which is equal at every instant of time to the magnitude of the original time function.

Because the two phasors of Fig. 15–2 convey essentially the same information, it is common practice to draw only a single phasor, of length I_m instead of $\frac{1}{2}I_m$, and rather than considering it to rotate, fix its position at some arbitrary instant of time (usually $t = 0$). By representing all voltages and currents in a circuit in the same manner (using the same time reference, of course) the resulting phasor diagram gives all the essential information concerning the magnitude and phase of the various quantities but suppresses their time-varying characteristics.

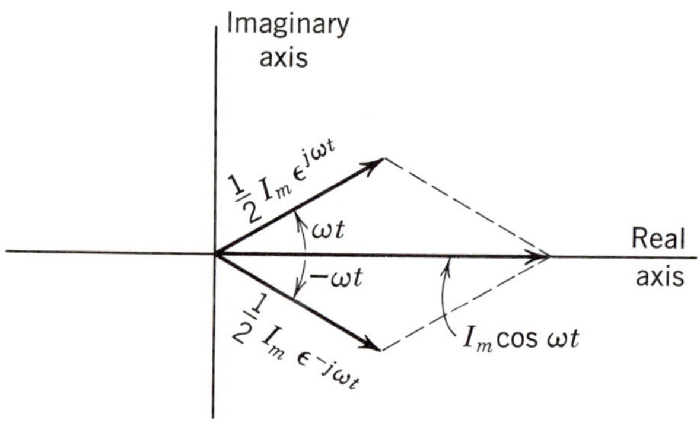

Fig. 15–2. Phasor representation of $I_m \cos \omega t$.

This review of rotating phasors may now be extended to consider the other time functions just discussed. In the case of a function of the form

$$I_m\epsilon^{-\alpha t} \cos \omega t = \tfrac{1}{2}I_m(\epsilon^{-\alpha t}\epsilon^{j\omega t} + \epsilon^{-\alpha t}\epsilon^{-j\omega t}) \tag{15–28}$$

it may be seen that each phasor not only rotates at a constant angular velocity but also decreases in length because of the factor $\epsilon^{-\alpha t}$. On the other hand, a function of the form

$$I_m\epsilon^{-\alpha t} = I_m\epsilon^{-\alpha t}\epsilon^{j(0)t} \tag{15–29}$$

changes in length but does not rotate since its angular velocity is zero. These results are summarized in Fig. 15–3, which shows the phasors and the resultant time functions for all three basic waveforms which have been discussed.

15–5. Other Time Functions. The discussion so far has indicated that the concept of complex frequencies can be applied directly to three different time functions. It has been stated, however, that a desirable attribute of any generalized frequency concept was that it be applicable to almost any time function, and there are certainly many more possible functions than these three.

It is not desirable, at this time, to enter into a complete discussion of how almost any time function may be represented in terms of its frequencies.

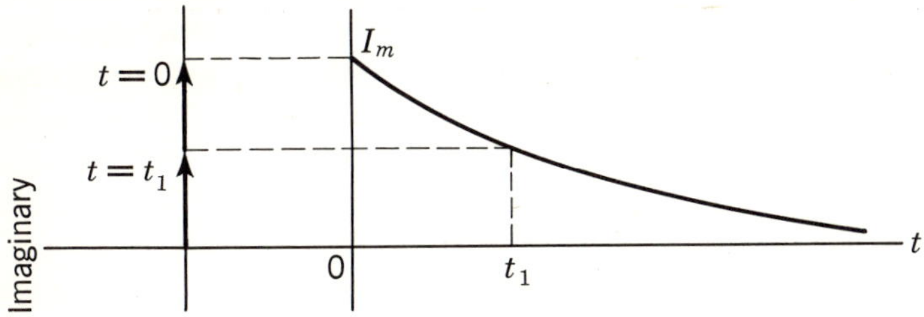

(a) $i(t) = I_m \epsilon^{-\alpha t}$
$s = -\alpha \pm j0$

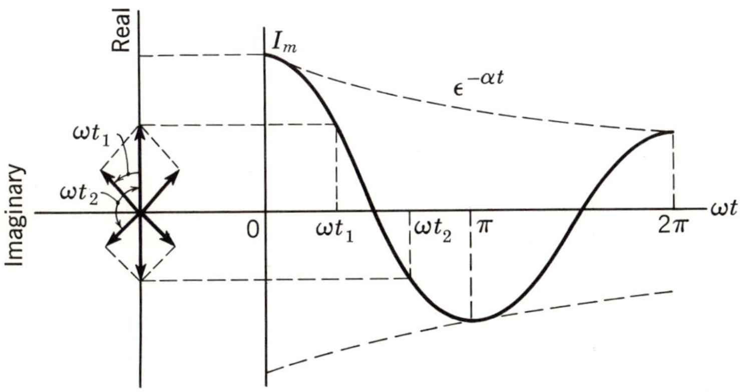

(b) $i(t) = I_m \epsilon^{-\alpha t} \cos \omega t$
$s = -\alpha \pm j\omega$

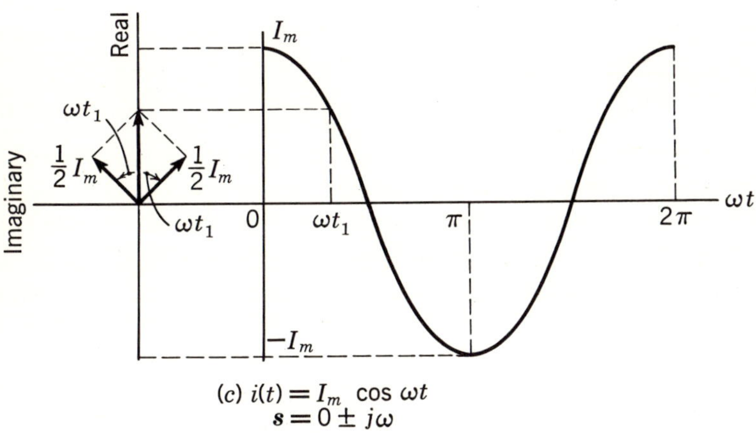

(c) $i(t) = I_m \cos \omega t$
$s = 0 \pm j\omega$

Fig. 15-3. Phasor representation of several time functions.

About all that can be done is to state that most time functions can be represented as the sum of a number of basic time functions of the form just discussed. For example, consider the time function shown in Fig. 15–4.

Although it is by no means obvious, this time function may be expressed as

$$i = -5\epsilon^{-t} + 15\epsilon^{-2t} - 10\epsilon^{-3t} \tag{15–30}$$

from which it may be seen that three real frequency components are present. These frequencies are

$$s_1 = -1, \qquad s_2 = -2, \qquad s_3 = -3 \tag{15–31}$$

In general, it may be stated that almost any time function may be created by combining any number of the three basic time functions that have been discussed with appropriate relative magnitudes. Further, it is possible to

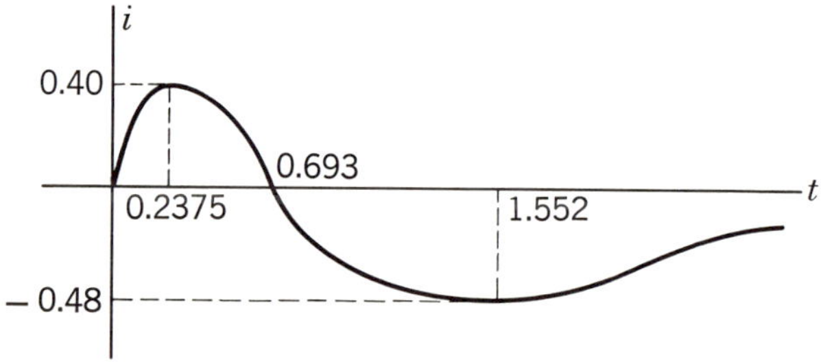

Fig. 15–4. Plot of $(-5\epsilon^{-t} + 15\epsilon^{-2t} - 10\epsilon^{-3t})$.

specify the time function uniquely, simply by specifying the frequency components and their relative amplitudes. For example, in the illustration just given, a statement to the effect that we have a frequency component of -1 with an amplitude of -5 amp, another frequency component of -2 and an amplitude of 15 amp, and a third frequency component of -3 and an amplitude of -10 amp is just as complete a description of the time function as is Eq. 15–30.

It may be shown that any time function may be specified in either of two equivalent ways. In the first way, it may be described by means of an equation which gives its magnitude at any instant of time. If this is done, the function is said to be specified in the *time domain*. Alternatively, the function may be described by giving its frequency components and their amplitudes. In this case, the function is said to be specified in the *frequency domain*. In advanced courses techniques are discussed whereby a description in either domain may be transformed into the other. To a large extent the ability of an electrical engineer to solve a wide range of circuit problems easily depends upon the facility with which he can interpret his problem in *either* the time domain or the frequency domain and transform from one to the other as the occasion demands.

PROBLEMS

15–1. The electrical quantities that are ordinarily used in circuit theory are commonly assigned symbols and units of measurement as in Table 15–1. The

TABLE 15–1
Units for Electrical Quantities

Quantity	Symbol	Unit
Charge	Q, q	Coulomb
Current	I, i	Ampere
Potential	V, v, E, e	Volt
Capacitance	C	Farad
Inductance	L	Henry
Magnetic flux linkage	λ	Weber-turn
Resistance	R	Ohm
Power	P	Watt
Energy	W	Joule or watt-second
Angular velocity	ω	Radian/second

dimensions of electrical quantities can be expressed in terms of four fundamental dimensions. One combination that is often used consists of mass, length, time, and charge (see Art. 1–3). It is often convenient to use potential, current, time, and length (with the units of the volt, the ampere, the second, and the meter respectively) as the fundamental dimensions. Determine the dimensions of the quantities listed in terms of the latter four dimensions.

15–2. Determine the dimensions of the following combinations of electrical quantities in terms of the four fundamental dimensions of Prob. 15–1.

(a) R/L (b) $1/RC$ (c) $1/LC$
(d) L/RC (e) E/L (f) I/C
(g) ωL (h) $1/\omega C$ (i) EC
(j) IL (k) EC/Q (l) IL/λ
(m) R^2C (n) $\sqrt{L/C}$

15–3. Show, by means of a sketch, how each of the following quantities varies with time. Determine the frequency components, with their amplitudes, of each quantity.

(a) $e = 5\epsilon^{-50t}$
(b) $e = 10 \cos (100t + 30°)$
(c) $e = 3 \cosh 10t$
(d) $e = 5\epsilon^{-100t} + 10 \sin 1000t + 20 \cos 1000t$
(e) $i = 0.1\epsilon^{-1000t} + 0.2\epsilon^{-2000t}$
(f) $i = 5 \sinh 100t$
(g) $e = 100 \sin (1000t - 45°)$
(h) $i = 2\epsilon^{-100t} \cos (10{,}000t + 60°)$
(i) $e = 2\epsilon^{-500t} + 10\epsilon^{-500t} \sin (10{,}000t + 45°)$

15–4. (a) Evaluate $\epsilon^{-(50/\pi)t}$ for $t = 0$, $t = \pi/200$, $t = 3\pi/400$, $t = \pi/100$, $t = 3\pi/200$, $t = 7\pi/200$, and $t = \pi/50$. (b) Plot the magnitude of the function versus time. (c) Represent this quantity as a phasor in a complex plane and show how it varies with time.

15–5. (a) Evaluate cos $(100t + 45°)$ for the values of t used in Prob. 15–4. (b) Plot the magnitude of the function versus time. (c) Consider this function to be an appropriate sum of two exponential terms, and represent these exponential terms as phasors in a complex plane, showing their positions at the values of t used in (a). (d) Verify the fact that the sum of the two phasors at any instant of time is a real quantity that is the magnitude of the original time function.

15–6. A time function $\epsilon^{-(50/\pi)t} \cos (100t + 45°)$ is given. (a) Evaluate this function for the values of t specified in Prob. 15–4. (b) Plot the magnitude of the function versus time. (c) Consider the function to be an appropriate sum of two exponential terms and represent these exponential terms as phasors in a complex plane, showing their positions at the values of t used in (a). Describe what happens to these phasors as t increases to ∞. (d) Verify the fact that the sum of the two phasors at any instant of time is a real quantity that is the magnitude of the original time function.

15–7. A certain current is described in the frequency domain as having frequency components of 0, $-1 - j1$, and $-1 + j1$ with amplitudes of $\frac{1}{2}$, $\epsilon^{-j\pi/2}$, and $\epsilon^{j\pi/2}$ respectively. Describe this current in the time domain and sketch a plot of its magnitude versus time.

15–8. Determine the complete solution for $i(t)$ of the figure and separate it into its two components, the particular integral (steady-state) and complementary function (transient). What frequencies does the voltage contain? What frequencies does the steady-state current contain? What frequencies does the transient current contain? What determines the frequencies of the steady-state and transient currents?

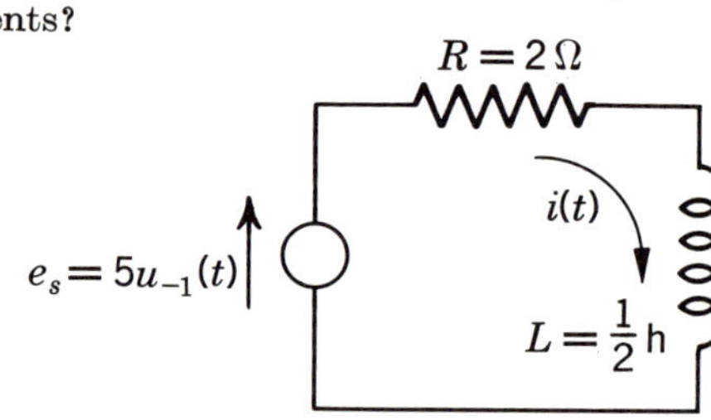

Prob. 15–8.

15–9. Determine $e(t)$ of the figure. What frequencies does the current i_s contain? What frequencies does the voltage contain? Does the voltage have any

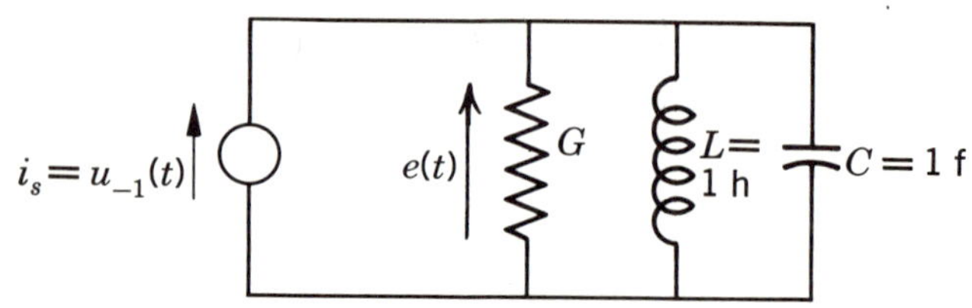

$G = 1$ mho

Prob. 15–9.

frequencies differing from those of the current? If so, what determines those frequencies?

15–10. Generalize on the results of Prob. 15–9 by considering the solution for $e(t)$ for any values of G, L, and C. Under what conditions will $e(t)$ contain complex frequencies? Under what conditions will $e(t)$ contain real frequencies only?

16

IMPEDANCE IN THE COMPLEX FREQUENCY PLANE

The preceding chapter introduced the concept of complex frequencies and attempted to illustrate their physical nature through the use of various time functions. This discussion emphasized the possibility of describing any time function in either the time domain or the frequency domain. This concept is extremely important and is considered in detail in advanced courses. For the present, however, the major interest lies in employing complex frequencies as an aid to the understanding of the steady-state sinusoidal behavior of circuits. In doing this it will be found useful to introduce the complex frequency plane and consider the graphical representation of *impedance* on this plane. By so doing, it becomes possible to describe, in a relatively simple way, the behavior of quite complex circuits and make evident by inspection relationships which could otherwise be obtained only as a result of tedious mathematical developments. The degree of circumspection which the complex frequency plane lends to a multitude of steady-state sinusoidal problems may be considered as one of its outstanding advantages.

16–1. Impedance in Terms of Complex Frequency. It is now appropriate to consider the representation of impedance in terms of complex frequencies and this will be done by combining certain ideas presented in Chapters 14 and 15. In Chapter 14 the impedance of circuits operating in the sinusoidal steady state was shown to be the ratio of positive-frequency components of voltage and current. In Chapter 15 it was shown that a steady-state sinusoid was composed of two frequency components, $\pm j\omega$, and represented a special case of a more general frequency which was defined as

$$s = \frac{di/dt}{i} \tag{16–1}$$

or

$$s = \frac{de/dt}{e} \tag{16–2}$$

for any time function $i(t)$ or $e(t)$.

This definition of impedance will be extended still further as follows:

The impedance of any linear element or circuit to a single frequency component is the ratio at every instant of time of the voltage at that frequency to the current at that frequency, where the frequencies considered may be real, imaginary, or complex.

The impedance of the three types of circuit elements in terms of the complex frequency s may now be obtained from their corresponding volt-ampere characteristics.

Resistance:

$$e = iR$$

and therefore

$$\mathbf{Z} = \frac{e}{i} = R \tag{16–3}$$

at any frequency.

Inductance:

$$e = L\frac{di}{dt}$$

and therefore

$$\mathbf{Z} = \frac{e}{i} = \frac{L(di/dt)}{i} = sL \tag{16–4}$$

Capacitance:

$$i = C\frac{de}{dt}$$

and therefore

$$\mathbf{Z} = \frac{e}{i} = \frac{e}{C(de/dt)} = \frac{1}{sC} \tag{16–5}$$

These impedances, and their corresponding admittances, are summarized in Table 16–1.

TABLE 16–1

Impedance and Admittance for Passive Circuit Elements

Element	Impedance	Admittance
Resistance	R	$1/R$
Inductance	sL	$1/sL$
Capacitance . . .	$1/sC$	sC

If several elements are connected together the resulting impedance may be obtained in the same manner as for the sinusoidal steady state. Impedances in series add directly, as do admittances in parallel. Several examples of the impedance of combinations of elements are illustrated in Fig. 16–1. It is suggested that the student verify these results.

There are several points worth noting in connection with writing these impedance expressions:

1. The impedance is often designated as $\mathbf{Z}(s)$, as shown in Fig. 16–1, in order to indicate more clearly that it is a function of s.

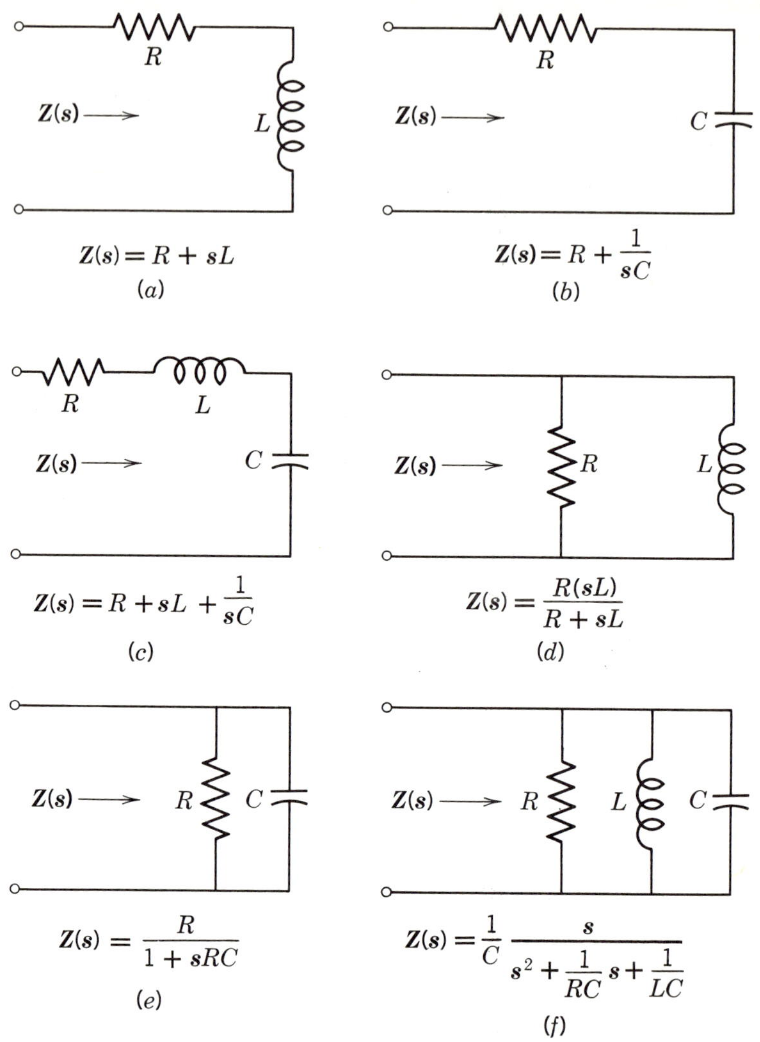

Fig. 16–1. Element combinations and their corresponding impedances.

2. The impedance is *not* designated as a complex number, as was done previously, since it will *always* be a *real* function of s. In fact, there is a rather subtle but very significant distinction between the present expressions for impedance and those previously obtained. The present form, $\mathbf{Z}(s)$, may be considered as a *real* function of a *complex* variable s. The previous form $\mathbf{Z}$ is a *complex* function of a *real* variable ω.
3. The impedance function $\mathbf{Z}(s)$ could have been obtained from the previous sinusoidal steady-state expressions for impedance simply by replacing $j\omega$ with s. Conversely, if we start with $\mathbf{Z}(s)$, the sinusoidal steady-state impedance may be obtained simply by replacing s with $j\omega$.

16–2. The Complex Frequency Plane. The next step in obtaining a geometrical representation for impedance is to introduce the complex frequency plane. This plane is simply a rectangular coordinate system in which the horizontal and vertical axes represent the real and imaginary parts of the complex frequency variable s. In fact, there is no difference between this complex plane and any plane on which complex numbers are plotted other than the interpretation given the axes.

For purposes of notation it is desirable to adopt standardized symbols for the two axes of the complex frequency plane. The symbols to be used throughout this book are

$$s = \sigma + j\omega \tag{16–6}$$

and this notation is consistent with the majority of present-day technical literature. The horizontal, or real, axis will be designated as the σ axis and the vertical, or imaginary, axis will be designated as the $j\omega$ axis. The entire complex frequency plane will usually be referred to as the s-plane, for the sake of brevity. It is often desirable to be able to describe various portions of the s-plane in a concise manner and for this purpose it is usually divided into two major areas. All portions of the s-plane lying to the left of the $j\omega$ axis are referred to as the left-half-plane (LHP) and all portions to the right of the $j\omega$

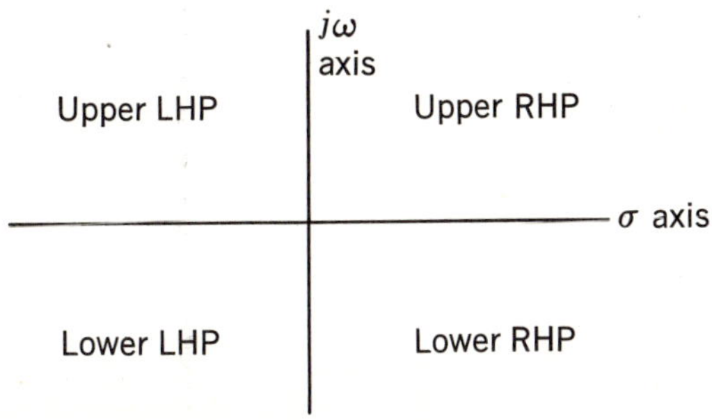

Fig. 16–2. The s-plane.

axis as the right-half-plane (RHP). Hence, the LHP corresponds entirely to negative values of σ and the RHP to positive values of σ. Occasionally it may be necessary to further subdivide each half-plane into its positive and negative parts with respect to ω. The quadrant corresponding to positive values of ω and negative values of σ would be designated as the upper LHP and a similar appropriate designation would apply to the other quadrants. All of the ideas concerning notation are illustrated in Fig. 16–2.

An additional point of consideration is the location of the point $s = \infty$. It is, of course, not possible to plot this point on a finite piece of paper but the question may arise as to the path in the s-plane that one would follow in order to approach this value. From a mathematical standpoint, $s = \infty$ is a single point, hence any straight path leading radially from the origin of the s-plane will eventually arrive at infinity. From a geometrical standpoint, this may be interpreted to mean that the s-plane is really not a plane at all but a gigantic sphere, such as the earth. The origin of the s-plane could be considered to be the North Pole, and the South Pole would then become the point of infinity. For most purposes it is sufficient to ignore this refinement and consider the s-plane as a plane, but the sphere interpretation is helpful in certain specialized cases.

Before proceeding with the representation of impedance in the complex frequency plane it is desirable to consider how some of the time functions discussed in Chapter 15 might be represented. Figure 16–3 indicates several time functions and their corresponding frequency components plotted in the s-plane.

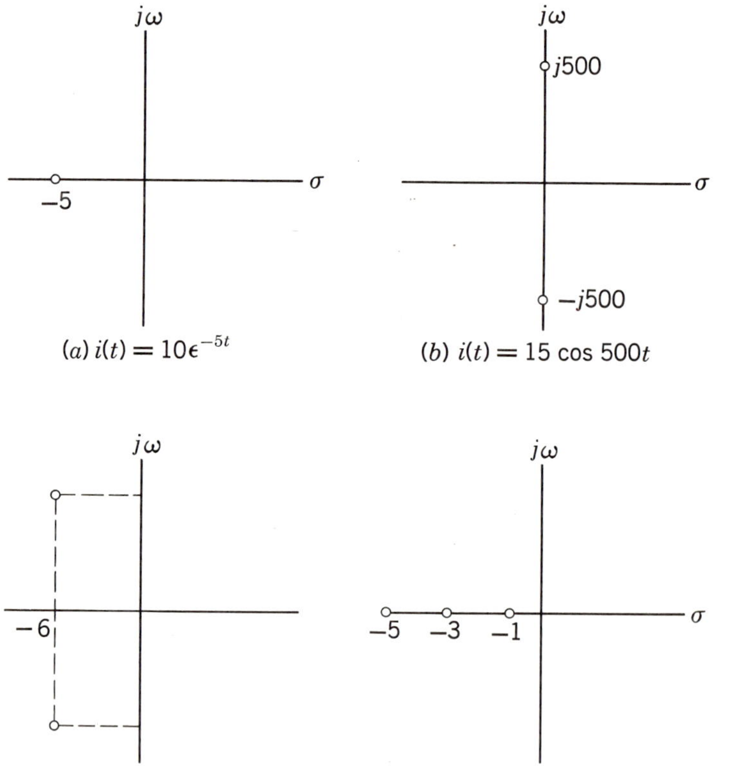

(a) $i(t) = 10\epsilon^{-5t}$ (b) $i(t) = 15 \cos 500t$

(c) $i(t) = 20\epsilon^{-6t} \cos 200t$ (d) $i(t) = -5\epsilon^{-t} + 15\epsilon^{-3t} - 10\epsilon^{-5t}$

Fig. 16–3. s**-plane plots of several time functions.**

Two points should be noted in connection with these figures. In the first place, the plots show only the values of the frequency components involved and indicate nothing concerning the magnitude or algebraic sign of the coefficients multiplying these components. Although it is possible to remove this difficulty by means of a suitable notation, this is generally not done because this method of representation is so seldom used. Secondly, it should be noted that the conjugate relationship existing between imaginary or complex pairs of frequency components implies a symmetry with respect to the σ axis. This will always be the case. There is no such symmetry with respect to the $j\omega$ axis.

The representation of time functions in the s-plane in the manner of Fig. 16–3 is not commonly used. Its presentation here is solely for the purpose of our acquiring additional familiarity with the complex frequency plane.

16–3. Impedance in the Complex Frequency Plane. One of the most common applications of the complex frequency plane is its use in the portrayal of impedance functions. In order to illustrate the nature of this technique consider the series R-L circuit shown in Fig. 16–1a. As indicated there, the impedance function for this circuit is

$$Z(s) = R + sL \tag{16–7}$$

which may be rewritten as

$$Z(s) = L\left(s + \frac{R}{L}\right) \tag{16–8}$$

It has been indicated that the objective of this section is the representation of impedance functions in the complex frequency plane. In order to accomplish this there must be designated on the s-plane frequencies which are particularly descriptive of this impedance. Such frequencies are those at which the impedance becomes zero or infinite. The reasons for this selection will be discussed in detail in a later section.

Examination of Eq. 16–8 reveals that there is one, and only one, value of s which will make the impedance zero. This value is designated s_1 and is

$$s_1 = -\frac{R}{L} \tag{16–9}$$

If Eq. 16–8 is written in terms of s_1 it becomes

$$Z(s) = L(s - s_1) \tag{16–10}$$

The geometrical representation of this impedance function may be accomplished by plotting the frequency s_1 on the s-plane, as shown in Fig. 16–4. Note that this plot does not show the multiplicative constant L appearing in Eq. 16–10; it only designates the frequency at which the impedance is zero.

It was stated earlier that frequencies at which the impedance is infinite are also of special interest. In this particular example the only frequency at which the impedance is infinite is $s_2 = \infty$, and this cannot be designated on a finite portion of the s-plane. Examples will be given later in which the impedance is infinite at finite frequencies.

Before proceeding with an interpretation of this geometrical representation, or further examples, it is advisable to introduce certain terminology and notation as an aid to this discussion. In order to present this terminology as concisely as possible, and facilitate future reference to it, the various items

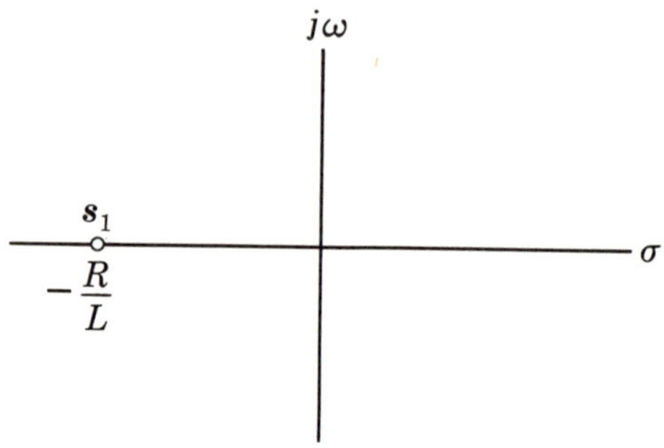

Fig. 16–4. *s*-plane representation of series *R-L* impedance.

of interest are tabulated in Table 16–2. Since these terms will be used repeatedly in future discussions it is essential that the student familiarize himself with them before proceeding further.

TABLE 16–2

Summary of *s*-Plane Terminology

Critical frequency	Any frequency at which an impedance function is *zero* or *infinite*. These frequencies may be real, imaginary, or complex and may occur at $s = 0$ and $s = \infty$.
Impedance zero	Any critical frequency at which the value of the *impedance* is zero. These will be designated with *odd-numbered* subscripts only, i.e., s_1, s_3, s_5, etc.
Impedance pole	Any critical frequency at which the value of the *impedance* is infinite. These will be designated with *even-numbered* subscripts only, i.e., s_2, s_4, s_6, etc.
Admittance zero	Any critical frequency at which the value of the *admittance* is zero. For a given circuit, these frequencies are identical to the *impedance poles*.
Admittance pole	Any critical frequency at which the value of the *admittance* is infinite. For a given circuit, these frequencies are identical to the *impedance zeros*.
Scale factor	The multiplicative constant associated with any impedance or admittance function. This constant involves some combination of the elements in the circuit.

As a second example of the geometrical representation of impedance functions consider the series *R-C* circuit shown in Fig. 16–1*b*. The impedance of this circuit is

$$Z(s) = R + \frac{1}{sC} \tag{16–11}$$

After algebraic manipulation this may be written

$$Z(s) = R\frac{\left(s + \frac{1}{RC}\right)}{s} \tag{16–12}$$

Since the impedance will be zero whenever the numerator is zero, it is seen that this function has an impedance zero at

$$s_1 = -\frac{1}{RC} \tag{16–13}$$

Similarly, the impedance will be infinite whenever the denominator of the impedance function is zero. Hence, there will be an impedance pole at

$$s_2 = 0 \tag{16–14}$$

In terms of the critical frequencies s_1 and s_2, the impedance function may be written as

$$Z(s) = R\frac{(s - s_1)}{(s - s_2)} \tag{16–15}$$

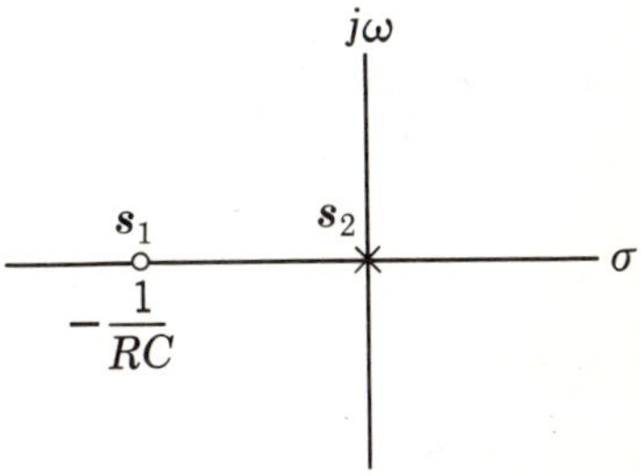

Fig. 16–5. *s*-plane representation of series *R-C* impedance.

The scale factor for this impedance is R. The critical frequencies are shown in Fig. 16–5. *Note that the pole is designated by a cross and the zero by a circle.*

It is also possible to consider the corresponding admittance function, which is

$$Y(s) = \frac{1}{Z(s)} = \frac{1}{R}\frac{s}{\left(s + \frac{1}{RC}\right)} = \frac{1}{R}\frac{(s - s_2)}{(s - s_1)} \tag{16–16}$$

From this it may be seen that the admittance zero is at $s_2 = 0$ and the admittance pole at $s_1 = -1/RC$. The scale factor is $1/R$.

The examples discussed so far have had only one pole and one zero because there were only two elements in each circuit. If a circuit contains more elements there will in general be more critical frequencies. Such circuits will be discussed shortly. A brief discussion of the general case will also be included.

16–4. Interpretation of the Pole-Zero Configuration. The preceding section indicated how the poles and zeros of simple impedance functions could be plotted on the complex frequency plane. This plot is referred to as a pole-zero configuration or, occasionally, a pole-zero constellation. There would be little point in making such a plot if it could not be used advantageously in gaining additional knowledge about the impedance function.

The first application that will be discussed, therefore, is the use of the pole-zero configuration to determine graphically the value of the impedance at any frequency.

Returning now to the series R-L circuit, it was found that its impedance function could be written in the form

$$Z(s) = L(s - s_1) \tag{16–10}$$

where

$$s_1 = -\frac{R}{L} \tag{16–9}$$

This is shown again in Fig. 16–6.

Since any special value of s is simply a complex number, the *difference* between any two s-values may be interpreted as a phasor in the s-plane. Hence, the factor $(s - s_1)$ in Eq. 16–10 is simply the phasor directed from s_1 to s as shown in Fig. 16–6a. The magnitude of the impedance is seen to be

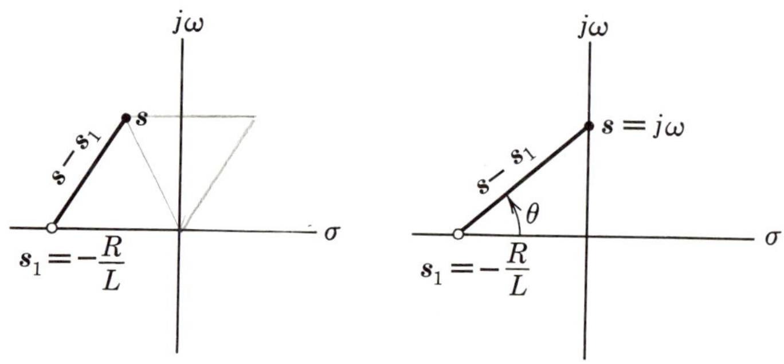

(a) Value of impedance at any frequency

(b) Value of impedance to steady-state sinusoids

Fig. 16–6. s-plane plot for the determination of the impedance of an R-L series circuit.

the *length* of this phasor multiplied by the scale factor. Likewise, the angle of the impedance is simply the angle which this phasor makes with the σ axis.

Although it is frequently necessary to know the value of an impedance function for almost any value of s, the major interest at this time is in frequencies lying on the $j\omega$ axis since these are the values corresponding to steady-state sinusoidal functions. Therefore, this study will be restricted to the evaluation of the impedance function along the $j\omega$ axis. In this case, the phasor is the hypotenuse of a right triangle having a base of R/L and an altitude of ω. The length of the phasor is, therefore,

$$|s - s_1| = \sqrt{\left(\frac{R}{L}\right)^2 + \omega^2} \tag{16–17}$$

and the magnitude of the impedance becomes

$$|Z| = L|s - s_1| = L\sqrt{\left(\frac{R}{L}\right)^2 + \omega^2} = \sqrt{R^2 + \omega^2 L^2} \tag{16-18}$$

The angle which the phasor makes with the σ axis, and therefore the angle of the impedance, is

$$\theta = \tan^{-1}\frac{\omega}{R/L} = \tan^{-1}\frac{\omega L}{R} \tag{16-19}$$

Comparison of these results with those already established in earlier work for the impedance of a series R-L circuit reveals complete correspondence. Note that the pole at infinity does not enter into the computations because it does not appear explicitly as a factor in the impedance function.

Next consider the series R-C circuit for which the impedance function was shown to be

$$Z(s) = R\frac{(s - s_1)}{(s - s_2)} \tag{16-15}$$

where

$$s_1 = -\frac{1}{RC} \tag{16-13}$$

and

$$s_2 = 0 \tag{16-14}$$

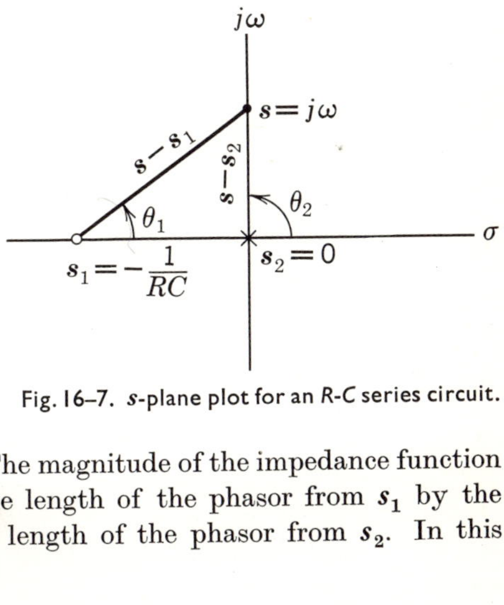

Fig. 16-7. s-plane plot for an R-C series circuit.

The pole-zero configuration is illustrated again in Fig. 16-7.

In this case there are two critical frequencies to be considered. The numerator term $(s - s_1)$ is the phasor from s_1 to s; the denominator term $(s - s_2)$ is the phasor from s_2 to s. The magnitude of the impedance function may be obtained by multiplying the length of the phasor from s_1 by the scale factor R and dividing by the length of the phasor from s_2. In this example the magnitude becomes

$$|Z| = R\frac{|s - s_1|}{|s - s_2|} = \frac{R\sqrt{(1/RC)^2 + \omega^2}}{\omega} = \sqrt{R^2 + \frac{1}{\omega^2 C^2}} \tag{16-20}$$

The angle of the impedance may be determined by remembering that when one complex number is divided by another the angle of the quotient is obtained by subtracting the denominator angle from the numerator angle. For this case

$$\theta = \theta_1 - \theta_2 = \tan^{-1}\frac{\omega}{1/RC} - \frac{\pi}{2} = \tan^{-1}\omega RC - \frac{\pi}{2} \tag{16-21}$$

Note that the scale factor does not enter into the computation of the angle.

16–5. The Generalized Impedance Function. The discussion so far has considered only very simple circuits involving one pole and one zero. In order to present a complete picture of this application of the complex frequency plane it is necessary to consider more general impedance functions. Unfortunately, it is not possible at this stage to justify everything that should be said concerning generalized impedance functions. Therefore, as an interim measure, the necessary facts concerning such functions will be stated without proof with the understanding that they will be justified in advanced courses.

In general, any impedance function may be written as a scale factor times the ratio of two polynomials in s as shown in Eq. 16–22:

$$Z(s) = H\frac{s^n + a_{n-1}s^{n-1} + a_{n-2}s^{n-2} + \cdots + a_1 s + a_0}{s^m + b_{m-1}s^{m-1} + b_{m-2}s^{m-2} + \cdots + b_1 s + b_0} \tag{16–22}$$

There are certain restrictions that are imposed on these polynomials, however. The ones that are important are:

1. The degree of the polynomials cannot differ by more than 1. That is, $(m - n)$ may be $+1$ or 0 or -1, but nothing else.
2. All the coefficients of s must be positive in both polynomials.
3. The scale factor H must be a positive, real constant.

Impedance zeros occur at those values of s which make the numerator zero. These values are obtained by finding the roots of the equation

$$0 = s^n + a_{n-1}s^{n-1} + a_{n-2}s^{n-2} + \cdots + a_1 s + a_0 \tag{16–23}$$

As shown in college algebra, such an equation has n roots and may be expressed in terms of these roots by writing the product of factors of the form $(s - s_1)$, where s_1 is any root. Hence, Eq. 16–23 may be written in factored form as

$$(s - s_1)(s - s_3)(s - s_5) \cdots (s - s_{2n-1}) = 0 \tag{16–24}$$

It is apparent in this form that the polynomial is zero at any of its roots since that particular factor will be zero. It should also be noted that the zeros of Eq. 16–24 are designated by odd subscripts in accordance with the previously established convention.

Impedance poles occur at those values of s which make the denominator of Eq. 16–22 zero. This polynomial may also be written in factored form. Hence, the impedance function may always be written in factored form as

$$Z(s) = H\frac{(s - s_1)(s - s_3)(s - s_5) \cdots (s - s_{2n-1})}{(s - s_2)(s - s_4)(s - s_6) \cdots (s - s_{2m})} \tag{16–25}$$

Certain restrictions exist on the possible locations of the critical frequencies of Eq. 16–25. These restrictions are not independent of those just listed but are a consequence of them. For use here, the most important restrictions are:

1. The critical frequencies for any impedance (or admittance) function must lie in the LHP or on the $j\omega$ axis.

2. The critical frequencies must appear as negative, real values or in conjugate, complex pairs.
3. There must be the same number of poles as there are zeros, including any at infinity. If infinity is a critical frequency, it does not appear explicitly as a factor in the impedance function but must be inferred from the degree of the numerator and denominator. That is, $n = m + 1$ corresponds to a pole at infinity, $n = m - 1$ is a zero at infinity, and $n = m$ indicates there is no critical frequency at infinity.

All of these restrictions are discussed in detail in advanced courses, as well as some additional ones not considered here.

In order to obtain the impedance function in factored form it is usually necessary to determine the roots of the numerator and denominator polynomials. If these are of degree higher than 3, there is no straightforward method of accomplishing this. There are, however, some special trial-and-error methods which reduce the labor required to a minimum, but unless one has many polynomials to solve it is probably just as satisfactory to utilize the standard procedure of trial and error by synthetic division as discussed in most college algebra texts.

16-6. Interpretation of Generalized Pole-Zero Configurations. The determination of the value of an impedance function at some particular value of $s = j\omega$ is basically the same as was illustrated in Art. 16-4; the only difference is in the number of factors that must be considered. Therefore, this procedure may now be stated in general form as follows:

1. The *magnitude* of any impedance or admittance function at any value of s may be determined by multiplying the scale factor by the *product* of the lengths of the phasors from all of the *zeros* to s and dividing by the product of the lengths of the phasors from all of the *poles* to s.
2. The *angle* of any impedance or admittance function at any value of s may be determined by *adding* the angles of the phasors from all of the *zeros* to s and *subtracting* the angles of the phasors from all of the *poles* to s.

These rules may be illustrated by reference to the arbitrary pole-zero configuration shown in Fig. 16-8. It will be noted that this configuration has the same number of poles as zeros and thus does *not* have a critical frequency at infinity. The impedance function may be written immediately by inspection of the configuration and is

$$Z(s) = H\frac{(s - s_1)(s - s_3)(s - s_5)}{(s - s_2)(s - s_4)(s - s_6)} \tag{16-26}$$

where the scale factor H is either unspecified or must be specified independently of the pole-zero configuration.

From the first rule given in this section, the magnitude of the impedance at any value of ω may be written as

$$|Z| = H\frac{|s - s_1|\,|s - s_3|\,|s - s_5|}{|s - s_2|\,|s - s_4|\,|s - s_6|} \tag{16-27}$$

where each factor is simply the length of the corresponding phasor. These lengths are, of course, all considered to be positive.

The second rule indicates that the angle of the impedance may be written as

$$\theta = (\theta_1 + \theta_3 + \theta_5) - (\theta_2 + \theta_4 + \theta_6) \tag{16–28}$$

From Fig. 16–8 it may be seen that, according to the usual convention, θ_1, θ_2, θ_5, and θ_6 are positive angles, and θ_3 and θ_4 are negative angles. Also, it may be observed that $\theta_5 = \pi/2$ and $\theta_3 = -\pi/2$ whereas all other angles are less than $\pi/2$. It may be stated in general that the angle associated with each phasor will *always* be equal to or less than $\pi/2$ and the total angle on the impedance will also *always* be less than or equal to $\pi/2$.

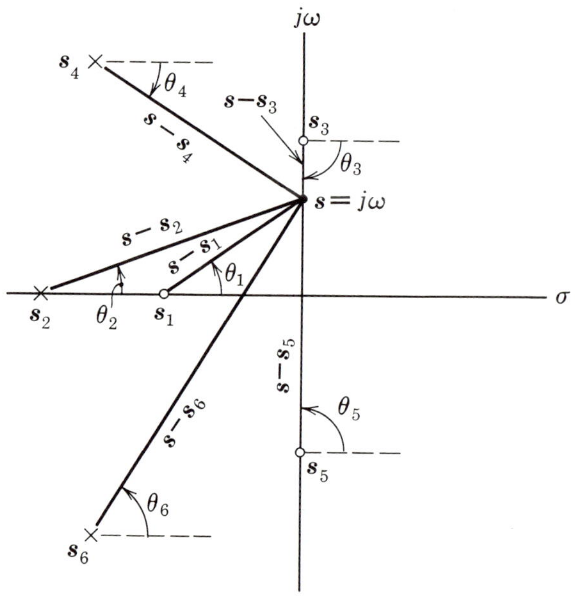

Fig. 16–8. Illustration of the evaluation of $|Z|$ and θ from the s-plane plot.

If it is imagined that the point $s = j\omega$ is being moved slowly along the $j\omega$ axis from zero to infinity, it may be visualized how the lengths and angles of the various phasors change with respect to this motion and it becomes possible to visualize the variation with frequency of the total impedance and total angle. The ability to carry out this visualization rapidly and accurately is a tremendous advantage to the circuit analyst and should be practiced as often as possible by the student. The numerical examples considered in the following sections afford an excellent opportunity to commence this practice.

16–7. Examples of More General Impedance Functions. Consider the circuit of Fig. 16–9 and obtain the impedance function as

$$Z(s) = R + sL + \frac{1}{sC}$$

$$= \frac{RCs + LCs^2 + 1}{sC}$$

$$= L\frac{s^2 + \frac{R}{L}s + \frac{1}{LC}}{s} \tag{16–29}$$

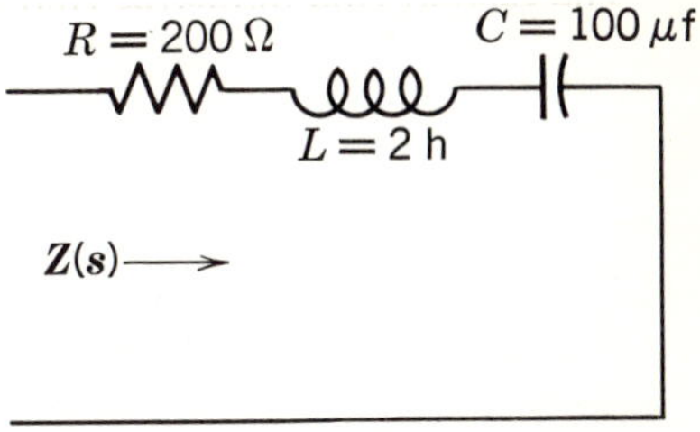

Fig. 16–9. *R-L-C* series circuit for analysis.

It will be observed that this function has two zeros and one pole appearing explicitly, hence, must also have a pole at infinity. The indicated pole is at $s_2 = 0$.

The zero location may be determined by using the quadratic formula to obtain the roots of the equation as

$$s^2 + \frac{R}{L}s + \frac{1}{LC} = 0 \tag{16–30}$$

from which

$$s_{1,3} = -\frac{R}{2L} \pm \sqrt{\left(\frac{R}{2L}\right)^2 - \frac{1}{LC}} \tag{16–31}$$

These zero positions may be either real or complex, depending upon whether the quantity inside the radical is positive or negative. The form of Eq. 16–31 is used if the roots are real, whereas complex roots can be obtained by factoring $j = \sqrt{-1}$ out of the radical, or

$$s_{1,3} = -\frac{R}{2L} \pm j\sqrt{\frac{1}{LC} - \left(\frac{R}{2L}\right)^2} \tag{16–32}$$

If the numerical values specified in Fig. 16–9 are utilized, the result is

$$Z(s) = 2\,\frac{s^2 + 100s + 5000}{s} \tag{16–33}$$

and

$$s_{1,3} = -50 \pm j\sqrt{5000 - 2500} = -50 \pm j50 \tag{16–34}$$

The pole-zero configuration is indicated in Fig. 16–10.

Evaluation of the magnitude and angle of this impedance at $\omega = 100$ may be accomplished as shown in Fig. 16–11. The magnitude of the impedance is obtained as

$$|Z| = 2\,\frac{\sqrt{\overline{50}^2 + \overline{50}^2}\,\sqrt{\overline{50}^2 + \overline{150}^2}}{100} = 223.4 \text{ ohms} \tag{16–35}$$

and the angle as

$$\theta = \tan^{-1}\frac{50}{50} + \tan^{-1}\frac{150}{50} - \frac{\pi}{2}$$

$$= \frac{\pi}{4} + 1.249 - \frac{\pi}{2} = 0.4636 \text{ radian} \quad (16\text{–}36)$$

The magnitude and angle may be calculated for any value of ω but it is also possible, in this case, to make rough sketches indicating how these quantities vary with ω without making any calculations. By observing the

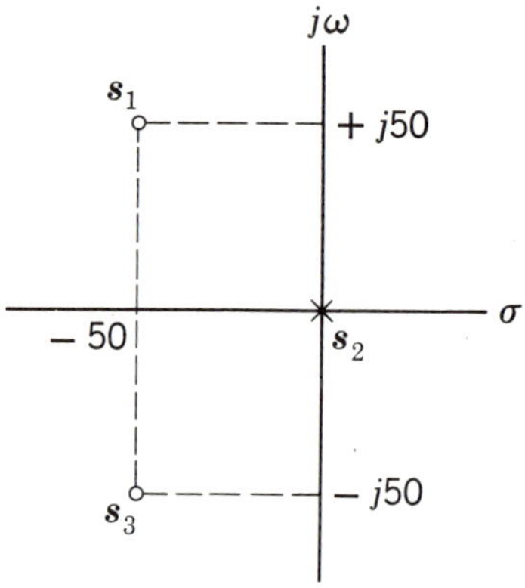

Fig. 16–10. *s*-plane pole-zero plot for *R-L-C* network of Fig. 16–9.

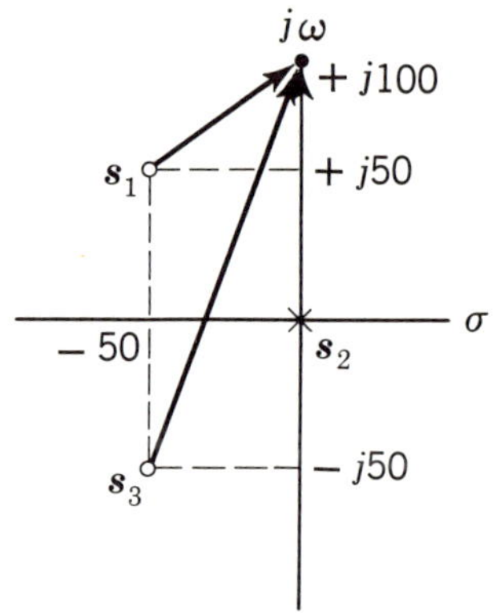

Fig. 16–11. Evaluation of $\mathbf{Z}$ for Fig. 16–9 at $\omega = 100$.

changes in phasor length and angle as ω is varied from zero to infinity, the curves of Fig. 16–12 may be rapidly sketched.

As a second example consider the circuit of Fig. 16–13, which involves series and parallel combinations of elements. As an aid in obtaining the impedance function it is convenient to designate the parallel equivalent impedance of R_1 and C_1 as $\mathbf{Z}_1$ and the series equivalent impedance of R_2 and C_2 as $\mathbf{Z}_2$. Hence,

$$\mathbf{Z}(s) = \mathbf{Z}_1(s) + \mathbf{Z}_2(s) \quad (16\text{–}37)$$

where

$$\mathbf{Z}_1(s) = \frac{R_1\left(\dfrac{1}{sC_1}\right)}{R_1 + \dfrac{1}{sC_1}} = \frac{R_1}{sR_1C_1 + 1}$$

and

$$\mathbf{Z}_2(s) = R_2 + \frac{1}{sC_2} = \frac{sR_2C_2 + 1}{sC_2}$$

Combining impedances gives

$$\mathbf{Z}(s) = \frac{R_1}{sR_1C_1 + 1} + \frac{sR_2C_2 + 1}{sC_2}$$

$$= \frac{R_1C_2s + R_1C_1R_2C_2s^2 + R_1C_1s + R_2C_2s + 1}{sC_2(R_1C_1s + 1)}$$

$$= R_2 \frac{s^2 + \left(\dfrac{1}{R_1C_1} + \dfrac{1}{R_2C_2} + \dfrac{1}{R_2C_1}\right)s + \dfrac{1}{R_1C_1R_2C_2}}{s\left(s + \dfrac{1}{R_1C_1}\right)} \tag{16–38}$$

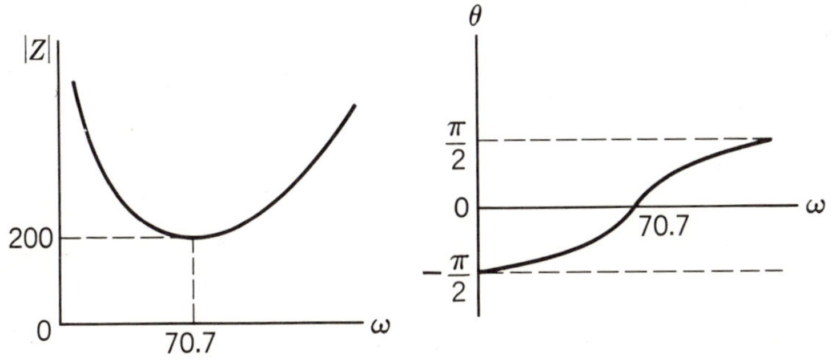

(a) Magnitude (b) Angle

Fig. 16–12. $|Z|$ and θ for network of Fig. 16–9.

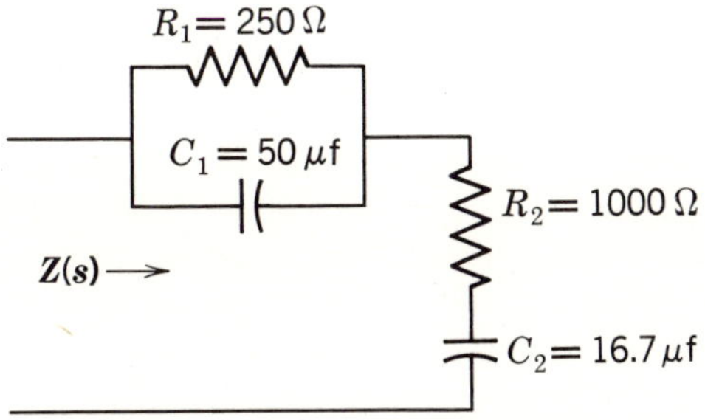

Fig. 16–13. Series-parallel network.

It should be noted that the poles of $\mathbf{Z}(s)$ are simply the pole of $\mathbf{Z}_1(s)$ and the pole of $\mathbf{Z}_2(s)$. It is apparent from Eq. 16–38 that this must be true since $\mathbf{Z}(s)$ will be infinite if *either* $\mathbf{Z}_1(s)$ or $\mathbf{Z}_2(s)$ is infinite and *cannot* be infinite at any other frequency. Hence, the poles of $\mathbf{Z}(s)$ occur at

$$s_2 = 0$$

$$s_4 = -\frac{1}{R_1C_1} = -\frac{1}{250 \times 50 \times 10^{-6}}$$

The zeros of $Z(s)$ may be determined by first substituting in the specified numerical values, or

$$Z(s) = 1000\frac{(s^2 + 160s + 4800)}{s(s+80)} \qquad (16\text{–}39)$$

and the numerator factored by inspection into two real roots:

$$Z(s) = 1000\frac{(s+120)(s+40)}{s(s+80)} \qquad (16\text{–}40)$$

The resulting pole-zero configuration is shown in Fig. 16–14.

It may be noted that all of the critical frequencies lie on the negative σ axis. This will always be true of circuits involving only resistance and capacitance or only resistance and inductance.

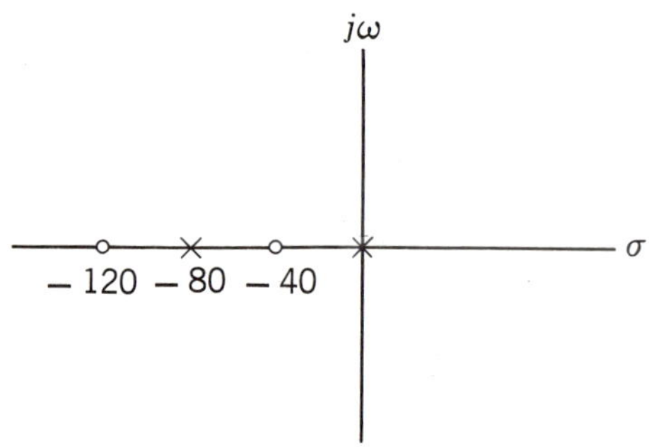

Fig. 16–14. s-plane pole-zero configuration for network of Fig. 16–13.

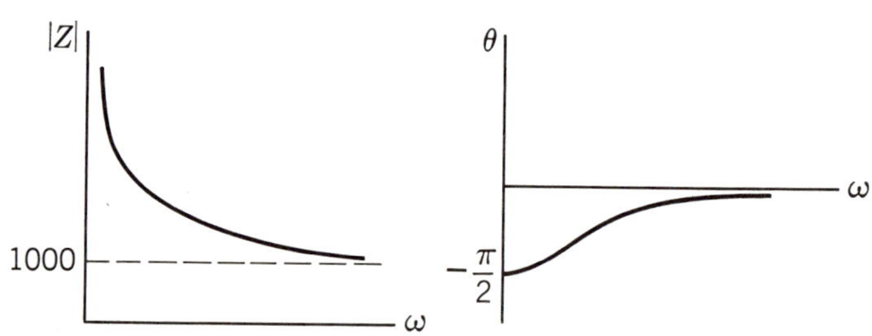

(*a*) Magnitude (*b*) Angle

Fig. 16–15. $|Z|$ and θ for network of Fig. 16–13.

The magnitude and angle of the impedance may be evaluated for any value of ω by a procedure essentially identical to that discussed in the previous example and will not be repeated here. It is of interest, however, to sketch the variation in magnitude and angle with respect to ω. This may be done by inspection and the results are shown in Fig. 16–15.

16–8. The Effect of Element Values on the Pole-Zero Configuration. It is often desirable to be able to study the effect of varying element sizes upon the operation of a circuit. Observing the changes in pole and zero positions with the changes in element values provides a convenient method of accomplishing this. This procedure is frequently referred to as a *root-locus technique* since it involves plotting the path, or locus, of the critical frequencies as one or more elements are varied in value. There are many ramifications of this technique which are of value to a circuit designer but all that will be done at the present time is to discuss a single example.

The example to be considered here examines the effect of changing the resistance in the series R-L-C circuit discussed in the previous section. At that time it was shown that the impedance function was

$$Z(s) = L\,\frac{s^2 + \frac{R}{L}s + \frac{1}{LC}}{s} \tag{16–29}$$

which has a pole at zero and at infinity and zeros given by

$$s_{1,3} = -\frac{R}{2L} \pm j\sqrt{\frac{1}{LC} - \left(\frac{R}{2L}\right)^2} \tag{16–32}$$

It may be noted that only the zero positions will be affected by changing the resistance.

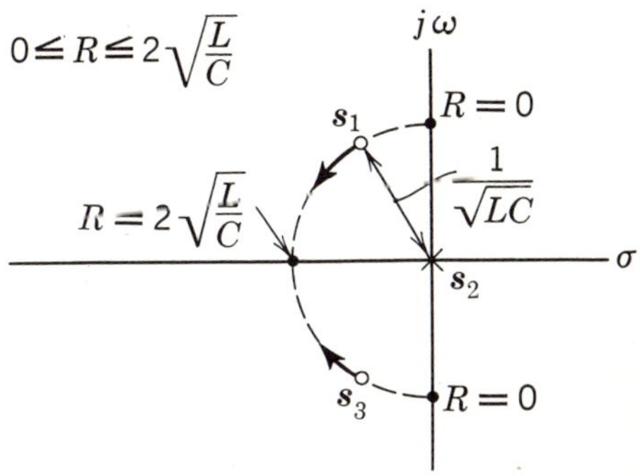

Fig. 16–16. Zero locus plot for R-L-C series circuit with varying R.

The difficulty with which the locus of the zeros is determined may be eased considerably by noting that, when they are complex, they remain the same distance from the origin of the s-plane regardless of the value of the resistance. This fact is made apparent by obtaining the magnitude of s_1 or s_3 for complex values as

$$|s_1| = |s_3| = \sqrt{\left(\frac{R}{2L}\right)^2 + \left(\frac{1}{LC} - \left(\frac{R}{2L}\right)^2\right)} = \frac{1}{\sqrt{LC}} \tag{16–41}$$

which depends on the values of inductance and capacitance only. This condition implies that the locus follows the arc of a circle as shown in Fig. 16–16.

There are several particular values of R which are of special interest and these are considered now:

$R = 0;$	$s_{1,3} = 0 \pm j\sqrt{\frac{1}{LC}}$	Zeros on $j\omega$ axis
$R = \sqrt{\frac{2L}{C}};$	$s_{1,3} = -\sqrt{\frac{1}{2LC}} \pm j\sqrt{\frac{1}{2LC}}$	Zeros at 45° angles with respect to negative σ axis
$R = 2\sqrt{\frac{L}{C}};$	$s_{1,3} = -\sqrt{\frac{1}{LC}} \pm j0$	Zeros coincide on negative σ axis

If the value of R is made greater than $2\sqrt{L/C}$, the zeros become purely real and are given by

$$s_{1,3} = -\frac{R}{2L} \pm \sqrt{\left(\frac{R}{2L}\right)^2 - \frac{1}{LC}} \tag{16–31}$$

as shown previously. The pole-zero configuration is now as shown in Fig. 16–17.

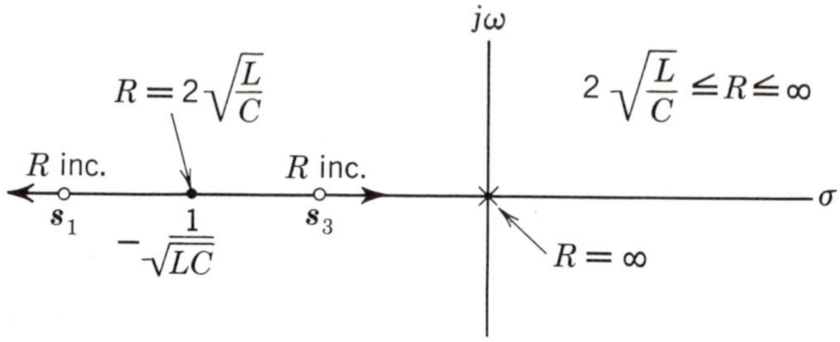

Fig. 16–17. Continuation of zero locus plot of Fig. 16–16.

Again some special values may be considered:

$R = 2\sqrt{\frac{L}{C}}$;	$s_{1,3} = -\frac{1}{\sqrt{LC}}$	Zeros coincide on σ axis
$R = 2.5\sqrt{\frac{L}{C}}$;	$s_1 = -\frac{2}{\sqrt{LC}}$, $s_3 = -\frac{1}{2\sqrt{LC}}$	Zeros at $\frac{1}{2}(1/\sqrt{LC})$ and twice $1/\sqrt{LC}$
$R = 10.1\sqrt{\frac{L}{C}}$;	$s_1 = -\frac{10}{\sqrt{LC}}$, $s_3 = -\frac{1}{10\sqrt{LC}}$	Zeros at $\frac{1}{10}(1/\sqrt{LC})$ and 10 times $1/\sqrt{LC}$
$R = \infty$;	$s_1 = \infty$, $s_3 = 0$	Zeros have coincided with poles

It is also of interest to note that the zeros exhibit geometric symmetry with respect to $-1/\sqrt{LC}$, that is,

$$s_1 s_3 = \frac{1}{LC} \tag{16–42}$$

This may be proved by direct substitution as

$$s_1 s_3 = \left[-\frac{R}{2L} - \sqrt{\left(\frac{R}{2L}\right)^2 - \frac{1}{LC}}\right]\left[-\frac{R}{2L} + \sqrt{\left(\frac{R}{2L}\right)^2 - \frac{1}{LC}}\right]$$

$$= \frac{R^2}{4L^2} - \frac{R^2}{4L^2} + \frac{1}{LC} = \frac{1}{LC} \tag{16–43}$$

This result is true for any value of R, including those which make s_1 and s_3 complex.

The first discussion of series R-L-C circuits occurred in the preliminary study of the transient response. At that time it was found convenient to divide the solution of the differential equation up into three parts as follows:

$R < 2\sqrt{\dfrac{L}{C}}$ Underdamped case—oscillatory response

$R = 2\sqrt{\dfrac{L}{C}}$ Critically damped case—exponentially damped ramp

$R > 2\sqrt{\dfrac{L}{C}}$ Overdamped case—non-oscillatory response

It may be observed that these three parts correspond exactly to the cases of complex zeros, coincident zeros, and real zeros respectively. Hence, it is possible to establish a correlation between the pole-zero configuration and the transient response of the circuit as well as its steady-state sinusoidal response. This correlation will not be discussed further at this time. Indeed, the ability of the pole-zero configuration to be useful in many types of problems constitutes the principal advantage that this technique has to offer.

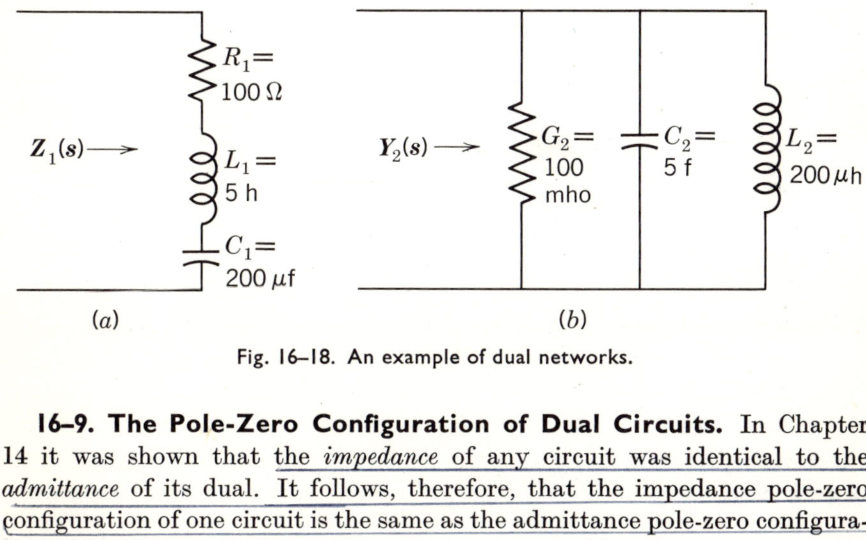

Fig. 16–18. An example of dual networks.

16–9. The Pole-Zero Configuration of Dual Circuits. In Chapter 14 it was shown that the *impedance* of any circuit was identical to the *admittance* of its dual. It follows, therefore, that the impedance pole-zero configuration of one circuit is the same as the admittance pole-zero configuration of its dual. As an illustration of this point, consider the dual of the series R-L-C circuit shown in Fig. 16–18*a*.

The impedance function of the series R-L-C circuit has been shown to be

$$\mathbf{Z}_1(s) = L_1 \frac{s^2 + \dfrac{R_1}{L_1}s + \dfrac{1}{L_1C_1}}{s} \tag{16–29}$$

The admittance of the circuit in Fig. 16–18b may be obtained by adding admittances, or

$$Y_2(s) = G_2 + sC_2 + \frac{1}{sL_2} \tag{16–44}$$

$$= \frac{sL_2G_2 + s^2L_2C_2 + 1}{sL_2}$$

$$= C_2 \frac{s^2 + \dfrac{G_2}{C_2}s + \dfrac{1}{L_2C_2}}{s} \tag{16–45}$$

Comparison of Eqs. 16–45 and 16–29 indicates the similarity of functions and the interchange of elements.

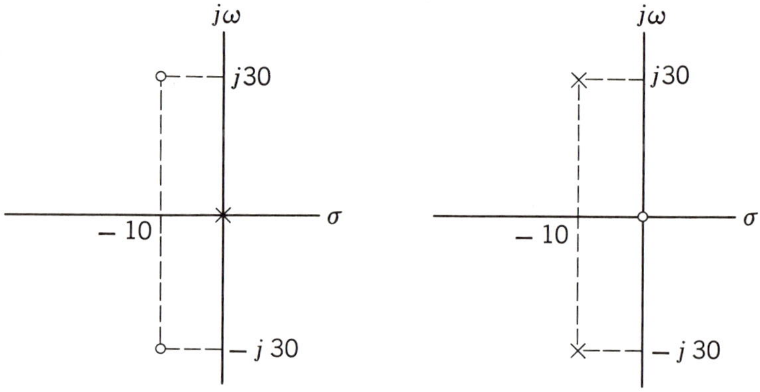

(a) $Z_1(s) = Y_2(s)$ (b) $Z_2(s) = Y_1(s)$

Fig. 16–19. Pole-zero configuration for circuits of Fig. 16–18.

If numerical values are substituted in both equations, they become

$$Z_1(s) = 5\,\frac{s^2 + 20s + 1000}{s}, \qquad Y_2(s) = 5\,\frac{s^2 + 20s + 1000}{s} \tag{16–46}$$

which makes their duality even more evident. The pole-zero configuration for each may be established by determining the roots of the numerator from

$$Z_1(s) = Y_2(s) = 5\,\frac{(s + 10 - j30)(s + 10 + j30)}{s} \tag{16–47}$$

Figure 16–19 shows the pole-zero configurations for the impedance and the admittance of both circuits.

16–10. Scaling of Impedance Functions. Many occasions arise in circuit analysis and design in which it is convenient to be able to modify an existing impedance function to make it easier to manipulate mathematically or to make it correspond to a desired function. There are, of course, many ways in which an impedance may be changed but the simplest are those which preserve the poles and zeros in the same *relative* positions. There are three ways in which a change of this sort may be accomplished:

1. The positions of the poles and zeros remain unchanged but the *magnitude* of the impedance function at any *other* frequency is modified by the same positive real factor. A change of this sort is designated as *magnitude scaling*.
2. The positions of the poles and zeros are all moved by the same positive, real factor but the magnitude of the impedance function (on the new frequency scale) is the same as it was on the previous frequency before modification. This type of change is designated as *frequency scaling*.
3. The circuit may be scaled in *both* magnitude and frequency simultaneously. This may always be handled as two independent scaling operations of the types just discussed.

The simplest scaling operation is that of magnitude scaling. In terms of the impedance function, this operation amounts to changing the scale factor only. If an impedance function has the form

$$Z(s) = H\frac{(s - s_1)(s - s_3)\cdots(s - s_{2n-1})}{(s - s_2)(s - s_4)\cdots(s - s_{2m})} \tag{16–25}$$

and it is desired to change the magnitude at every frequency by a factor K_m, then the scaled impedance function is

$$Z_k(s) = K_m H\frac{(s - s_1)(s - s_3)\cdots(s - s_{2n-1})}{(s - s_2)(s - s_4)\cdots(s - s_{2m})} \tag{16–48}$$

The constant K_m must be a positive, real number but may be either larger or smaller than 1. The critical frequencies and, therefore, the pole-zero configurations do not change at all.

It is just as simple to carry out magnitude scaling in terms of modifying the element values of a given circuit. If the impedance of the entire circuit is to change by a factor K_m, then it is necessary that the impedance of *each element* change by the same factor. Hence, resistance and inductance values are multiplied by K_m and capacitance values are divided by K_m (since the impedance of a capacitor is inversely proportional to the capacitance).

As an illustration of magnitude scaling consider the circuit shown in Fig. 16–20*a*. The impedance function for this circuit is

$$Z(s) = 2\frac{s^3 + s^2 + 3s + 1}{s^2 + s + 2} \tag{16–49}$$

If it is desired to raise the impedance level by a factor of 8, the impedance function becomes

$$Z_8(s) = 8 \times 2 \frac{s^3 + s^2 + 3s + 1}{s^2 + s + 2} = 16 \frac{s^3 + s^2 + 3s + 1}{s^2 + s + 2} \qquad (16\text{–}50)$$

The modified circuit is obtained by multiplying the resistance and inductance by 8 and dividing the capacitance by 8. The resulting circuit is shown in Fig. 16–20*b*.

In the operation of frequency scaling it is desired to obtain the same range of impedance values but in a different range of frequencies. This amounts to changing the frequency scale on a plot of impedance as a function of frequency.

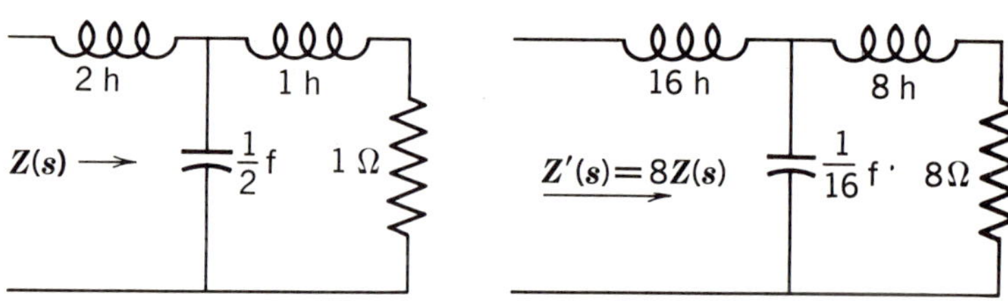

(*a*) Series-parallel circuit (*b*) Magnitude-scaled by a factor of 8

Fig. 16–20. An example of magnitude scaling.

Again, it is possible to accomplish frequency scaling from either the impedance function or from the circuit diagram. If the impedance function is of the form

$$Z(s) = H \frac{(s - s_1)(s - s_3) \cdots (s - s_{2n-1})}{(s - s_2)(s - s_4) \cdots (s - s_{2m})} \qquad (16\text{–}25)$$

and it is desired to scale the frequency by a factor of K_f, then all of the critical frequencies will be multiplied by this factor. At first thought, it might appear that this is the only change needed in the impedance function. However, it should be noted that the value of s at which a given impedance is to occur also increases by K_f; hence, each pole or zero factor in Eq. 6–25 is multiplied by K_f. If there are the same number of factors in both the numerator and denominator ($n = m$), then each is increased by the same amount and the scale factor of $Z(s)$ is unchanged. However, if the number of factors in the numerator and denominator are not the same ($n = m \pm 1$), then there will be an additional K_f left over in one or the other and the scale factor will change. The value of H must then be modified to compensate for this. The scaled impedance function may, therefore, be written generally as

$$Z_k(s) = K_f^{(m-n)} H \frac{(s - K_f s_1)(s - K_f s_3) \cdots (s - K_f s_{2n-1})}{(s - K_f s_2)(s - K_f s_4) \cdots (s - K_f s_{2m})} \qquad (16\text{–}51)$$

As an illustration of all possibilities consider three different impedance functions and scale each one in frequency by a factor 5. The results, with $K_f = 5$, are:

$$\mathbf{Z}(s) = 15\frac{(s+1)(s+3)}{(s+2)(s+4)}, \qquad \mathbf{Z}_k(s) = 15\frac{(s+5)(s+15)}{(s+10)(s+20)}$$

$$\mathbf{Z}(s) = 15\frac{(s+1)(s+3)}{(s+2)}, \qquad \mathbf{Z}_k(s) = 3\frac{(s+5)(s+15)}{(s+10)}$$

$$\mathbf{Z}(s) = 15\frac{(s+3)}{(s+2)(s+4)}, \qquad \mathbf{Z}_k(s) = 75\frac{(s+15)}{(s+10)(s+20)}$$

A simple rule which accomplishes the same result as Eq. 16–51 is merely to multiply all critical frequencies by K_f and insert an additional factor of K_f in the polynomial of *lower degree.*

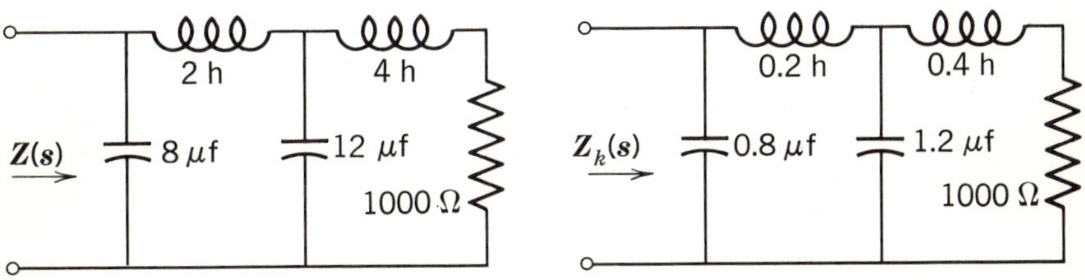

(*a*) Original circuit (*b*) Frequency-scaled by a factor of 10

Fig. 16–21. An example of frequency scaling.

Frequency scaling of an existing circuit is more straightforward. All that needs to be remembered is that each element is to have the same impedance at a frequency $K_f s$ as it originally had at a frequency s. This means that resistors do not change in value and inductance and capacitance values are divided by K_f. This is immediately apparent from the following expressions:

Resistance	*Inductance*	*Capacitance*
$\mathbf{Z}(s) = R$	$\mathbf{Z}(s) = sL$	$\mathbf{Z}(s) = 1/sC$
$\mathbf{Z}_k(s) = R_k$	$\mathbf{Z}_k(s) = K_f s L_k$	$\mathbf{Z}_k(s) = 1/K_f s C_k$
$\therefore\ R_k = R$	$\therefore\ L_k = L/K_f$	$\therefore\ C_k = C/K_f$

As an illustration of this procedure, consider the circuit of Fig. 16–21*a*. If this impedance is to be scaled in frequency by a factor of 10, the resulting circuit is shown in Fig. 16–21*b*.

If both frequency and magnitude scaling are to be accomplished, it is generally desirable to carry out the scaling in two steps—one for magnitude and one for frequency—and thus avoid much of the possibility for error that arises if combination scaling is attempted in a single step.

As far as this book is concerned, the use of scaling makes it possible to discuss analysis methods in terms of simple element values and then modify the results to obtain practical values.

PROBLEMS

16–1. Determine the impedance $\boldsymbol{Z}(s)$ for each of the circuits shown.

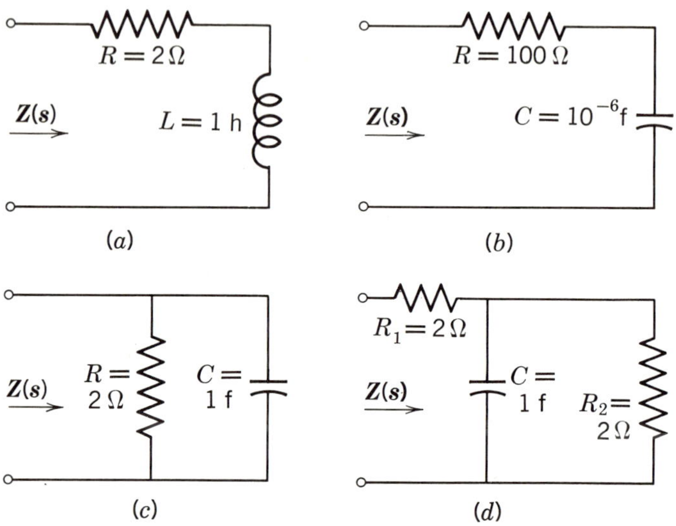

Prob. 16–1.

16–2. Determine the admittance $\boldsymbol{Y}(s)$ for each of the circuits illustrated.

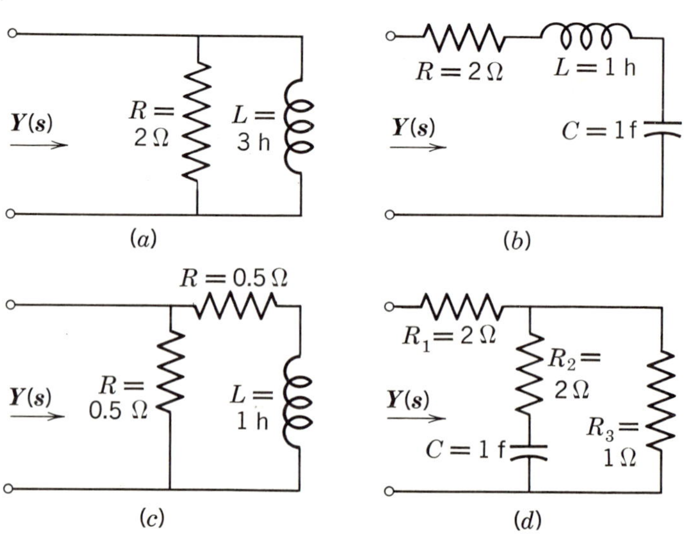

Prob. 16–2.

16–3. A voltage $e_s = 10 \cos (10^4 t + 30°)$ is applied to the circuit of Prob. 16–1*b*. Determine $i(t)$.

16–4. Display the frequency components of each of the following time functions on a complex frequency plane and comment on the physical significance of the various locations for the frequency components.

(a) $10\epsilon^{-5t}$ (b) $10\epsilon^{10t}$ (c) $5 \cos 10t$

(d) 100 (e) $5\epsilon^{-5t} \sin 10t$ (f) $10\epsilon^{10t} \cos 20t$

16–5. Determine the scale factor and the critical frequencies for each of the impedance and admittance functions of Probs. 16–1 and 16–2. Plot the pole-zero configuration for each circuit.

16–6. Compare the scale factors and pole-zero configurations for the circuits of Probs. 16–1*d* and 16–2*c*. How are these circuits related?

16–7. Determine by inspection whether the impedance of each of the circuits shown has a pole or zero (or neither) at $s = 0$. Repeat for $s = \infty$.

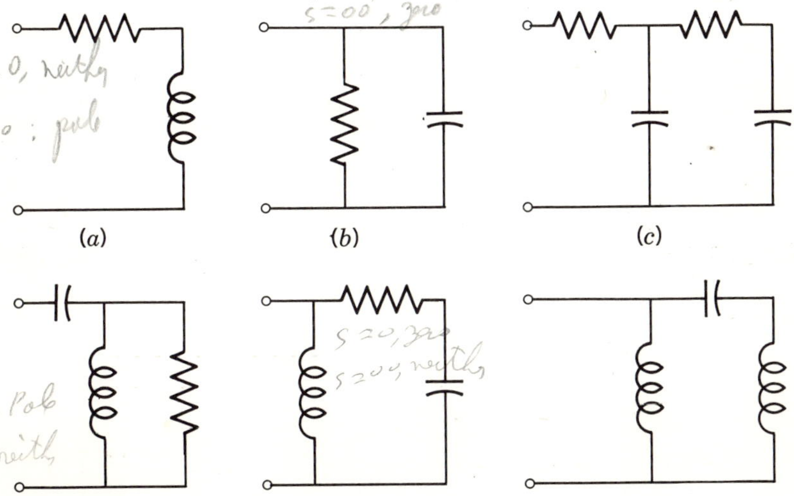

Prob. 16–7.

16–8. Does Prob. 16–7 contain examples of all possible combinations of zero and infinite frequency characteristics? If not, devise simple circuits with the characteristics of the missing combinations.

16–9. Determine $Z(s)$ of the figure on p. 244 and plot the pole-zero configuration on the s-plane. Let $s = j\omega$, with ω ranging from $-\infty$ to $+\infty$, and plot curves of the magnitude and angle of the impedance versus ω. A graphical method of obtaining data is suggested. Comment on these curves. At what points do the maximum and minimum magnitudes occur and what relation do these points have to the pole-zero configuration? How do the magnitude and angle for negative values of ω compare with the magnitude and angle for positive values of ω?

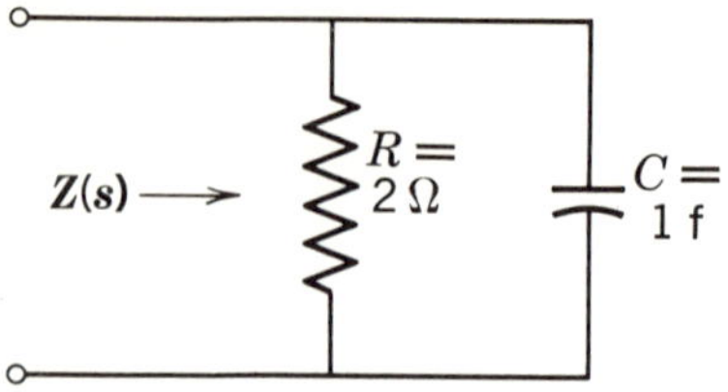

Prob. 16–9.

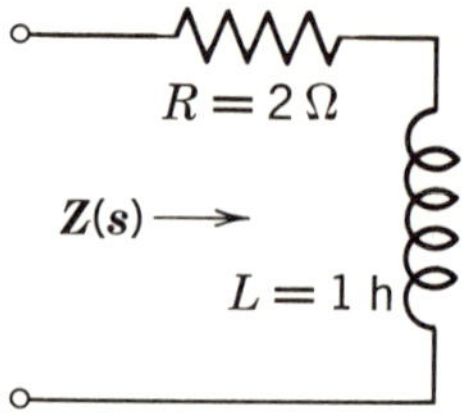

Prob. 16–10.

16–10. Repeat Prob. 16–9 for the circuit shown.

16–11. The admittance of this circuit is given as

$$Y(s) = \frac{1}{R}\frac{(s - s_2)}{(s - s_1)} = \frac{1}{2}\frac{(s + 2)}{s}$$

Plot the pole-zero configuration. Let $s = j\omega$, with ω varying from $-\infty$ to $+\infty$. On the same set of coordinate axes plot curves showing how $|s - s_2|$ and $|s - s_1|$ vary with ω. Use the information contained in these two curves to produce a curve of $|Y|$ versus ω. Plot curves showing how the angles of $(s - s_2)$ and $(s - s_1)$ vary with ω and use these to produce a curve of the angle of the admittance versus ω. Comment on your results, covering such things as the points of minimum and maximum magnitude, the influence of the pole and the zero in determining the magnitude and angle, any symmetry exhibited by the curves, and so forth.

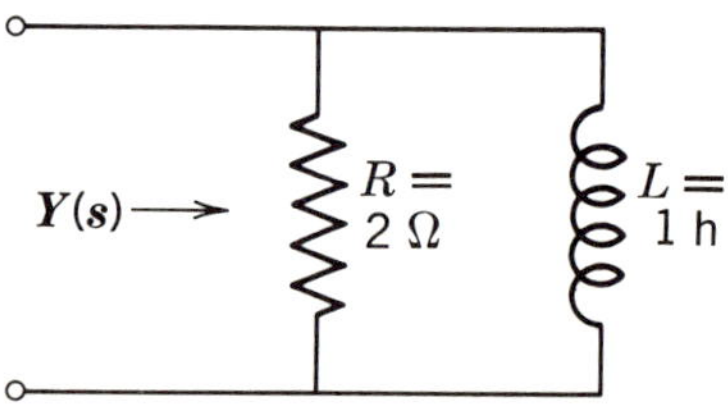

Prob. 16–11.

16–12. A voltage source with $e_s = 3 \cos (3t + 60°)$ is applied to the circuit of Prob. 16–9. Determine the equation for the steady-state current as a function of time.

16–13. Justify the statement made with respect to Eq. 16–22 that all the coefficients of s must be positive in both polynomials.

16–14. Factor the following polynomials and write them in the form of Eq. 16–24.

(a) $s^2 + s$ (b) $s^2 + 1$ (c) $s^2 + 5$

(d) $s^2 + 5s + 6$ (e) $s^2 + s + 1$ (f) $s^2 + 3s + 1$

(g) $s^3 + 6s^2 + 5s$ (h) $s^4 + 5s^2 + 4$ (i) $s^3 + 9s^2 + 26s + 24$

16–15. Inspect the following impedance functions and decide whether $s = \infty$ is a pole, a zero, or neither.

(a) $Z(s) = \dfrac{s^4 + 3s^2 + 1}{s^3 + 2s}$ (b) $Z(s) = 2\dfrac{s}{s^2 + 1}$

(c) $Z(s) = 2\dfrac{s^2 + \frac{3}{2}s + 1}{s^2 + 3s + 1}$ (d) $Z(s) = \dfrac{s^2 + 2s + 17}{s^2 + 3s + 2}$

16–16. Express the impedance of Prob. 16–15, part (d), in the form

$$Z(s) = H\frac{(s - s_1)(s - s_3)}{(s - s_2)(s - s_4)}$$

and plot the pole-zero configuration. Let $s = j\omega$ and plot curves of $|s - s_1|$, $|s - s_3|$, $|s - s_2|$, and $|s - s_4|$ versus ω with ω taking on values from -8 to $+8$. Also plot curves of the angles of $(s - s_1)$, $(s - s_3)$, $(s - s_2)$, and $(s - s_4)$ versus ω.

16–17. Review the results of Probs. 16–9, 16–10, 16–11, and 16–16. Can you draw any conclusions that will apply to all circuits? Answer the following questions. (a) How do the magnitude and angle for negative values of ω compare with those for positive values of ω? (b) What is the derivative of the magnitude with respect to ω at $\omega = 0$? (c) What possible values can the angle have at $\omega = 0$? At $\omega = \infty$? (d) The plot of a single factor $|s - s_2|$ versus ω produces what kind of curve? What determines the value of ω where the minimum value of $|s - s_2|$ occurs? What determines the minimum value?

16–18. Determine $Z(s)$ for each of the circuits illustrated and plot the pole-zero configuration.

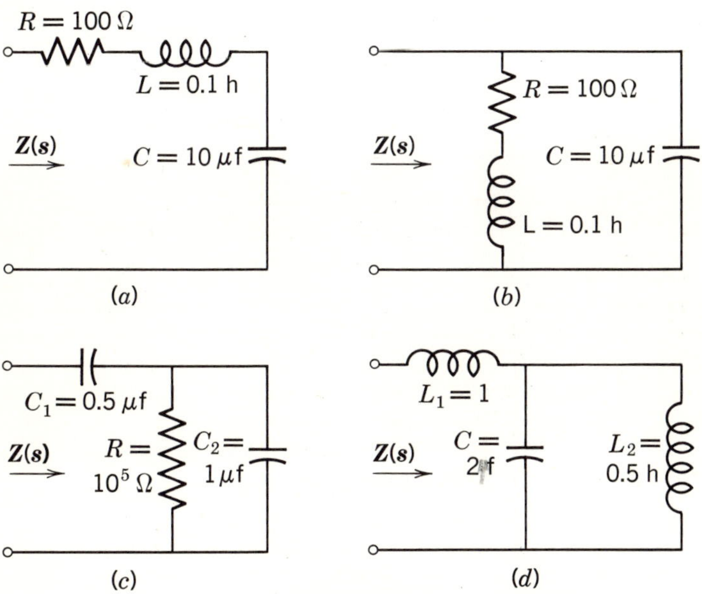

Prob. 16–18.

16–19. Sketch curves showing how the magnitude and angle of the impedance vary as ω varies from 0 to ∞ for each of the circuits of Prob. 16–18. This should be done by inspection.

16–20. Describe the loci of the poles and zeros of the impedance of the figure as R and L are held constant and C is varied. Indicate specifically the points corresponding to $C = 0$ and $C = \infty$.

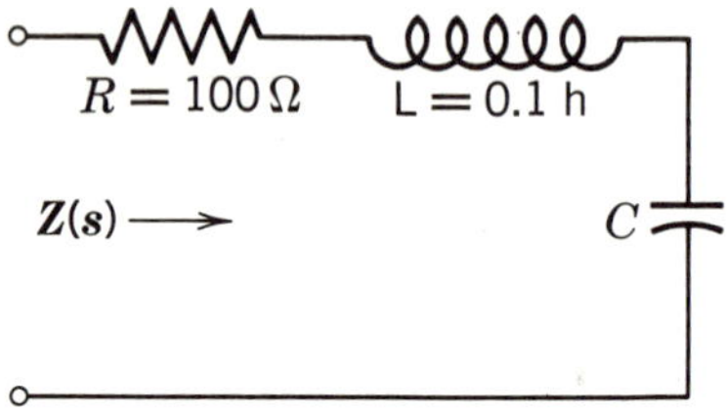

Prob. 16–20.

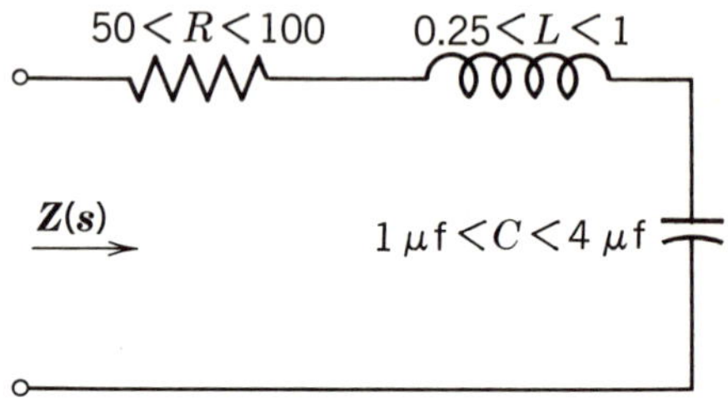

Prob. 16–21.

16–21. Describe the regions in the s-plane in which the poles and zeros of the impedance can lie with the indicated ranges of values for R, L, and C.

16–22. Describe the loci of the poles and zeros of the admittance of this circuit as R is varied. Indicate specifically the points corresponding to $R = 0$, $R = \infty$, and any value of R where the poles or zeros change from real to complex values or vice versa.

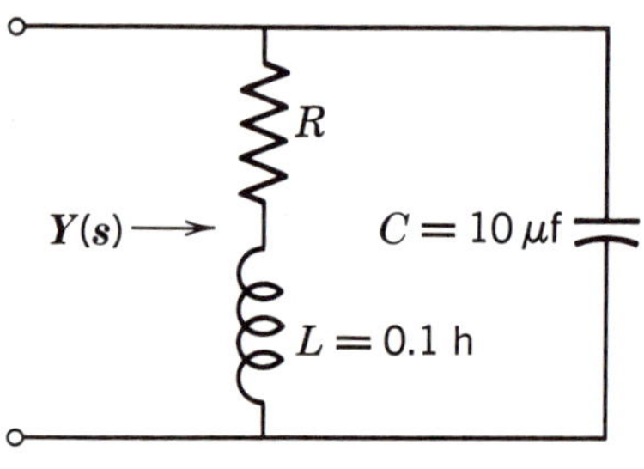

Prob. 16–22.

16–23. Which pair of the three circuits illustrated are duals?

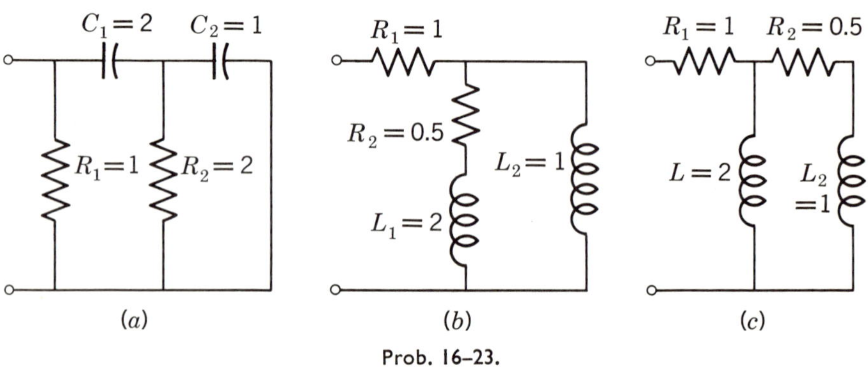

Prob. 16–23.

16–24. Test this expression

$$Z(s) = L_1 \frac{s^3 + s^2 \dfrac{R_2}{L_2} + s\left(\dfrac{R_1R_2}{L_1L_2} + \dfrac{1}{L_1C_1} + \dfrac{1}{L_2C_1}\right) + \dfrac{R_1 + R_2}{L_1L_2C_1}}{s^2 + s\dfrac{R_2}{L_2} + \dfrac{1}{L_2C_1}}$$

as the impedance of the circuit shown to see if it is dimensionally correct.

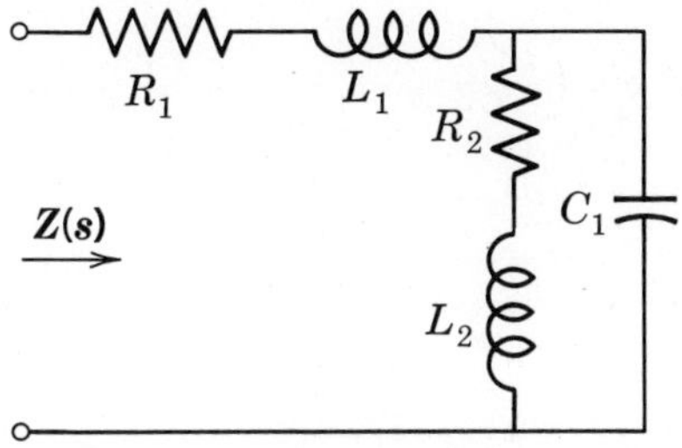

Prob. 16–24.

16–25. Several impedance and admittance functions are given. Determine the dimensions of the scale factor for each of these. Do the dimensions in each case correspond to resistance, inductance, capacitance, or reciprocals of these?

(a) $Z(s) = 2\dfrac{s^2 + 3s + 1}{s + 2}$ (b) $Y(s) = 3\dfrac{s + 1}{s + 2}$

(c) $Z(s) = 5\dfrac{s + 3}{(s + 1)(s + 4)}$ (d) $Y(s) = \dfrac{1}{2}\dfrac{s^2 + 5s + 4}{s + 2}$

(e) $Z(s) = 4\dfrac{s^2 + 4s + 3}{s^2 + 6s + 8}$ (f) $Y(s) = \dfrac{1}{4}\dfrac{s^2 + 4s + 3}{s^3 + 6s^2 + 8s}$

16–26. From the results of Prob. 16–25 formulate a general rule for determining the nature of the scale factor from a knowledge of the degree of the numerator polynomial and the degree of the denominator polynomial.

16–27. Each of the circuits of Prob. 16–1 is to have its impedance magnitude scaled by a factor $K_m = 100$. Determine the new element values. Determine the scale factor and impedance poles and zeros of each circuit after scaling and compare them with the values obtained before scaling.

16–28. Each of the circuits of Prob. 16–18 is to have the frequency scaled by a factor $K_f = 1000$. Determine the scale factor and the poles and zeros of the impedance of each circuit after scaling and compare them with the values obtained before scaling.

16–29. Determine the impedance of the circuit shown on p. 248 and plot the pole-zero configuration. The impedance magnitude is to be scaled by a factor of 100 and the frequency is to be scaled by a factor of 10^4. Determine the new element values and plot the new pole-zero configuration. What has happened to the scale factor of the impedance?

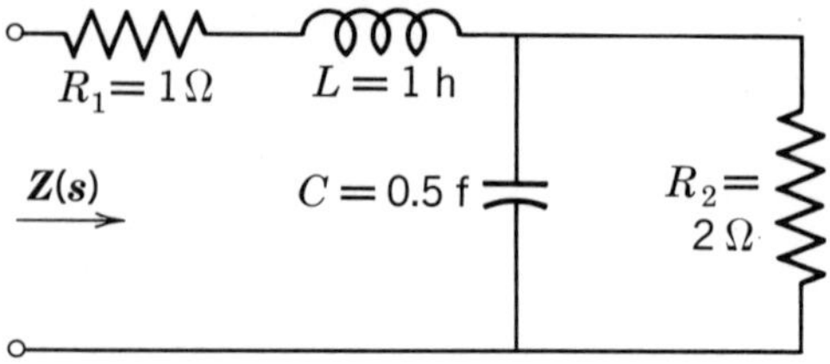

Prob. 16–29.

16–30. By means of magnitude and frequency scaling, change the circuit illustrated to one in which the voltage e_s has a magnitude of unity and a radian frequency of unity. Solve for the current and interpret the result in terms of the original circuit.

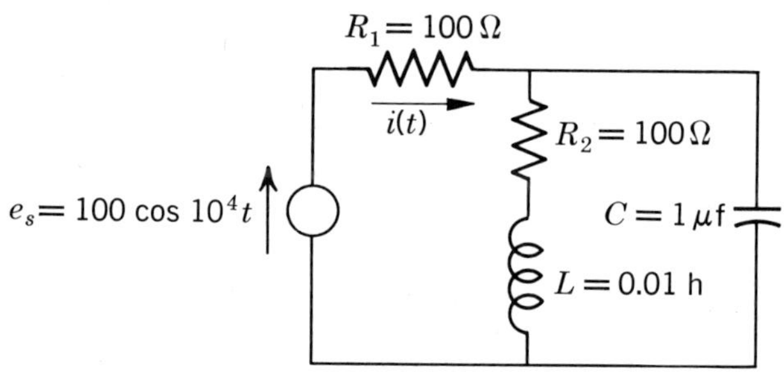

Prob. 16–30.

16–31. The admittance of the adjacent circuit is given as

$$\mathbf{Y}(s) = 2\,\frac{s(s + 0.75)}{(s + 0.5)}$$

It is desired to change the values of resistance and capacitance so that the admittance becomes

$$\mathbf{Y}(s) = 2 \times 10^{-6}\,\frac{s(s + 75 \times 10^3)}{(s + 50 \times 10^3)}$$

Determine the required element values.

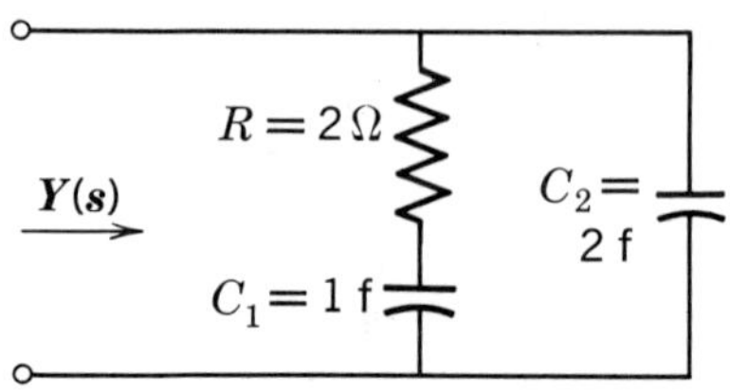

Prob. 16–31.

16–32. The network impedance illustrated is described as follows: (a) The scale factor is 20. (b) There is one pole, located at -30. (c) There is one zero, located at -10. Determine $i(t)$, the steady-state current.

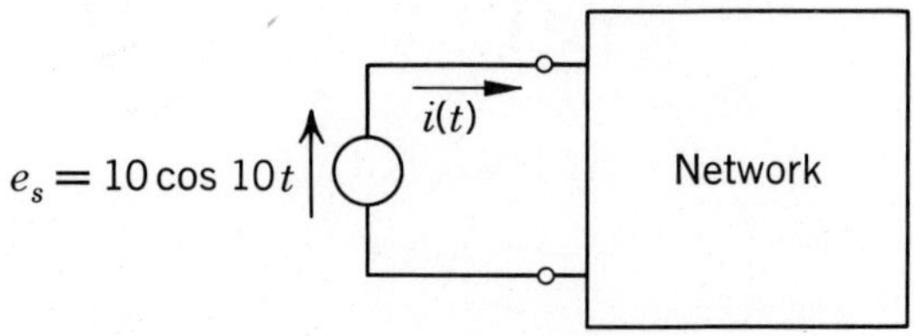

Prob. 16–32.

16–33. The impedance of the network shown is given as

$$Z(s) = 10^6 \frac{s + 100}{s^2 + 100s + 10^6}$$

Determine the steady-state current $i(t)$ for $e_s = 10 + 10 \cos 1000t + 10 \cos 2000t$.

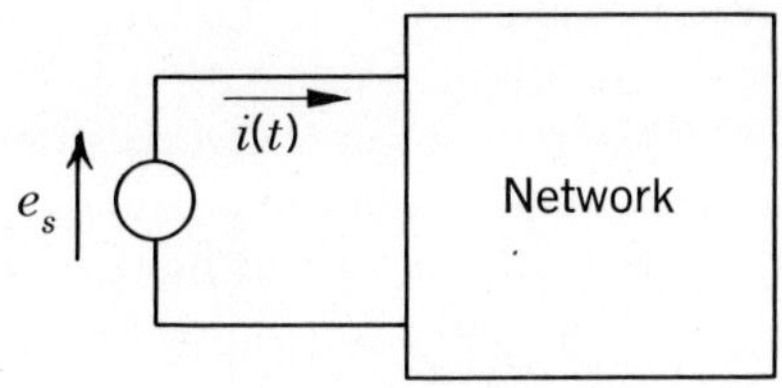

Prob. 16–33.

17

RESONANCE

The phenomenon of resonance is a possibility in any circuit involving both inductance and capacitance. A special case of this, simple series resonance, was discussed in the earlier part of this book from a conventional impedance standpoint and also in terms of stored energies. This chapter will review and extend this discussion in terms of the pole-zero configuration on the complex frequency plane.

The interpretation of resonance phenomena in the complex frequency plane is successful because it makes evident, almost by inspection, relationships which could otherwise be obtained only as a result of extensive algebraic manipulations. The primary reason that this simplification is possible is that the approximations which are necessary to achieve simple end results are more obvious in the geometrical representation than in the algebraic form.

This present chapter will also introduce some applications of resonance phenomena in order to emphasize that it is more than a matter of academic interest.

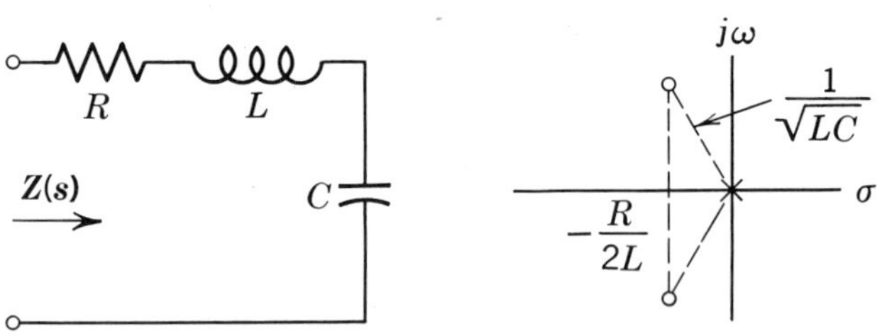

(*a*) R-L-C series circuit (*b*) Pole-zero configuration

Fig. 17–1. Series *R-L-C* circuit and *s*-plane plot of $Z(s)$.

17–1. Series Resonance in the Complex Frequency Plane. The series *R-L-C* circuit, illustrated in Fig. 17–1*a*, has been shown to have the impedance function

$$Z(s) = L\frac{s^2 + \frac{R}{L}s + \frac{1}{LC}}{s} \tag{17–1}$$

which exhibits poles at zero and infinity and zeros at

$$s_{1,3} = -\frac{R}{2L} \pm j\sqrt{\frac{1}{LC} - \left(\frac{R}{2L}\right)^2} \tag{17–2}$$

(Fig. 17–1*b*). Further, it has been demonstrated that as long as $R < 2\sqrt{LC}$ the zeros are complex and at a distance of $1/\sqrt{LC}$ from the origin regardless of the value of R. This condition is satisfied for practically all cases in which resonance is of interest and will be assumed to be true throughout this section.

The impedance at frequencies along the $j\omega$ axis, which correspond to steady-state sinusoidal functions, is the item of major interest. *The condition of resonance is said to exist if this impedance is pure resistance.* Examination of the impedance function of Eq. 17–1 reveals that resonance will occur when $s = \pm j(1/\sqrt{LC})$ since at this frequency

$$Z(s) = L\frac{[j(1/\sqrt{LC})]^2 + \frac{R}{L}[j(1/\sqrt{LC})] + \frac{1}{LC}}{j(1/\sqrt{LC})}$$

$$Z(s) = L\frac{-\frac{1}{LC} + j\frac{R}{L}\cdot\frac{1}{\sqrt{LC}} + \frac{1}{LC}}{j(1/\sqrt{LC})} = R \tag{17–3}$$

This particular frequency will appear so often in the work that follows that it is convenient to represent it by means of a special symbol, or $1/\sqrt{LC} = \omega_0$. The impedance function may now be written in terms of ω_0 as

$$Z(s) = L\frac{s^2 + \frac{R}{L}s + \omega_0^2}{s} \tag{17–4}$$

$$= L\frac{s^2 + \frac{\omega_0 R}{\omega_0 L}s + \omega_0^2}{s} \tag{17–5}$$

The term Q has already been introduced to designate the ratio of inductive reactance of a coil to its resistance. If the resistance in this circuit is assumed to be entirely associated with the inductance, then it becomes appropriate to utilize this terminology.* Since Q is a function of frequency it becomes desirable to specify it at one particular frequency, and a logical choice is the resonant frequency, or

$$Q_0 = \frac{\omega_0 L}{R} \tag{17–6}$$

* The terminology is mathematically correct regardless of the actual source of the resistance—it is only a matter of interpretation.

If this is introduced into the impedance expression of Eq. 17–5,

$$Z(s) = L\frac{s^2 + \dfrac{\omega_0}{Q_0}s + \omega_0^{\,2}}{s} \tag{17–7}$$

from which the zero positions may be expressed as

$$\begin{aligned} s_{1,3} &= -\frac{\omega_0}{2Q_0} \pm j\sqrt{\omega_0^{\,2} - \left(\frac{\omega_0}{2Q_0}\right)^2} \\ &= \omega_0\left[-\frac{1}{2Q_0} \pm j\sqrt{1 - \left(\frac{1}{2Q_0}\right)^2}\right] \end{aligned} \tag{17–8}$$

The pole-zero configuration is now redrawn as Fig. 17–2 in order to emphasize the change in notation.

In most circuits of interest Q_0 will be larger than 10 and is more likely to be in the range from 50 to 200. Therefore the zeros are actually very close to the $j\omega$ axis and the relative length of their real components in Fig. 17–2 has been exaggerated for the sake of clarity.

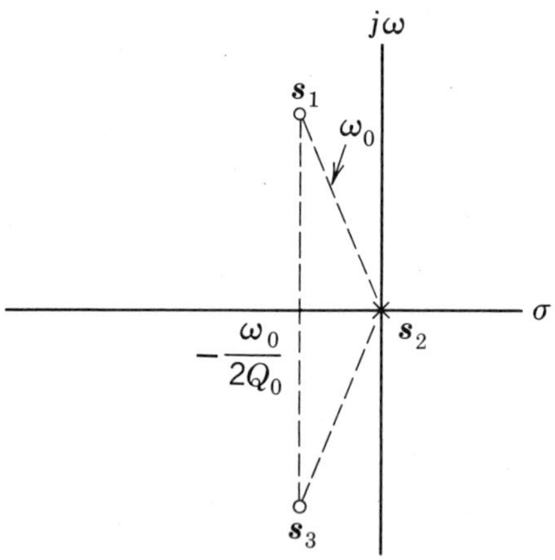

Fig. 17–2. Pole-zero plot for *R-L-C* series circuit.

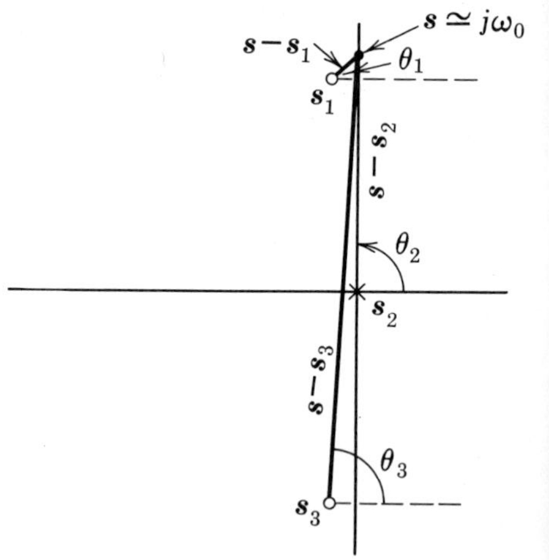

Fig. 17–3. *s*-plane phasor computation of $Z(j\omega)$.

17–2. Series Resonance in High-*Q* Circuits ($Q_0 > 10$). Before proceeding with a general analysis of series resonance it is instructive to consider the simplifications that are possible when a high-Q circuit may be assumed. These simplifications are actually approximations that are introduced but the resulting error is less than $\frac{1}{2}$% if $Q_0 > 10$. Since this is usually the case, the high-Q approximations are more commonly used than the more exact results which are discussed in the next section.

Although there are many ways in which suitable approximations may be arrived at, probably the simplest and most effective approach is to utilize the geometrical relationships evident in the pole-zero configuration. For this purpose consider Fig. 17–3, which again portrays the pole-zero configuration for the series R-L-C circuit and also indicates the phasors that would be involved in computing the magnitude and angle of the impedance at some frequency near the resonant frequency ω_0. In terms of these phasors the magnitude of the impedance may be written as

$$|\boldsymbol{Z}| = L\frac{|\boldsymbol{s} - \boldsymbol{s}_1|\,|\boldsymbol{s} - \boldsymbol{s}_3|}{|\boldsymbol{s} - \boldsymbol{s}_2|} \tag{17–9}$$

and the angle as

$$\theta = \theta_1 + \theta_3 - \theta_2 \tag{17–10}$$

If it is assumed at this time that the value of impedance is of interest only for frequencies in the vicinity of $\boldsymbol{s} = j\omega_0$ (which would generally be the case for high-Q circuits), then it is evident from Fig. 17–3 that $|\boldsymbol{s} - \boldsymbol{s}_1|$ will be the length mostly dependent upon the value of $\boldsymbol{s}$ since the other two lengths will not change much with small changes in $\boldsymbol{s}$. For example, a 1% change in $\boldsymbol{s}$ along the $j\omega$ axis may result in a 100% change in $|\boldsymbol{s} - \boldsymbol{s}_1|$ with only a 1% change in $|\boldsymbol{s} - \boldsymbol{s}_2|$ and a $\frac{1}{2}$% change in $|\boldsymbol{s} - \boldsymbol{s}_3|$. In addition, the two nearly constant lengths appear as a ratio and the percentage change of this ratio is even less. As a result, it is very nearly correct to assume that the ratio of $|\boldsymbol{s} - \boldsymbol{s}_3|$ to $|\boldsymbol{s} - \boldsymbol{s}_2|$ remains constant at 2. The magnitude of the impedance then becomes approximately

$$|Z| = 2L\,|\boldsymbol{s} - \boldsymbol{s}_1| \tag{17–11}$$

In considering the angle it is observed that the angle θ_2 is exactly $\pi/2$ for any value of $\boldsymbol{s}$ on the $j\omega$ axis and the angle θ_3 is very nearly $\pi/2$. Hence, the total angle of the impedance is approximately

$$\theta = \theta_1 \tag{17–12}$$

As a result of these approximations it may be seen that the effects of the critical frequencies $\boldsymbol{s}_2$ and $\boldsymbol{s}_3$ have been completely removed so far as variation in $\boldsymbol{Z}(\boldsymbol{s})$ is concerned and $\boldsymbol{s}_1$ plays the predominant role. In view of this it is possible to write the approximate complete impedance function as

$$\boldsymbol{Z}(\boldsymbol{s}) = 2L(\boldsymbol{s} - \boldsymbol{s}_1) \tag{17–13}$$

The critical frequency $\boldsymbol{s}_1$ was shown by Eq. 17–8 to be

$$\boldsymbol{s}_1 = \omega_0\left[-\frac{1}{2Q_0} + j\sqrt{1 - \left(\frac{1}{2Q_0}\right)^2}\right] \tag{17–14}$$

which for high Q values is approximately

$$\boldsymbol{s}_1 \simeq \omega_0\left(-\frac{1}{2Q_0} + j\right) \tag{17–15}$$

This approximation amounts to saying that the imaginary component of the zero's position is essentially the same as its distance from the origin, ω_0. If this value of s_1 is substituted into Eq. 17–13 the resulting approximate impedance function becomes

$$Z(s) = 2L\left(s + \frac{\omega_0}{2Q_0} - j\omega_0\right) \tag{17–16}$$

This impedance function looks peculiar because it violates a previously stated requirement that there be no imaginary terms in any physically realizable impedance function. This difficulty is removed, however, if it is remembered that Eq. 17–16 need not, and does not, *exactly* represent any actual impedance but is only a good *approximation* to an actual impedance in a small portion of the s-plane, near $s = j\omega_0$. For purposes of computation it does not matter that the impedance of Eq. 17–16 is not realizable.

Since the critical frequencies s_2 and s_3 do not contribute to the approximate impedance function, it becomes possible to redraw the pole-zero configuration and include only the portion of the s-plane near s_1. By this expansion of a small portion of the s-plane, the nature of the impedance function near $j\omega_0$ may be examined in more detail. This expansion is illustrated in Fig. 17–4.

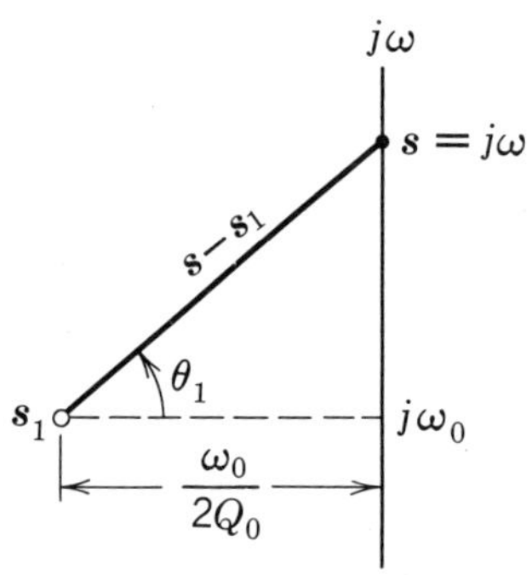

Fig. 17–4. Approximate s-plane plot for the *R-L-C* series circuit near resonance.

The steps taken in arriving at the approximate impedance function and the approximate pole-zero configuration have been discussed in considerable detail in order that they may be understood clearly. As a result of this detail it may be thought that obtaining the approximation is more trouble than it is worth. This is only a temporary difficulty, however, and after gaining experience it becomes possible to obtain approximations of this sort almost by inspection. When this degree of facility with the complex frequency plane has been attained, this type of approach to high-Q circuit problems can result in significant savings in time and effort. It is, therefore, highly recommended that the student be sure that he understands the nature of, and the reason for, every step of the approximation just completed.

Obtaining the approximate forms for the impedance function and the pole-zero configuration is, of course, just the first step in carrying out any analysis of the *R-L-C* circuit near resonance. There are many items which might be investigated in such an analysis; only a few will be considered here.

Magnitude of Impedance. Since the magnitude of impedance is given by

$$Z = 2L\,|s - s_1| \tag{17–17}$$

it may be seen by inspection of Fig. 17–4 that this will be

$$Z = 2L\sqrt{(\omega - \omega_0)^2 + \left(\frac{\omega_0}{2Q_0}\right)^2} \tag{17-18}$$

Generally, the difference between the actual frequency and the resonant frequency is the variable of primary interest and it is convenient to represent this by

$$\Delta\omega = \omega - \omega_0$$

from which

$$|Z| = 2L\sqrt{(\Delta\omega)^2 + \left(\frac{\omega_0}{2Q_0}\right)^2} = \frac{\omega_0 L}{Q_0}\sqrt{1 + \left(\frac{2Q_0\,\Delta\omega}{\omega_0}\right)^2}$$

$$= R\sqrt{1 + \left(\frac{2Q_0\,\Delta\omega}{\omega_0}\right)^2} = R\sqrt{1 + \left(\frac{2L}{R}\,\Delta\omega\right)^2} \tag{17-19}$$

This impedance is sketched in Fig. 17–5*a* as a function of $\Delta\omega$ for several values of R. If a constant voltage source of variable frequency is applied to this circuit, the resulting current magnitude will be

$$|I| = \frac{|E|}{|Z|} = \frac{E}{R\sqrt{1 + \left(\frac{2L}{R}\,\Delta\omega\right)^2}} \tag{17-20}$$

This current is sketched in Fig. 17–5*b*.

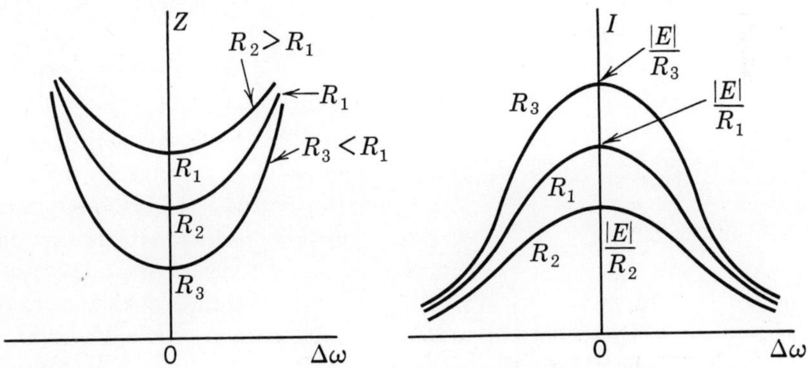

(*a*) Magnitude of impedance (*b*) Magnitude of current

Fig. 17–5. Z and I versus $\Delta\omega$ for various R values in the series *R-L-C* circuit.

Angle of Impedance. Inspection of Fig. 17–4 reveals that the angle of the impedance is

$$\theta = \theta_1 = \tan^{-1}\frac{2Q_0(\Delta\omega)}{\omega_0} = \tan^{-1}\left(\frac{2L}{R}\,\Delta\omega\right) \tag{17-21}$$

This angle is illustrated versus $\Delta\omega$ in Fig. 17–6 for various values of R.

BANDWIDTH. The bandwidth of a resonant circuit is usually an item of considerable interest in any practical application. The implication of the term *bandwidth* is that it designates the range of frequencies to which the circuit will respond but it is apparent from the preceding figures that there is some

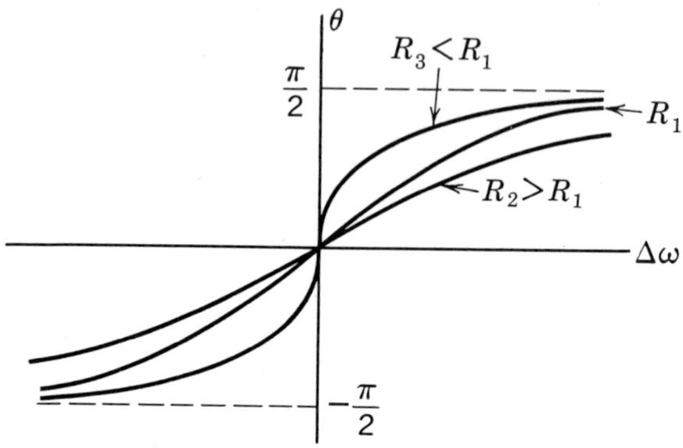

Fig. 17–6. θ versus $\Delta\omega$ for various R values in the series R-L-C circuit.

response at all frequencies. Hence, it is necessary to make an arbitrary definition as to the exact specification of bandwidth. An often used definition, which is applicable in many different types of circuits, designates the bandwidth as the interval between frequencies at which the response is 0.707 of the maximum response. The figure 0.707 ($1/\sqrt{2}$) is selected to give simple equations for the bandwidth of most circuits.

For the series R-L-C circuit, to which a constant voltage source is applied, the current will be $1/\sqrt{2}$ times its maximum value when the impedance $|Z|$ is $\sqrt{2}$ times its minimum value. From Fig. 17–4 it is apparent that this will occur when

$$\omega - \omega_0 = \frac{\omega_0}{2Q_0} \qquad (17\text{–}22)$$

and the frequency increment between such points is

$$W = 2\frac{\omega_0}{2Q_0} = \frac{\omega_0}{Q_0} \qquad (17\text{–}23)$$

where W is the bandwidth in radians per second. This result is indicated in Fig. 17–7.

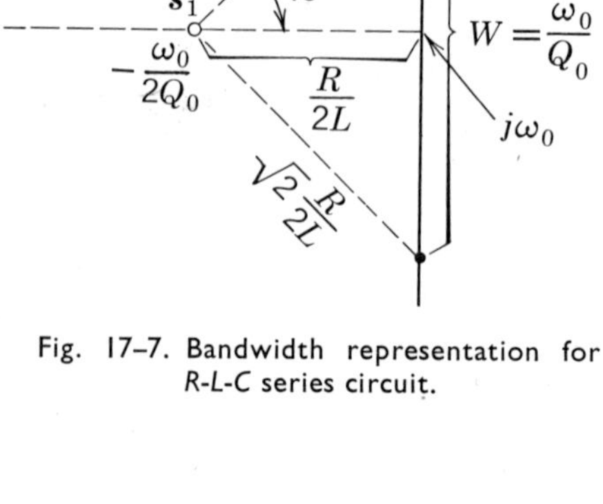

Fig. 17–7. Bandwidth representation for R-L-C series circuit.

17–3. Exact Analysis of Series Resonance. In the preceding section it was found that the high-Q circuit could be analyzed very easily from an approximate pole-zero configuration which consisted of a single complex zero. If the Q of the circuit is not high enough to permit this approximation, more exact results may be obtained from the entire pole-zero configuration. It is of interest to compare these exact results with the approximate ones just obtained.

Magnitude of Impedance. The exact expression for $\mathbf{Z}(s)$ was given by Eq. 17–7 as

$$\mathbf{Z}(s) = L\frac{s^2 + \dfrac{\omega_0}{Q_0}s + {\omega_0}^2}{s} \tag{17–7}$$

which, for $s = j\omega$, becomes

$$\mathbf{Z}(j\omega) = L\frac{-\omega^2 + j\dfrac{\omega_0}{Q_0}\omega + {\omega_0}^2}{j\omega} \tag{17–24}$$

The magnitude of the impedance is

$$|Z| = L\frac{\sqrt{({\omega_0}^2 - \omega^2)^2 + \left(\dfrac{\omega_0}{Q_0}\omega\right)^2}}{\omega} \tag{17–25}$$

which may be rewritten as

$$\begin{aligned} |Z| &= L\sqrt{\left(\frac{\omega_0}{Q_0}\right)^2 + \left(\frac{{\omega_0}^2}{\omega} - \omega\right)^2} \\ &= L\sqrt{\left(\frac{\omega_0}{Q_0}\right)^2 + {\omega_0}^2\left(\frac{\omega_0}{\omega} - \frac{\omega}{\omega_0}\right)^2} \\ &= \frac{\omega_0 L}{Q_0}\sqrt{1 + {Q_0}^2\left(\frac{\omega_0}{\omega} - \frac{\omega}{\omega_0}\right)^2} \\ &= R\sqrt{1 + \left(\frac{\omega_0 L}{R}\right)^2\left(\frac{\omega_0}{\omega} - \frac{\omega}{\omega_0}\right)^2} \end{aligned} \tag{17–26}$$

This result appears to be considerably different from the one obtained with the high-Q approximation, but actually the difference is not great. In order to compare them more easily, it is convenient to introduce a new frequency variable which is defined as

$$\Delta\beta = \frac{\omega_0}{2}\left(\frac{\omega}{\omega_0} - \frac{\omega_0}{\omega}\right) \tag{17–27}$$

Substituting this variable into Eq. 17–26 results in

$$|Z| = R\sqrt{1 + \left(\frac{2L}{R}\Delta\beta\right)^2} \tag{17–28}$$

which is identical in form with Eq. 17–19. Curves for the magnitude of the impedance, as a function of $\Delta\beta$ instead of $\Delta\omega$, would be identical to those plotted in Fig. 17–5*a*. Therefore, the only comparison that needs to be made is that of $\Delta\beta$ to $\Delta\omega$. This comparison is facilitated by rewriting $\Delta\beta$ as

$$\Delta\beta = \frac{\omega_0}{2}\left(\frac{\omega^2 - \omega_0^2}{\omega\omega_0}\right) = \frac{1}{2}\left(\frac{\omega + \omega_0}{\omega}\right)(\omega - \omega_0) = \frac{1}{2}\left(1 + \frac{\omega_0}{\omega}\right)\Delta\omega \tag{17–29}$$

From this expression it may be seen that when ω is very close to ω_0, then $\Delta\beta$ and $\Delta\omega$ are nearly the same.

An item that is, perhaps, of even greater interest than the relative sizes of $\Delta\beta$ and $\Delta\omega$ is the nature of the symmetry they exhibit. The high-Q approximations exhibit *arithmetic symmetry*—a term which implies that the resonant frequency ω_0 is the arithmetic mean of any pair of frequencies for which $|\Delta\omega|$, and therefore $|Z|$, have the same value. Graphically, arithmetic symmetry means that curves of $|Z|$ are symmetrical about ω_0 when plotted on a *linear ω scale.*

The exact expressions, involving $\Delta\beta$, exhibit *geometric symmetry*—which implies that the resonant frequency ω_0 is the geometric mean of any pair of frequencies for which $|\Delta\omega|$ and $|Z|$ have the same value. This may be demonstrated by letting $\Delta\beta_1$ be the value of $\Delta\beta$ for a frequency ω_1. The frequency ω_2 at which $\Delta\beta$ is $-\Delta\beta_1$ (hence, has the same value of $|Z|$) is

$$\frac{\omega_0}{2}\left(\frac{\omega_2}{\omega_0} - \frac{\omega_0}{\omega_2}\right) = -\Delta\beta_1 = -\frac{\omega_0}{2}\left(\frac{\omega_1}{\omega_0} - \frac{\omega_0}{\omega_1}\right) \tag{17–30}$$

from which

$$\frac{\omega_2}{\omega_0} - \frac{\omega_0}{\omega_2} = \frac{\omega_0}{\omega_1} - \frac{\omega_1}{\omega_0} \tag{17–31}$$

Multiplying both sides by $\omega_1\omega_2\omega_0$ gives

$$\omega_1\omega_2^2 - \omega_1\omega_0^2 = \omega_2\omega_0^2 - \omega_1^2\omega_2$$

$$\omega_1\omega_2^2 + \omega_1^2\omega_2 = \omega_2\omega_0^2 + \omega_1\omega_0^2$$

$$\omega_1\omega_2(\omega_2 + \omega_1) = \omega_0^2(\omega_2 + \omega_1)$$

$$\omega_1\omega_2 = \omega_0^2$$

or

$$\omega_0 = \sqrt{\omega_1\omega_2} \tag{17–32}$$

Graphically, geometric symmetry means that curves of $|Z|$ are symmetrical about ω_0 when plotted on a *logarithmic ω scale.* Some further implications of this type of symmetry are discussed later.

Angle of Impedance. The exact angle of the impedance may be obtained directly from Eq. 17–18 as

$$\theta = \tan^{-1}\frac{\omega_0\omega/Q_0}{\omega_0^2 - \omega^2} - \frac{\pi}{2} \tag{17–33}$$

$$= \tan^{-1}\frac{\omega^2 - \omega_0^2}{\omega_0\omega/Q_0} = \tan^{-1}\left[Q_0\left(\frac{\omega}{\omega_0} - \frac{\omega_0}{\omega}\right)\right]$$

$$= \tan^{-1}\left[\frac{\omega_0 L}{R}\left(\frac{\omega}{\omega_0} - \frac{\omega_0}{\omega}\right)\right]$$

$$= \tan^{-1}\left(\frac{2L}{R}\,\Delta\beta\right) \tag{17–34}$$

It is seen again that the exact angle may be specified in terms of $\Delta\beta$ by exactly the same expression as was used to obtain the approximate angle in terms of $\Delta\omega$.

Bandwidth. In the preceding section it was specified that the bandwidth of a series resonant circuit was the frequency increment between points at which the magnitude of the impedance was 1.414 times its minimum value. Applying this same definition to the exact case makes it necessary to find the frequencies at which

$$Q_0\left(\frac{\omega}{\omega_0} - \frac{\omega_0}{\omega}\right) = 1 \tag{17–35}$$

as may be seen from Eq. 17–26. This may be written, after dividing both sides by Q_0/ω_0, as

$$\omega - \frac{\omega_0^2}{\omega} = \frac{\omega_0}{Q_0} \tag{17–36}$$

If one of the frequencies at which this condition occurs is designated as ω_1 and the other one as ω_2, then Eq. 17–36 may be expressed as

$$\omega_2 - \frac{\omega_0^2}{\omega_2} = \frac{\omega_0}{Q_0} \tag{17–37}$$

but

$$\omega_1 = \frac{\omega_0^2}{\omega_2} \tag{17–38}$$

so it becomes

$$\omega_2 - \omega_1 = \frac{\omega_0}{Q_0} \tag{17–39}$$

But $(\omega_2 - \omega_1)$ is the desired frequency increment, hence, is the bandwidth W. Therefore,

$$W = \frac{\omega_0}{Q_0} \tag{17–40}$$

It is somewhat surprising to note that the *approximate high-Q* bandwidth was really no approximation at all but was an *exact* result for any Q_0 even though it was obtained from an approximate pole-zero configuration. In view of the different types of symmetry involved in the two cases it is of interest to compare them and see how they can have the same bandwidth. This comparison is indicated in Fig. 17–8, which shows the variation in current magnitude with frequency for the exact and approximate cases for a Q_0 of 1. This

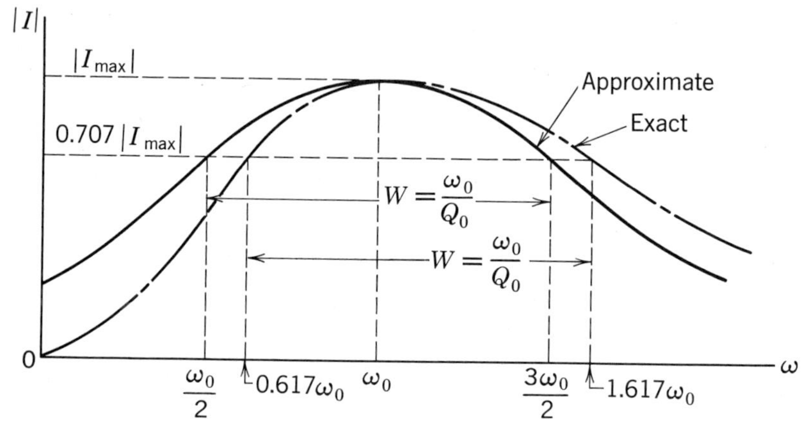

Fig. 17–8. Comparison between approximate and exact values of $|I|$ versus ω for $Q_0 = 1$.

value of Q_0 is far too low to expect the approximate result to be very close to the exact result but the curves do indicate how the bandwidths might be the same. If Q_0 were greater than 10, the approximate and exact curves would be very close together.

The significant aspects of the comparison between the exact and the approximate high-Q analysis may now be summarized as follows:

1. The same form of expression for the magnitude and angle of the impedance may be used by selecting the appropriate frequency variable.
2. The frequency variable for the approximate case exhibits arithmetic symmetry whereas that for the exact case exhibits geometric symmetry.
3. The expressions for bandwidth and resonant frequency are exact in either case.

The foregoing results were obtained specifically for the series R-L-C circuit, but they are typical of what might be expected from any resonant circuit.

17–4. Voltages in the Series R-L-C Circuit. In all of the discussions so far pole-zero configurations have been used to represent impedance functions—which are, of course, the ratio of a voltage to a current. There is no reason, however, why the ratio of two voltages cannot also be represented by means of poles and zeros in the complex frequency plane. Actually, this application of the pole-zero configuration is probably more common than the impedance representation.

As an illustration of this application and as a means of further investigating the series *R-L-C* circuit, consider the pole-zero configurations associated with the ratios of resistor voltage $\mathbf{E}_R$, capacitor voltage $\mathbf{E}_C$, and inductor voltage

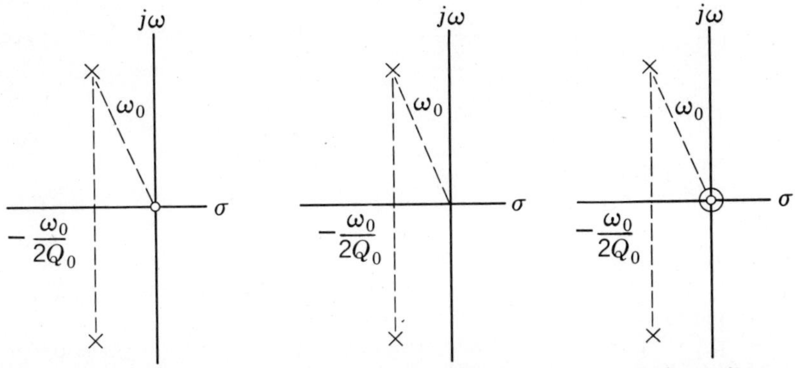

(*a*) Resistor voltage E_R/E_s (*b*) Capacitor voltage E_C/E_s (*c*) Inductor voltage E_L/E_s

Fig. 17–9. Pole-zero configurations for the ratio of resistor, capacitor, and inductor voltage to the applied voltage.

$\mathbf{E}_L$, all with respect to the source voltage $\mathbf{E}_s$. These ratios may be obtained by taking the ratio of the impedance of the appropriate element to the total impedance, and the following expressions result:

$$\frac{\mathbf{E}_R}{\mathbf{E}_s}(s) = \frac{R}{\mathbf{Z}(s)} = \frac{R}{L}\,\frac{s}{s^2 + \dfrac{\omega_0}{Q_0}s + {\omega_0}^2} \tag{17–41}$$

$$\frac{\mathbf{E}_C}{\mathbf{E}_s}(s) = \frac{1/sC}{\mathbf{Z}(s)} = \frac{{\omega_0}^2}{s^2 + \dfrac{\omega_0}{Q_0}s + {\omega_0}^2} \tag{17–42}$$

$$\frac{\mathbf{E}_L}{\mathbf{E}_s}(s) = \frac{sL}{\mathbf{Z}(s)} = \frac{s^2}{s^2 + \dfrac{\omega_0}{Q_0}s + {\omega_0}^2} \tag{17–43}$$

The resulting pole-zero configurations are shown in Fig. 17–9.

Examination of these pole-zero configurations yields much of the significant information concerning these voltages with little or no computation.

For example,

1. For high-Q circuits in which the poles would be close to the $j\omega$ axis, all three voltages should have a maximum in the vicinity of $\boldsymbol{s} = j\omega_0$.
2. At $\boldsymbol{s} = j\omega_0$ the exact voltage ratios become

$$\frac{\boldsymbol{E}_R}{\boldsymbol{E}_s} = \frac{R}{L} \cdot \frac{j\omega_0}{-\omega_0^2 + j\dfrac{\omega_0^2}{Q_0} + \omega_0^2} = \frac{R}{L}\frac{Q_0}{\omega_0} = 1 \tag{17-44}$$

$$\frac{\boldsymbol{E}_C}{\boldsymbol{E}_s} = \frac{\omega_0^2}{-\omega_0^2 + j\dfrac{\omega_0^2}{Q_0} + \omega_0^2} = -jQ_0 \tag{17-45}$$

$$\frac{\boldsymbol{E}_L}{\boldsymbol{E}_s} = \frac{-\omega_0^2}{-\omega_0^2 + j\dfrac{\omega_0^2}{Q_0} + \omega_0^2} = jQ_0 \tag{17-46}$$

3. The ratio $\boldsymbol{E}_R/\boldsymbol{E}_s$ will have its maximum value exactly at $j\omega_0$ since its pole-zero configuration is simply the reciprocal of that for the impedance, which is a minimum at this value.
4. The other two ratios will *not* have their maxima exactly at $j\omega_0$ because they have different zeros from $\boldsymbol{E}_R/\boldsymbol{E}_s$. In a high-$Q$ circuit, however, the deviation will be negligibly small. An exact analysis for low-Q circuits is left to the student.

One of the most interesting aspects of series resonance is the large voltages appearing across the inductance and capacitance at resonance. As indicated before, this voltage is Q_0 times the source voltage, hence, may become very large indeed. Whether this result is desirable or undesirable depends upon the application of series resonance being considered, as will be seen in the following section.

17–5. Applications of Series Resonant Circuits. There are many useful applications of series resonance and the purpose of this section is to discuss a few of these as a means of further emphasizing some of the properties which have been presented. These applications will be discussed very briefly with just enough detail to explain the major reasons for utilizing series resonance. By now, the student should be able to supply the analytical details of any of these circuits.

As a first application consider the input circuit of a typical broadcast-band radio receiver. Most receivers of this type now use a loop antenna as a means of extracting a signal from electromagnetic waves. Such a loop antenna usually consists of a coil having a few turns and enclosing a large area. This coil is then connected to a capacitor, whose value can be varied. The voltage appearing across the capacitor is applied to the first vacuum tube in the receiver. A diagram of this arrangement is shown in Fig. 17–10a. The loop antenna may be considered to be electrically equivalent to a voltage source in series with inductance and resistance and so the entire equivalent circuit will be as shown in Fig. 17–10*b*. The capacitor C may be adjusted to make the circuit resonant at any desired frequency.

This resonant condition accomplishes two desirable results. First, the voltage $\mathbf{E}_C$ which is applied to the first vacuum tube is greater than $\mathbf{E}_s$, the voltage induced in the loop, by the factor Q_0. Since Q_0 may well be in the order of 100, an appreciable voltage gain has been achieved very simply. Second, voltages of other frequencies induced in the loop will not be increased by this amount because the circuit is not in resonance with these other frequencies. Therefore, the circuit has separated the desired signal from many

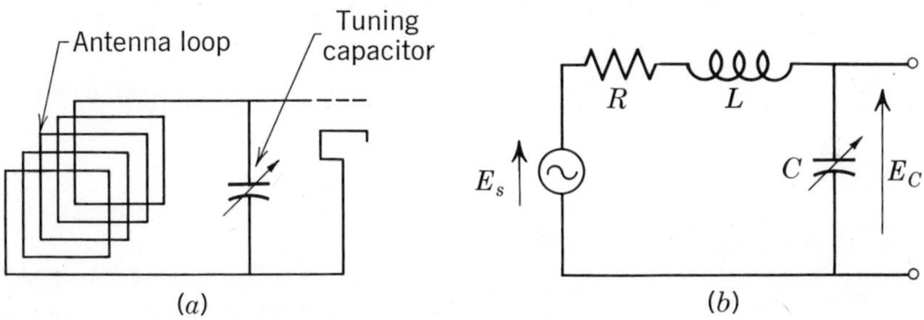

Fig. 17–10. Input circuit of a typical broadcast-band receiver.

undesired signals. How well it can accomplish this separation for frequencies close to the resonant frequency depends upon the *bandwidth* of the circuit, which was also shown to be a function of Q_0. In general, therefore, a high value of Q_0 is desired.

The second application to be considered involves a method for measuring the Q of any coil. This may be done by having a voltage source of small known amplitude and of variable frequency, a variable capacitor, and a vacuum-tube voltmeter (VTVM), all connected as shown in Fig. 17–11. The

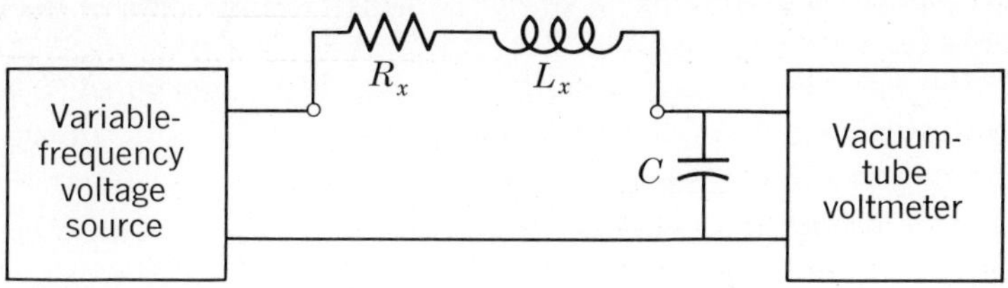

Fig. 17–11. A method of measuring Q.

coil whose Q is to be measured is designated as L_x, R_x. The measurement consists of adjusting the capacitor C for maximum deflection on the VTVM for the frequency at which it is desired to know the Q. The reading of the VTVM is just Q times the applied voltage and, if this applied voltage is held constant, the scale of the VTVM may be calibrated to read directly in units of Q. It is also possible to obtain the inductance of the coil from a knowledge of the frequency and the value of capacitance needed to tune a circuit to resonance. The entire device is known as a *Q-meter* and is commercially available in a convenient form.

A third application of series resonance involves the grounding of a power system. It is generally desirable in a power system to have one or more points connected to the earth in order to establish a definite set of voltages in the system with respect to ground or zero voltage. Generally, this grounding is accomplished by means of a solid conductor tying the desired point to ground, as shown in Fig. 17–12a. Occasionally, however, it is desirable to have this

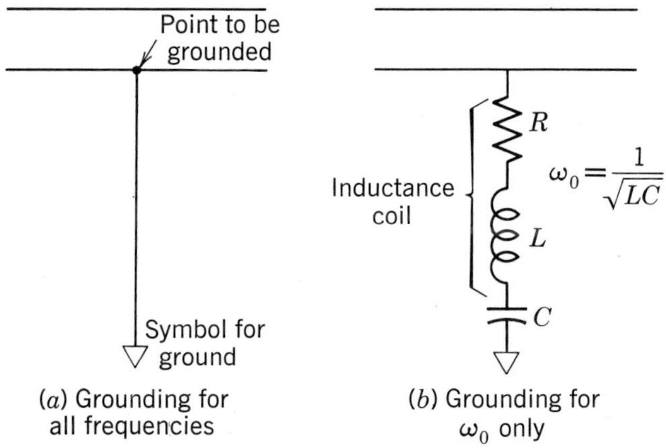

Fig. 17–12. Selective grounding of a power system.

grounding effective only for the principal frequency in the power system—usually 60 cps—and have no ground for other frequencies, such as 180 cps, which may have arisen in the system as a result of some nonlinearity. In such instances the grounding may be accomplished by means of a series circuit, resonant at 60 cps, as shown in Fig. 17–12b, but having a high impedance at 180 cps. From the previous work it will be remembered that the impedance at resonance is simply the coil resistance and this, of course, should be as low as possible. However, a low value of R means a large value of Q, which in turn may result in large voltages appearing across the inductance and capacitance. In this instance, the large voltage is undesirable—but unavoidable. The desired characteristic of the circuit is its low impedance at the resonant frequency.

17–6. Parallel Resonance in the Complex Frequency Plane. There is really very little point in attempting to distinguish between series and parallel resonance because they are only slightly different aspects of the same phenomenon. In fact, it is often possible to interpret the resonance behavior of a given circuit as *either* series or parallel resonance and arrive at the same result. However, a distinction is often made between the two as a convenience in specifying the characteristic of major interest for a given application. Such a distinction is made in this chapter but it should be emphasized that this does *not* imply two different types of resonance; the resonance phenomenon is the same and is simply being interpreted from two different viewpoints.

Consider first the dual of the series R-L-C circuit with a voltage source. In accordance with the previous discussion of duals, this will be a parallel G-L-C circuit with a current source as shown in Fig. 17–13. The admittance func-

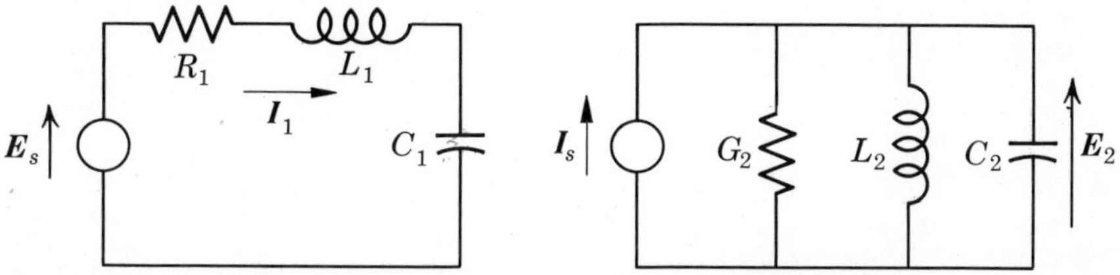

(*a*) Series R-L-C circuit (*b*) Dual of (*a*)

Fig. 17–13. Equivalent series and parallel circuits (duals).

tion of the dual will be identical to the impedance function of the series R-L-C circuit if each symbol is replaced by its dual. By analogy to Eq. 17–1:

$$Y(s) = C_2 \frac{s^2 + \dfrac{G_2}{C_2} s + \dfrac{1}{L_2 C_2}}{s} \tag{17–47}$$

This may be converted into the form involving resonant frequency and Q by noting that

$$\frac{1}{L_2 C_2} = \frac{1}{L_1 C_1} = \omega_0^{\,2} \tag{17–48}$$

and

$$Q_0 = \frac{\omega_0 L_1}{R_1} = \frac{\omega_0 C_2}{G_2} \tag{17–49}$$

from which Eq. 17–47 may be written as

$$Y(s) = C_2 \frac{s^2 + \dfrac{\omega_0}{Q_0} s + \omega_0^{\,2}}{s} \tag{17–50}$$

The pole-zero configuration for this admittance, shown in Fig. 17–14*a*, is identical to that previously drawn for the series R-L-C circuit. For some purposes it may be more convenient to utilize the impedance function $\boldsymbol{Z}(s)$

which is associated with this circuit. This impedance function is shown in Fig. 17–14*b*.

It is now possible to utilize all of the results obtained for the series *R-L-C* circuit by interpreting them in their dual fashion. Sketches of impedance magnitude may be interpreted as admittance magnitude, sketches of current interpreted as voltage, and so on. Because of this duality it is not necessary to repeat the analysis or even tabulate the end results.

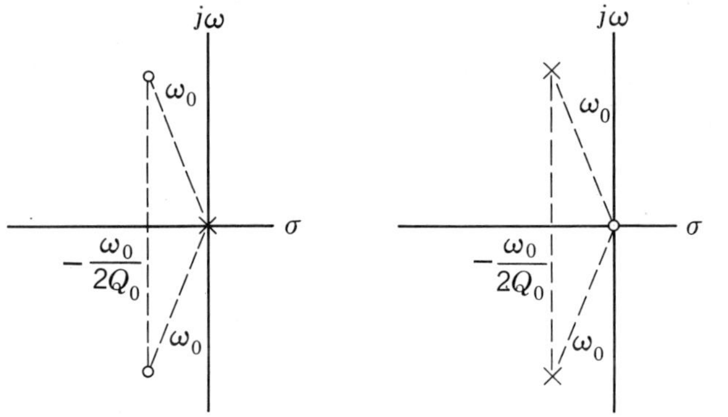

(*a*) Pole-zero configuration of ***Y*(*s*)**

(*b*) Pole-zero configuration of ***Z*(*s*)**

Fig. 17–14. Comparison of s-plane plots of *Y*(s) and *Z*(s) for Fig. 17–13.

There is one point of terminology that should be clarified, however. In the series *R-L-C* circuit it was found that the resonant condition corresponded to a minimum value of impedance and this minimum was R. In the present case, therefore, resonance will correspond to a *minimum* value of *admittance* and this minimum will be G_2. However, it is perhaps more common to consider this as a *maximum* value of *impedance* and designate it as the *antiresonant resistance* R_{ar}, or

$$R_{ar} = \frac{1}{G_2} = \frac{Q_0}{\omega_0 C_2} = Q_0 \omega_0 L_2 \tag{17–51}$$

As would be expected, a large value of Q_0 will correspond to a large value of R_{ar}.

17–7. Modifications of Parallel Resonant Circuits. In the preceding section the circuit considered consisted of all three elements in parallel. If all of the loss in the circuit is associated with coil resistance it may be felt that this method of representation is somewhat artificial in that the d-c resistance of the coil is not accounted for. Since it has long been common practice to represent coil resistance as being in series with the inductance, a more realistic circuit for discussing parallel resonance might be that shown in Fig. 17–15a. The admittance function for this circuit may be obtained as

$$Y(s) = \frac{1}{R + sL} + sC = \frac{1 + RCs + LCs^2}{R + Ls}$$

$$= C\frac{s^2 + \dfrac{R}{L}s + \dfrac{1}{LC}}{s + \dfrac{R}{L}} \tag{17–52}$$

Introducing the quantities ω_0 and Q_0 as previously defined yields

$$Y(s) = C\frac{s^2 + \dfrac{\omega_0}{Q_0}s + {\omega_0}^2}{s + \dfrac{\omega_0}{Q_0}} \tag{17–53}$$

and the corresponding pole-zero configuration is shown in Fig. 17–15b. Comparison of this pole-zero configuration with that in Fig. 17–14a reveals that the only change is that the pole has moved from the origin to a point on the negative σ axis.

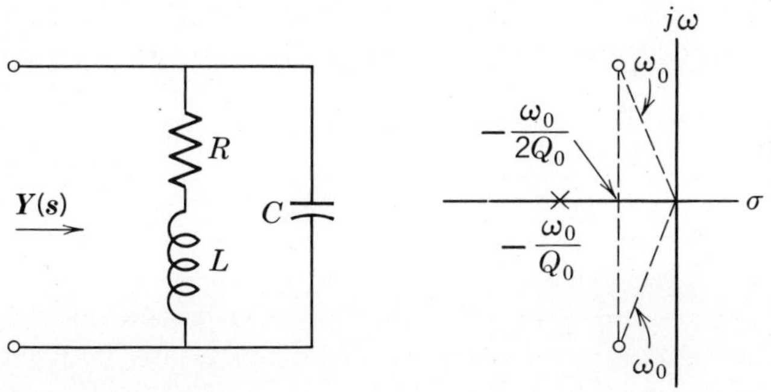

(a) Modified parallel R-L-C circuit

(b) Pole-zero configuration for (a)

Fig. 17–15. Another R-L-C parallel configuration.

If the Q of the circuit is high this shift in pole position will have a negligible effect on the circuit response in the vicinity of $j\omega_0$. For such cases, the results indicated in the previous section may be assumed to apply and either representation of the resonant circuit is adequate. In this connection, it is of interest to relate the coil resistance of Fig. 17–15a to the antiresonant resistance of the circuit. As indicated in Eq. 17–51, this resistance is

$$R_{\text{ar}} = Q_0\omega_0 L = Q_0(Q_0 R) = {Q_0}^2 R \tag{17–54}$$

If the Q of the circuit is not large the shift in the pole position will have a noticeable effect. In particular, it is found that the minimum admittance and the point of zero angle do not occur at the same frequency, and neither occurs at ω_0. The question then arises as to which condition is to be designated as resonance. In keeping with the previous definition of resonance, the answer to this question is that resonance occurs when the admittance is purely real. This definition is also in complete agreement with the concept that resonance occurs when the maximum stored energy in the inductance equals the maximum stored energy in the capacitance. The determination of the exact locations of resonance and minimum admittance are left as student exercises.

Another form of a parallel R-L-C circuit is that shown in Fig. 17–16a in which there is resistance in both branches of the circuit. The admittance function for this circuit is obtained as follows:

$$\mathbf{Y}(s) = \frac{1}{R_L + sL} + \frac{1}{R_C + \dfrac{1}{sC}} = \frac{1}{R_L + sL} + \frac{sC}{1 + R_C sC}$$

$$= \frac{1 + R_C sC + R_L sC + s^2 LC}{(R_L + sL)(1 + R_C sC)} = \frac{1}{R_C} \frac{s^2 + \dfrac{R_C + R_L}{L} s + \dfrac{1}{LC}}{\left(s + \dfrac{R_L}{L}\right)\left(s + \dfrac{1}{R_C C}\right)} \qquad (17\text{–}55)$$

In order to express this admittance function in a form similar to that used previously, it is convenient to make the following definitions of terms:

$$\omega_0^{\,2} = \frac{1}{LC}$$

$$Q_0 = \frac{\omega_0 L}{R_L + R_C} \qquad \text{(equivalent } Q \text{ of entire circuit)}$$

$$Q_L = \frac{\omega_0 L}{R_L} \qquad (Q \text{ of the inductive branch only})$$

$$Q_C = \frac{1}{\omega_0 R_C C} \qquad (Q \text{ of the capacitive branch only})$$

Using these symbols, Eq. 17–55 becomes

$$\mathbf{Y}(s) = \frac{1}{R_C} \frac{s^2 + \dfrac{\omega_0}{Q_0} s + \omega_0^{\,2}}{\left(s + \dfrac{\omega_0}{Q_L}\right)(s + Q_C \omega_0)} \qquad (17\text{–}56)$$

The pole-zero configuration for this admittance function is illustrated in Fig. 17–16b. Note that in this case the pole at the origin has again moved its position on the negative σ axis and that, in addition, the pole at infinity has moved in to a finite position on the σ axis.

If the Q of the circuit is high its response near $j\omega_0$ will still be essentially the same as that discussed in the previous section. If Q is *not* large there will be, in general, a noticeable shift in the locations of the frequencies of minimum admittance and zero angle. However, as shown in the first part of this book, this shift does not occur if the resistances are equal.

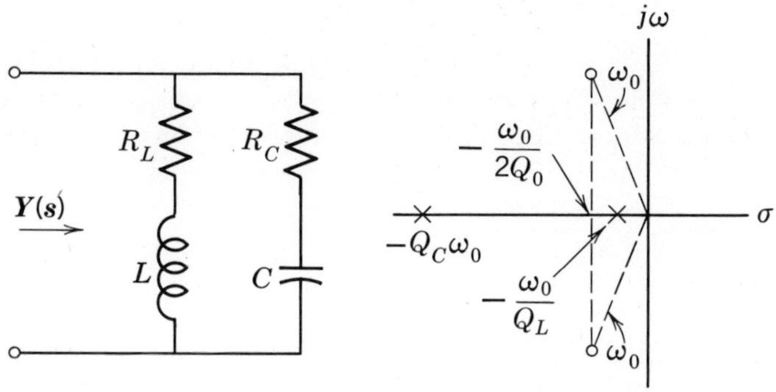

(*a*) Parallel R-L-C circuit (*b*) Pole-zero configuration

Fig. 17–16. *R-L-C* parallel circuit with resistance in both branches.

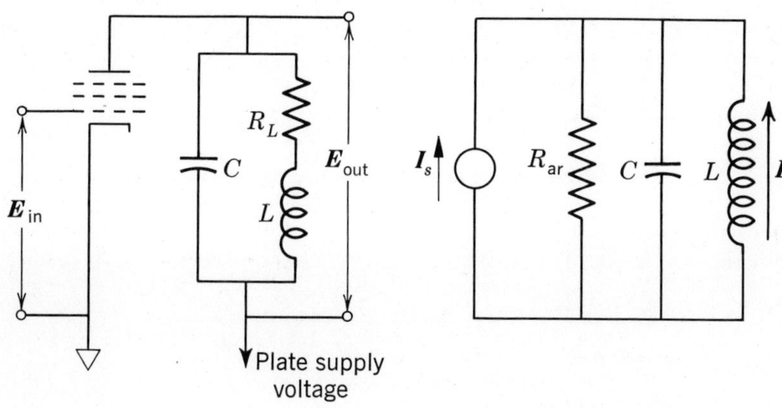

(*a*) A tuned amplifier (*b*) Equivalent circuit

Fig. 17–17. A tuned amplifier and its equivalent circuit.

17–8. Applications of Parallel Resonance. The characteristic of greatest importance of a parallel resonant circuit in most applications is its high impedance at resonance. One example of this is the *tuned amplifier* such as might be used in a radio receiver. Many vacuum tubes behave very nearly as an ideal current source, and the circuit of Fig. 17–17*a* may be represented quite satisfactorily by the idealized circuit of Fig. 17–17*b*.

From Eq. 17–50 the output voltage of this circuit is

$$\boldsymbol{E}_{\text{out}} = \boldsymbol{I}_s \boldsymbol{Z}(s) = \frac{\boldsymbol{I}_s}{C} \frac{s}{s^2 + \dfrac{\omega_0}{Q_0} s + \omega_0^2} \qquad (17\text{–}57)$$

This voltage is a maximum at the resonant frequency ω_0, where

$$\boldsymbol{E}_{\text{out}} = \frac{\boldsymbol{I}_s}{C} \frac{Q_0}{\omega_0} = \boldsymbol{I}_s R_{\text{ar}} \qquad (17\text{–}58)$$

As an illustration of the order of magnitudes involved, assume that the

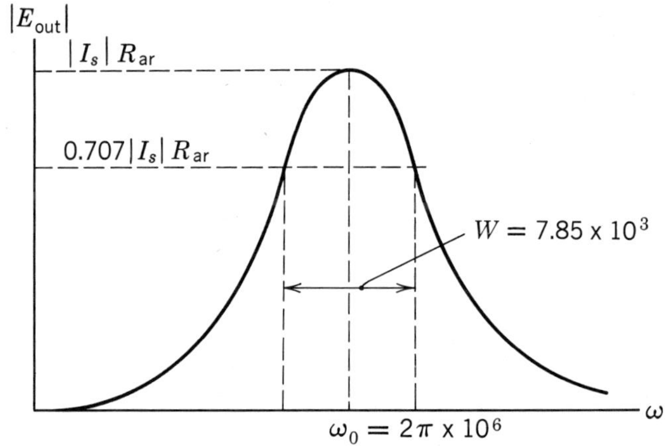

Fig. 17–18. Voltage output versus ω for the circuit of Fig. 17–17.

resonant frequency is to be 1000 kc ($\omega_0 = 2\pi \times 10^6$), the capacitance 100 $\mu\mu$f, and the coil Q 80. The value of inductance required for these values is

$$L = \frac{1}{\omega_0^2 C} = \frac{1}{(2\pi)^2 \times 10^{12} \times 100 \times 10^{-12}} = 253\ \mu\text{h} \qquad (17\text{–}59)$$

The antiresonant impedance of the circuit becomes

$$R_{\text{ar}} = Q_0 \omega_0 L = 80 \times 2\pi \times 10^6 \times 253 \times 10^{-6} = 127{,}300 \text{ ohms} \qquad (17\text{–}60)$$

The bandwidth of tuned amplifiers is also of interest and for this case is

$$W = \frac{\omega_0}{Q_0} = \frac{2\pi \times 10^6}{80} = 78{,}500 \text{ radians/sec} = 12.5 \text{ kc} \qquad (17\text{–}61)$$

The magnitude of the output voltage as a function of frequency is shown in Fig. 17–18.

Not all applications of parallel resonance involve the use of high-Q circuits. For example, in the first part of this text, the problem of power factor correction was discussed. When this correction is carried to the point of unity power factor, the resulting system is resonant but generally the Q is very low.

17-9. Tuning of Resonant Circuits. The previous chapter discussed the motion of the impedance zeros of a series R-L-C circuit in response to changes in circuit resistance. In this section, the effects of changing the inductance and capacitance will be discussed. In general, any resonant circuit must have its resonant frequency adjusted to the desired value by varying either the inductance, or the capacitance, or both. This process is designated as *tuning.*

The present discussion will be restricted to the series R-L-C circuit but it should be remembered that the same results will be applicable to the parallel R-L-C circuit by utilizing the concept of duality. It has been shown that the zeros of the series R-L-C circuits may be specified by

$$s_{1,3} = -\frac{R}{2L} \pm j\sqrt{\frac{1}{LC} - \left(\frac{R}{2L}\right)^2} \tag{17-2}$$

or

$$s_{1,3} = -\frac{\omega_0}{2Q_0} \pm j\omega_0\sqrt{1 - \left(\frac{1}{2Q_0}\right)^2} \tag{17-8}$$

For high-Q circuits these may be closely approximated by

$$s_{1,3} = -\frac{R}{2L} \pm j\sqrt{1/LC} \tag{17-62}$$

$$= -\frac{\omega_0}{2Q_0} \pm j\omega_0 \tag{17-63}$$

Consider first how the zeros are affected by changing the capacitance. There are actually two different conditions that should be considered in connection with this. One condition involves the assumption that the resistance remains constant. At low frequencies this is nearly true. However, if the circuit is to be used at high frequencies then the resistance tends to increase with frequency as the result of a phenomenon known as *skin effect.* Under these circumstances, a much more realistic assumption (although again not exactly correct) is that Q remains constant at its resonant frequency value Q_0.

In the case of constant resistance it may be seen from either Eq. 17-2 or Eq. 17-62 that an increase in C moves the zeros along a vertical path as shown in Fig. 17-19a until they reach the negative σ axis at

$$C = \frac{4L}{R^2} \tag{17-64}$$

Further increases in C cause them to move in opposite directions along the negative σ axis, one approaching the origin and the other approaching $-R/L$.

If Q_0 is considered to remain constant, the zeros follow the paths indicated in Fig. 17–19*b*, which are lines of constant slope approaching the origin at $C = \infty$.

It should be emphasized that neither of the root-locus plots of Fig. 17–19 are exactly correct since the actual change in resistance with frequency is more complicated than has been indicated here. The situation shown in Fig. 17–19*a*, however, is reasonably close for resonant frequencies in the order of a

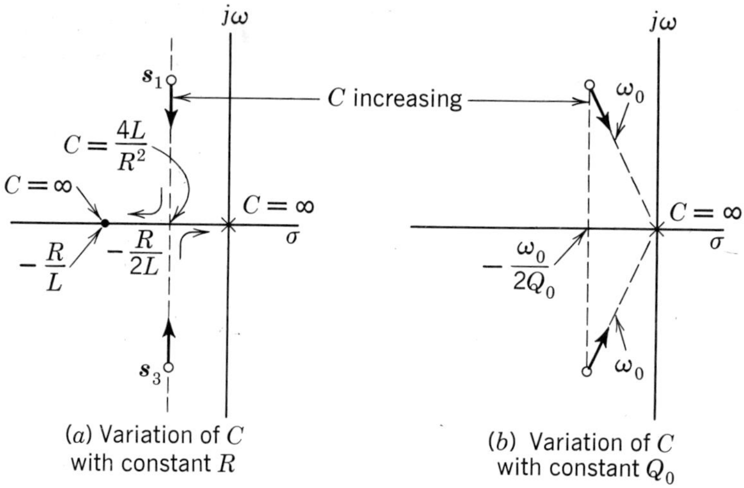

Fig. 17–19. Variation of pole-zero location as a function of varying *C* for an *R-L-C* series circuit.

few thousand cycles per second or less, and the situation shown in Fig. 17–19*b* is usually adequate for resonant frequencies in excess of several hundred kilocycles per second.

Similar locus diagrams may be drawn for tuning accomplished by varying the inductance. This is left as a student exercise.

PROBLEMS

17–1. A series *R-L-C* circuit has parameters with the values $R = 10$, $L = 0.01$, and $C = 10^{-6}$. (a) What is the resonant frequency ω_0? (b) What is Q_0? (c) Determine $\boldsymbol{Z}(\boldsymbol{s})$ and plot the poles and zeros of the impedance in the $\boldsymbol{s}$-plane.

17–2. Repeat Prob. 17–1 for the values $R = 20$, $L = 0.01$, and $C = 10^{-6}$.

17–3. Show the areas in the $\boldsymbol{s}$-plane in which the impedance poles and zeros of a series *R-L-C* circuit can be located if $10 < Q_0 < 15$ and $1000 < \omega_0 < 1500$.

17–4. Determine $i(t)$, $e_R(t)$, $e_L(t)$, and $e_C(t)$ for the circuit shown. Plot these quantities as a function of $\omega_0 t$ on the same set of coordinate axes and comment on the significance of the applied voltage being at the resonant frequency of the circuit.

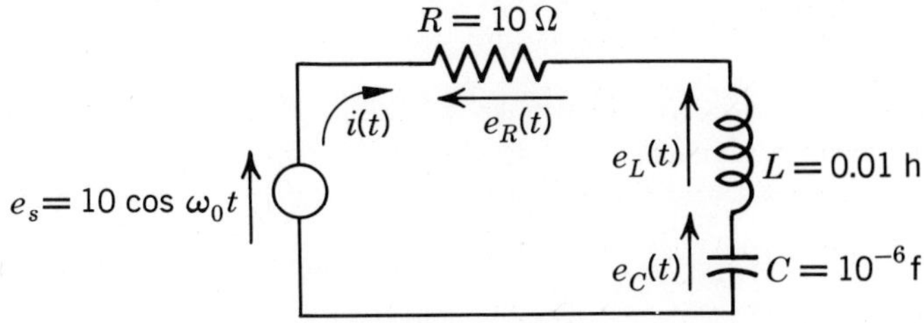

Prob. 17–4.

17–5. Determine, for Prob. 17–4, equations for the energy stored in the inductance and the energy stored in the capacitance as functions of time. Compare these equations. What is the total stored energy at any instant of time? What is the physical significance of your results?

17–6. Start with the relations

$$|Z| = \sqrt{R^2 + \left(\omega L - \frac{1}{\omega C}\right)^2}, \qquad \theta = \tan^{-1} \frac{\omega L - (1/\omega C)}{R}$$

and derive Eqs. 17–18 and 17–19. Note that suitable approximations will be necessary. Compare the approximations made here with those used in obtaining Eq. 17–17.

17–7. Use Eqs. 17–19 and 17–21 to determine data for curves of the approximate values of Z and θ versus ω for the circuit illustrated. Plot the curves accurately and to a sufficiently large scale to permit their use for the graphical determination of data. Demonstrate how these curves can be used for any values of C and ω by determining Z and θ for $C = 100\ \mu$f and $\omega = 99.7$.

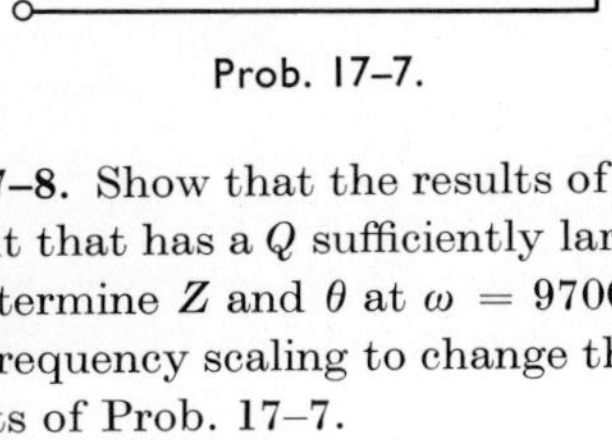

Prob. 17–7.

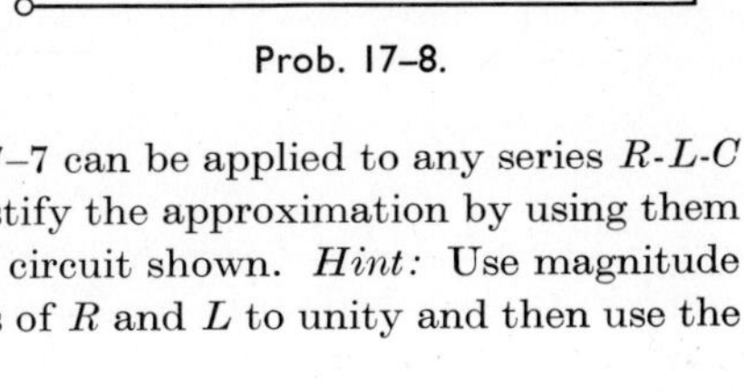

Prob. 17–8.

17–8. Show that the results of Prob. 17–7 can be applied to any series R-L-C circuit that has a Q sufficiently large to justify the approximation by using them to determine Z and θ at $\omega = 9700$ for the circuit shown. *Hint:* Use magnitude and frequency scaling to change the values of R and L to unity and then use the results of Prob. 17–7.

17–9. Show that the results of Prob. 17–7, with $\Delta\omega$ becoming $\Delta\beta$, as defined by Eq. 17–29, can be used to determine the exact values of Z and θ at $\omega = 8000$ for the circuit of Prob. 17–8.

17–10. (a) Determine $\boldsymbol{Z}(\boldsymbol{s})$ and plot the impedance poles and zeros in the $\boldsymbol{s}$-plane for the circuit shown. (b) Use the results of Prob. 17–7 to obtain data and plot the approximate values of Z and θ versus ω for values of ω between 3000 and 7000. (c) Use the results of Prob. 17–8 to obtain data and plot the exact values of Z and θ versus ω for values of ω between 3000 and 7000. (d) Compare the results of parts (b) and (c). What are the resonant frequency, the upper and lower half-power frequencies, the bandwidth, and Q_0 of the circuit in each case? Discuss the significance of the errors in the approximate values.

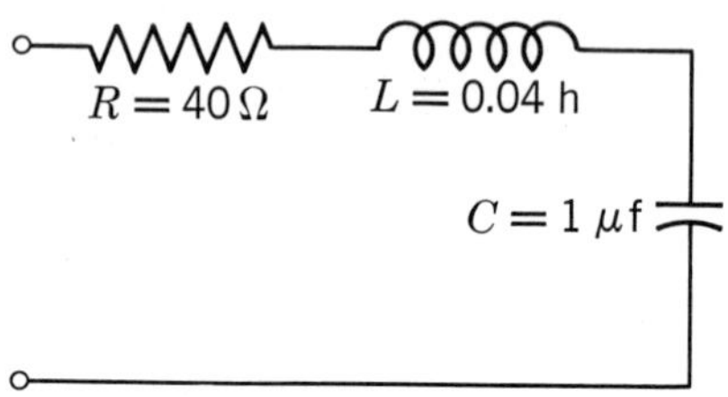

Prob. 17–10.

17–11. Two coils are available, coil No. 1 having a resistance of 8.1 ohms and an inductance of 160 μh and coil No. 2 having a resistance of 9.4 ohms and an inductance of 210 μh. Each of these coils is to be placed in series with a capacitor and the circuit is to be tuned to resonance at 1500 kc. What value of capacitance is required in each case? Which coil will produce the most selective circuit?

17–12. Show that in a series R-L-C circuit the ratio $|E_C|/|E_s|$ is a maximum at

$$\omega = \omega_0\sqrt{1 - \frac{1}{2Q_0^2}}$$

and that the maximum value of this ratio is

$$\left|\frac{E_C}{E_s}\right|_{\max} = \frac{Q_0}{\sqrt{1 - (1/4Q_0^2)}}$$

17–13. Show that in a series R-L-C circuit the ratio $|E_L|/|E_s|$ is a maximum at

$$\omega = \frac{\omega_0}{\sqrt{1 - (1/2Q_0^2)}}$$

and that the maximum value of this ratio is

$$\left|\frac{E_L}{E_s}\right|_{\max} = \frac{Q_0}{\sqrt{1 - (1/4Q_0^2)}}$$

17–14. A series R-L-C circuit is to have a resonant frequency of 10,000 cps and is to have a bandwidth of 500 cps. Determine the values of R and C required if the inductance is 10 μh.

17–15. Design a series R-L-C circuit that can be tuned by a variable capacitor to have the resonant frequency anywhere between 50,000 cps and 100,000 cps. The coil is to have an inductance of 1 mh and the minimum permissible Q is to be 50. Determine the range over which the capacitor must be variable. Determine the maximum resistance that can be present in the circuit.

17–16. A series R-L-C circuit is to have a resonant frequency of 10,000 radians/sec and a bandwidth of 500 radians/sec. The impedance at resonance is to be 200 ohms. Determine the values of resistance, inductance, and capacitance required.

17–17. The circuit shown is an equivalent circuit of the loop antenna circuit of a radio receiver. The voltage $e_s(t)$ represents the signal received by the antenna; the voltage $e_c(t)$ is the useful response of the circuit. The antenna is receiving two signals, one at a frequency of 920 kc and the other at a frequency of 890 kc. This results in

$$e_s = 100 \cos 1840\pi t + 100 \cos 1780\pi t$$

It is desired to tune the circuit, by varying the capacitance, to have a resonant frequency of 920 kc. (a) Determine the value of C required. (b) Using this value of C, make a complete analysis of the circuit for each of the frequencies involved, determining the complex amplitudes $\boldsymbol{I}$, $\boldsymbol{E}_R$, $\boldsymbol{E}_L$, and $\boldsymbol{E}_C$ and sketching a phasor diagram. (c) What is $e_c(t)$? Compare $e_c(t)$ to $e_s(t)$ and discuss the significance of your results.

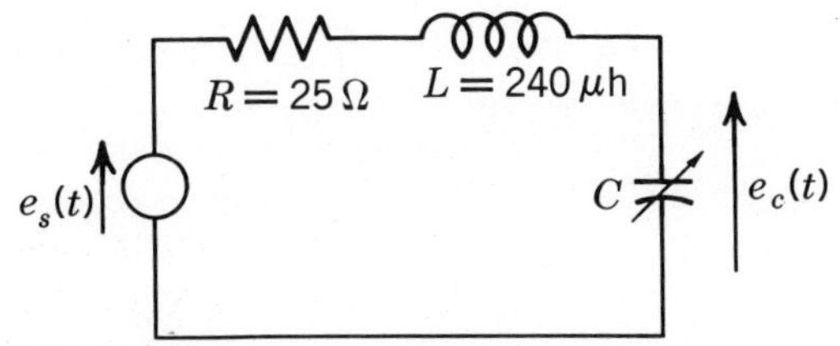

Prob. 17–17.

17–18. The impedance of a series R-L-C circuit is given as

$$\boldsymbol{Z}(\boldsymbol{s}) = \frac{(\boldsymbol{s} + 0.05 + j1)(\boldsymbol{s} + 0.05 - j1)}{\boldsymbol{s}}$$

(a) What are the values of R, L, and C? (b) What is Q_0? (c) What is the bandwidth?

17–19. How do the parameter values in Prob. 17–18 change if the frequency scale is changed to make resonance occur at 10^4 radians/sec? What is the effect on Q_0? What is the effect on the bandwidth?

17–20. How do the parameter values in Prob. 17–19 change if a magnitude scale change is made to make the impedance at resonance equal to 50 ohms? What is the effect on the bandwidth and Q_0?

17–21. (a) What is the resonant frequency of the circuit illustrated on p. 276? (b) What is Q_0? (c) Determine $\boldsymbol{Y}(\boldsymbol{s})$ and plot the admittance poles and zeros in the $\boldsymbol{s}$-plane. (d) What is $|Y|$ at resonance? (e) Sketch a curve showing how the magnitude and angle of the admittance vary with ω.

17–22. Determine equations similar to Eqs. 17–19 and 17–21 that will apply to the admittance of parallel G-L-C circuits.

17–23. Determine equations similar to Eqs. 17–28 and 17–29 that will apply to the admittance of parallel G-L-C circuits.

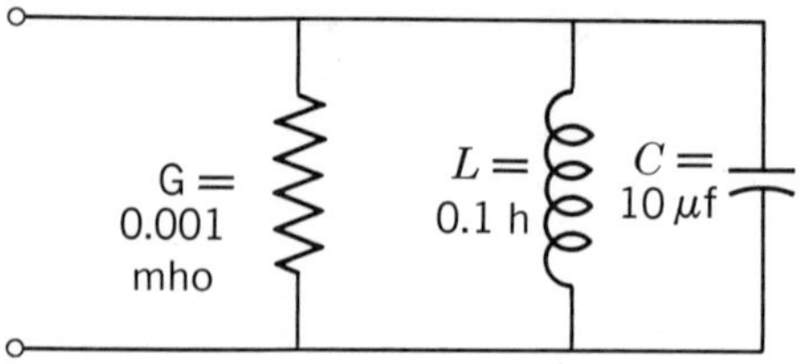

Prob. 17–21.

17–24. Show that the results of Prob. 17–7 can also be interpreted to give the approximate and exact values of the magnitude and angle of the admittance of the circuit shown at $\omega = 10{,}500$.

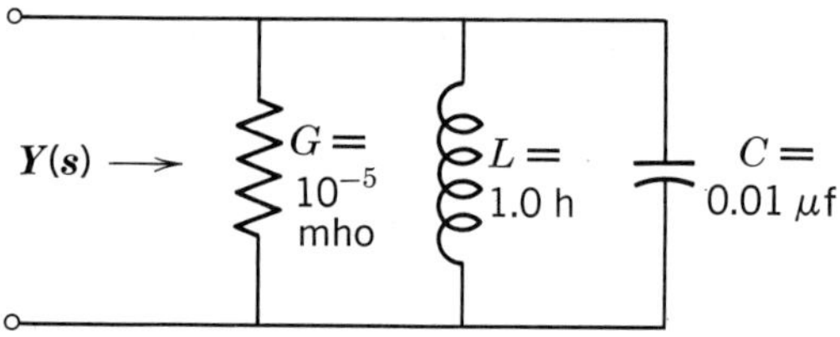

Prob. 17–24.

17–25. (a) At what frequency will the ratio $|I_G|/|I_s|$ of the illustrated figure be maximum and what is the ratio at this frequency? (b) At what frequency will the ratio $|I_L|/|I_s|$ be maximum and what is the ratio at this frequency? (c) At what frequency will the ratio $|I_C|/|I_s|$ be maximum and what is the ratio at this frequency? (d) At what frequency will $|E|$ be maximum? What is $|E|$ at this frequency?

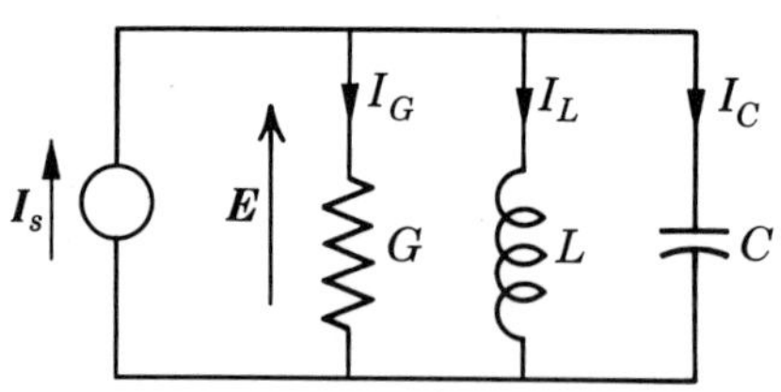

Prob. 17–25.

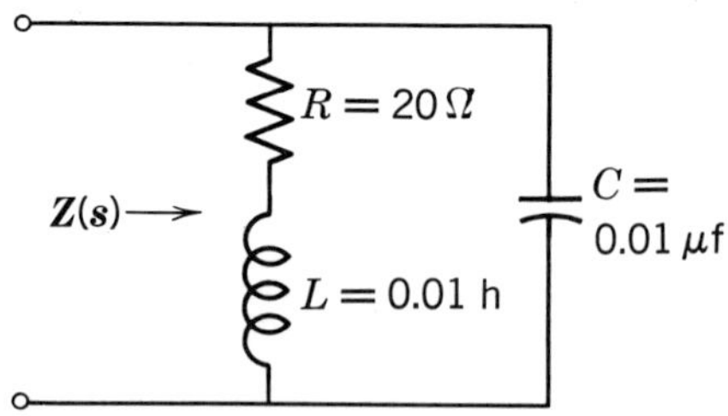

Prob. 17–26.

17–26. (a) Determine $\boldsymbol{Z}(\boldsymbol{s})$ of the circuit shown and plot the impedance poles and zeros in the $\boldsymbol{s}$-plane. (b) Determine ω_0 and Q_0. (c) What is the impedance of the circuit at resonance? (d) Sketch curves of $|Z|$ and θ versus ω. (e) What is the bandwidth? (f) Determine a parallel G-L-C circuit with the same resonant frequency, the same bandwidth, and the same impedance at resonance.

17–27. (a) Prove that the exact resonant frequency of the circuit illustrated is

$$\omega = \omega_0\left(1 - \frac{1}{Q_0{}^2}\right)$$

(b) Prove that $|Z|$ is a maximum at

$$\omega = \omega_0 \sqrt{\left(1 + \frac{2}{Q_0^2}\right)^{1/2} - \frac{1}{Q_0^2}}$$

(c) If $Q_0 = 10$, how much error is introduced by considering the resonant frequency and the frequency of maximum $|Z|$ to be ω_0?

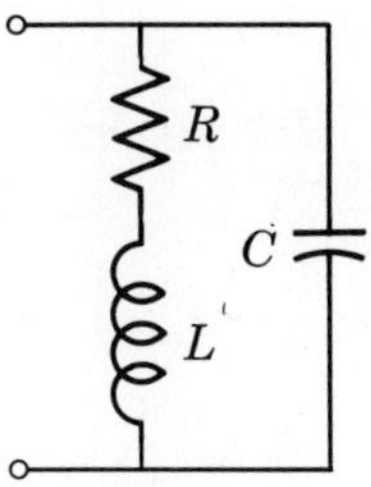

Prob. 17–27.

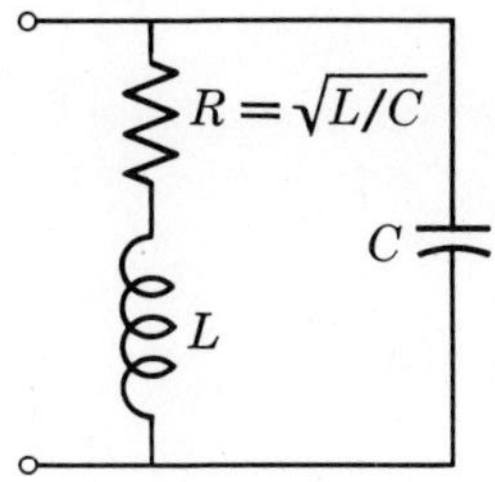

Prob. 17–28.

17–28. Determine the pole-zero configuration of the admittance of the adjacent circuit and comment on the significance of this particular value of R.

17–29. Show what happens to the pole-zero configuration of Fig. 17–16 when

(a) $R_L = R_C = \sqrt{L/C}$ (b) $R_L = R_C > \sqrt{L/C}$

(c) $R_L = R_C < \sqrt{L/C}$ (d) $R_L > \sqrt{R/C} > R_C$

(e) $R_L < \sqrt{L/C} < R_C$

In each case discuss the implications of the resulting pole-zero configuration as it affects the resonant frequency and the plot of $|Z|$ versus ω.

17–30. A circuit of the form indicated in Prob. 17–27 is to be designed to meet the following specifications: (a) The impedance is to have a maximum value of 50,000 ohms at a frequency of 10^6 cps. (b) The bandwidth is to be 2000 cps. Determine the appropriate values of R, L, and C.

17–31. If a capacitor has appreciable energy dissipation it can be represented as an ideal capacitance in parallel with an appropriate conductance. The circuit of Fig. 17–15*a* is thus modified as shown. Determine the resonant frequency, the bandwidth, $|Z|$ at the resonant frequency, and Q_0 for the circuit for the values $G = 10^{-5}$, $R = 100$, $L = 0.01$, and $C = 0.01 \times 10^{-6}$. Make any justifiable approximations.

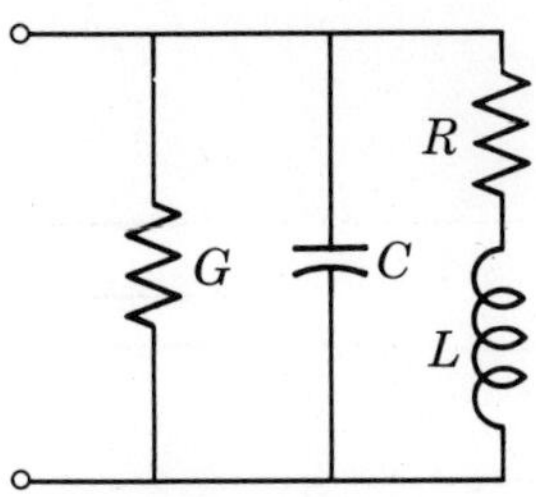

Prob. 17–31.

17–32. For the circuit shown, use any justifiable approximations and determine: (a) The maximum value of the ratio $|E_C|/|E_s|$ as ω is varied and the frequency at which it occurs. (b) The frequencies at which the ratio $|E_C|/|E_s|$ is the maximum value divided by $\sqrt{2}$.

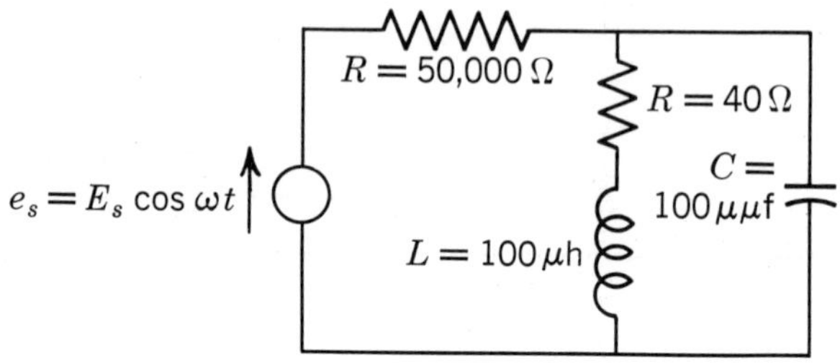

Prob. 17–32.

17–33. Determine the root-locus plots for the impedance poles and zeros of a series R-L-C circuit for each of the following conditions: (a) $R = 10$, $L = 1.0$, and C varied. (b) $R = 100$, $L = 1.0$, and C varied. (c) $Q_0 = 10$ and C varied. (d) $Q_0 = 1$ and C varied. (e) $Q_0 = 1/\sqrt{2}$ and C varied.

17–34. Repeat Prob. 17–33 for the following conditions: (a) $R = 5, C = 10^{-6}$, and L varied. (b) $R = 100, C = 10^{-6}$, and L varied. (c) $Q_0 = 10, C = 10^{-6}$, and L varied. (d) $Q_0 = \frac{1}{2}, C = 10^{-6}$, and L varied. (e) $R/L = 100, C = 10^{-6}$, and L varied.

17–35. For the circuit of Prob. 17–27, determine the root-locus plots for the admittance poles and zeros for the following conditions: (a) $R = 10$, $L = 1.0$, and C varied. (b) $Q_0 = 10$ and C varied. (c) $L = 1, C = 10^{-6}$, and R varied.

18

IMPEDANCE AND ADMITTANCE LOCI

The preceding chapters discussed the representation of impedance and admittance on the complex frequency plane (s-plane) and illustrated this with examples. Also discussed was the use of the complex frequency plane for root-locus plots; that is, it was shown how the poles and zeros of the impedance and transfer functions shifted as various element values were changed.

The complex frequency plane is not the only way in which impedances and admittances may be represented graphically. This chapter will discuss another technique which is useful in visualizing the sinusoidal steady-state behavior of these functions. Unlike the s-plane approach, this technique does not apply to generalized time functions but it does utilize a complex plane and portrays loci in that plane.

18–1. The Complex Impedance or Admittance Plane. The value of any impedance function along the $j\omega$ axis can be obtained by replacing s by $j\omega$. If this is done the resulting function of ω will in general be complex and have a real and an imaginary part. As a simple example of this, consider the impedance function corresponding to the parallel R-C circuit shown in Fig. 18–1.

The impedance function for this network is

$$Z(s) = \frac{1}{\dfrac{1}{R} + sC} = \frac{1}{C}\frac{1}{s + \dfrac{1}{RC}} \tag{18–1}$$

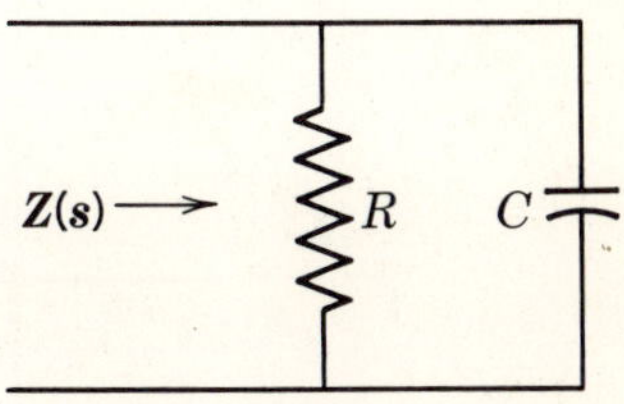

Fig. 18–1. Parallel R-C circuit.

If s is replaced by $j\omega$

$$Z(j\omega) = \frac{1}{C}\frac{1}{j\omega + \dfrac{1}{RC}} \tag{18–2}$$

which may be rationalized to yield

$$Z(j\omega) = \frac{1}{C}\frac{\dfrac{1}{RC} - j\omega}{\dfrac{1}{R^2C^2} + \omega^2}$$

$$= R\frac{1 - j\omega RC}{1 + \omega^2R^2C^2} \tag{18–3}$$

If the real part of $\boldsymbol{Z}(j\omega)$ is designated as $R(\omega)$ and the imaginary part as $X(\omega)$, then

$$R(\omega) = \frac{R}{1 + \omega^2 R^2 C^2} \tag{18–4}$$

and

$$X(\omega) = -\frac{\omega R^2 C}{1 + \omega^2 R^2 C^2} \tag{18–5}$$

There are several points in connection with these expressions which are worth noting although they are not particularly pertinent to the present discussion. These points apply to any set of such functions which arise from passive physically possible impedances, and are presented without proof.

1. Any $R(\omega)$ will be positive for any real value of ω, and will be an *even* function of ω, that is, $R(\omega) = R(-\omega)$.
2. Any $X(\omega)$ will be an *odd* function of ω, that is, $X(\omega) = -X(-\omega)$.
3. The magnitude of the impedance at any frequency is given by

$$|Z| = \sqrt{R^2(\omega) + X^2(\omega)} \tag{18–6}$$

4. The angle of the impedance at any frequency is given by

$$\theta = \tan^{-1} \frac{X(\omega)}{R(\omega)} \tag{18–7}$$

It is now possible to consider a *complex impedance plane* in which the real component represents the resistive part of any impedance and the imaginary

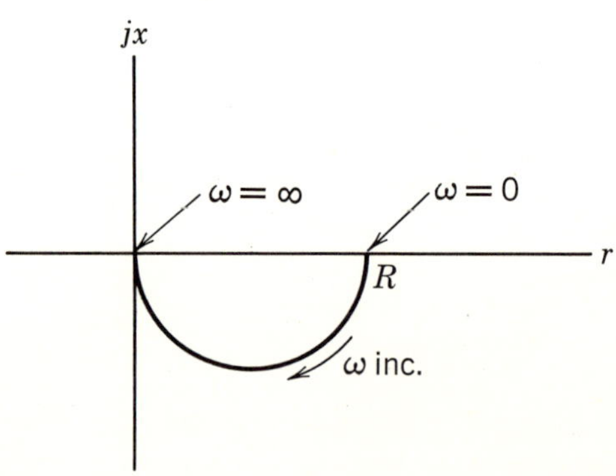

Fig. 18–2. Impedance locus on complex plane for the parallel *R-C* circuit.

component the reactive part of the impedance. In the present example these are $R(\omega)$ and $X(\omega)$ respectively. For each value of ω there is a specific point in the impedance plane. If ω is allowed to vary from 0 to ∞, this point traces out a path which is designated the *impedance locus*. The impedance locus for the parallel *R-C* circuit is shown in Fig. 18–2. Note that the axes have been labeled r and x rather than $R(\omega)$ and $X(\omega)$. This has been done in order to generalize the impedance plane so it will be applicable to variables other than ω.

It is possible to plot this locus by evaluating $R(\omega)$ and $X(\omega)$ for a number of specific values of ω, but this procedure is both time-consuming and unnecessary. A more direct approach, particularly for a simple circuit such as this, is to obtain an equation for the locus and then sketch it by identifying the equation. In order to obtain an equation it is necessary to relate $R(\omega)$ and $X(\omega)$ by eliminating the variable ω. This may be done as follows: Let

$$R(\omega) = \frac{R}{1 + \omega^2 R^2 C^2} = r \tag{18-8}$$

and

$$X(\omega) = -\frac{\omega R^2 C}{1 + \omega^2 R^2 C^2} = x \tag{18-9}$$

Then

$$\frac{r}{x} = -\frac{R}{\omega R^2 C}$$

from which

$$\omega = -\frac{X}{rRC} \tag{18-10}$$

Substituting Eq. 18–10 into Eq. 18–8 yields

$$r = \frac{R}{1 + \dfrac{X^2 R^2 C^2}{r^2 R^2 C^2}} = \frac{Rr^2}{r^2 + x^2} \tag{18-11}$$

or

$$r^2 + x^2 - Rr = 0 \tag{18-12}$$

Completing the square for the r terms results in

$$\left(r^2 - Rr + \frac{R^2}{4}\right) + x^2 = \frac{R^2}{4}$$

or

$$\left(r - \frac{R}{2}\right)^2 + x^2 = \left(\frac{R}{2}\right)^2 \tag{18-13}$$

which may be recognized as the equation of a circle of radius $R/2$ centered at $R/2$ on the r axis. The entire circle corresponds to values of ω ranging from $-\infty$ to $+\infty$. Since only positive values of ω have been plotted in Fig. 18–2, only a semicircle appears.

It is also possible to consider the admittance of a circuit in exactly the same way as has been done for impedance. Again for the parallel R-C circuit

$$\mathbf{Y}(s) = \frac{1}{R} + sC \tag{18-14}$$

from which

$$\mathbf{Y}(j\omega) = \frac{1}{R} + j\omega C \tag{18-15}$$

Since this function is already in rectangular form, the two components may be identified as

$$G(\omega) = \frac{1}{R} \tag{18–16}$$

and

$$B(\omega) = \omega C \tag{18–17}$$

The restrictions on these functions are exactly the same as those previously mentioned for $R(\omega)$ and $X(\omega)$.

An admittance plane having a real axis g and an imaginary axis b may now be defined and the locus plotted as shown in Fig. 18–3. In this case the functions are sufficiently simple that there is no need to obtain the equation of the locus. Again the portion contributed by negative values of ω has been eliminated.

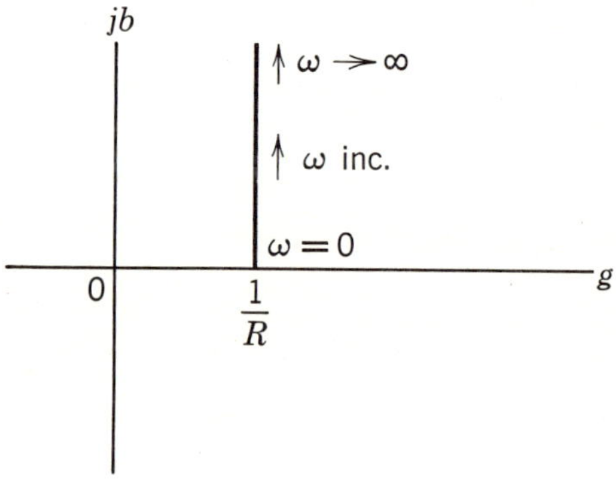

Fig. 18–3. Admittance locus on complex plane for parallel *R-C* circuit.

18–2. General Locus Plots. It was suggested in the preceding section that obtaining an equation for the locus by eliminating ω between the two equations was a more straightforward procedure than point-by-point evaluation. But even in the simple example of Fig. 18–1 some algebraic manipulation was necessary in order to obtain the desired equation and, after obtaining it, there was still the necessity of recognizing the form of the equation. In a more general case the algebra would be even more involved and the resulting equation would probably not be recognizable as any simple curve. Therefore, it appears that if the impedance or admittance locus is to be useful in any but the simplest cases, there should be an easier procedure for sketching the locus. This section will outline such a procedure.

It should be noted that every point on the locus has a distance from the origin which is equal to the magnitude of the impedance, and the angle that a line from any point to the origin makes with the positive real axis is the angle of the impedance. It should also be recalled that a general impedance function can be expressed as

$$Z(s) = H\frac{(s - s_1)(s - s_3)\cdots(s - s_{2n-1})}{(s - s_2)(s - s_4)\cdots(s - s_{2m})} \tag{18–18}$$

and that the magnitude and angle of the impedance may be obtained from phasors drawn from the critical frequencies to any point on the $j\omega$ axis. If the impedance locus associated with each critical frequency can be obtained, the composite locus involving all critical frequencies is also determined. As will be seen, there are at least two ways in which the loci from individual poles and zeros may be obtained.

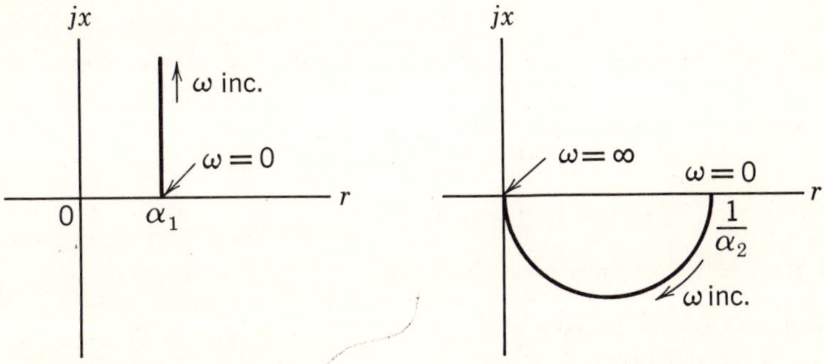

(a) Locus for a real zero (b) Locus for a real pole

Fig. 18–4. Complex plane loci for a single real zero and a single real pole.

The impedance locus for a single, real impedance zero may be obtained by consideration of the following function:

$$Z_1(s) = (s - s_1) = s + \alpha_1 \tag{18–19}$$

Letting $s = j\omega$, this becomes

$$Z_1(j\omega) = \alpha_1 + j\omega \tag{18–20}$$

from which

$$R_1(\omega) = \alpha_1 \tag{18–21}$$

and

$$X_1(\omega) = \omega \tag{18–22}$$

The locus corresponding to these equations is a straight line as shown in Fig. 18–4a for positive ω values only.

The impedance locus corresponding to a single, real pole may be obtained by considering

$$Z(s) = \frac{1}{(s - s_2)} = \frac{1}{s + \alpha_2} \tag{18–23}$$

Letting $s = j\omega$ results in

$$Z_2(j\omega) = \frac{1}{\alpha_2 + j\omega} = \frac{\alpha_2 - j\omega}{\alpha_2^2 + \omega^2} \tag{18–24}$$

from which

$$R_2(\omega) = \frac{\alpha_2}{\alpha_2^2 + \omega^2} \tag{18–25}$$

and

$$X_2(\omega) = -\frac{\omega}{\alpha_2^2 + \omega^2} \tag{18–26}$$

Comparing these expressions with those obtained in the preceding section indicates that the locus will be a semicircle as shown in Fig. 18–4*b*.

In order to visualize how these results may be utilized in constructing loci for more elaborate circuits, consider the circuit shown in Fig. 18–5. The impedance function for this circuit may be written as

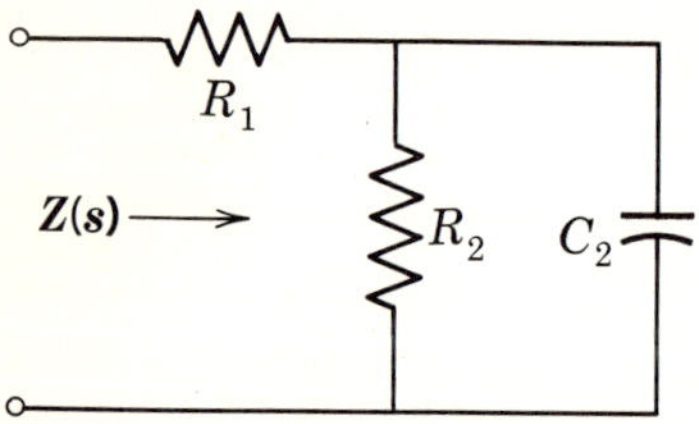

Fig. 18–5. Series-parallel circuit to illustrate the determination of a more complex locus.

$$Z(s) = R_1 + \frac{1}{\dfrac{1}{R_2} + sC_2}$$

$$= R_1 + \frac{1}{C_2\left(s + \dfrac{1}{R_2C_2}\right)} \tag{18–27}$$

Although these two terms could be combined to produce an impedance function having a zero as well as a pole, the form of Eq. 18–27 is more convenient. The second term of Eq. 18–27 has a single real pole and its impedance locus is a semicircle as shown in Fig. 18–6*a*. The first term is simply a

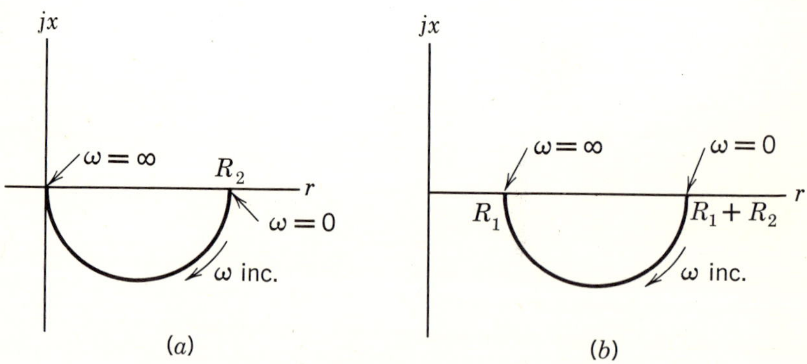

Fig. 18–6. Partial and complete locus for network of Fig. 18–5.

constant real value which is added and has the effect of moving the preceding locus to the right by an amount R_1. The complete locus is illustrated in Fig. 18–6*b*.

As a second example of combining loci consider the circuit of Fig. 18-7. The impedance function for this circuit is

$$Z(s) = R_1 + sL_1 + \frac{1}{\dfrac{1}{R_2} + sC_2}.$$

$$= L_1\left(s + \frac{R_1}{L_1}\right) + \frac{1}{C_2}\frac{1}{s + \dfrac{1}{R_2C_2}} \tag{18-28}$$

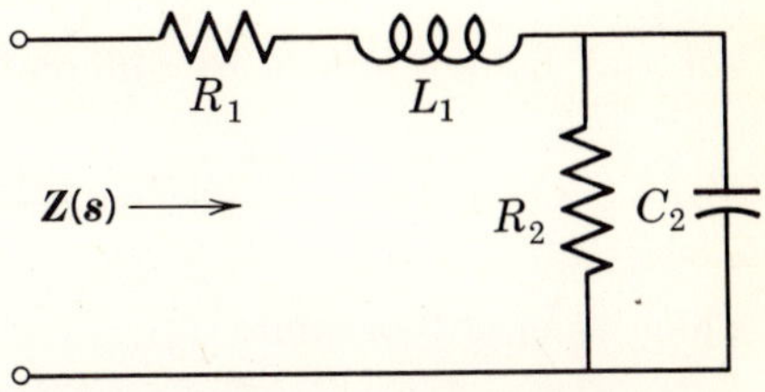

Fig. 18-7. Another series-parallel circuit to illustrate locus techniques.

Again, it is convenient to leave the function in this form rather than combine terms. The first term of Eq. 18-28 is a zero and has a straight-line locus as indicated in Fig. 18-8*a*. The second term is a pole and its locus is a semicircle as shown in Fig. 18-8*b*.

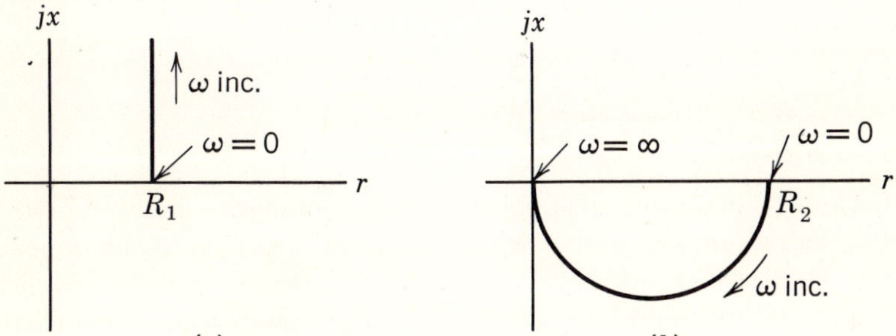

(*a*) (*b*)

Fig. 18-8. Partial loci for network of Fig. 18-7.

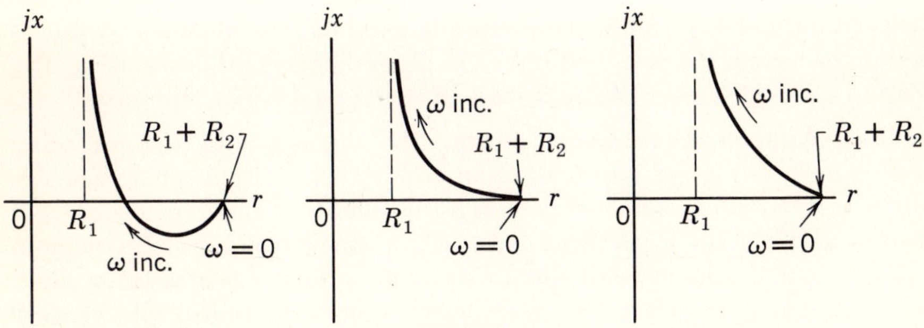

(*a*) $L_1 < R_2^2C_2$ (*b*) $L_1 = R_2^2C_2$ (*c*) $L_1 > R_2^2C_2$

Fig. 18-9. Possible complete loci for network of Fig. 18-7.

The composite locus is the sum of these two partial loci and might have one of several forms, depending upon the relative rates of traversal along the two. Some of the possibilities are shown in Fig. 18-9.

In order to determine which of these three cases is applicable it is necessary to investigate the relative values of $dX(\omega)/(d\omega)$ at $\omega = 0$. For the *R-L* term this is simply the value of L_1. For the *R-C* term it is somewhat more involved but may be determined by differentiating Eq. 18–26 with respect to ω,

$$\frac{dX_2(\omega)}{d\omega} = \frac{-(\alpha^2 + \omega^2) - \omega(2\omega)}{(\alpha^2 + \omega^2)^2} \tag{18–29}$$

which at $\omega = 0$ becomes

$$\left.\frac{dX_2(\omega)}{d\omega}\right|_{\omega=0} = \frac{1}{\alpha_2^{\,2}} \tag{18–30}$$

For the particular example being considered this becomes

$$\left.\frac{dX_2(\omega)}{d\omega}\right|_{\omega=0} = -\frac{1}{C_2} R_2^{\,2}C_2^{\,2} = -R_2^{\,2}C_2 \tag{18–31}$$

The three cases illustrated in Fig. 18–9 correspond to the following limiting conditions:

1. $L_1 < R_2^2C_2$ (locus goes negative first)
2. $L_1 = R_2^2C_2$ (locus starts with zero slope)
3. $L_1 > R_2^2C_2$ (locus positive at all times)

In both of the examples just considered the composite loci were obtained by *adding* individual loci parts corresponding to each part of the impedance function. Had the examples involved combinations in parallel, essentially the same procedure could be applied to the admittance loci. As a matter of fact, the *addition* of a number of simpler loci can be applied to most situations in which the circuit diagram is given.

If the circuit is not specified but an impedance function is given in the factored form of Eq. 18–18, the composite locus may be obtained by multiplying the loci for the poles and zeros. In general, this multiplication is not as easy to accomplish as is the addition and therefore it is not often used.

18–3. Approximate Locus Plots. The discussion of the preceding section on determining impedance and admittance loci indicated that this might well be a fairly involved process if the number of circuit elements were increased. Since the locus plot is primarily a conceptual aid rather than an end result, it is seldom justifiable to spend an appreciable amount of labor on it. As a matter of fact, the impedance and admittance loci achieve their greatest usefulness if they can be sketched by inspection from the circuit diagram with little or no computation. Therefore, it seems desirable to consider ways of making such sketches, even if some accuracy of result has to be sacrificed in the process.

Basically, the procedure for obtaining these approximate locus diagrams is to locate such points on the locus as are obtainable by inspection. Two such points always occur at zero and infinite frequency. In addition, points of resonance are also usually easy to obtain.

As an illustration of this technique, consider the series R-L-C circuit shown in Fig. 18–10a. If the resistance is assumed to remain constant, and since the impedance is infinite at both zero and infinite frequency, the resulting impedance locus is a straight line as shown in Fig. 18–10b. This locus clearly indicates that the point of resonance (zero angle) and the point of minimum impedance coincide, as was mentioned earlier in connection with series resonance.

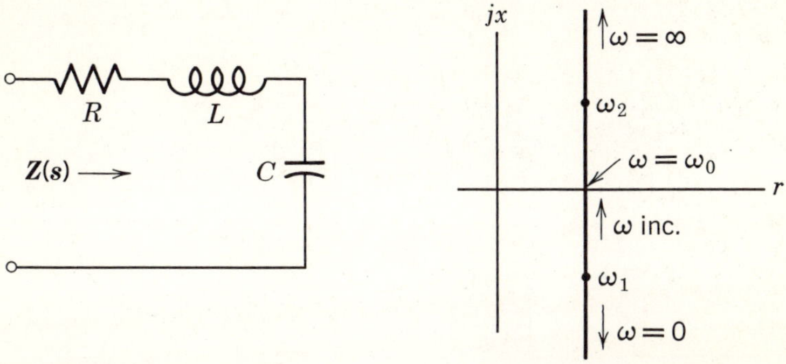

(*a*) Series R-L-C circuit (*b*) Impedance locus

Fig. 18–10. *R-L-C* series circuit and impedance locus for varying ω.

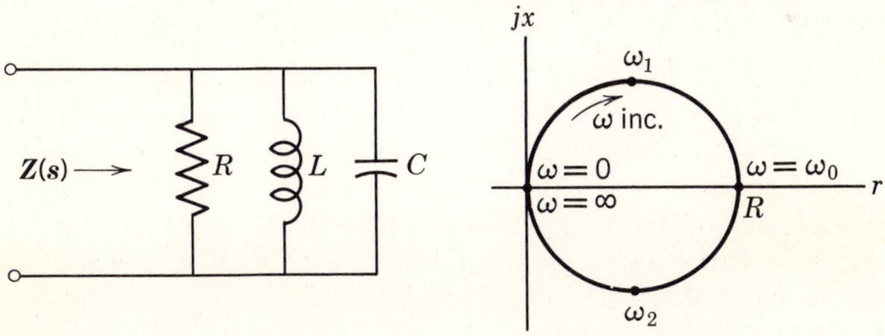

(*a*) (*b*)

Fig. 18–11. *R-L-C* parallel circuit and impedance locus.

There are two other points that can be located easily on this locus although they are not needed in order to plot the locus. These points are the half-power frequencies which determine the *bandwidth* of the circuit. These occur when $|Z| = \sqrt{2}\,R$. Phasors to these points make an angle of 45° with the r axis. Such points are designated as ω_1 and ω_2 on the locus diagram.

As a second example consider the parallel R-L-C circuit shown in Fig. 18–11a and note that the impedance is *zero* at zero and infinite frequency. Since the impedance of this type of circuit at resonance is large, the resulting locus should appear as in Fig. 18–11*b*. This locus has been shown as a circle although this fact has not been proved. In actuality, it is a circle. The half-power frequencies are located at the top and bottom of the circle. Examination of the locus reveals that the half-power frequencies are also the frequencies of maximum reactance and the magnitude of this reactance is $R/2$. These ideas were not so immediately evident in the previous discussion of parallel resonance.

For a third example of loci by inspection consider the circuit shown in Fig. 18–12*a*. It is observed that at zero frequency the value of the impedance is

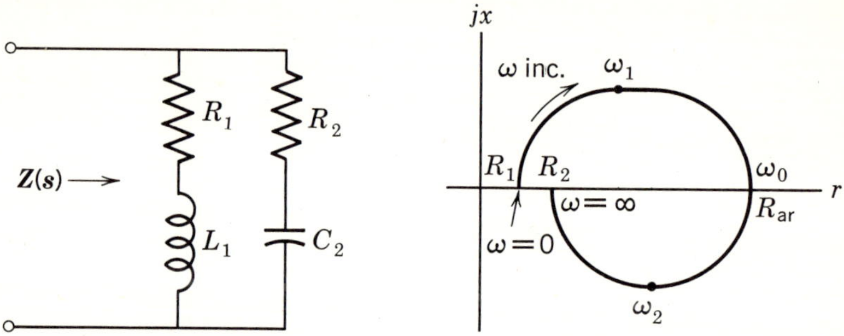

(*a*) Modified parallel *R-L-C* circuit

(*b*) Impedance locus for $R_1 < R_2$

Fig. 18–12. Another *R-L-C* parallel circuit and its impedance locus.

R_1 and that at infinite frequency it is R_2. Assuming the resistances are not too large, there will be a resonance corresponding to some much larger value of resistance occurring at a frequency ω_0. The impedance locus is shown in Fig. 18–12*b*. This locus is obviously not a circle (unless $R_1 = R_2$), but if R_1 and R_2 are very small compared to R_{ar} (that is, a high-Q circuit) it is very close to being a circle and may be so assumed for most applications. Under this assumption the half-power frequencies ω_1 and ω_2 are approximately as indicated.

In all of the examples just discussed, the locus was obtained by locating certain specific points but nothing was said concerning the direction the locus takes for positively increasing frequencies although this direction was indicated in all cases. The following rules, which are given without proof at this time, make it possible to determine this direction quite easily:

1. If the impedance or admittance locus is a straight line it will always proceed *vertically upward* for increasing frequencies.
2. If the impedance or admittance locus is a curved path it will always proceed *clockwise* around this path for increasing frequency. Therefore, the radius of curvature will always be to the *right* of the locus when proceeding in the direction of increasing frequency.

If the circuit whose locus is desired has several resonant frequencies, then the locus may make several complete revolutions in the impedance plane. Although well-established methods for determining the loci are available for these cases, they will not be discussed here because they involve circuits having a greater degree of complexity than it is desired to consider at this time.

18–4. Impedance and Admittance Loci for Varying Parameters. The discussion so far has been limited to cases in which the frequency was the parameter varied, but there are, of course, many other possibilities. In this section it is desired to consider some situations in which one of the element values is changed. It is again advantageous to be able to sketch the locus by inspection and only examples in which this is possible will be considered.

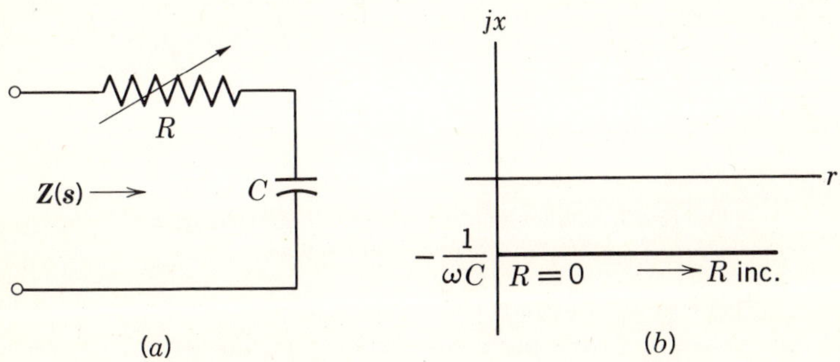

Fig. 18–13. *R-C* series circuit and impedance locus for varying *R* and fixed ω.

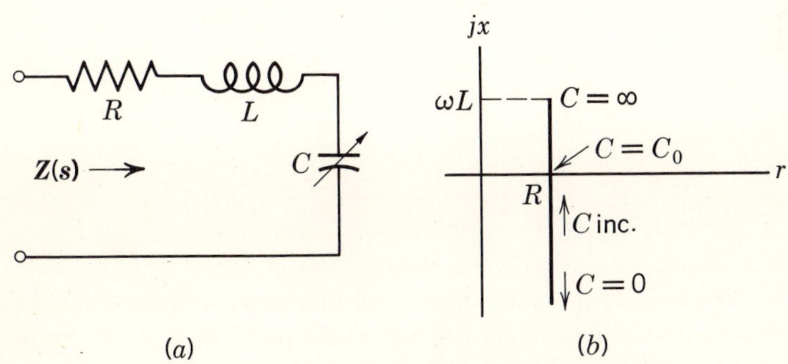

Fig. 18–14. Series *R-L-C* circuit and impedance locus for varying *C* and fixed ω.

For the first example, consider the series *R-C* circuit as shown in Fig. 18–13*a* in which the resistance is varied. If the resistance is zero the impedance is simply that of the capacitor. This reactive impedance remains constant as R is increased, and the impedance locus will be as shown in Fig. 18–13*b*.

The second example, shown in Fig. 18–14*a*, is a series *R-L-C* circuit in which the capacitance is changed. At zero capacitance the impedance is infinite. As C is increased the impedance decreases but with a constant real part since R is not changing. At infinite capacitance the impedance is simply that of R and L. This locus is illustrated in Fig. 18–14*b*.

The third example is the parallel R-L-C circuit shown in Fig. 18–15*a* in which the inductance is varied and, therefore, its resistance also. In order to obtain a unique locus it is necessary to specify the manner in which R is

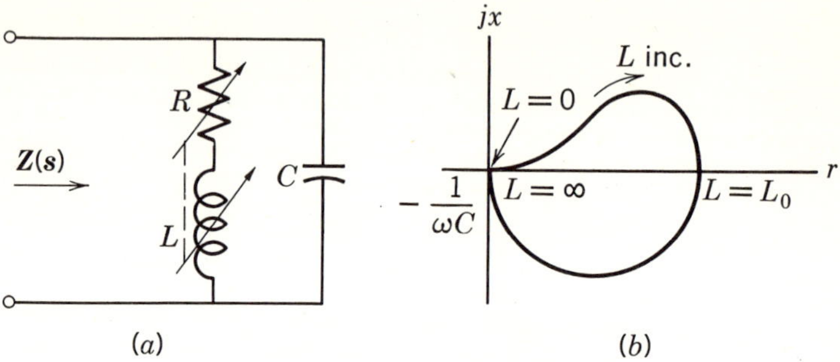

Fig. 18–15. Parallel R-L-C circuit and impedance locus for varying R and L and fixed ω.

varied with respect to L. For purposes of illustration it will be assumed that the resistance varies in such a manner as to keep Q constant at the resonant frequency value Q_0. Since

$$Q_0 = \frac{\omega_0 L}{R} \qquad \text{or} \qquad R = \frac{\omega_0 L}{Q_0} \tag{18–32}$$

and

$$\omega_0 = \frac{1}{\sqrt{LC}} \tag{18–33}$$

it follows that

$$R = \frac{L}{Q_0}\frac{1}{\sqrt{LC}} = \frac{1}{Q_0}\sqrt{\frac{L}{C}} \tag{18–34}$$

from which it may be seen that the resistance varies as the square root of the inductance. The impedance is zero at zero inductance. At infinite inductance the impedance is simply that of the capacitor and when the applied frequency of the circuit is equal to the resonant frequency the impedance will be $Q_0\omega_0 L$ (a large value for high-Q circuits). The resulting approximate locus will be as shown in Fig. 18–15*b*.

From the preceding examples it may be seen that the loci resulting from varying element sizes may be obtained in essentially the same manner as those resulting from varying frequency. It should be emphasized, however, that the rules governing the direction of the locus for increasing frequency do *not* apply to loci in which element sizes are changing. It is not possible to establish a set of rules concerning direction which will apply in all cases.

PROBLEMS

18–1. (a) Determine $\boldsymbol{Z}(s)$ and plot the impedance poles and zeros in the s-plane for the circuit shown. (b) Plot $|Z|$ and θ versus ω. (c) Plot $R(\omega)$ and $X(\omega)$ versus ω. (d) Establish a complex impedance plane and plot the impedance locus as ω is varied. (e) Determine $\boldsymbol{Y}(s)$ and plot the admittance poles and zeros in the s-plane. (f) Plot the magnitude and angle of the admittance versus ω. (g) Plot $G(\omega)$ and $B(\omega)$ versus ω. (h) Establish a complex admittance plane and plot the admittance locus as ω is varied. (i) Compare, in a general way, the information concerning the circuit characteristics that is available in parts (a) through (h).

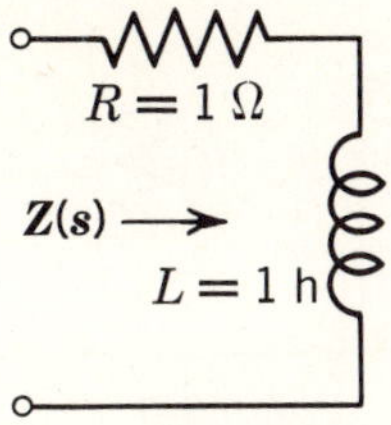

Prob. 18–1.

18–2. Show, by means of magnitude and frequency scaling, that the results of Prob. 18–1 can be used for any combination of R and L in series. Demonstrate this for the values $R = 100$ and $L = 0.01$.

18–3. Sketch the impedance and admittance loci for a series R-C circuit when ω is varied. Indicate the points corresponding to $\omega = 0$ and $\omega = \infty$ and the direction of traversal as ω increases.

18–4. Repeat Prob. 18–3 for resistance and inductance in parallel.

18–5. Determine the admittance locus for the circuit shown with ω varying. Consider the various forms the locus may have and determine any critical relations involving the values of R, L, and C.

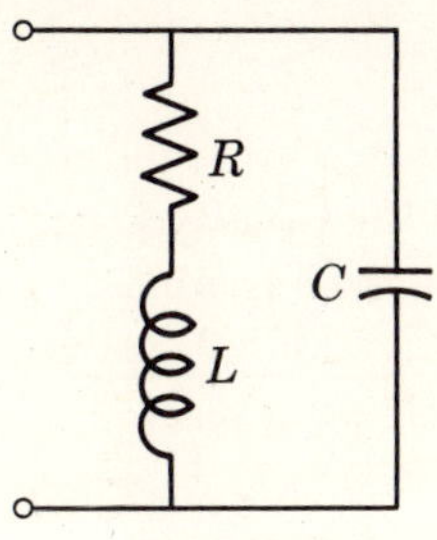

Prob. 18–5.

18–6. Plot the locus of the admittance of Prob. 18–5 for the values $R = 100$, $L = 1.0$, and $C = 10^{-6}$. Identify the points corresponding to resonance and to minimum magnitude of admittance.

18–7. An impedance is given as

$$Z(s) = \frac{(s + 1 + j5)(s + 1 - j5)}{s + 2}$$

Determine the locus of the impedance with ω varying from 0 to ∞ by first determining the locus for each pole and zero and combining these to produce the composite locus. Is there any frequency that produces a resonant condition?

18–8. Determine the locus of the admittance for the circuit illustrated with ω varying from 0 to ∞. Consider the various forms that the locus may have and determine any critical relations involving the values of R, L, and C.

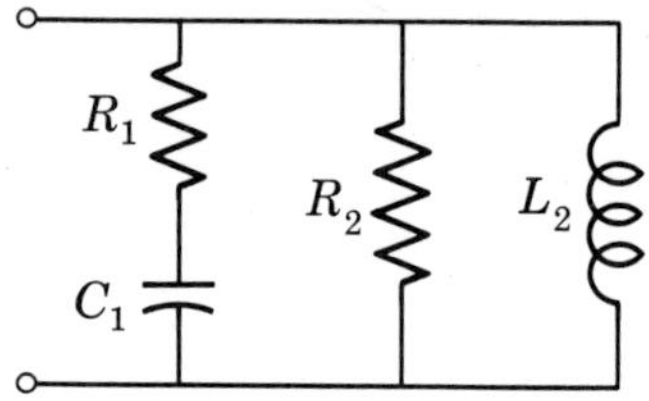

Prob. 18–8.

18–9. Determine the impedance locus for the circuit of Prob. 18–5.

18–10. Determine the admittance locus for a series R-L-C circuit with $R = 200$, $L = 0.01$, $C = 0.01 \times 10^{-6}$, and with ω varying from 0 to ∞. Identify the points corresponding to the resonant frequency, the frequency of maximum admittance, the half-power frequencies, $\omega = 0$, and $\omega = \infty$.

18–11. The admittance locus for the circuit shown can have several possible forms, depending upon the relations existing among the parameter values. Determine as many of these forms as you can together with the appropriate relations involving the parameter values. In each case comment on the significance insofar as possible resonant frequencies are concerned.

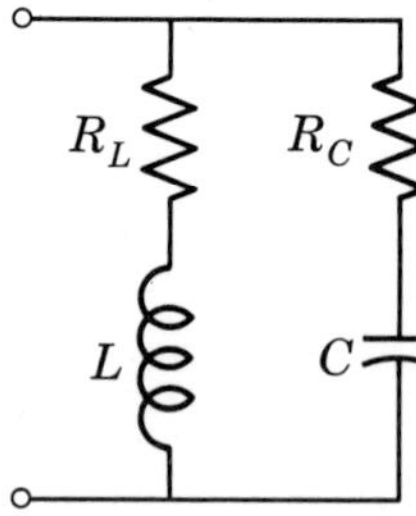

Prob. 18–11.

18–12. Determine the admittance locus for a series R-C circuit for the following conditions: (a) R and ω constant, C varied. (b) C and ω constant, R varied.

18–13. Determine the admittance locus for a series R-L circuit for the following conditions: (a) R and ω constant, L varied. (b) L and ω constant, R varied.

18–14. Determine the admittance locus for a series R-L-C circuit for the following conditions: (a) R, L, and ω constant, C varied. (b) R, C, and ω constant, L varied. (c) L, C, and ω constant, R varied.

18–15. Determine the admittance locus of the figure shown as C is varied. Is there any value of C that will produce resonance? If so, what is that value?

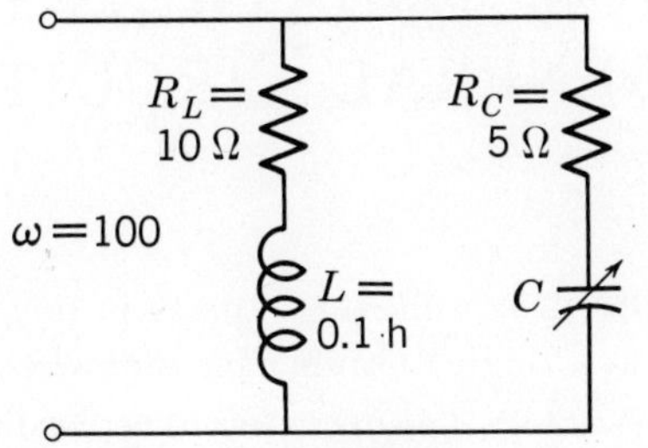

Prob. 18–15.

18–16. Determine the impedance locus of the adjacent figure as C is varied. Is there any value of C that will produce resonance? If so, what is that value?

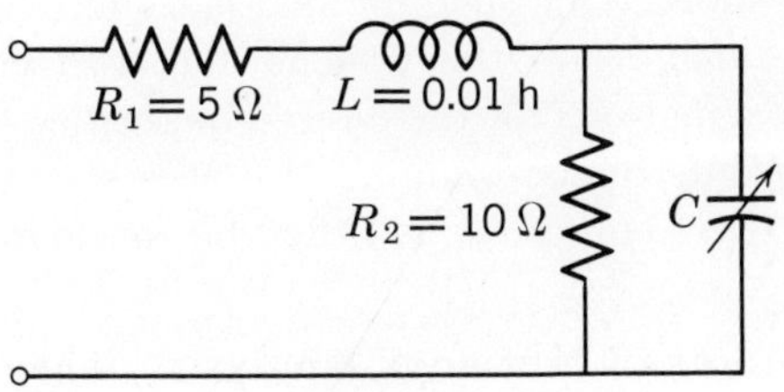

Prob. 18–16.

19

EQUILIBRIUM EQUATIONS FOR MORE GENERAL CIRCUITS

The voltage and current equilibrium equations discussed so far have been those in which there was a single unknown current or voltage. More elaborate circuits have been analyzed by making series and parallel combinations of elements in order to reduce the complexity. There are many circuits, however, having more than one unknown voltage or current in which this technique is either not possible or requires far too much labor. It is desirable, therefore, to consider ways of analyzing such circuits in a straightforward manner and with a minimum of effort. One such method is to determine a set of simultaneous equations which relate the various unknown voltages or currents and then solve these equations for the desired quantities. The present chapter will discuss methods of determining the necessary equations. Several methods of solving these equations will then be considered in the following chapter.

19–1. The Problems of Network Analysis. When a large number of elements are connected together the resulting combination is often referred to as a *network,* a term which is somewhat descriptive of the manner of connection. For our present purposes the words "network" and "circuit" will be used interchangeably, although many authorities would make a distinction between them.

The network usually has one or more sources connected to it and the basic problem of network analysis is that of computing the voltage or current at one or more points throughout the network. Since there are a number of elements, and a voltage and current associated with each element, it should be evident that there will be a number of unknown variables, that is, unknown voltages or currents. In fact, it is only in the case of series circuits in which all elements have the same current, or parallel circuits in which all elements have the same voltage, that the analysis problem reduces to a single unknown.

The relationships that exist among the variables are exactly prescribed by the types of elements involved and the manner in which these elements are connected. The type of element determines the volt-ampere characteristic for that particular element and the manner of connection determines the relationships among the voltages and currents of different elements. In addition to knowing these relationships, it is also necessary to specify the sources if it is desired to obtain the actual time function for any voltage or current.

From the preceding discussion it is evident that a knowledge of *three* different aspects of the circuit is necessary in order to complete the analysis. These are:

1. The energy sources—their location in the circuit, their nature and time variation, and their magnitude. All these items have been discussed previously.
2. The types of elements involved and the volt-ampere characteristics for these elements. These items have also been considered at length.
3. The manner in which elements are connected—that is, the network geometry—and the relationships implied by these connections. These items have *not* been considered in any detail.

For any network, having prescribed energy sources, there is a set of relationships which describe all of the voltages and currents in the network. For any *linear* network these relationships may *always* be described mathematically by a set of linear differential equations. These are the equilibrium equations. The solution to the basic network problem, therefore, is a matter of solving these equations for the desired voltages and currents.

It appears trivial to suggest that before the equilibrium equations can be solved they must first be obtained, but it is unfortunately often true that too little attention is given to the problems associated with writing the equations. A little more careful thought on these problems may ease the problem of solving the equations tremendously. The purpose of the present chapter is to discuss some methods for obtaining the necessary equilibrium equations in a complete, concise, and straightforward manner.

19–2. Network Geometry. The preceding section pointed out that of the three items necessary for general circuit analysis the only one that has not been previously considered is that of network geometry. The term *network geometry* refers to the manner in which the various elements are connected without regard to the nature of the element. There are numerous ways in which the geometry might be characterized, but a purely graphical approach has the advantages of being easily visualized and making apparent many relationships among the variables.

In order to portray the geometry of a network in terms of a graph it is only necessary to replace each element of the circuit diagram by a single line which is designated as a *branch*. Figure 19–1 shows a network and its graph. The graph also carries arrowheads on each branch to indicate the assumed positive direction for each *branch current* and a number to identify each branch.

In the graph of Fig. 19–1 there are six branches, hence, six branch currents and six branch voltages. Since the voltage and current of any branch are related by the nature and size of the element in that branch, it is apparent that the network response may be *completely* specified by *either* the six branch currents *or* the six branch voltages. However, it is also apparent that a complete specification of network response can be made with fewer than six currents or six voltages because there are some relationships among these voltages and currents that are imposed by the manner in which the branches are connected. Hence, there are fewer than six *independent* currents or

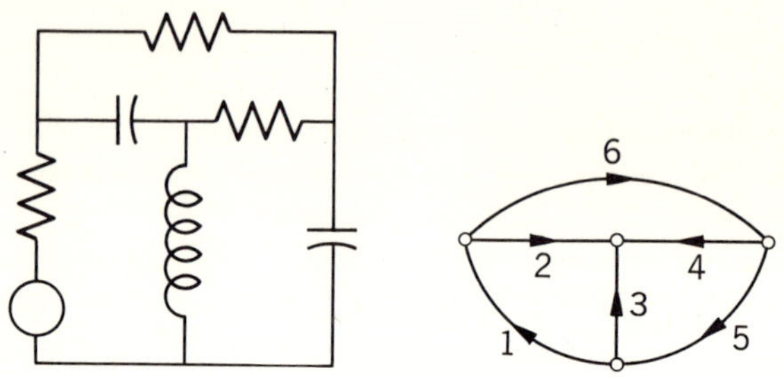

Fig. 19–1. A network and its graph.

independent voltages. The graph provides a convenient way of determining exactly how many independent currents or voltages are present in any circuit and which combinations form an independent set.

In order to utilize the graph for the objective described, it is convenient to introduce the concepts of *trees* and *nodes*. A node is any point in the network where two or more branches connect together. In the graph they are usually

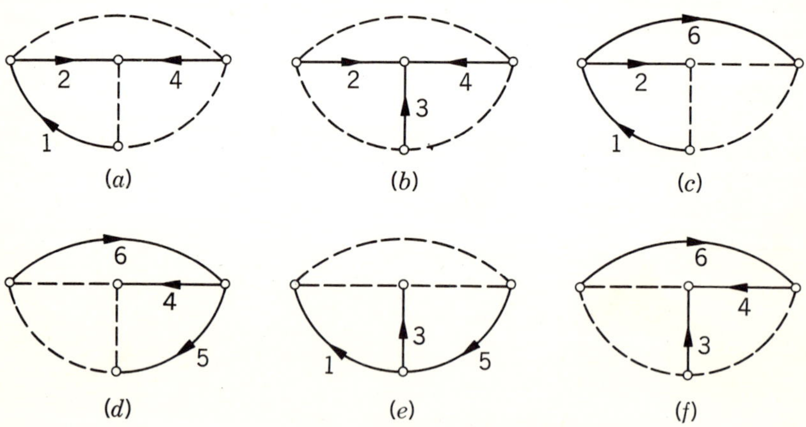

Fig. 19–2. Possible trees for the graph of Fig. 19–1.

designated by solid dots as shown in Fig. 19–1. A tree is a reduced form of graph in which all of the nodes are connected by the smallest number of branches and in such a way that no closed paths are formed. For any given graph there are generally a number of possible trees. Figure 19–2 indicates, by solid lines, some of the possible trees for the graph of Fig. 19–1. The lines which form any given tree are designated as *tree branches* and the lines which are not used in a tree (shown dashed) are designated as *links*.

Inspection of the trees in Fig. 19–2 reveals that regardless of how the tree is formed it always has the same number of tree branches, hence, the same number of links. (There need *not* be the same number of tree branches as links, although in this case there are.) This result leads one to suspect that the number of tree branches and links must be somewhow related to the number of independent voltage or current variables. This relationship may be seen by the following reasoning.

Imagine first that all of the tree branches of any given tree are *short-circuited*, thus making the voltages between any pair of nodes zero. Since all nodes are included in the tree, and they are now all at the same potential (say, zero), there can be no voltages existing anywhere in the circuit and therefore the remaining link voltages cannot be independent of the tree branch voltages. Furthermore, if any tree branch were not short-circuited there would be one node at a different potential and thus at least one tree branch voltage *not* dependent upon the others. Therefore, it can be concluded that there are *exactly* as many independent voltage variables as there are tree branches and that *any* set of *tree branch voltages* is independent.

Since there are no closed paths in any tree it is apparent that *open-circuiting* all links will result in zero current in *every* branch of the network. Hence, there can be *no* branch currents independent of the link currents. Furthermore, if any link is closed there will be a path for current independent of the other links. Therefore, it can be concluded that there are *exactly* as many independent branch currents as there are links and that *any* set of *link currents* is independent.

Thus, it may be seen that the tree provides a simple, graphical method of determining the *number* of independent variables needed and precisely *which* variables form an independent set. It is now necessary to consider ways of writing equilibrium equations in terms of these variables.

19–3. The Voltage Equilibrium Equation. When dealing with simple series circuits the voltage equilibrium equation is established by considering a *simple* current variable and expressing the voltage across each element in terms of this current. Thus, in the simple series R-L-C circuit shown in Fig. 19–3 the voltage equilibrium equation could be written as

$$e_R(t) + e_L(t) + e_C(t) = e_s(t) \tag{19–1}$$

or

$$Ri(t) + L\frac{di(t)}{dt} + \frac{1}{C}\int i(t)\,dt = e_s(t) \tag{19–2}$$

Thus, it may be seen that the basic concept employed in writing any such voltage equilibrium equation is to proceed around a closed path in the circuit and equate the sum of element voltages, expressed in terms of the element currents, to the source voltages. Exactly the same concept is employed in writing the voltage equilibrium equations for general networks. However, since there are a number of closed paths in a general network there will be several different ways in which this procedure may be carried out.

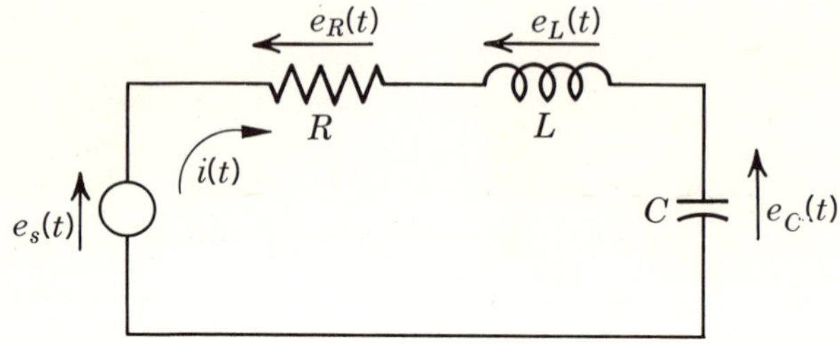

Fig. 19–3. *R-L-C* series circuit.

In order to illustrate several of these possibilities, and thereby emphasize the simplicity of the *one* procedure which will be recommended, consider the circuit of Fig. 19–4*a* with the graph of Fig. 19–4*b*. In order to minimize complications from unessential details, all elements in this circuit are chosen to be resistances.

Since this network has exactly the same graph as that of Fig. 19–1, any of the trees shown in Fig. 19–2 could be used. For purposes of illustration we arbitrarily select Fig. 19–2*c* as the one to be considered. This tree is redrawn

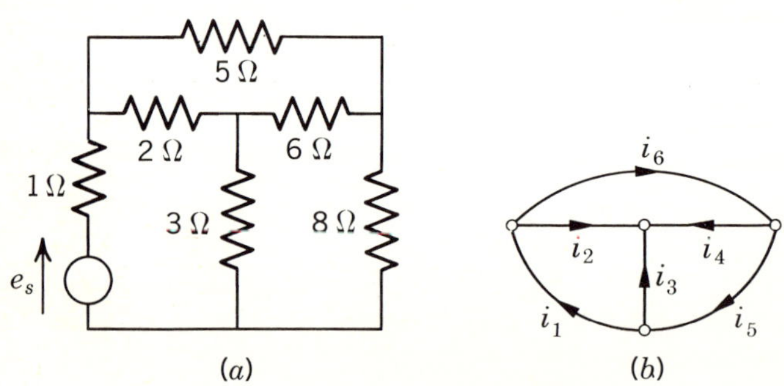

Fig. 19–4. A network and its graph.

in Fig. 19–5*a* and the branch currents are designated. There are six branch currents shown but there are only three *independent* currents since there are three links. It is necessary, therefore, to specify the three independent currents which are to be used as variables in terms of the branch currents. Although there are many ways in which this may be done, the method to be outlined has the advantage of utilizing the information available from the tree and thus provides a visual aid to the mathematical techniques.

First it should be recalled that any tree is drawn so that there are no closed paths. Furthermore, closing any *one* link will always produce *one and only one* closed path. If these closed paths are designated as *loops*, and a *loop current* assigned to each, then there will be as many loop currents as links; hence, the loop currents may be taken as the independent variables of the equilibrium

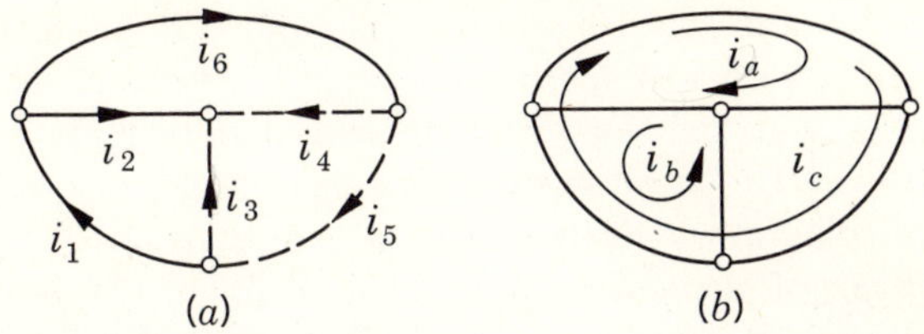

Fig. 19–5. A possible tree and loop set for the graph of Fig. 19–4*b*.

equations. The loop currents for the tree selected in this example are shown in Fig. 19–5*b*. Note that each *loop current* is identical to the *link current* which defines that loop. That is,

$$i_a = i_4$$
$$i_b = i_3$$
$$i_c = i_5$$

The *tree branch currents* may also be related to the *loop currents* by

$$i_6 = i_a + i_c$$
$$i_2 = -i_a - i_b$$
$$i_1 = -i_b + i_c$$

Hence, the three loop currents constitute an independent set from which all of the branch currents may be determined.

If a different tree had been considered a different set of loop currents would have resulted. For example, suppose that the tree shown in Fig. 19–6*a* is considered. The loop currents established by closing the links of this tree are shown in Fig. 19–6*b*. The relationships between the loop currents and the branch currents now become

$$i_1 = i_a$$
$$i_2 = i_a - i_c$$
$$i_3 = -i_a + i_b$$
$$i_4 = i_c - i_b$$
$$i_5 = i_b$$
$$i_6 = i_c$$

The particular set of loop currents indicated in Fig. 19–6 are usually designated as *mesh currents*. A *mesh* in a circuit diagram is a closed path which does not encircle any circuit elements or sources. Hence, it is a special form of loop and the current in it is a mesh current, which is a special form of loop current. There will always be the same number of meshes as links; hence, the mesh currents will always form a complete, independent set of variables.

The mesh currents form a particularly useful set of variables because they may be obtained directly from the circuit diagram without the necessity of drawing a graph and selecting a tree. Therefore, if one writes the equilibrium equations in terms of mesh currents, the problem of selecting the current variables is straightforward and easily handled. It is worth noting, however, that even the mesh currents are not unique but depend upon the way the circuit diagram is drawn. This is indicated clearly in Fig. 19–7 in which a given set of elements connected in a given way is shown in three different forms so as to give rise to three different sets of mesh currents. Once the form of the circuit diagram has been established, however, the mesh currents are uniquely determined.

In all of the discussion which follows the current variables will be selected as mesh currents. On this basis, reconsider the circuit of Fig. 19–4*a*, which is redrawn in Fig. 19–8 with the mesh currents indicated. Note that all the mesh currents have been assumed to circulate in the same direction. This assumption is not essential but is usually convenient in that the resulting equilibrium equations will exhibit a greater degree of symmetry and thus reduce the possibility of errors.

Once a suitable set of current variables has been selected, the voltage equilibrium equations may be written immediately. One procedure for doing this (not the only one, however) is to proceed around each mesh in the direction of the assumed mesh current and determine the voltage drops for every mesh current associated with that mesh. A voltage drop encountered in proceeding through an element in the direction of the mesh current involved is assumed positive and one encountered in going opposite to a mesh current is assumed negative. Enter all voltage drops on the left side of a set of simultaneous equations. The voltage rises due to sources may be entered on the right side of the simultaneous equations. The voltage rise is considered positive if one proceeds through the source in the direction of the source arrow; otherwise it is considered negative.

On the basis of this procedure the voltage equilibrium equations for Fig. 19–8 become

$$i_1 + 2i_1 - 2i_3 + 3i_1 - 3i_2 = e_s \quad \text{for mesh 1} \tag{19–3}$$

$$3i_2 - 3i_1 + 6i_2 - 6i_3 + 8i_2 = 0 \quad \text{for mesh 2} \tag{19–4}$$

and

$$2i_3 - 2i_1 + 5i_3 + 6i_3 - 6i_2 = 0 \quad \text{for mesh 3} \tag{19–5}$$

If the coefficients are combined the equations become

$$6i_1 - 3i_2 - 2i_3 = e_s \tag{19–6}$$

$$-3i_1 + 17i_2 - 6i_3 = 0 \tag{19–7}$$

$$-2i_1 - 6i_2 + 13i_3 = 0 \tag{19–8}$$

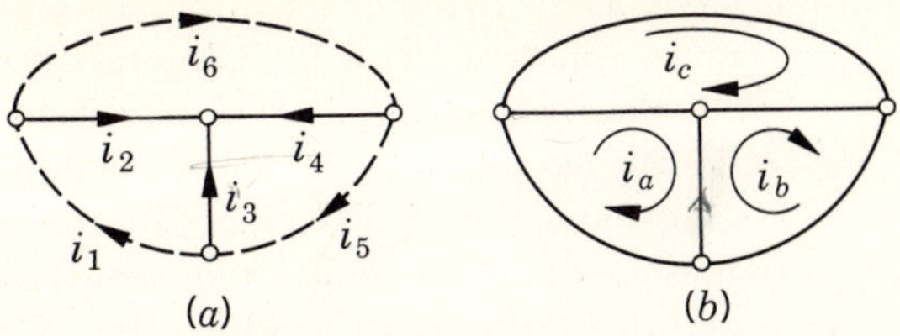

Fig. 19-6. Another possible tree with its loop currents for the graph of Fig. 19-4*b*.

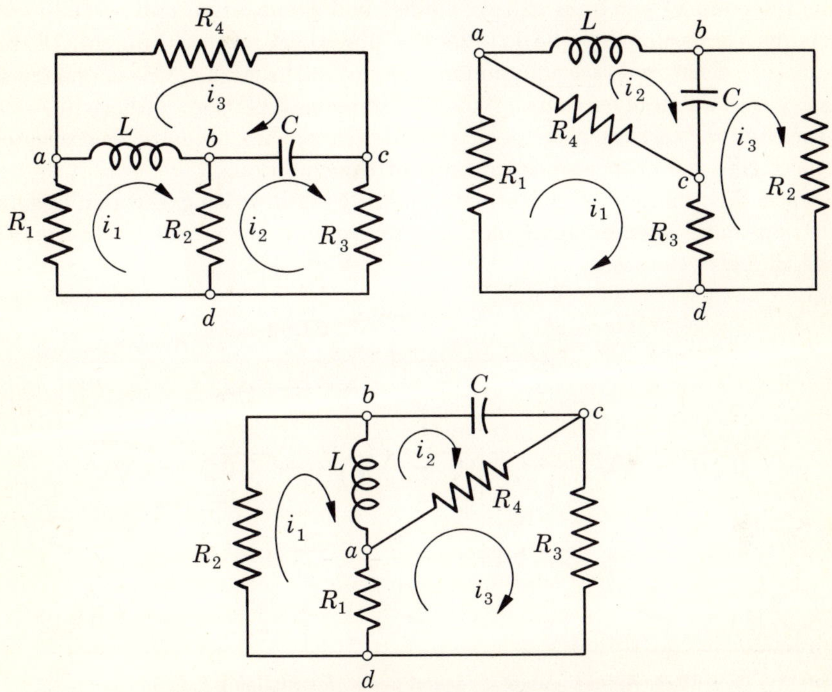

Fig. 19-7. Three different forms of the same network.

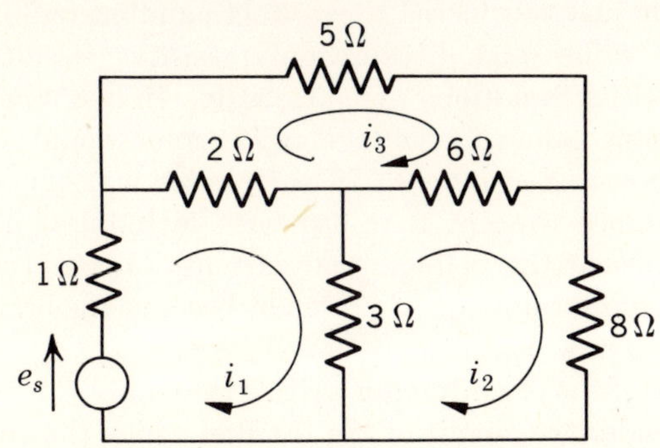

Fig. 19-8. The network of Fig. 19-4*a* and its mesh currents.

Once the equations have been written in this form, a simplified procedure for obtaining them becomes apparent. It will be observed that in Eq. 19–6, which pertains to mesh 1, the coefficient of i_1 is simply the total impedance around mesh 1, the coefficient of i_2 is the negative of the impedance which is common to meshes 1 and 2, and the coefficient of i_3 is the negative of the impedance common to meshes 1 and 3. Similar considerations apply to Eqs. 19–7 and 19–8, which pertain to the other meshes. Hence, it is possible to write the equilibrium equations in this combined form immediately without any preliminary steps. It should be noted, however, that the terms resulting from impedances common to two meshes will be negative only if the mesh currents are assumed to be in opposite directions in that branch. If this result is to occur for *every* common impedance in the circuit, it is necessary to assume that all mesh currents circulate in the *same* direction. Otherwise, *some* of the common terms will be positive and greater care needs to be exercised in specifying the proper signs throughout the equations.

Since this chapter is concerned primarily with writing the equilibrium equations, any discussion of methods for solving them will be postponed until the next chapter.

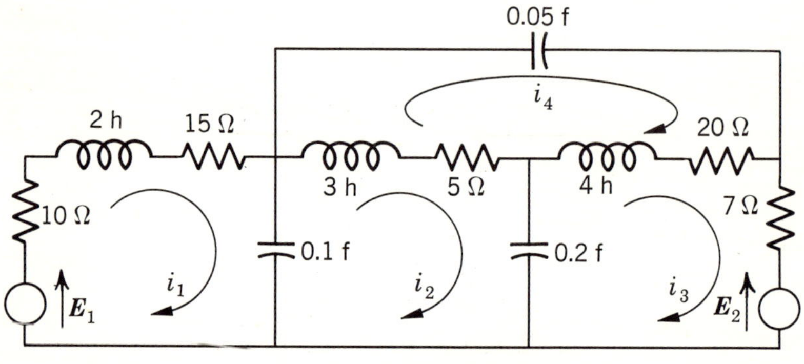

Fig. 19–9. An example of a general network containing *R*, *L*, and *C*.

In the example just considered the voltages and currents were expressed as instantaneous values and, because only resistive circuit elements were involved, the resulting equations were algebraic. In a more general case, the use of instantaneous values for voltage and current would result in a set of simultaneous *differential equations*. In such cases it is more convenient to utilize the impedance concept that has been introduced in the preceding chapters and represent the voltages and currents as complex quantities. If this is done the equilibrium equations again become algebraic, although the terms involved are now complex.

As an example of the procedure for writing the equilibrium equations in the general case, consider the circuit of Fig. 19–9 in which the voltages are specified to be at the same frequency and the assumed mesh currents are indicated.

Following the simplified procedure suggested in the previous example, the voltage equilibrium equations may be written immediately as

$$\left(25 + 2s + \frac{10}{s}\right)I_1 - \frac{10}{s}I_2 = E_1 \tag{19–9}$$

$$-\frac{10}{s}I_1 + \left(5 + 3s + \frac{15}{s}\right)I_2 - \frac{5}{s}I_3 - (5 + 3s)I_4 = 0 \tag{19–10}$$

$$-\frac{5}{s}I_2 + \left(27 + 4s + \frac{5}{s}\right)I_3 - (20 + 4s)I_4 = -E_2 \tag{19–11}$$

$$-(5 + 3s)I_2 - (20 + 4s)I_3 + \left(2s + 7s + \frac{20}{s}\right)I_4 = 0 \tag{19–12}$$

Methods for solving equations of this sort will be discussed in the next chapter.

In both of the preceding examples the sources have been considered to be voltage sources. If the circuit to be analyzed contains one or more current sources, it is convenient to transform them to equivalent voltage sources by application of Thévenin's theorem.

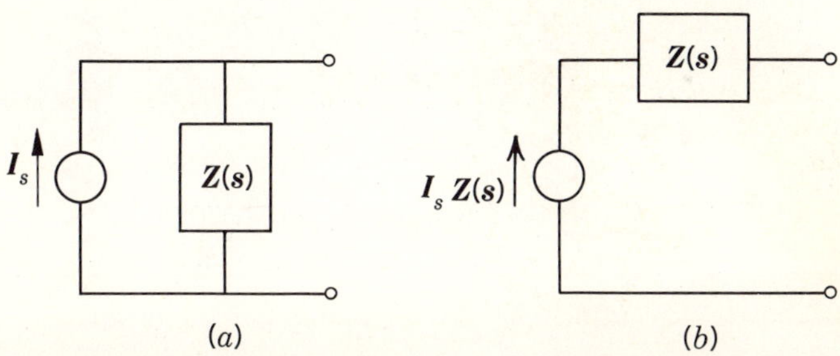

Fig. 19–10. A current source and Thévenin's equivalent voltage source.

If such transformations are not made it may be necessary to write more equilibrium equations than would otherwise be needed. As an example of such source transformations in the most general case consider the circuit of Fig. 19–10*a*, which contains a current source and an arbitrary impedance in parallel. The equivalent voltage source may be obtained by recalling that it consists of the *open-circuit voltage* in *series* with the same impedance. This is illustrated in Fig. 19–10*b*.

As a more specific example of this procedure consider Fig. 19–11. A circuit containing three current sources is shown in Fig. 19–11*a*; in Fig. 19–11*b* these sources have been replaced by equivalent voltage sources. In particular, note the directions of the arrows assigned to each voltage source.

Having made these transformations, the voltage equilibrium equations for mesh currents may be written as

$$(14 + 5s)I_1 - (4 + 3s)I_2 - 2sI_3 = 10I_a + (4 + 3s)I_b \quad (19\text{–}13)$$

$$-(4 + 3s)I_1 + \left(24 + 3s + \frac{10}{s}\right)I_2 - \frac{10}{s}I_3 = -(4 + 3s)I_b \quad (19\text{–}14)$$

$$-2sI_1 - \frac{10}{s}I_2 + \left(5 + 2s + \frac{15}{s}\right)I_3 = \left(5 + \frac{5}{s}\right)I_c \quad (19\text{–}15)$$

The student should verify these equations and pay particular attention to the manner in which the voltage sources are introduced.

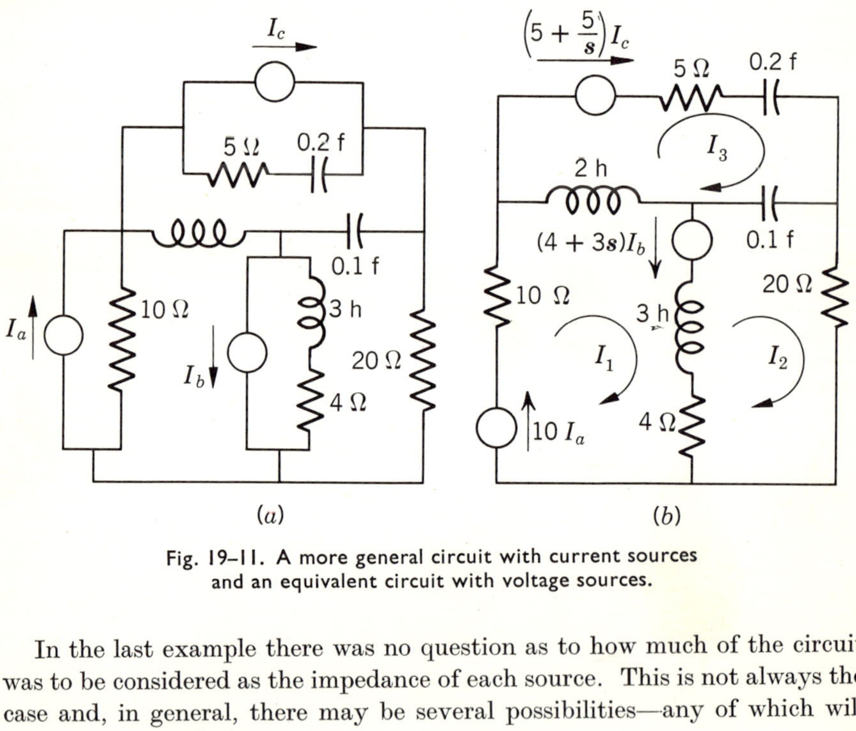

Fig. 19–11. A more general circuit with current sources and an equivalent circuit with voltage sources.

In the last example there was no question as to how much of the circuit was to be considered as the impedance of each source. This is not always the case and, in general, there may be several possibilities—any of which will yield a correct solution.

A final comment is in order before leaving the subject of voltage equilibrium equations. The last two sets of equations have been written in terms of the complex frequency variable s. When a particular sinusoidal frequency is specified, s becomes a purely imaginary numerical constant and the coefficients of the current variables become the resistive and reactive components of appropriate branch impedances. In such cases it is generally desirable to insert the reactance values on the circuit diagram and write the equilibrium equations immediately with these values. As an example, assume that the voltages in Fig. 19-9 are given as

$$e_1(t) = 100 \cos (2t + 30°)$$
$$e_2(t) = 200 \cos (2t - 45°)$$

from which

$$\boldsymbol{E}_1 = 100\underline{/30°}$$
$$\boldsymbol{E}_2 = 200\underline{/-45°}$$

Since $s = j2$ in this case the circuit diagram may be drawn as shown in Fig. 19-12 and the equilibrium equations written immediately as

$$(25 - j1)\boldsymbol{I}_1 + j5\boldsymbol{I}_2 = 100\underline{/30°} \qquad (19\text{-}16)$$
$$j5\boldsymbol{I}_1 + (5 - j1.5)\boldsymbol{I}_2 + j2.5\boldsymbol{I}_3 - (5 + j6)\boldsymbol{I}_4 = 0 \qquad (19\text{-}17)$$
$$j2.5\boldsymbol{I}_2 + (27 + j5.5)\boldsymbol{I}_3 - (20 + j8)\boldsymbol{I}_4 = -200\underline{/-45°} \qquad (19\text{-}18)$$
$$-(5 + j6)\boldsymbol{I}_2 - (20 + j8)\boldsymbol{I}_3 + (25 + j4)\boldsymbol{I}_4 = 0 \qquad (19\text{-}19)$$

When writing equations in this manner it is imperative that all sources have the same frequency. If they did not have the same frequency, the reactance values would be different for each source. In this case it would be necessary

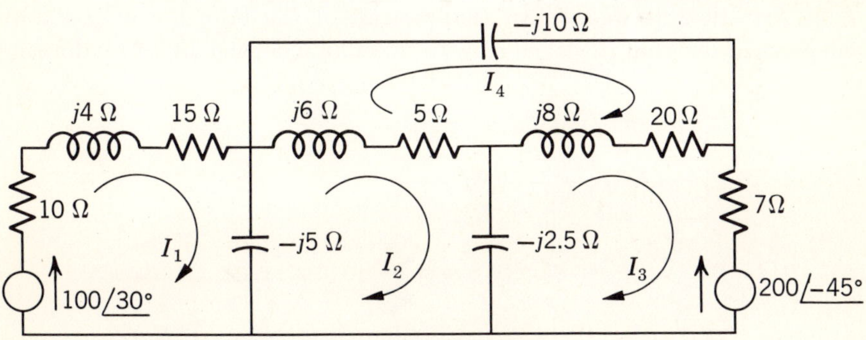

Fig. 19-12. The network of Fig. 19-9 with sinusoidal sources of the same frequency.

to obtain a solution for each source independently and combine the results according to the superposition theorem.

19-4. The Current Equilibrium Equations. It has been shown in previous work that when dealing with circuit elements connected in parallel it is convenient to write a current equilibrium equation rather than a voltage equilibrium equation. In the simplest case this involves a *single* voltage variable and the current in each element is expressed in terms of this voltage.

Thus, in the simple parallel R-L-C circuit shown in Fig. 19–13 the current equilibrium equation may be written as

$$i_R(t) + i_L(t) + i_C(t) = i_s(t) \tag{19–20}$$

or

$$\frac{1}{R}\,e(t) + \frac{1}{L}\int e(t)\,dt + C\,\frac{de(t)}{dt} = i_s(t) \tag{19–21}$$

The basic concept employed in writing any such current equilibrium equation is to equate the sum of the currents leaving any junction to the sum of the currents entering that junction. This same concept may be used in any

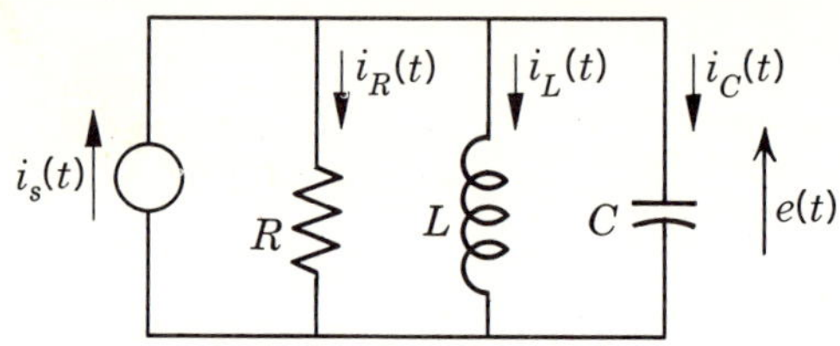

Fig. 19–13. An R-l-C parallel circuit.

general network except that there are a number of junctions that must be considered instead of just one. In any case, however, the currents in the elements are expressed in terms of the branch voltages and these voltages become the variables of the equilibrium equations.

In Art. 19–2 it was shown that not all of the branch voltages were independent but that it was possible to select an independent set by drawing

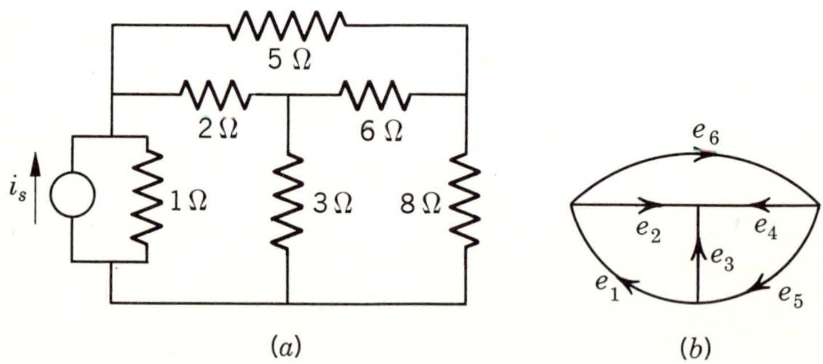

Fig. 19–14. A network and its graph.

one of the possible trees for the network. As an illustration of some of the possibilities consider the circuit of Fig. 19–14*a* which has the graph of Fig. 19–14*b*. (Note that the current source and its associated resistance are considered to form a single branch.)

Since this network has exactly the same graph as that of Fig. 19–1, any of the trees of Fig. 19–2 could be used for selecting the voltage variables. As an example, consider the tree of Fig. 19–2*a*, which is redrawn in Fig. 19–15 with the tree branch voltages designated. Since each voltage appears between two nodes of the network they may also be designated as *node-pair voltages*. The indicated node-pair voltages of Fig. 19–15 may be related to *all* of the branch voltages by reference to Fig. 19–14*b*. Thus, it is possible to write

$$\begin{aligned}
e_1 &= e_a \\
e_2 &= e_b \\
e_3 &= e_a + e_b \\
e_4 &= -e_c \\
e_5 &= -e_a - e_b - e_c \\
e_6 &= e_b + e_c
\end{aligned}$$

and it is seen that e_a, e_b, and e_c form an independent set from which the branch voltages may be determined.

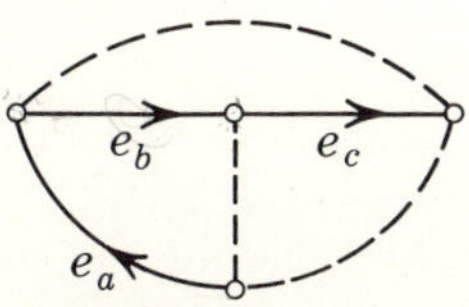

Fig. 19–15. A possible tree for the network of Fig. 19–14.

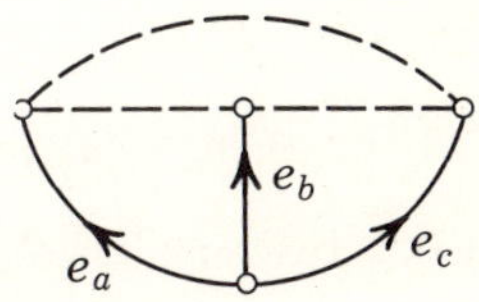

Fig. 19–16. Another possible tree, known as a datum tree, for the network of Fig. 19–14.

If a different tree had been selected, a different set of node-pair voltages would have resulted. For example, consider the tree of Fig. 19–16. By referring to Fig. 19–14*b*, the node-pair voltages of this tree may be related to the branch voltages by

$$\begin{aligned}
e_1 &= e_a \\
e_2 &= -e_a + e_b \\
e_3 &= e_b \\
e_4 &= e_b - e_c \\
e_5 &= -e_c \\
e_6 &= -e_a + e_c
\end{aligned}$$

This particular set of node-pair voltages has one node which is common to all three voltages. Such a node is referred to as a *datum node* or *reference node* and the voltages e_a, e_b, and e_c as *node-datum voltages*. This notation is often shortened to simply *node voltages* and as long as the reference node is clearly specified no uncertainty will arise.

The node-datum voltages form a particularly useful set of variables because they may be obtained from the circuit diagram without the necessity of drawing a graph and selecting a tree. Therefore, if one writes the equilibrium equations in terms of node-datum voltages the problem of selecting an independent set of voltage variables is straightforward and easily handled. It should be noted, however, that the node-datum voltages are not unique but depend upon which node is selected as the reference. This is clearly indicated in Fig. 19–17, where the reference node is designated by a "ground" symbol and the appropriate node-datum voltages by arrows. Once the reference node has been established, the node-datum voltages are uniquely

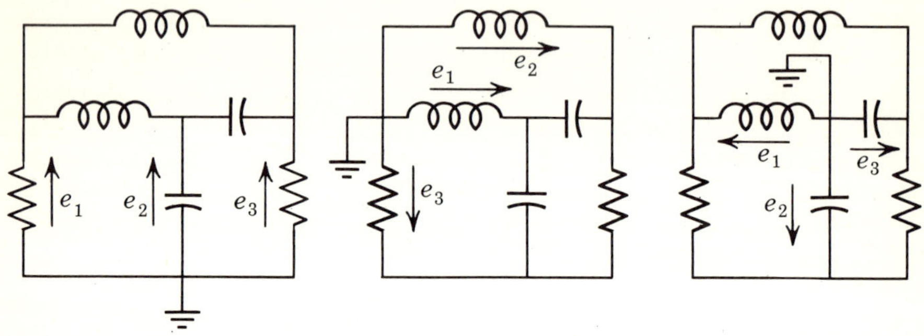

Fig. 19–17. Three possible datum-node selections for the same network.

determined, however. There are no definite rules for selecting the reference node but it is usually convenient to select either the node which would normally be at "zero" or "ground" potential or the node which has the greatest number of branches connected to it.

In the discussion which follows the voltage variables will always be selected as node-datum voltages. In addition, the reference node will be assumed

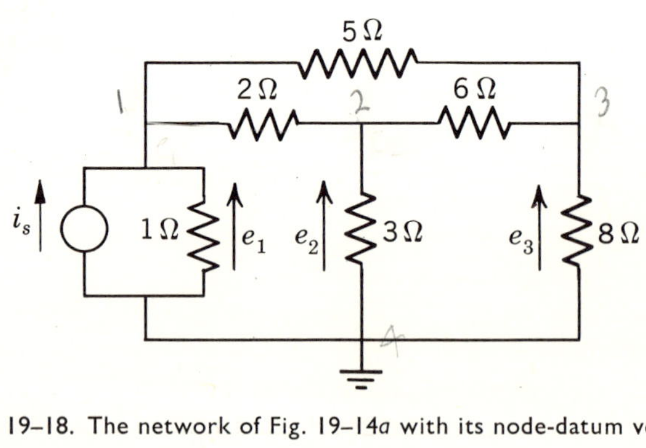

Fig. 19–18. The network of Fig. 19–14*a* with its node-datum voltages.

to be at zero potential; hence, the potentials of all other nodes will be equal to the *node-datum voltages* and will be designated simply as *node voltages*. On this basis, Fig. 19–14*a* is redrawn as in Fig. 19–18 and the node voltages designated by arrows. The significance of assuming all node voltages to be directed away from the reference node will be discussed later.

Having selected the voltage variables, the next step is to write equations expressing the current equilibrium at each of the nodes except the reference node. A possible procedure for doing this is to determine the currents entering or leaving each node by way of the circuit elements in terms of the node voltages. If the node-voltage arrow points toward the node, then all leaving currents are assumed positive and all entering currents are assumed negative. If the node voltage arrow points away from the node, the convention is reversed. All of the currents in the elements are then entered on the left side of a set of simultaneous equations. The currents due to sources may be entered on the right side of the simultaneous equations. If the node-voltage arrow points toward the node, entering source currents are assumed positive and leaving source currents negative. If the node-voltage arrow points away from the node, the convention is reversed.

On the basis of this procedure the current equilibrium equations for Fig. 19–18 become

$$\tfrac{1}{1}e_1 + \tfrac{1}{2}(e_1 - e_2) + \tfrac{1}{5}(e_1 - e_3) = i_s \quad \text{for node 1} \tag{19–22}$$

$$\tfrac{1}{3}e_2 + \tfrac{1}{2}(e_2 - e_1) + \tfrac{1}{6}(e_2 - e_3) = 0 \quad \text{for node 2} \tag{19–23}$$

$$\tfrac{1}{8}e_3 + \tfrac{1}{6}(e_3 - e_2) + \tfrac{1}{5}(e_3 - e_1) = 0 \quad \text{for node 3} \tag{19–24}$$

If the coefficients are combined the equations become

$$(1 + \tfrac{1}{2} + \tfrac{1}{5})e_1 - \tfrac{1}{2}e_2 - \tfrac{1}{5}e_3 = i_s \tag{19–25}$$

$$-\tfrac{1}{2}e_1 + (\tfrac{1}{3} + \tfrac{1}{2} + \tfrac{1}{6})e_2 - \tfrac{1}{6}e_3 = 0 \tag{19–26}$$

$$-\tfrac{1}{5}e_1 - \tfrac{1}{6}e_2 + (\tfrac{1}{8} + \tfrac{1}{6} + \tfrac{1}{5})e_3 = 0 \tag{19–27}$$

When written in this form, a simplified procedure for obtaining the equations becomes apparent. It will be observed that in Eq. 19–22, which pertains to node 1, the coefficient of e_1 is simply the total admittance connected to node 1, the coefficient of e_2 is the negative of the admittance connected between nodes 1 and 2, and the coefficient of e_3 is the negative of the admittance connected between nodes 1 and 3. Similar considerations apply to Eqs. 19–26 and 19–27, which pertain to the other nodes. Hence, it is possible to write the equilibrium equations immediately in this combined form without any preliminary steps. It should be noted, however, that the terms resulting from admittances connected between two nodes will be negative only if the node voltages are assumed to be in the same direction (with respect to the reference node) for every node. If this is not done, some of the coefficients pertaining to admittances *between* nodes will be positive, with the result that greater care is needed to insure the correct signs throughout.

Since the admittances in the foregoing example are purely conductive, it is possible to write algebraic equations in terms of instantaneous values of voltage and current. In a more general case the admittances would be expressed as functions of s and the voltages and currents represented by complex quantities. As an example of this, consider the circuit of Fig. 19–19 in which the reference node and assumed node voltages are indicated. Following the simplified procedure suggested in the preceding example, the equilibrium equations may be written immediately as

$$\left(\frac{1}{10}+\frac{1}{15+2s}\right)E_1-\left(\frac{1}{15+2s}\right)E_2 \qquad = I_1 \tag{19–28}$$

$$-\left(\frac{1}{15+2s}\right)E_1+\left(\frac{1}{15+2s}+0.1s+\frac{1}{5+3s}+0.05s\right)E_2 - \left(\frac{1}{5+3s}\right)E_3-0.05sE_4=0 \tag{19–29}$$

$$-\left(\frac{1}{5+3s}\right)E_2+\left(\frac{1}{5+3s}+0.2s+\frac{1}{20+4s}\right)E_3-\left(\frac{1}{20+4s}\right)E_4=0 \tag{19–30}$$

$$-0.05sE_2-\left(\frac{1}{20+4s}\right)E_3+\left(\frac{1}{20+4s}+\frac{1}{7}+0.05s\right)E_4=0 \tag{19–31}$$

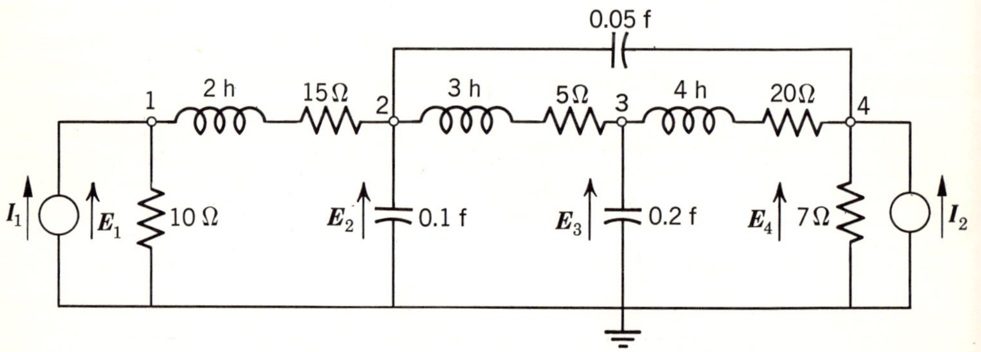

Fig. 19–19. A general network containing *R*, *L*, and *C*.

In both of the preceding examples the sources have been considered to be current sources. If the circuit to be analyzed contains one or more voltage sources it is convenient to transform them to equivalent current sources by means of Norton's theorem. As a matter of fact, both of these examples have been obtained by just this procedure. The circuit of Fig. 19–18 is the same as that of Fig. 19–8 with the source transformed. To complete the equivalence it is necessary to specify that

$$i_s = \frac{e_s}{1} \tag{19–32}$$

Likewise, the circuit of Fig. 19–19 is identical to that of Fig. 19–9 except for the source transformations. The equivalences needed here are

$$I_1 = \frac{E_1}{10} \tag{19–33}$$

and
$$I_2 = \frac{E_2}{7} \tag{19–34}$$

where E_1 and E_2 are the source voltages of Fig. 19–9.

In connection with the second transformation mentioned, it is worth noting that a different transformation would have been more desirable. If all of the impedance which was in mesh 1 only had been considered as part of the source, the transformed circuit would have been as shown in Fig. 19–20.

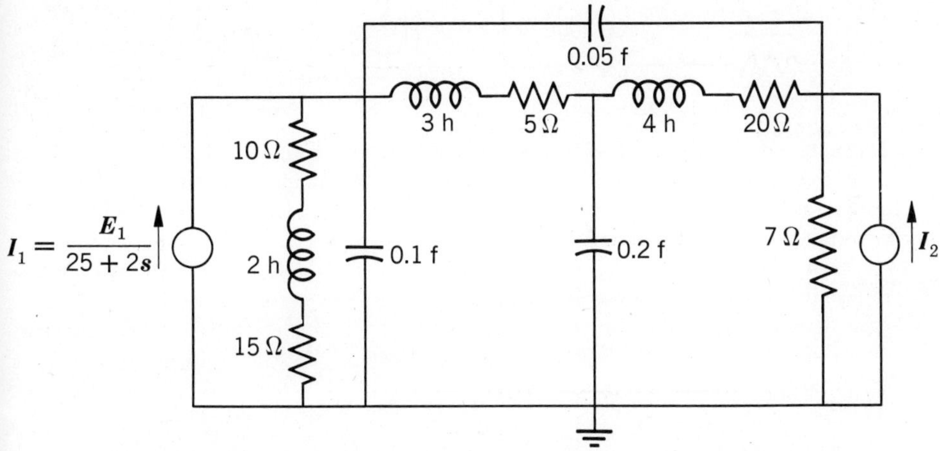

Fig. 19–20. Another possible transformation of the network of Fig. 19–9.

The advantage here is that there are only three node pairs, hence, only three equations to solve simultaneously. It is often possible to reduce the number of variables by making appropriate source transformations and it is generally desirable to exploit this possibility whenever the occasion arises.

PROBLEMS

19–1. Show that any combination of two of the three branch currents illustrated forms an independent set. In each case demonstrate that the third current is not independent by expressing it in terms of the other two currents.

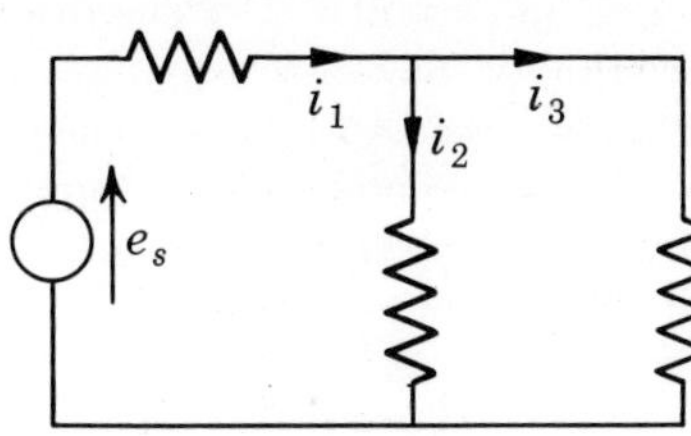

Prob. 19–1.

19–2. Show that any combination of five of the six branch voltages illustrated forms an independent set. In each case demonstrate that the sixth voltage is not independent by expressing it in terms of the other five voltages.

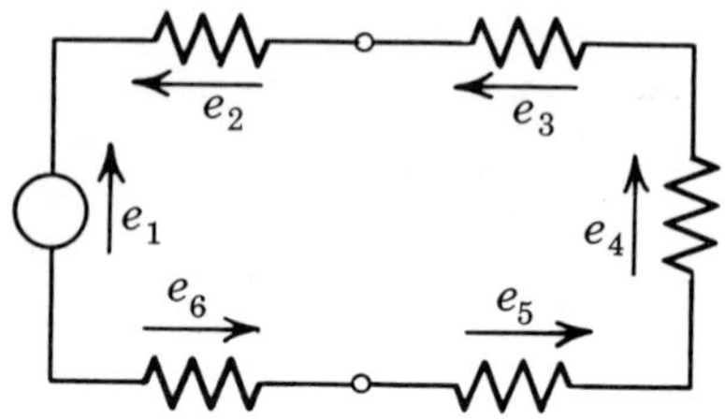

Prob. 19–2.

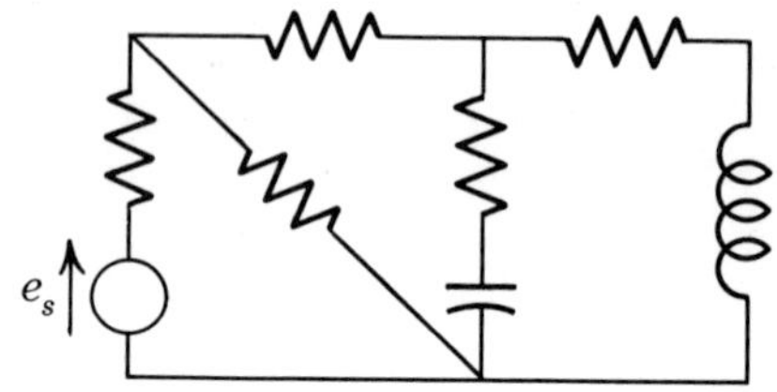

Prob. 19–3.

19–3. Draw the graph of the network shown and identify the branches and branch currents. Determine at least four different combinations of branch currents that form independent sets. In each case demonstrate that the remaining currents are not independent by expressing them in terms of the independent set.

19–4. Draw the graph of the network shown and determine at least four different combinations of branch voltages that form independent sets. In each case demonstrate that the remaining branch voltages are not independent by expressing them in terms of the independent set.

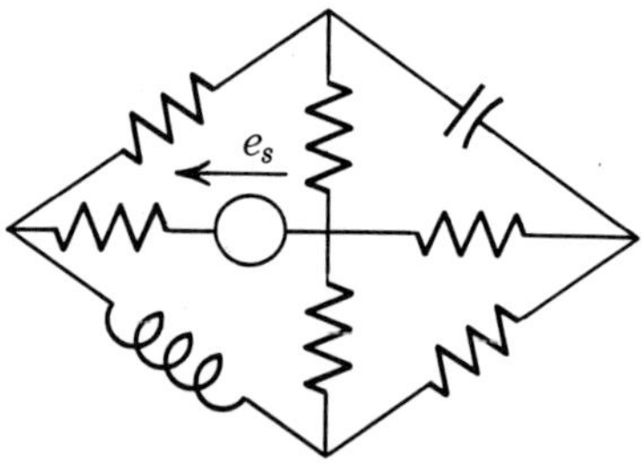

Prob. 19–4.

19–5. If the number of branches in a network is designated as b and the number of nodes as n there will be $(n - 1)$ independent branch voltages and $(b + 1 - n)$ independent branch currents. Use the results of Probs. 19–1, 19–2, 19–3, and 19–4 to confirm these statements.

19–6. Shown here are a network and two possible sets of loop currents (the last one corresponding to mesh currents). Write equations relating the branch currents to the loop currents for each of the two specified sets of loop currents. Write voltage equilibrium equations in terms of the loop currents for each of the two sets.

19–7. Write voltage equilibrium equations in terms of the specified mesh currents of the figure.

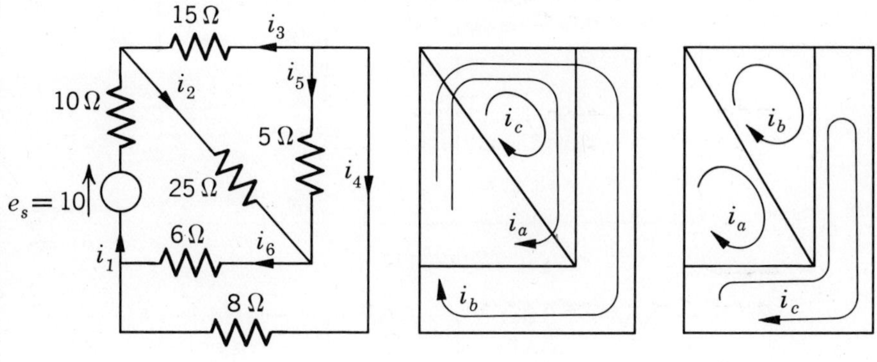

(a) (b) (c)

Prob. 19–6.

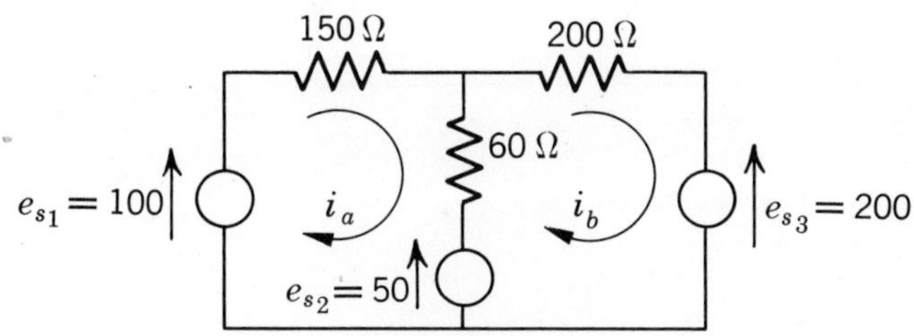

Prob. 19–7.

19–8. Write voltage equilibrium equations in terms of the specified mesh currents of the figure. Note that this problem is identical to Prob. 19–7 with the exception of the direction of i_b. Comment on the effects of reversing the direction of i_b.

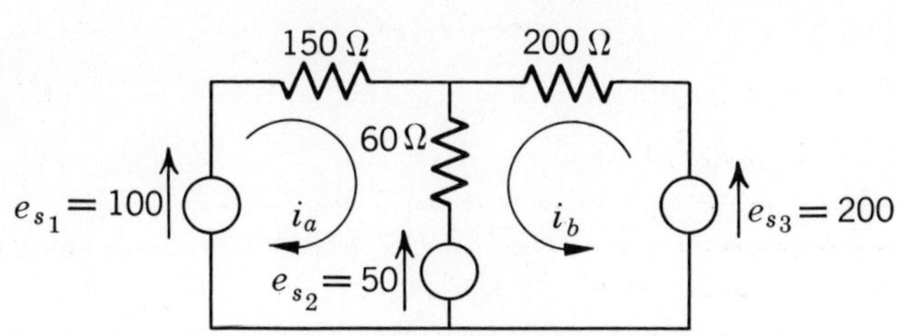

Prob. 19–8.

19–9. Write voltage equilibrium equations in terms of the specified mesh currents of the figure on p. 314. Instantaneous values of voltage and current are to be used, thus requiring integrodifferential equations.

19–10. Select mesh currents for the circuit shown on p. 314 and write the voltage equilibrium equations, using impedances as functions of s.

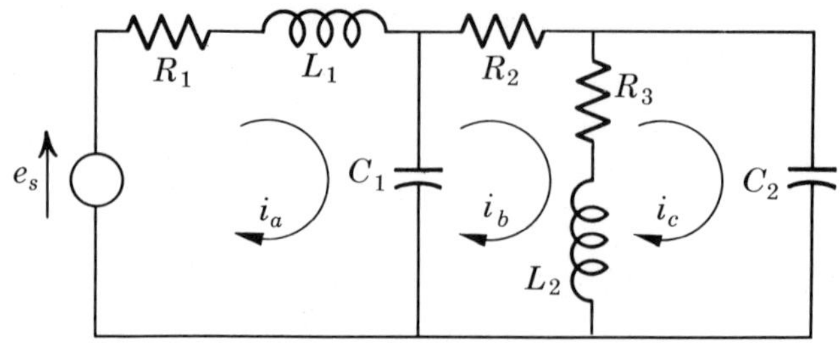

Prob. 19–9.

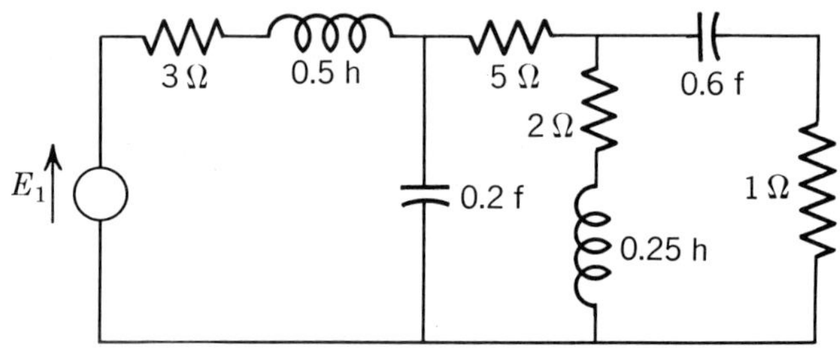

Prob. 19–10.

19–11. Select mesh currents for the network illustrated and write voltage equilibrium equations.

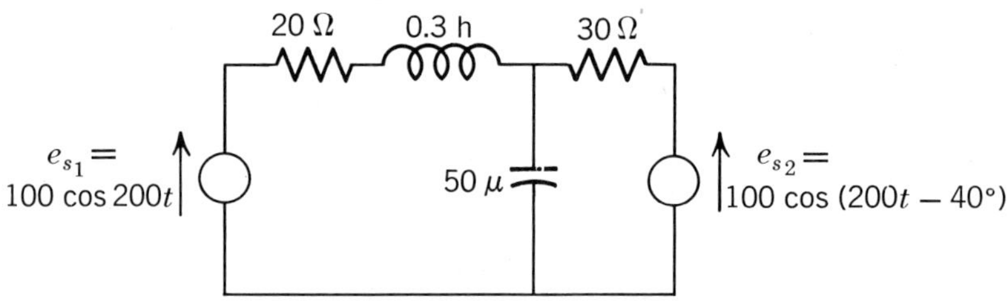

Prob. 19–11.

19–12. Select mesh currents for the network shown and write voltage equilibrium equations.

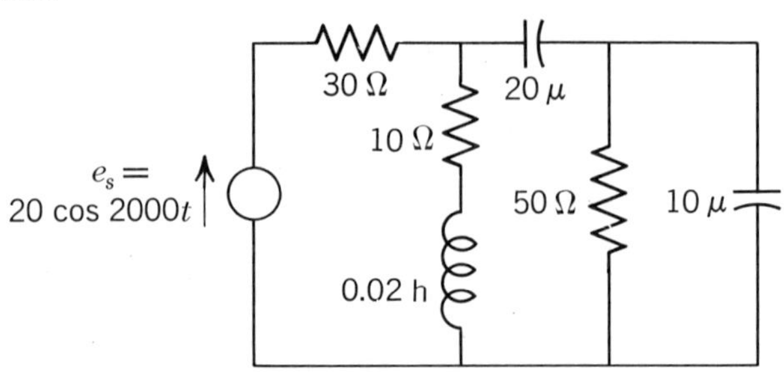

Prob. 19–12.

19–13. Make the required source transformation for the figure and write voltage equilibrium equations.

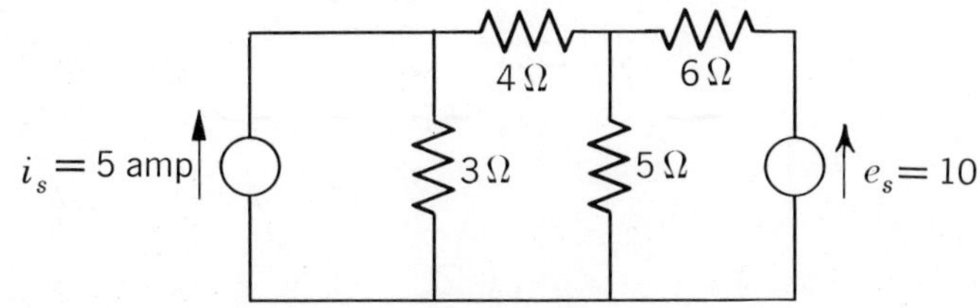

Prob. 19–13.

19–14. Make the required source transformation for the figure and write voltage equilibrium equations.

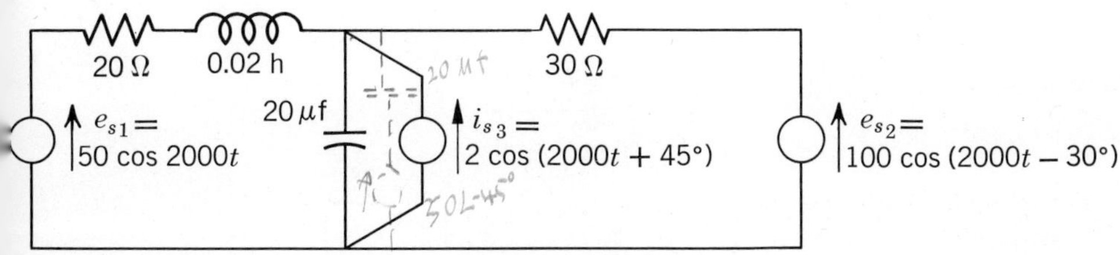

Prob. 19–14.

19–15. Express each of the branch currents shown in terms of the voltages e_1, e_2, and e_3. Write current equations at nodes 1, 2, and 3. Show that a fourth equation written at node 4 would not be independent.

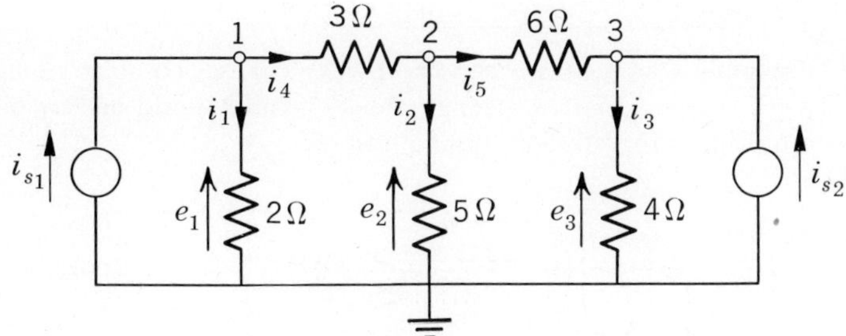

Prob. 19–15.

19–16. Write current equilibrium equations for each of the three possible choices of datum nodes indicated in the figure on p. 316.

19–17. Write current equilibrium equations in terms of the node-datum voltages specified in the figure on p. 316.

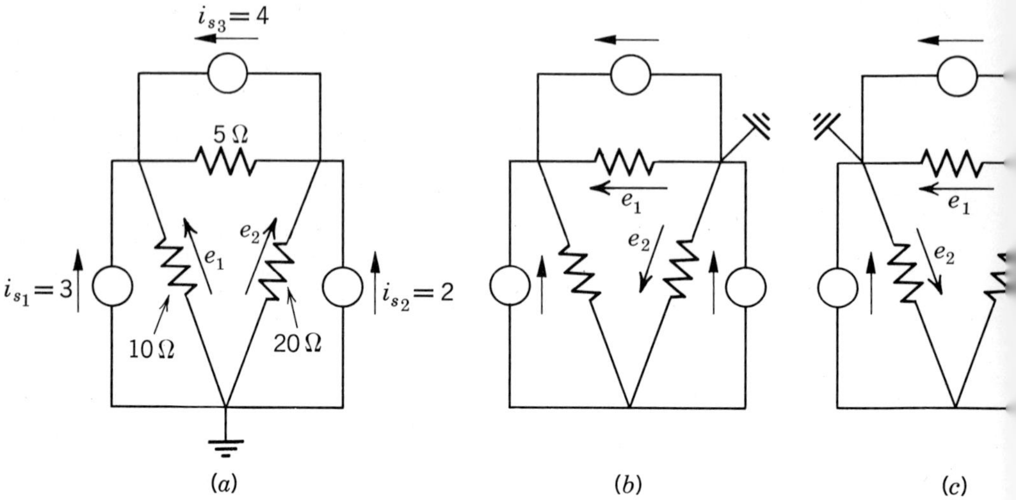

Prob. 19–16.

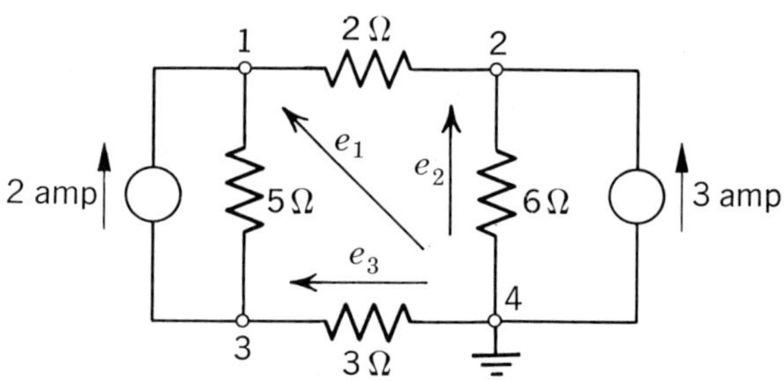

Prob. 19–17.

19–18. Use node 4 as a datum node and write current equilibrium equations for the network shown. Instantaneous values of voltage and current are to be used, thus requiring integrodifferential equations.

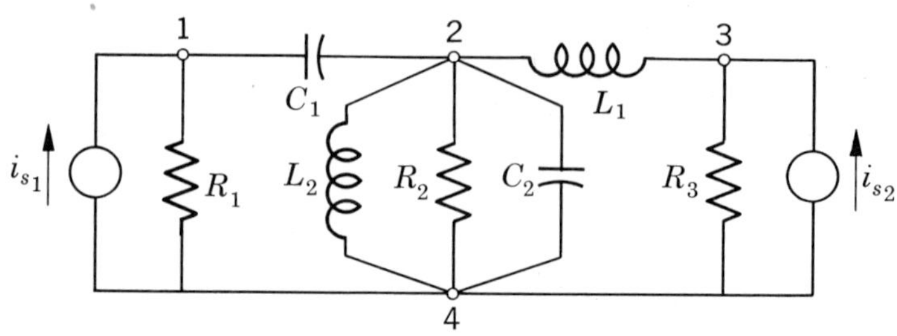

Prob. 19–18.

19–19. Write current equilibrium equations for the circuit illustrated, using node 4 as the datum node. Use the complex representation of voltages, currents, and admittances.

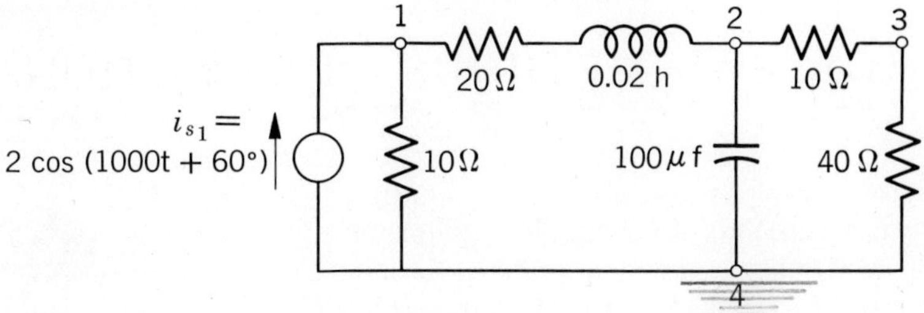

Prob. 19–19.

19–20. Write current equilibrium equations for the network shown, with node 4 as the datum node. Express the admittances as functions of s.

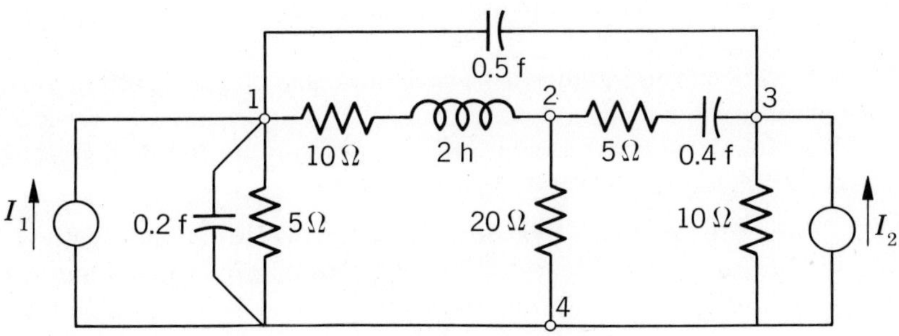

Prob. 19–20.

19–21. Make the necessary source transformations with the circuits of Probs. 19–7, 19–10, and 19–14 and write current equilibrium equations.

20

GENERAL METHODS OF SOLUTION

Now that methods of writing the equilibrium equations for general circuits have been discussed, the next problem is that of finding solutions for these equations. The student is probably familiar with one or more techniques for solving a set of simultaneous algebraic equations from previous work in college algebra. If so, he is already aware of the large amount of time required to carry out the solution if the number of equations is large. Since this process is so time-consuming it is desirable to be aware of as many time-saving short cuts as possible. This chapter will attempt to outline the most efficient methods for obtaining numerical solutions to the general problem. The next chapter will discuss additional short cuts that are possible in certain specialized cases. The ability to use any or all of these techniques wisely is an extremely valuable asset which not only saves time in circuit analysis but is also a definite aid in circuit synthesis.

20–1. The Method of Systematic Elimination. One method of solving simultaneous equations is the familiar one of reducing the number of variables by addition and subtraction. This method has the advantage of requiring very little specialized knowledge concerning the theory of equations and also lends itself to physical interpretation in terms of removing certain loops or nodes from the circuit. Its major disadvantage is that unnecessary calculations are often required if the work is not carried out in an organized fashion. The method of systematic elimination is simply an organized procedure for carrying out the additions and subtractions in the most concise manner possible and with the least amount of lost motion.

This procedure is best illustrated by considering a specific example. For this purpose, rewrite Eqs. 19–6, 19–7, and 19–8, assuming the source voltage e_s to be 50 v. These equations now become

$$6i_1 - 3i_2 - 2i_3 = 50 \tag{20–1}$$

$$-3i_1 + 17i_2 - 6i_3 = 0 \tag{20–2}$$

$$-2i_1 - 6i_2 + 13i_3 = 0 \tag{20–3}$$

Since the computations to be performed involve only the numerical coefficients of these equations, a more compact manner of presenting these coefficients is to arrange them in an array known as a *matrix*. In doing this it is only necessary to specify which columns of the matrix correspond to the coefficients of which variable. A convenient and logical method of doing this is to arrange the columns in the same sequence as they appear in the equations. Thus, a suitable matrix for specifying the numerical values in the above set of equations would be

$$\begin{bmatrix} 6 & -3 & -2 & 50 \\ -3 & 17 & -6 & 0 \\ -2 & -6 & 13 & 0 \end{bmatrix}$$

Although the rows of this matrix are written in the same sequence as the equations they came from, this is by no means necessary. Occasionally some other sequence may be preferable for reasons to be discussed later.

The process of eliminating variables corresponds to making certain entries in the matrix zero. A systematic way of doing this is to make all of the entries below the dashed line zero with as few manipulations as possible. The reasons for doing this will become apparent as the example proceeds. When this has been accomplished the matrix is said to be in *triangular form*.

There are many ways in which the triangular form might be accomplished. One possible way involves the following steps:

1. Multiply row 3 by 3 and add row 1 to it.
2. Multiply row 2 by 2 and add row 1 to it. These two steps may be carried out simultaneously and the resulting matrix will be

$$\begin{bmatrix} 6 & -3 & -2 & 50 \\ 0 & 31 & -14 & 50 \\ 0 & -21 & 37 & 50 \end{bmatrix}$$

3. Multiply row 3 by $\frac{31}{21}$ and add row 2 to it to yield the matrix

$$\begin{bmatrix} 6 & -3 & -2 & 50 \\ 0 & 31 & -14 & 50 \\ 0 & 0 & \frac{31}{21} \times 37 - 14 & \frac{31}{21} \times 50 + 50 \end{bmatrix}$$

which is seen to be in triangular form.

The third row of the final matrix represents the algebraic equation

$$\left(\tfrac{31}{21} \times 37 - 14\right) i_3 = \tfrac{31}{21} \times 50 + 50 \tag{20-4}$$

since the coefficients of both i_1 and i_2 are zero. This equation may be solved immediately to give

$$(31 \times 37 - 14 \times 21) i_3 = 31 \times 50 + 21 \times 50$$

$$i_3 = \frac{50 \times 52}{1147 - 294} = 3.05 \text{ amp}$$

Likewise, the second row of the final matrix represents the equation

$$31i_2 - 14i_3 = 50 \tag{20–5}$$

Since i_3 is already known, this equation may be solved for i_2. Hence,

$$31i_2 = 50 + 14 \times 3.05$$

$$i_2 = \frac{50 + 14 \times 3.05}{31} = 2.99 \text{ amp}$$

Finally, the first row of the matrix represents the equation

$$6i_1 - 3i_2 - 2i_3 = 50 \tag{20–6}$$

which may be solved for i_1 since both i_2 and i_3 are now known. Thus,

$$6i_1 = 50 + 3 \times 2.99 + 2 \times 3.05 = 64.87$$

$$i_1 = \frac{65.07}{6} = 10.85 \text{ amp}$$

The significance of putting the matrix into triangular form is now apparent. By so doing, the row with the most zeros will always represent an equation with a single variable in it. The remaining rows will represent equations with one additional variable in each. Hence, this form results in the *least* amount of elimination that is possible. It should also be apparent that the triangular form just discussed is not the only form that will give the same results. For example, in a set of three equations any of the following forms would be equally good:

$$\underset{(a)}{\begin{bmatrix} a & b & c & d \\ 0 & e & f & g \\ 0 & 0 & h & i \end{bmatrix}} \qquad \underset{(b)}{\begin{bmatrix} 0 & 0 & a & b \\ 0 & c & d & e \\ f & g & h & i \end{bmatrix}} \qquad \underset{(c)}{\begin{bmatrix} 0 & a & b & c \\ 0 & 0 & d & e \\ f & g & h & i \end{bmatrix}}$$

As a matter of fact any of these forms can be made into a triangular form simply by rearranging the sequence of rows.

In the example just given, all of the coefficients were pure real numbers since the circuit from which they came was purely resistive. In the case of more general circuits the coefficients will be complex and the process of systematic elimination will be correspondingly more difficult. The elimination procedure may be carried out in essentially the same manner, however, and as an illustration of this consider the following set of equations:

$$(5 + j10)\boldsymbol{I}_1 - (3 - j3)\boldsymbol{I}_2 = 100 + j0 \tag{20–7}$$

$$-(3 - j3)\boldsymbol{I}_1 + (8 + j12)\boldsymbol{I}_2 - (4 + j2)\boldsymbol{I}_3 = 0 \tag{20–8}$$

$$-(4 + j2)\boldsymbol{I}_2 + (20 - j10)\boldsymbol{I}_3 = 0 \tag{20–9}$$

The matrix of the coefficients, therefore, is

$$\begin{bmatrix} 5+j10 & -3+j3 & 0 & 100+j0 \\ -3+j3 & 8+j12 & -4-j2 & 0 \\ 0 & -4-j2 & 20-j10 & 0 \end{bmatrix}$$

In order to obtain a zero in the first column of the second row, multiply row 2 by $(5+j10)$ and row 1 by $(3-j3)$ and add. The resulting matrix is

$$\begin{bmatrix} 5+j0 & -3+j3 & 0 & 100+j0 \\ 0 & -80+j158 & -j50 & 300-j300 \\ 0 & -4-j2 & 20-j10 & 0 \end{bmatrix}$$

The zero in the second column of row 3 is obtained by multiplying that row by $(-80+j158)$ and row 2 by $(4+j2)$ and adding. The matrix now becomes

$$\begin{bmatrix} 5+j10 & -3+j3 & 0 & 100+j0 \\ 0 & -80+j158 & -j50 & 300-j300 \\ 0 & 0 & 80+j3760 & 1800-j600 \end{bmatrix}$$

The third row of this matrix represents the equation

$$(80+j3760)\boldsymbol{I}_3 = 1800 - j600$$

$$\boldsymbol{I}_3 = \frac{1800-j600}{80+j3760} = \frac{1898\angle{-18.42^\circ}}{3760\angle{88.78^\circ}} = 0.505\angle{-107.2^\circ}$$

The second row may now be solved for $\boldsymbol{I}_2$:

$$(-80+j158)\boldsymbol{I}_2 - j50(0.505\angle{-107.2^\circ}) = 300 - j300$$

$$\boldsymbol{I}_2 = \frac{324.1-j307.5}{-80+j158} = \frac{447\angle{-43.7^\circ}}{-177\angle{-63.15^\circ}} = -2.525\angle{19.65^\circ}$$

Finally, $\boldsymbol{I}_1$ is obtained from row 1:

$$(5+j10)\boldsymbol{I}_1 + (-3+j3)(-2.525\angle{19.65^\circ}) = 100 + j0$$

$$\boldsymbol{I}_1 = \frac{90.3+j4.58}{5+j10} = \frac{90.3\angle{2.88^\circ}}{11.17\angle{63.4^\circ}} = 8.08\angle{-60.5^\circ}$$

An alternative method of solving such a set of equations may be obtained by recognizing that the three complex variables that are being sought actually consist of *six* unknowns. Hence, if each current is expressed in terms of its real and imaginary components, it should be possible to write a set of *six* simultaneous equations with pure real coefficients. Thus, letting

$$\begin{aligned} \boldsymbol{I}_1 &= I_{1A} + jI_{1B} \\ \boldsymbol{I}_2 &= I_{2A} + jI_{2B} \\ \boldsymbol{I}_3 &= I_{3A} + jI_{3B} \end{aligned}$$

Eqs. 20–7, 20–8, and 20–9 may be written as

$$\begin{aligned} &5I_{1A} - 10I_{1B} + j10I_{1A} + j5I_{1B} - 3I_{2A} - 3I_{2B} - j3I_{2B} \\ &\qquad + j3I_{2A} = 100 + j0 \\ &-3I_{1A} - 3I_{1B} - j3I_{1B} + j3I_{1A} + 8I_{2A} - 12I_{2B} + j8I_{2B} \\ &\qquad + j12I_{2A} - 4I_{3A} + 2I_{3B} - j4I_{3B} - j2I_{3A} = 0 \\ &-4I_{2A} + 2I_{2B} - j4I_{2B} - j2I_{2A} + 20I_{3A} + 10I_{3B} - j10I_{3A} \\ &\qquad + j20I_{3B} = 0 \end{aligned}$$

Since the real and imaginary parts of these equations must be satisfied independently, they may be written as six equations as follows:

$$\begin{aligned} 5I_{1A} - 3I_{2A} \qquad\qquad - 10I_{1B} - 3I_{2B} \qquad\qquad &= 100 \\ -3I_{1A} + 8I_{2A} - 4I_{3A} - 3I_{1B} - 12I_{2B} + 2I_{3B} &= 0 \\ - 4I_{2A} + 20I_{3A} \qquad\qquad + 2I_{2B} + 10I_{3B} &= 0 \\ 10I_{1A} + 3I_{2A} \qquad\qquad + 5I_{1B} - 3I_{2B} \qquad\qquad &= 0 \\ 3I_{1A} + 12I_{2A} - 2I_{3A} - 3I_{1B} + 8I_{2B} - 4I_{3B} &= 0 \\ - 2I_{2A} - 10I_{3A} \qquad\qquad - 4I_{2B} + 20I_{3B} &= 0 \end{aligned}$$

The matrix for these equations is

$$\left[\begin{array}{rrr:rrr:r} 5 & -3 & 0 & -10 & -3 & 0 & 100 \\ -3 & 8 & -4 & -3 & -12 & 2 & 0 \\ 0 & -4 & 20 & 0 & 2 & 10 & 0 \\ \hdashline 10 & 3 & 0 & 5 & -3 & 0 & 0 \\ 3 & 12 & -2 & -3 & 8 & -4 & 0 \\ 0 & -2 & -10 & 0 & -4 & 20 & 0 \end{array}\right]$$

Although this matrix was obtained by a rather lengthy manipulation of the equations, when written in this form it becomes apparent that it could have been written immediately from the original set of equations without any intermediate steps. Note that the first six columns of the matrix have been divided into four submatrices by means of dashed lines. Note further that the upper-left matrix and the lower-right matrix are identical and are simply the real parts of the coefficients. The lower-left matrix is composed of the imaginary parts of the coefficients and the upper-right matrix contains the *negative* of these coefficients. The top three entries of the seventh column are the real parts of the constants on the right-hand side of the equations and the bottom three entries are the imaginary parts.

Once the matrix with real terms only has been obtained, the systematic elimination may proceed just as it did in the first example. The computation is more time-consuming in this case because there are six rows instead of three but it is probably no worse than the computation involving complex numbers.

Although the method of systematic elimination provides a straightforward method of solving any set of simultaneous equations, its use can be made even more effective by taking advantage of any short cuts that might result from rearranging the rows or columns to place the zeros in the most advantageous positions. Likewise, it is often possible to reduce the labor involved by spotting certain combinations that are easier to manipulate. To a large extent, the ingenuity and experience of the user play an important part in determining the over-all effectiveness of the method.

20-2. The Method of Determinants. A more elegant procedure for solving a set of simultaneous equations is the application of Cramer's rule, which expresses the solutions as ratios of determinants. This method is no shorter than that of systematic elimination when numerical coefficients are involved but is almost indispensable when the coefficients are represented by symbols. Furthermore, it will be shown later that the determinants have an additional significance that further increases their usefulness. Thus, a knowledge of the properties of determinants is particularly useful in the study of networks.

The present discussion assumes that the student has already studied the properties and uses of determinants in a course in college algebra; hence, it is intended to serve only as a review of the method and as a convenient reference to some of the properties. For the most part, the results are presented without proof and the student is referred to any college algebra text for formal justification.

The use of determinants in the solution of a set of linear equations is embodied in Cramer's rule, which may be stated as follows:

> If the determinant of the coefficients is different from zero, the value of each unknown is the quotient of two determinants. The determinant in the denominator is the determinant of the coefficients of the unknowns. The determinant in the numerator is the same as the determinant in the denominator except that the coefficients of the unknown whose value is being determined are replaced by the corresponding constant terms.

As an example of the application of this rule, reconsider Eqs. 20–1, 20–2, and 20–3:

$$6i_1 - 3i_2 - 2i_3 = 50 \quad (20\text{–}1)$$

$$-3i_1 + 17i_2 - 6i_3 = 0 \quad (20\text{–}2)$$

$$-2i_1 - 6i_2 + 13i_3 = 0 \quad (20\text{–}3)$$

The determinant of the coefficients, which appears in the denominator of each quotient, is

$$D = \begin{vmatrix} 6 & -3 & -2 \\ -3 & 17 & -6 \\ -2 & -6 & 13 \end{vmatrix} \quad (20\text{–}10)$$

The three unknown currents may then be expressed as

$$i_1 = \frac{\begin{vmatrix} 50 & -3 & -2 \\ 0 & 17 & -6 \\ 0 & -6 & 13 \end{vmatrix}}{D}; \quad i_2 = \frac{\begin{vmatrix} 6 & 50 & -2 \\ -3 & 0 & -6 \\ -2 & 0 & 13 \end{vmatrix}}{D}; \quad i_3 = \frac{\begin{vmatrix} 6 & -3 & 50 \\ -3 & 17 & 0 \\ -2 & -6 & 0 \end{vmatrix}}{D} \quad (20\text{–}11)$$

Before evaluating these determinants it is desirable to list some of the properties of determinants which are often helpful in reducing the labor involved. These properties are stated without proof.

1. A second-order determinant may be evaluated by cross-multiplication. Thus

$$\begin{vmatrix} a & b \\ c & d \end{vmatrix} = ad - bc \quad (20\text{–}12)$$

2. A third-order determinant may be evaluated by taking triple products with the proper sign. Thus,

$$\begin{vmatrix} a & b & c \\ d & e & f \\ g & h & i \end{vmatrix} = aei + bfg + cdh - gec - bdi - afh \quad (20\text{–}13)$$

3. The *minor* of any given element of a determinant is the determinant of next lower order obtained by deleting the row and column in which the element stands. Thus, for the determinant

$$\begin{vmatrix} a & b & c \\ d & e & f \\ g & h & i \end{vmatrix}$$

the minor of a is

$$\begin{vmatrix} e & f \\ h & i \end{vmatrix}$$

of e is

$$\begin{vmatrix} a & c \\ g & i \end{vmatrix}$$

of f is

$$\begin{vmatrix} a & b \\ g & h \end{vmatrix} \qquad \text{etc.}$$

4. The cofactor of any element in the mth row and nth column is $(-1)^{m+n}$ times the minor of that element. Thus, for the determinant just given the cofactor of a is

$$(-1)^2 \begin{vmatrix} e & f \\ h & i \end{vmatrix} = \begin{vmatrix} e & f \\ h & i \end{vmatrix}$$

of e is

$$(-1)^4 \begin{vmatrix} a & c \\ g & i \end{vmatrix} = \begin{vmatrix} a & c \\ g & i \end{vmatrix}$$

of f is

$$(-1)^5 \begin{vmatrix} a & b \\ g & h \end{vmatrix} = -\begin{vmatrix} a & b \\ g & h \end{vmatrix}$$

5. The value of a determinant is the *sum* of the results obtained by multiplying the elements of any row or column by their *cofactors*. Thus,

$$\begin{vmatrix} a & b & c \\ d & e & f \\ g & h & i \end{vmatrix} = a\begin{vmatrix} e & f \\ h & i \end{vmatrix} - d\begin{vmatrix} b & c \\ h & i \end{vmatrix} + g\begin{vmatrix} b & c \\ e & f \end{vmatrix}$$

6. The sign of a determinant is changed by interchanging two rows or two columns.
7. The value of a determinant is unchanged if each column is interchanged with the corresponding row.
8. If all the elements of a row or column are multiplied by the same number K, the value of the determinant is multiplied by K.
9. If the elements of any row (or column) are multiplied by any quantity and added to the corresponding elements of another row (or column) the value of the determinant is not changed.

Returning now to the numerical example, the determinant D of Eq. 20–10 may be evaluated as

$$D = \begin{vmatrix} 6 & -3 & -2 \\ -3 & 17 & -6 \\ -2 & -6 & 13 \end{vmatrix} = 6(17 \times 13 - 36) + 3(-39 - 12) - 2(18 + 34) = 853$$

The numerator determinants are evaluated most easily by expanding them into cofactors since there is one column in each which has only one element.

Hence,

$$i_1 = \frac{\begin{vmatrix} 50 & -3 & -2 \\ 0 & 17 & -6 \\ 0 & -6 & 13 \end{vmatrix}}{853} = \frac{5(17 \times 13 - 36)}{853} = 10.85 \text{ amp}$$

$$i_2 = \frac{\begin{vmatrix} 6 & 50 & -2 \\ -3 & 0 & -6 \\ -2 & 0 & 13 \end{vmatrix}}{853} = \frac{-50(-3 \times 13 - 12)}{853} = 2.99 \text{ amp}$$

$$i_3 = \frac{\begin{vmatrix} 6 & -3 & 50 \\ -3 & 17 & 0 \\ -2 & -6 & 0 \end{vmatrix}}{853} = \frac{50(18 + 34)}{853} = 3.05 \text{ amp}$$

Note that these values are the same as those obtained in the preceding section by means of systematic elimination.

If the coefficients of the equilibrium equations are complex, the evaluation of the determinants is more time-consuming but the methods are exactly the same. As an illustration of this, reconsider Eqs. 20–7, 20–8, and 20–9:

$$(5 + j10)\boldsymbol{I}_1 - (3 - j3)\boldsymbol{I}_2 = 100 + j0 \qquad (20\text{–}7)$$

$$-(3 - j3)\boldsymbol{I}_1 + (8 + j12)\boldsymbol{I}_2 - (4 + j2)\boldsymbol{I}_3 = 0 \qquad (20\text{–}8)$$

$$- (4 + j2)\boldsymbol{I}_2 + (20 - j10)\boldsymbol{I}_3 = 0 \qquad (20\text{–}9)$$

The determinant of the coefficients is

$$\boldsymbol{D} = \begin{vmatrix} 5 + j10 & -3 + j3 & 0 \\ -3 + j3 & 8 + j12 & -4 - j2 \\ 0 & -4 - j2 & 20 - j10 \end{vmatrix}$$

$$\boldsymbol{D} = (5 + j10)[(8 + j12)(20 - j10) - (-4 - j2)(-4 - j2)] + (3 - j3)[(-3 + j3)(20 - j10)] = 80 + j3760 = 3760\underline{/88.78°}$$

Application of Cramer's rule given for $\boldsymbol{I}_1$ is

$$\boldsymbol{I}_1 = \frac{\begin{vmatrix} 100 + j0 & -3 + j3 & 0 \\ 0 & 8 + j12 & -4 - j2 \\ 0 & -4 - j2 & 20 - j10 \end{vmatrix}}{3760\underline{/88.78°}}$$

$$= \frac{100[(8 + j12)(20 - j10) - (-4 - j2)(-4 - j2)]}{3760\underline{/88.78°}}$$

$$= \frac{100(268 + j144)}{3760\underline{/88.78°}} = \frac{30{,}400\underline{/28.2°}}{3760\underline{/88.78°}} = 8.08\underline{/-60.5°}$$

which is the result obtained in the previous section. The other two currents can be obtained in a similar manner.

One of the most useful applications of the determinant method of solving simultaneous equations arises when the coefficients are not numerical values but are algebraic symbols. This situation often results when one attempts to investigate the effects of certain circuit parameters upon the over-all performance of a circuit. As an example of this procedure, consider the circuit of Fig. 20–1 and assume that it is desired to find the effect of R_2 upon the output voltage $\boldsymbol{E}_3$.

If the voltage source is replaced by a current source of $\boldsymbol{I}_s = \boldsymbol{E}_s/R_s$, the current equilibrium equations may be written as

$$\begin{aligned}
\left(\frac{1}{R_s} + \frac{1}{R_1} + sC_1\right)\boldsymbol{E}_1 - sC_1\boldsymbol{E}_2 \qquad\qquad - \frac{1}{R_1}\boldsymbol{E}_3 &= \frac{\boldsymbol{E}_s}{R_s} \\
-sC_1\boldsymbol{E}_1 + \left(sC_1 + sC_2 + \frac{1}{R_2}\right)\boldsymbol{E}_2 - sC_2\boldsymbol{E}_3 &= 0 \\
-\frac{1}{R_1}\boldsymbol{E}_1 - sC_2\boldsymbol{E}_2 \qquad\qquad + \left(\frac{1}{R_1} + \frac{1}{R_3} + sC_2\right)\boldsymbol{E}_3 &= 0
\end{aligned} \tag{20–14}$$

The voltage $\boldsymbol{E}_3$ is, therefore,

$$\boldsymbol{E}_3 = \frac{\begin{vmatrix} \dfrac{1}{R_s} + \dfrac{1}{R_1} + sC_1 & -sC_1 & \boldsymbol{E}_s/R_s \\ -sC_1 & sC_1 + sC_2 + \dfrac{1}{R_2} & 0 \\ -\dfrac{1}{R_1} & -sC_2 & 0 \end{vmatrix}}{\begin{vmatrix} \dfrac{1}{R_s} + \dfrac{1}{R_1} + sC_1 & -sC_1 & -\dfrac{1}{R_1} \\ -sC_1 & sC_1 + sC_2 + \dfrac{1}{R_2} & -sC_2 \\ -\dfrac{1}{R_1} & -sC_2 & \dfrac{1}{R_1} + \dfrac{1}{R_3} + sC_2 \end{vmatrix}}$$

which may be expanded by cofactors to express $\boldsymbol{E}_3$ as a ratio of polynomials in s. This step is straightforward but time-consuming and will not be carried out here. The important thing is to note that it is not necessary to have numerical values in carrying out a solution by means of determinants.

The determinant method of solving simultaneous equations may also be extended to provide a direct procedure for obtaining impedance functions. This application will be discussed in the next chapter.

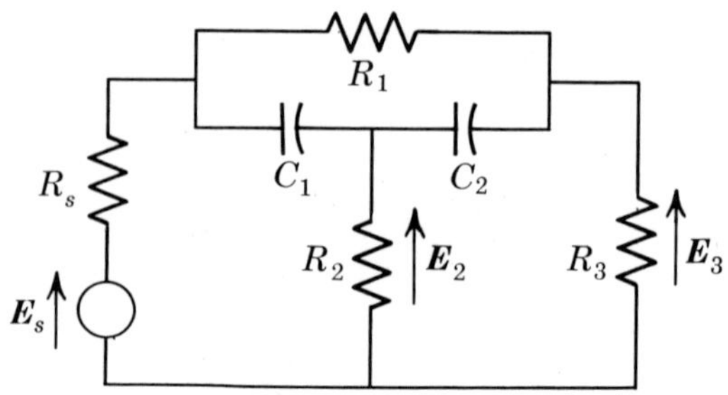

Fig. 20–1. A general *R-C* circuit.

PROBLEMS

20–1. Solve the following sets of simultaneous equations by systematic elimination of the variables:

(a)
$$\begin{aligned} 4i_1 - 2i_2 \qquad &= 20 \\ -2i_1 + 10i_2 - 5i_3 &= 0 \\ - 5i_2 + 8i_3 &= -30 \end{aligned}$$

(b)
$$\begin{aligned} 0.5e_1 - 0.2e_2 - 0.2e_3 &= 2 \\ -0.2e_1 + 0.75e_2 - 0.5e_3 &= 0 \\ -0.2e_1 - 0.5e_2 + 0.8e_3 &= 0 \end{aligned}$$

(c)
$$\begin{aligned} (4 + j6)\boldsymbol{I}_1 \qquad\qquad - (2 - j3)\boldsymbol{I}_3 &= 50 + j20 \\ (5 + j9)\boldsymbol{I}_2 - (3 + j4)\boldsymbol{I}_3 &= 0 \\ -(2 - j3)\boldsymbol{I}_1 - (3 + j4)\boldsymbol{I}_2 + (6 - j10)\boldsymbol{I}_3 &= 0 \end{aligned}$$

20–2. Evaluate the following determinants:

(a) $\begin{vmatrix} -2 & -1 \\ -1 & 4 \end{vmatrix}$ (b) $\begin{vmatrix} 4 & -1 & -2 \\ -1 & 5 & 0 \\ -2 & 0 & 7 \end{vmatrix}$ (c) $\begin{vmatrix} 6 & -2 & 0 & -1 \\ -2 & 5 & 3 & 0 \\ 0 & 3 & 8 & 0 \\ -1 & 0 & 0 & 4 \end{vmatrix}$

20–3. Solve for i_1 in Prob. 20–1a by means of determinants.

20–4. Solve for e_2 in Prob. 20–1b by means of determinants.

20–5. Make a complete analysis of the circuit shown to determine all branch currents.

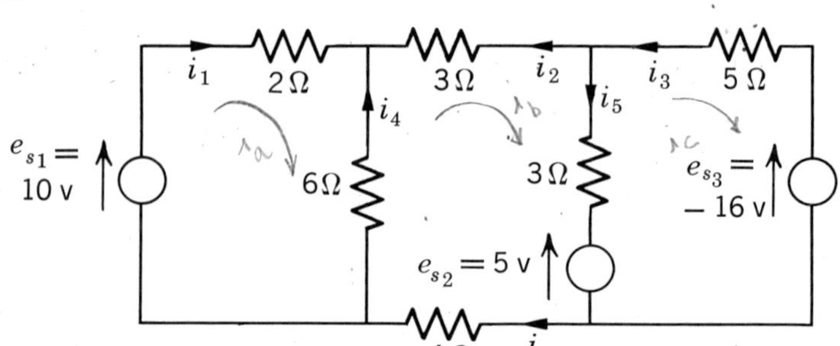

Prob. 20–5.

20–6. Determine all branch currents for the circuit of Prob. 19–11.

20–7. Determine the current i_1 for the network shown.

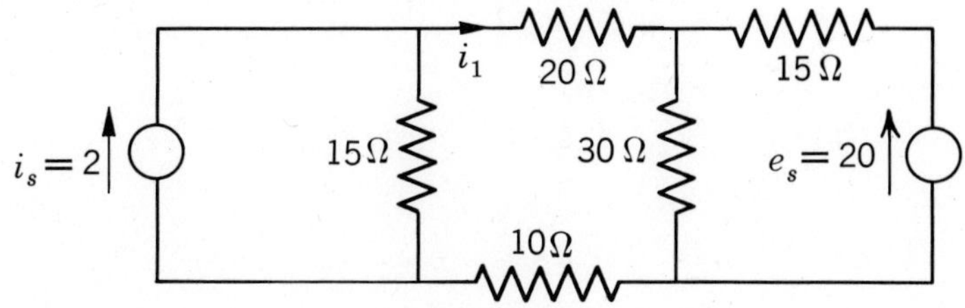

Prob. 20–7.

20–8. Determine the voltage $e_1(t)$ for the circuit illustrated.

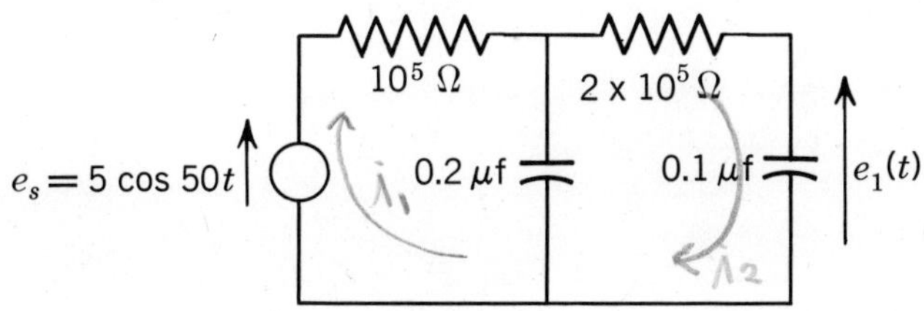

Prob. 20–8.

20–9. Determine the ratio $\boldsymbol{E}_2/\boldsymbol{E}_s$ of the figure as a function of $\boldsymbol{s}$

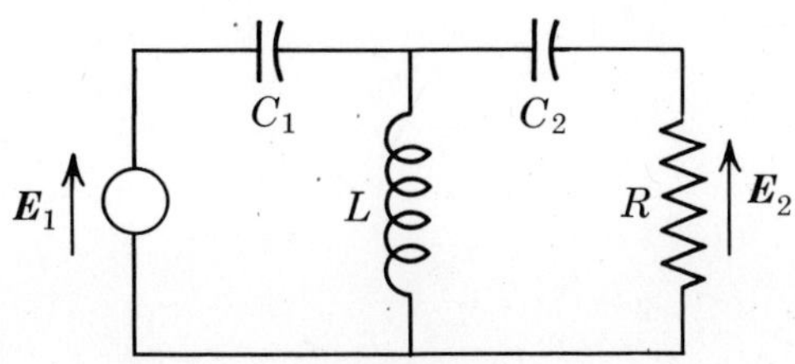

Prob. 20–9.

21

SIGNAL FLOW DIAGRAM

So far only classical methods have been used to solve simultaneous equations. In this chapter another technique, known as a signal flow diagram, will be studied. Although this technique can be used to give physical interpretations to the equations, its use here will be limited to the solution of equations.

21–1. The Signal Flow Diagram. Assume a circuit as shown in Fig. 21–1. The two mesh equations are:

$$E_s = Z_{11}I_1 - Z_{12}I_2 \tag{21–1}$$

$$0 = -Z_{12}I_1 + Z_{12}I_2 \tag{21–2}$$

where $Z_{11} = Z_1 + Z_{12}$

$Z_{22} = Z_2 + Z_{12}$

Let it be desired to find the current I_2. Equations 21–1 and 21–2 may be used to give the values of I_1 and I_2 thus:

$$I_1 = \frac{E_s}{Z_{11}} + \frac{Z_{12}}{Z_{11}} I_2 \tag{21–3}$$

and

$$I_2 = \frac{Z_{12}}{Z_{22}} I_1 \tag{21–4}$$

If a node is put to represent the source N_s, another to stand for I_1, a third for I_2, and a fourth labeled N_L (to be explained later), a graph of Eqs. 21–3 and 21–4 can be traced as shown in Fig. 21–2. Figure 21–2 shows arrows on the lines connecting the nodes. These arrows indicate the flow of signals

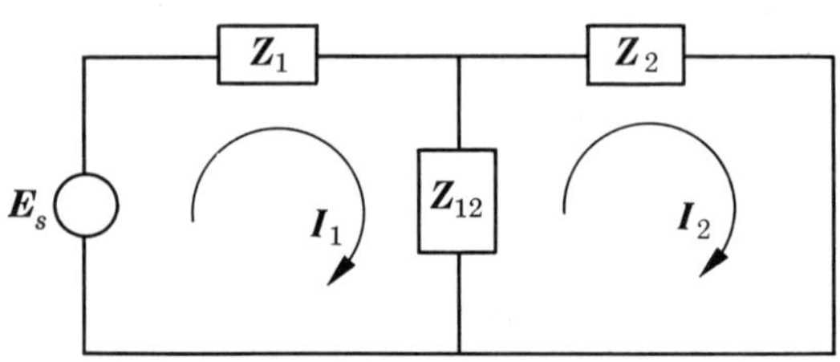

Fig. 21–1. Two-loop circuit.

between the nodes. The coefficients placed on the lines (E_1/Z_{11}, Z_{12}/Z_{22}, Z_{12}/Z_{11}, 1) are called *transmittances*. The extra node N_L is called the *output*.

The input or source node emits or transmits signals. It does not receive any signal from the circuit. For reasons which will become clear later on, its value is made unity.

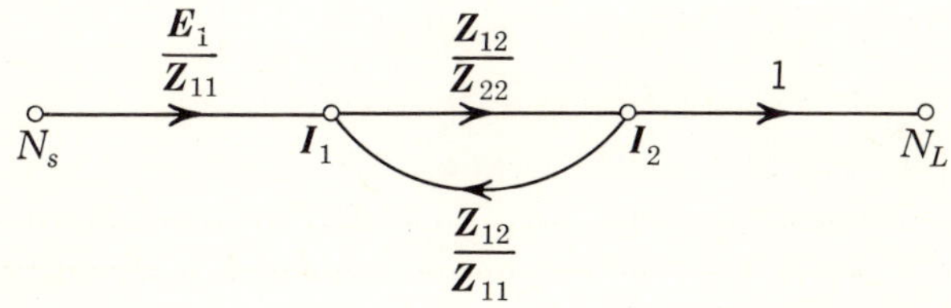

Fig. 21–2. Signal flow diagram for Eqs. 21–3 and 21–4.

The output represented by node N_L (also called *load* or *sink*) does not send any signal. As I_2, the value of current to be determined, sends a signal to I_1, it cannot represent the output and therefore the addition of N_L. Observe that I_2 is connected to N_L by a line unity transmittance.

The signal sent from the source to I_1 is a forward signal, hence the direction of the arrow. The signal from I_1 to I_2 is also a forward signal. However, the signal sent from I_2 to I_1 is a backward signal and is called *feedback*.

The transmittances are the value by which the sending node should be multiplied to give its contribution to the receiving node. The transmittances can be positive or negative but the sign does not change the direction of the arrow.

The source node is made to have a unity value; its actual magnitude is used in the transmittance. From the signal flow diagram one can write the equations at the nodes thus:

$$
\begin{aligned}
N_s &= 1 \\
I_1 &= \frac{E_1}{Z_{11}}\,1 + \frac{Z_{12}}{Z_{11}}\,I_2 \\
I_2 &= \frac{Z_{11}}{Z_{22}}\,I_1 \\
N_L &= 1I_2
\end{aligned}
\qquad (21\text{–}5)
$$

It should be emphasized that the source node N_s does not receive a signal. Its function is to send signals. The output node N_L does not send signals—either forward or backward. Its function is to receive signals.

At this point it should be mentioned that the flow diagram is not unique in this sense, that any equation can be represented by several different diagrams. However, the results given by these diagrams when reduced and solved are the same and give the same answer.

As a further introduction assume a circuit as shown in Fig. 21–3. The loop equations are:

$$10 = 8I_1 - 6I_2 \tag{21–6}$$

$$0 = -6I_1 + 12I_2 \tag{21–7}$$

Assume that it is desired to solve for I_2.

Step 1: From Eq. 21–6 solve for I_1 thus:

$$I_1 = 1.25 + 0.75I_2 \tag{21–8}$$

Step 2: From Eq. 21–7 find the value of I_2 thus:

$$I_2 = 0.5I_1 \tag{21–9}$$

Step 3: Draw four nodes N_s for the source with a magnitude 1, one for current I_1, one for current I_2, and one for the output labeled N_L.

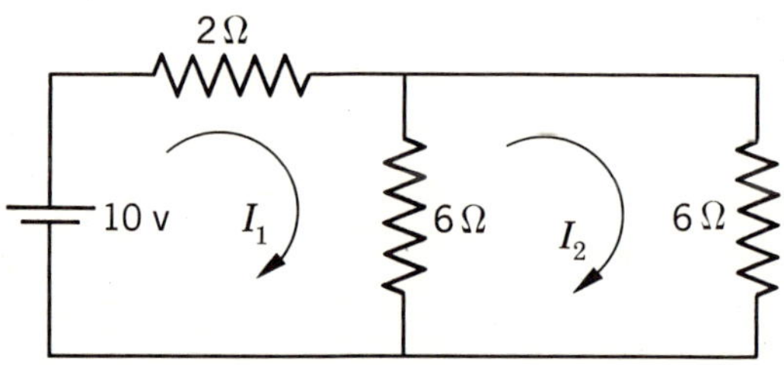

Fig. 21–3. Two-resistance mesh diagram.

Step 4: Connect the nodes according to Eqs. 21–8 and 21–9 and show on the connecting lines the direction of signal flow by arrows. Above the lines put the values of the transmittances. Connect I_2 to N_L. On this line put the arrow in the direction for I_2 to N_L and a transmittance of 1.

These four steps are shown in Fig. 21–4.

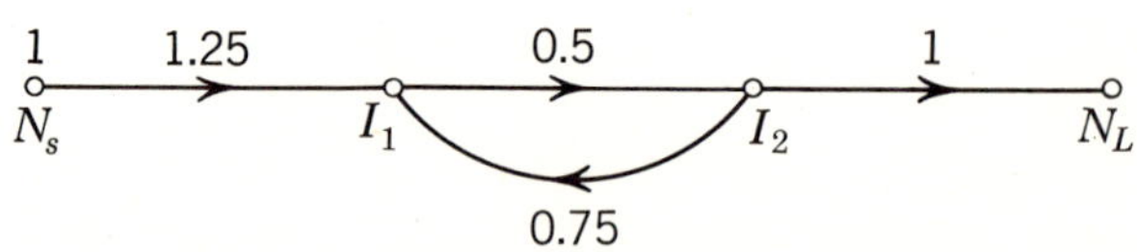

Fig. 21–4. Signal flow diagram for circuit of Fig. 21–3.

A more complex circuit is shown in Fig. 21–5. It is desired to find V_4. The loop equations are:

$$115 = 15I_1 - 10I_2 - 2I_3 \tag{21–10}$$

$$0 = -10I_1 + 15.5I_2 - I_3 \tag{21–11}$$

$$0 = -2I_1 - I_2 + 8I_3 \tag{21–12}$$

$$V_4 = 5I_3 \tag{21–13}$$

Find I_1 from Eq. 21–10, I_2 from Eq. 21–11, and I_3 from Eq. 21–12. Thus:

$$I_1 = \frac{23}{3} + \frac{2}{3} I_2 + \frac{2}{15} I_3 \tag{21-14}$$

$$I_2 = \frac{2}{3.1} I_1 + \frac{1}{15.5} I_3 \tag{21-15}$$

$$I_3 = \frac{1}{4} I_1 + \frac{1}{8} I_2 \tag{21-16}$$

$$V_4 = 5I_3 \tag{21-17}$$

Set five nodes and from Eqs. 21–14 through 21–17 draw the signal flow diagram as shown in Fig. 21–6.

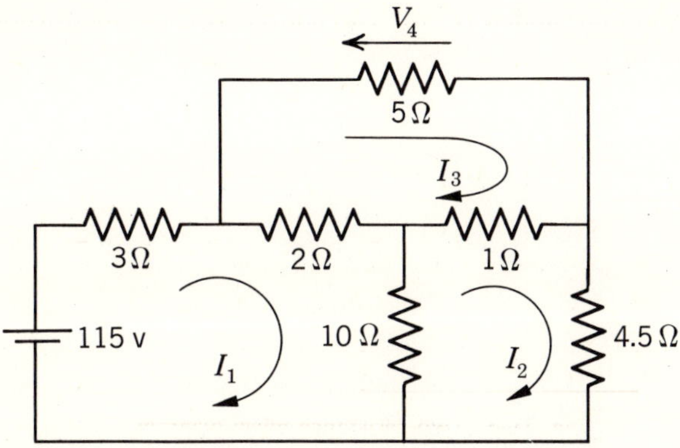

Fig. 21–5. Three-mesh circuit.

To solve the equations represented by the signal flow diagrams of Figs. 21–2, 21–4, and 21–6 is to eliminate all the nodes and the loops between N_s and N_L, namely between the source and the load. In other words, it is desired to find a straightforward simple transmittance between N_s and N_L.

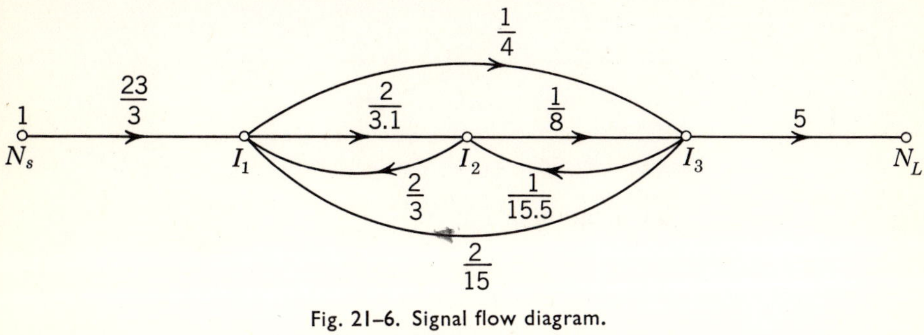

Fig. 21–6. Signal flow diagram.

Before proceeding to show how this elimination can be done, attention should be called to Figs. 21–2, 21–4, and 21–6. In these figures we observe signals fed forward and others fed back. These form closed loops. Thus in Fig. 21–2 the loop $\boldsymbol{I}_1\boldsymbol{I}_2\boldsymbol{I}_1$ is a closed loop with a forward signal $\boldsymbol{Z}_{12}/\boldsymbol{Z}_{22}$ from $\boldsymbol{I}_1$ to $\boldsymbol{I}_2$ and a backward signal $\boldsymbol{Z}_{12}/\boldsymbol{Z}_{11}$ from $\boldsymbol{I}_2$ to $\boldsymbol{I}_1$. In Fig. 21–6 there are several closed loops; namely, $I_1I_3I_1$, $I_1I_2I_3I_1$, $I_1I_2I_1$, and $I_2I_3I_2$. In the process of reducing the diagram to a single transmittance between N_s and N_L all these loops must be eliminated.

21–2. Circuit Reduction. To reduce the circuit to a straight transmission between source and load use is made of four theorems which are introduced and proved in this section.

THEOREM 1. If from one node a signal is transmitted by two or more parallel paths directly to another node, the parallel paths can be replaced by a single one having a transmittance equal to the sum of the transmittances of the parallel paths. For instance, in Fig. 21–7 between nodes n_1 and n_3 there

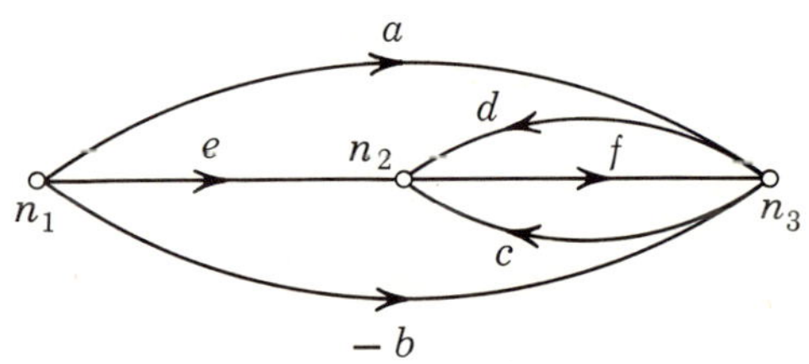

Fig. 21–7. Transmittances in parallel.

are two direct transmitting paths with transmittances a and $-b$. These could be replaced by a single path with a transmittance $a - b$ as shown in Fig. 21–8. Also between n_3 and n_2 there are two parallel feedback paths with transmittances c and d. These may also be replaced by one path with a transmittance $c + d$ as shown in Fig. 21–8.

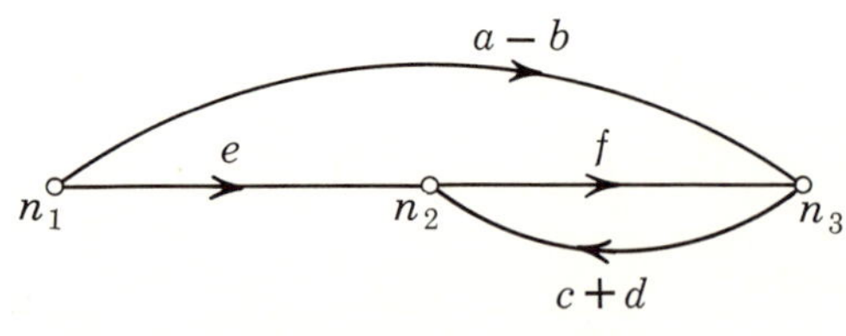

Fig. 21–8. Combination of parallel paths.

From the signal flow diagram of Fig. 21–7:

$$n_2 = n_1e + n_3(c + d) \tag{21–18}$$

$$n_3 = n_2f + n_1(a - b) \tag{21–19}$$

Equations 21–18 and 21–19 are also obtained from Fig. 21–8.

THEOREM 2. Two or more paths in cascade or tandem with transmittances a, b, and c respectively may be replaced by one path with a transmittance abc.

As proof of this consider Fig. 21–9, where

$$n_1 = an \tag{21-20}$$

$$n_2 = bn_1 = abn \tag{21-21}$$

$$n_3 = cn_2 = abcn \tag{21-22}$$

Therefore Fig. 21–9 can be reduced to Fig. 21–10.

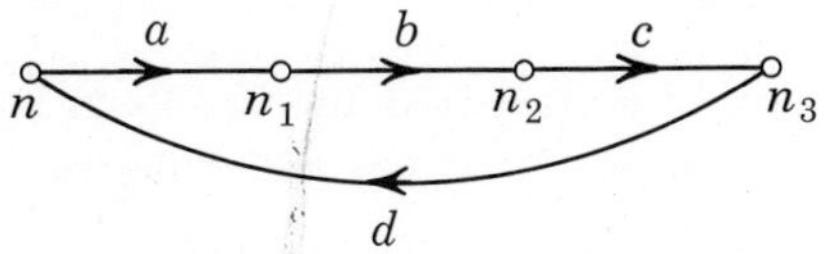

Fig. 21–9. Transmittances in cascade.

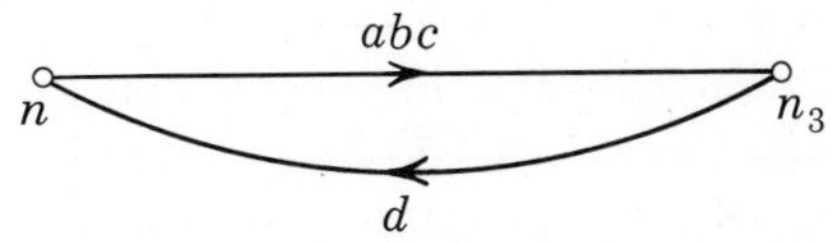

Fig. 21–10. Combination of transmittances in cascade.

THEOREM 3. A branch coming from a node n_x with transmittance a_x and terminating on node n_1 can be shifted from n_1 to an adjacent node n_2 receiving a transmittance signal b from n_1, provided the transmittance a_x of the branch n_x is multiplied by the transmittance b between n_1 and n_2 and also provided new branches are established from n_x and terminate on all nodes, except n_2, receiving signals from n_1. The transmittance of these new branches should be a_x times the transmittances between n_1 and the nodes under consideration.

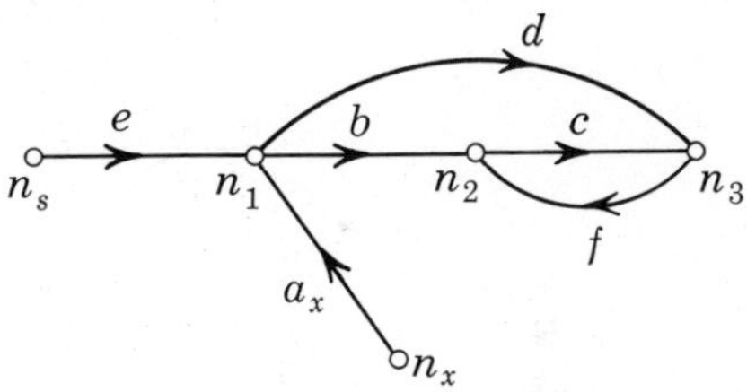

Fig. 21–11. Shifting of transmittance to an adjacent node.

To prove this theorem refer to Fig. 21–11. In Fig. 21–11 a branch n_x terminates on n_1. It is desired to shift this branch to n_2. What should the transmittance $n_x - n_2$ be? What other changes should be introduced in the circuit?

The equations of the circuit before the change is made are:

$$n_1 = n_se + n_xa_x \tag{21-23}$$

$$n_2 = n_1b + n_3f = n_sbe + n_xa_xb + n_3f \tag{21-24}$$

$$\begin{aligned} n_3 &= n_1d + n_2c \\ &= n_sde + n_xa_xd + n_sbce + n_xa_xbc + n_3fc \end{aligned} \tag{21-25}$$

The shift is made as shown in Fig. 21–12, and the values of n_2 and n_3 are not to change by this shift. The value of n_2 after the shift is:

$$\begin{aligned} n_2 &= n_1b + n_xk + n_3f \\ &= n_seb + n_xk + n_3f \end{aligned} \tag{21-26}$$

Equating n_2 before the shift to n_2 after the shift, or Eq. 21–24 to 21–26, we get

$$n_s eb + n_x k + n_3 f = n_s eb + n_x a_x b + n_3 f \tag{21–27}$$

From Eq. 21–27

$$k = a_x b \tag{21–28}$$

The value of n_3 after the shift is

$$\begin{aligned} n_3 &= n_1 d + n_2 c \\ &= n_s ed + n_s ebc + n_x a_x bc + n_3 fc \end{aligned} \tag{21–29}$$

But n_3 after the shift should equal n_3 before the shift, or Eq. 21–29 should be equal to Eq. 21–25. This cannot be correct unless a branch is added between n_x and n_3 with a transmittance $a_x d$ as shown in Fig. 21–13.

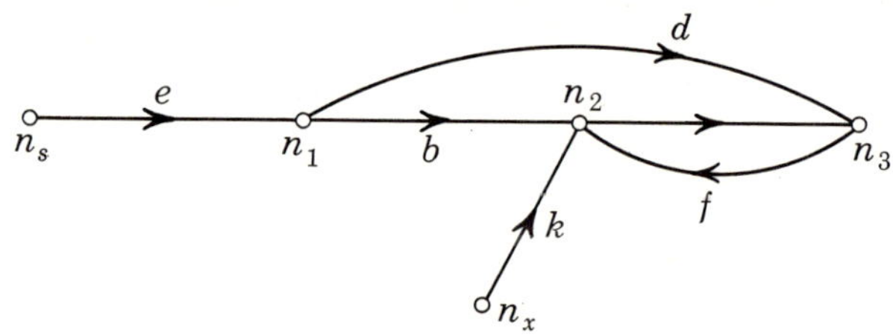

Fig. 21–12. Partial result of shift.

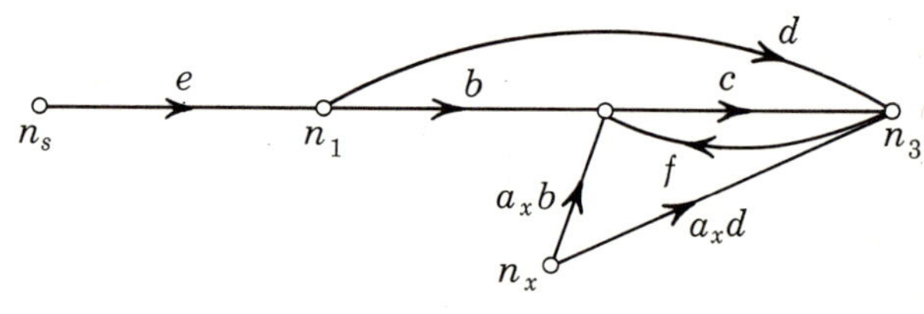

Fig. 21–13. Final result of shift.

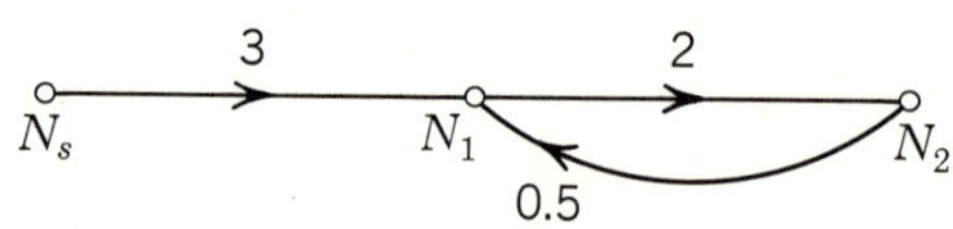

Fig. 21–14. Example for transmittance shifting.

To illustrate theorem 3 assume that in Fig. 21–14 it is desired to shift the feedback terminal from node N_1 to node N_2. Here N_x is N_2 and the feedback is between node N_2 and N_2 itself. The transmittance of the feedback path is $0.5 \times 2 = 1$. The circuit becomes as shown in Fig. 21–15 with a self-feedback. This circuit can be obtained directly by observation from Fig. 21–14.

Another illustration is given by Fig. 21–16. This can be reduced to Figs. 21–17 and 21–18.

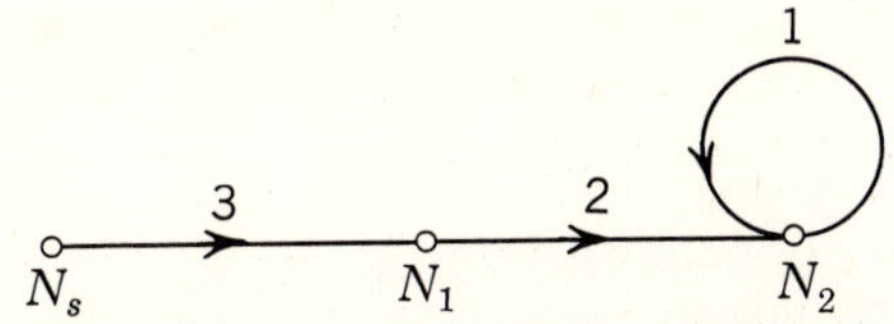

Fig. 21–15. Result of shifting in Fig. 21–14.

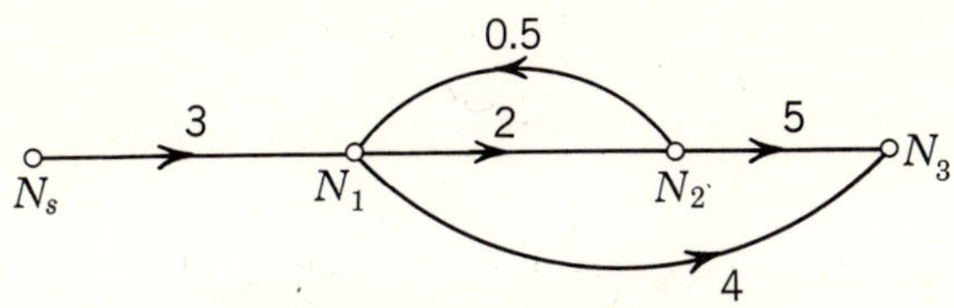

Fig. 21–16. Another signal flow diagram.

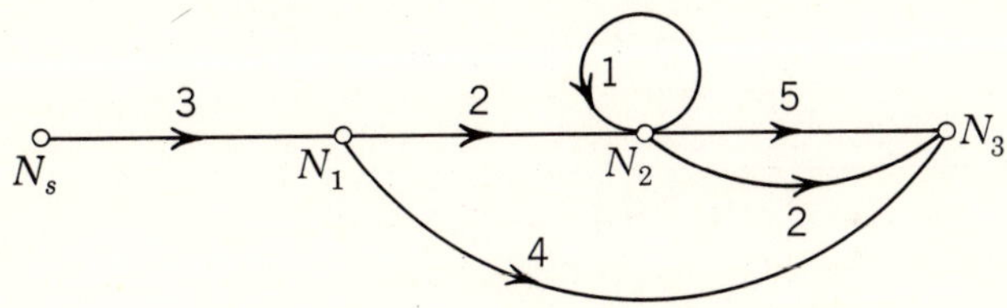

Fig. 21–17. Partial reduction of Fig. 21–16.

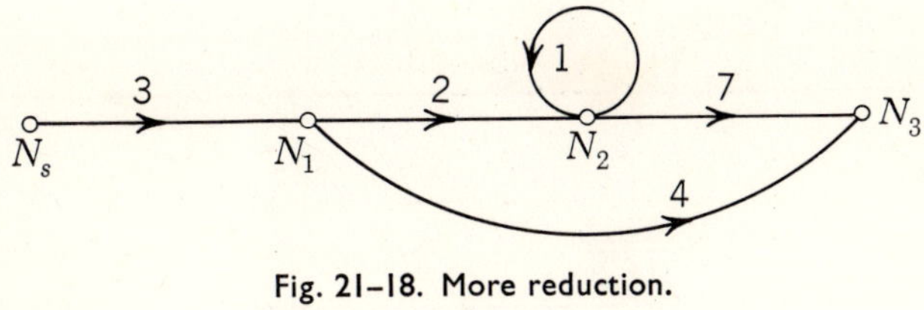

Fig. 21–18. More reduction.

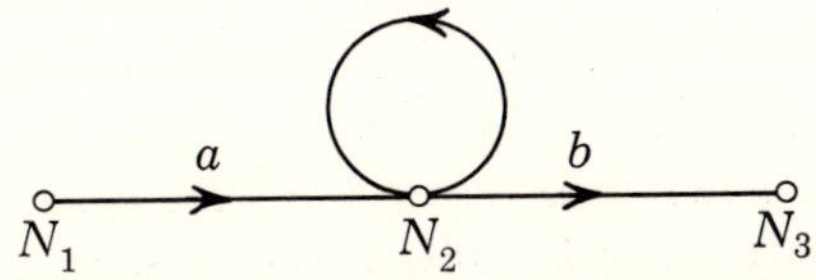

Fig. 21–19. Elimination of self-feedback.

THEOREM 4. The self-feedback can be eliminated as in Fig. 21–19.

$$N_2 = aN_1 + mN_2 \tag{21-30}$$

or

$$N_2(1 - m) = aN_1$$

and

$$\frac{N_2}{N_1} = \frac{a}{1 - m} \tag{21-31}$$

Figure 21–19 can now be redrawn as in Fig. 21–20. The N_2 node is split into N'_2 and N''_2. All the incoming signals to N_2 are connected to N'_2. All the outgoing signals from N_2 are emitted by N''_2. Between N'_2 and N''_2 there is a transmittance of $1/(1 - m)$. Figure 21–20 can now be reduced to Fig. 21–21.

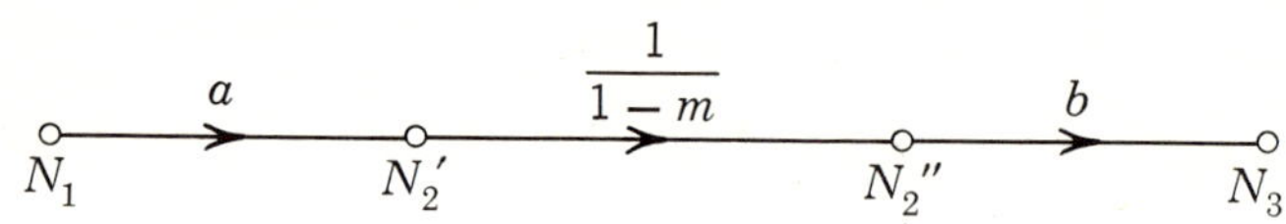

Fig. 21–20. Partial result.

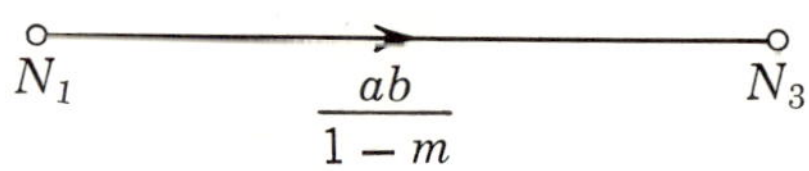

Fig. 21–21. Final result.

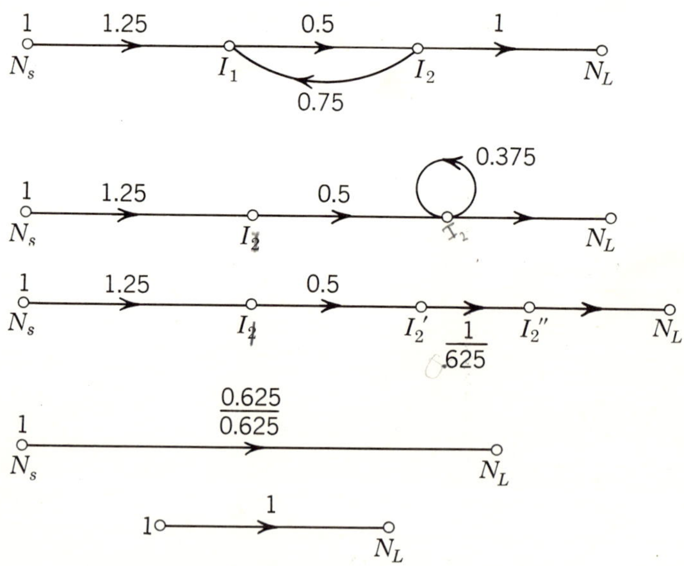

Fig. 21–22. Reproduction and reduction of Fig. 21–4.

21–3. Solution of Equation. By the application of the four theorems any simultaneous equation can be solved. For instance, Fig. 21–4 is reproduced in Fig. 21–22 and then reduced.

The second diagram to be reduced is that of Fig. 21-6, which is reproduced and reduced by steps, as shown in Fig. 21-23. Each step is explained in detail so that the student gets a good understanding of the technique.

STEP 1. The feedback between I_2 and I_1 is eliminated. To do this the branch coming from I_2 to I_1 is moved from I_1 to I_2. By theorem 3, when a branch is moved from one node to an adjacent one its transmittance is multiplied by the transmittance between the two adjacent nodes. Therefore there is a self-feedback on node I_2 equal to $\frac{2}{3} \times \frac{2}{3.1} = \frac{4}{9.3}$. Furthermore, as branch I_1 feeds I_3 a new branch between I_2 and I_3 is added with a transmittance between I_2 and I_3 of $\frac{2}{3} \times \frac{1}{4} = \frac{1}{6}$.

STEP 2. Between I_2 and I_3 there are two transmittances $\frac{1}{6}$ and $\frac{1}{8}$ in parallel. They can be replaced by a single transmittance having the value $\frac{1}{6} + \frac{1}{8} = \frac{7}{24}$.

STEP 3. The self-feedback of I_2 is replaced by a tandem transmittance according to theorem 4. Thus node I_2 is replaced by I'_2 and I''_2 with a transmittance from I'_2 to I''_2 of $\dfrac{1}{1 - \dfrac{4}{9.3}} = \dfrac{9.3}{5.3}$. All the transmittances coming to I_2 are connected to I'_2 and all the transmittances emitted by I_2 are emitted by I''_2.

STEP 4. The two cascade transmittances between I'_2 and I_3 are combined into one, thus: $\frac{9.3}{5.3} \times \frac{7}{24} = \frac{21.7}{42.4}$.

STEP 5. The feedback branch $I_3I'_2$ is moved one node to I_3, forming a self-feedback loop at I_3 of value $\frac{21.7}{42.4} \times \frac{1}{15.5} = \frac{21.7}{657.2}$. Notice that I'_2 does not send any signal to any other node beside I_3; therefore no new branch from I_3 is needed.

STEP 6. The two cascade transmittances between I_1 and I_3 are combined, thus eliminating I'_2. The value of the combined transmittance is $\frac{2}{3.1} \times \frac{21.7}{42.4} = \frac{21.7}{3.1 \times 21.2}$. Also, in step 6 the self-feedback around I_3 is changed according to theorem 4 to a transmittance between I'_3 and I''_3 of $\dfrac{1}{1 - \dfrac{21.7}{657.2}} = \dfrac{657.2}{635.5}$.

The incoming signal from I_1 feeds I'_3 and the feedback signal from I_3 is emitted from I''_3.

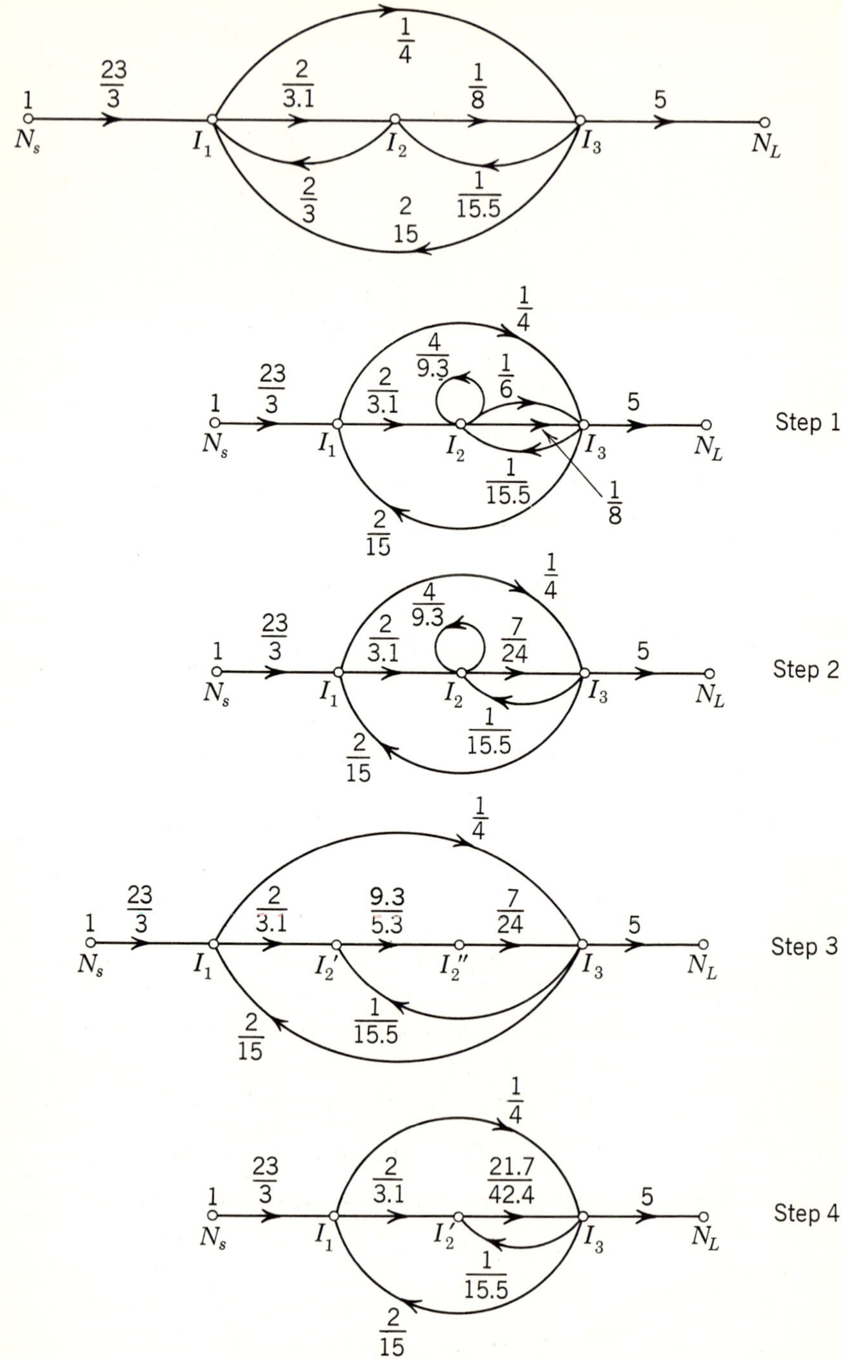

Fig. 21–23. Steps in reduction of Fig. 21–6.

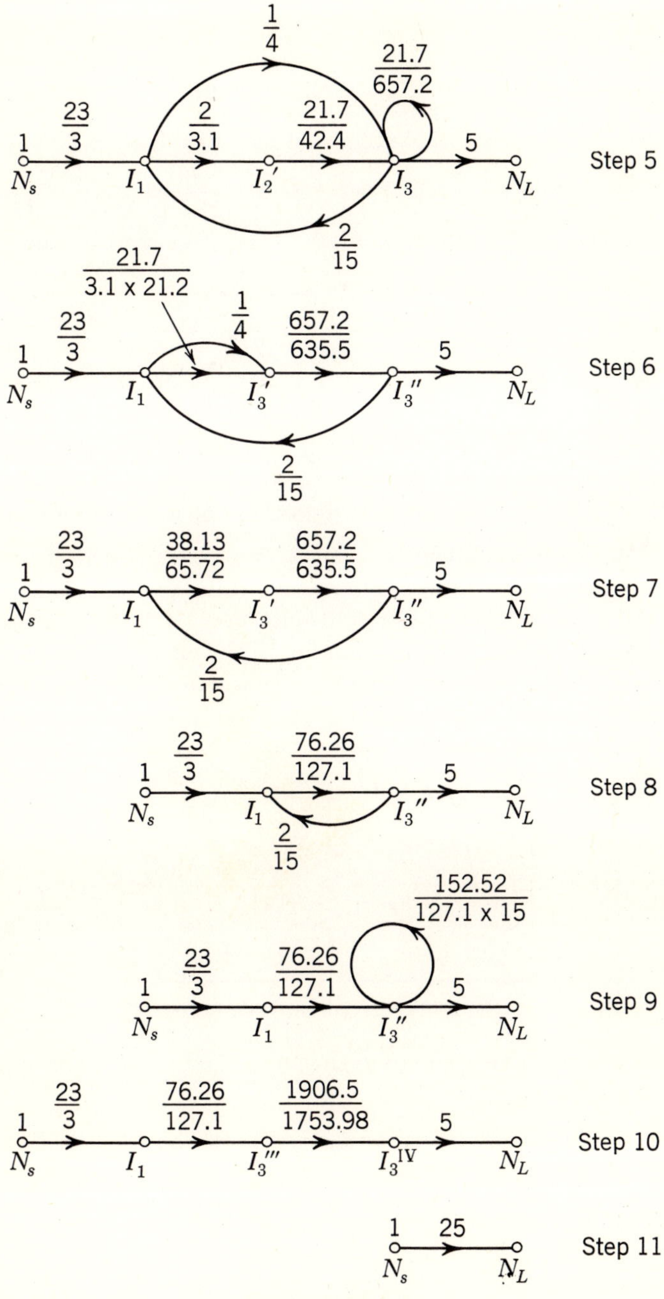

Fig. 21–23 (cont.).

STEP 7. The parallel admittances between I_1 and I'_3 are combined into an equivalent admittance of $\frac{1}{4} + \frac{21.7}{3.1 \times 21.2} = \frac{38.13}{65.72}$.

STEP 8. The cascade admittance between I_1 and I''_3 is found to be $\frac{38.13}{65.72} \times \frac{657.2}{635.5} = \frac{76.26}{127.1}$. Node I'_3 is eliminated.

STEP 9. The branch $I''_3 I_1$ is removed one node from I_1 to I''_3, causing a self-feedback around node I''_3 of value $\frac{2}{15} \times \frac{76.26}{127.1} = \frac{152.52}{127.1 \times 15}$.

STEP 10. The self-feedback around I''_3 is eliminated by the addition of I'''_3 and I^{iv}_3 with a transmittance of $\dfrac{1}{1 - \dfrac{152.52}{127.1 \times 15}} = \dfrac{1906.5}{1753.98}$.

STEP 11. The transmittance between N_s and N_L is the product of the cascade admittances and is equal to $\frac{23}{3} \times \frac{76.26}{127.1} \times \frac{1906.5}{1753.98} \times 5 = 25$. All the nodes between N_s and N_L are eliminated. Many steps could be eliminated in Fig. 21–23, but they are given in detail for the sake of the student.

21–4. Multisource Circuits. In the one-source circuit it was assumed that the value of N_s is unity. The actual value of the source voltage (or current) was used as part of the transmittance. If there is more than one source the same node N_s may be used for all the energy sources. The actual values of the voltages (or current) of the sources can be used as part of the transmittance.

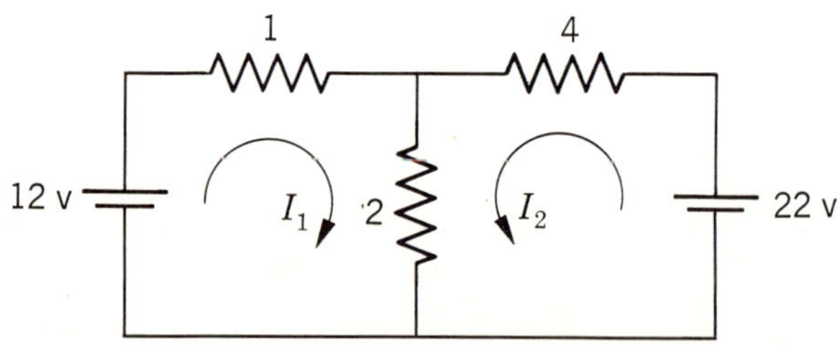

Fig. 21–24. Circuit with two sources.

To illustrate: Assume a circuit as shown in Fig. 21–24. The loop equations are given by

$$12 = 3I_1 + 2I_2 \tag{21–32}$$

$$22 = 2I_1 + 6I_2 \tag{21–33}$$

Assume that it is desired to find the value of I_2. Solve Eq. 21–32 for I_1 and Eq. 21–33 for I_2 and get:

$$I_1 = 4 - \frac{2}{3} I_2 \tag{21–34}$$

$$I_2 = \frac{11}{3} - \frac{1}{3} I_1 \tag{21–35}$$

Here put four nodes $N_s I_1 I_2$ and N_L. Assume the voltage at $N_s = 1$; then the transmittance between N_s and I_1 is 4. From Eq. 21-35 assume the source value to be 1 and use the same N_s; then the transmittance to I_2 is $\frac{11}{3}$. The flow diagram is shown in Fig. 21-25, and its reduction in Fig. 21-26.

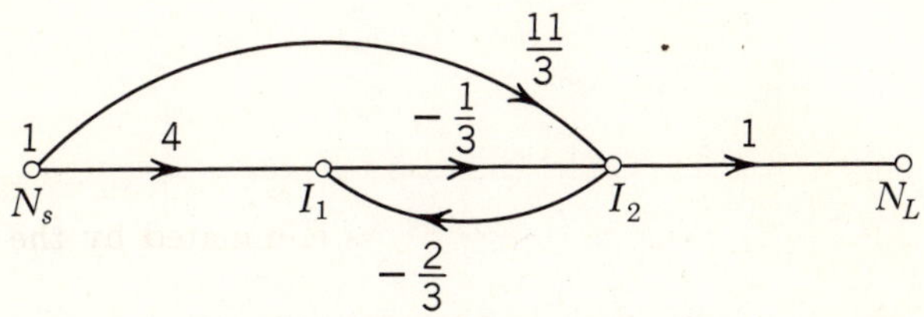

Fig. 21-25. Signal flow for Fig. 21-24.

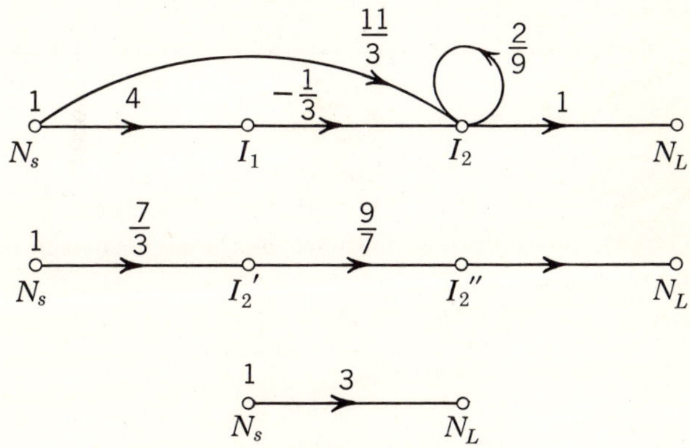

Fig. 21-26. Reduction of Fig. 21-25.

21-5. Multi-A-C Sources. The same procedure used in Art. 21-4 can be applied to a-c circuits with different phase angles between the source voltages (or current). The magnitude of the voltage (or current) and its phase angle can be used as part of the transmittance and one node N_s with a value of 1 used as the source.

PROBLEMS

21-1. Write the equations for the network illustrated and show the signal flow diagram. Solve for $\boldsymbol{I_4}$.

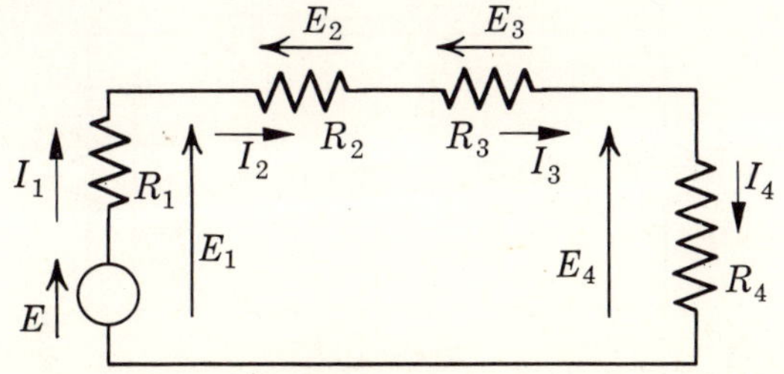

Prob. 21-1.

21–2. Write the equations for the figure illustrated and show the signal flow diagram. Solve for $\boldsymbol{E}_6$.

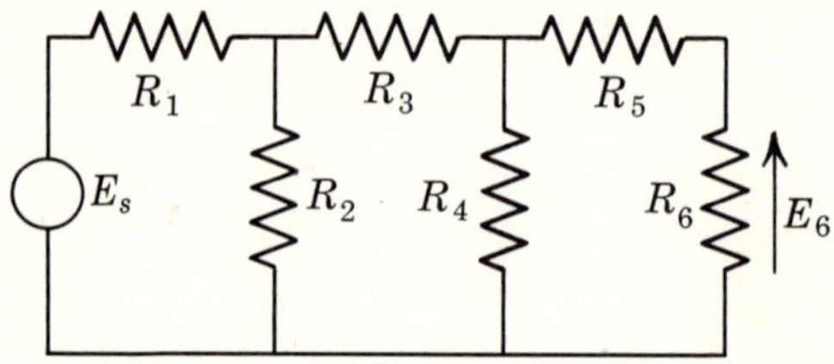

Prob. 21–2.

21–3. Write the equilibrium equations for the signal flow diagram shown.

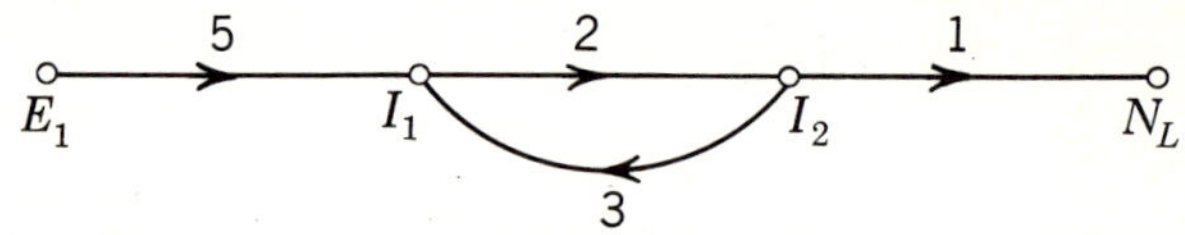

Prob. 21–3.

21–4. Write the equilibrium equations for the adjacent signal flow diagram and then find the corresponding electric circuit.

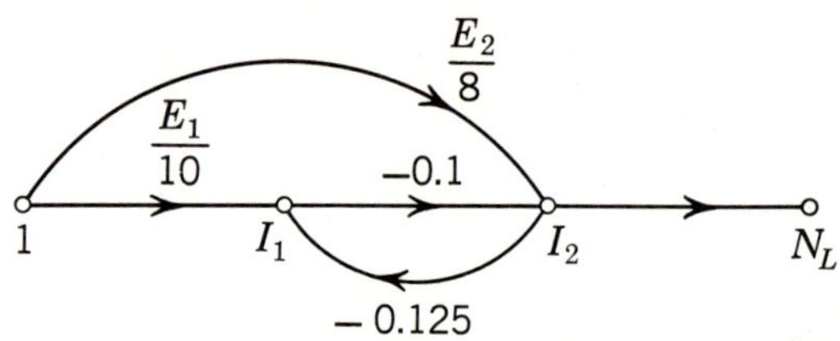

Prob. 21–4.

21–5. Write the node equations for the network illustrated. Draw the signal flow diagram and solve for $\boldsymbol{E}_2$.

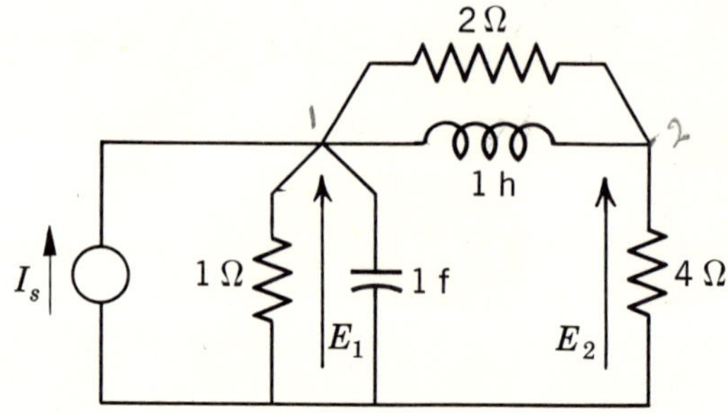

Prob. 21–5.

22

SPECIAL TECHNIQUES FOR CIRCUIT ANALYSIS

The preceding two chapters have been devoted to methods of writing and solving equilibrium equations for general linear circuits involving any number of elements and almost any geometry. In discussing these methods an effort was made to keep them as general as possible; hence, nothing was said concerning short-cut procedures which are sometimes possible if the circuit geometry is favorable. Because these short cuts may save a great deal of labor, they are worth knowing even if they are applicable to special cases only. This chapter will acquaint the student with several of these techniques.

22-1. The Ladder Network. The work required to solve a set of equilibrium equations increases rapidly as the number of equations increases. Therefore, any technique which eliminates the necessity of solving such a set of equations becomes extremely useful for analyzing large networks. It is not possible to find a technique that will work for any network, but one does exist for the special form of network designated as a *ladder*. In this case the technique is particularly valuable because ladder networks are commonly used in practice and, in addition, they are likely to have a large number of meshes.

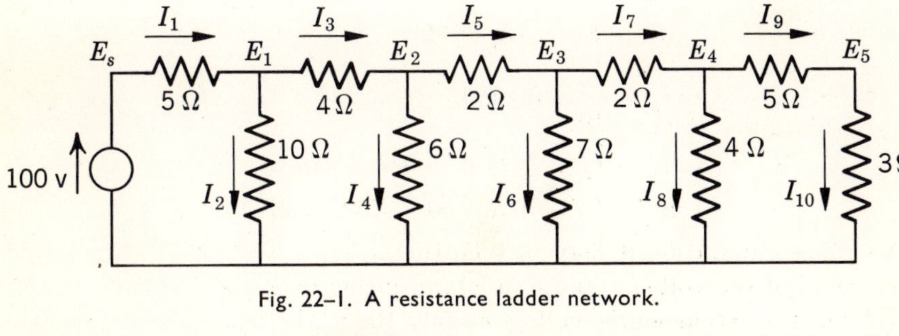

Fig. 22-1. A resistance ladder network.

The ladder network is so named because its graph resembles a ladder. Such a network, composed of resistances only, is shown in Fig. 22-1. Note that this circuit would require solving *five* voltage equilibrium equations if the general procedure were followed.

The simplified procedure starts by *assuming* a value of voltage or current in the resistor *farthest* from the source. This assumed value is then used to calculate all of the voltages and currents back to the source voltage. The value of source voltage thus calculated is *not* the actual source voltage but differs from it by a factor which can be easily determined. This factor may then be used to modify all of the computed voltages and currents. Note that this procedure utilizes the *linearity* of the circuit, which makes it possible to say that the response is directly proportional to the driving force.

The circuit of Fig. 22–1 will serve to illustrate the mechanics of the procedure. For simplicity, assume that the current I_{10} is 1 amp. The other voltages and currents may be calculated readily by the following systematic procedure:

$$\begin{aligned}
E_5 &= 3I_{10} = 3 \text{ v} \\
I_9 &= I_{10} = 1 \text{ amp} \\
E_4 &= E_5 + 5I_9 = 3 + 5 = 8 \text{ v} \\
I_8 &= \frac{E_4}{4} = 2 \text{ amp} \\
I_7 &= I_8 + I_9 = 2 + 1 = 3 \text{ amp} \\
E_3 &= E_4 + 2I_7 = 8 + 6 = 14 \text{ v} \\
I_6 &= \frac{E_3}{7} = 2 \text{ amp} \\
I_5 &= I_6 + I_7 = 2 + 3 = 5 \text{ amp} \qquad (22\text{–}1) \\
E_2 &= E_3 + 2I_5 = 14 + 10 = 24 \text{ v} \\
I_4 &= \frac{E_2}{6} = \frac{24}{6} = 4 \text{ amp} \\
I_3 &= I_4 + I_5 = 4 + 5 = 9 \text{ amp} \\
E_1 &= E_2 + 4I_3 = 60 \text{ v} \\
I_2 &= \frac{E_1}{10} = 6 \text{ amp} \\
I_1 &= I_2 + I_3 = 6 + 9 = 15 \text{ amp} \\
E_s &= E_1 + 5I_1 = 60 + 75 = 135 \text{ v}
\end{aligned}$$

From this computation it may be seen that a source voltage of 135 v would produce all of the voltages and currents resulting from the assumed current I_{10}. Since the actual source voltage is only 100 v, the true values of voltage and current may be obtained by multiplying the computed values by the ratio $\frac{100}{135}$. Hence, the true value of I_{10} is

$$I_{10} = 1 \times \tfrac{100}{135} = 0.741 \text{ amp}$$

and the voltage E_2 is

$$E_2 = 24 \times \tfrac{100}{135} = 17.78 \text{ v}$$

Any other voltage or current in the ladder network may be computed in a similar fashion.

There are several points worth noting in connection with this method of computation. In the first place, the method works only if there is a single source. If there were sources located at different points along the ladder it would not be possible to correct the assumed values of voltage and current by means of a simple multiplying factor. In the second place, the computation should always start in the element farthest from the source. If it were started at any other point it would be necessary to compute the equivalent impedance of the remaining portion of the circuit. Finally, although the ladder network shown in Fig. 22–1 has zero impedance in all of the bottom branches, it is not necessary that this be the case. Any, or all, of these branches may contain impedances and the general procedure is still applicable.

This method of solving ladder networks is still useful when the circuit contains reactive elements and the values of these elements may be specified either numerically or symbolically. As an example of the latter situation consider the low-pass R-C filter shown in Fig. 22–2. All resistances are

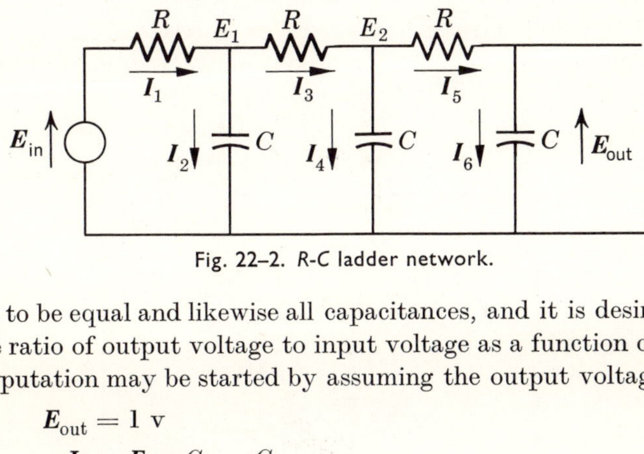

Fig. 22–2. R-C ladder network.

assumed to be equal and likewise all capacitances, and it is desired to determine the ratio of output voltage to input voltage as a function of frequency. The computation may be started by assuming the output voltage to be 1 v.

$$\begin{aligned}
E_{\text{out}} &= 1 \text{ v} \\
I_6 &= E_{\text{out}}sC = sC \\
I_5 &= I_6 = sC \\
E_2 &= E_{\text{out}} + RI_5 = 1 + RsC \\
I_4 &= E_2sC = sC + RC^2s^2 \\
I_3 &= I_4 + I_5 = 2sC + RC^2s^2 \\
E_1 &= E_2 + RI_3 = 1 + RCs + 2RCs + R^2C^2s^2 \\
I_2 &= E_1sC = sC + 3Rs^2C^2 + R^2C^3s^3 \\
I_1 &= I_2 + I_3 = 3sC + 4RC^2s^2 + R^2C^3s^3 \\
E_{\text{in}} &= E_1 + RI_1 = 1 + 6RCs + 5R^2C^2s^2 + R^3C^3s^3
\end{aligned} \tag{22-2}$$

Since this value of $\boldsymbol{E}_{\text{in}}$ will make $\boldsymbol{E}_{\text{out}}$ equal to 1 v, the ratio of output to input will be

$$\frac{\boldsymbol{E}_{\text{out}}}{\boldsymbol{E}_{\text{in}}} = \frac{1}{\boldsymbol{E}_{\text{in}}} = \frac{1}{1 + 6RCs + 5R^2C^2s^2 + R^3C^3s^3} \qquad (22\text{–}3)$$

In both of the examples just considered the major interest has been in determining values of voltage and current. It should be recognized, however, that exactly the same procedure is useful as a means of determining the input impedance to any ladder network. For example, in the circuit of Fig. 22–1 the input impedance would be

$$\boldsymbol{Z}_{\text{in}} = \frac{\boldsymbol{E}_s}{\boldsymbol{I}_1} = \frac{135}{15} = 9 \text{ ohms} \qquad (22\text{–}4)$$

and in the circuit of Fig. 22–2 it would be

$$\boldsymbol{Z}_{\text{in}} = \frac{\boldsymbol{E}_{\text{in}}}{\boldsymbol{I}_1} = \frac{1 + 6RCs + 5R^2C^2s^2 + R^3C^3s^3}{sC(3 + 4RCs + R^2C^2s^2)} \qquad (22\text{–}5)$$

Although this method of computation is most useful in connection with ladder networks, there are many modifications of it which may be useful with other network configurations. One such modification involves determining all of the voltages and currents in the network in terms of one *assumed value* and one *unknown variable*. This technique may be utilized when ladder

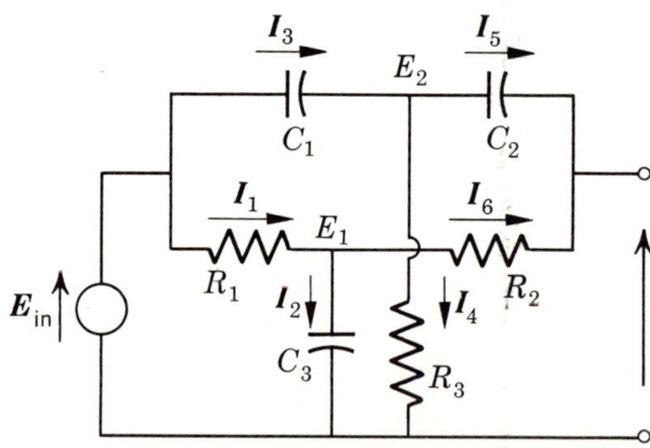

Fig. 22–3. A parallel-T *R-C* network.

networks are connected in parallel—a fairly common occurrence—so that a single assumed value does not uniquely determine the voltages and currents in *both* ladders.

As an example of this method of computation consider the circuit of Fig. 22–3, which represents a common form of *R-C* network known as a *parallel*-T or *twin*-T. It is again desired to find the ratio of output voltage to input voltage. In this case it is most convenient to assume a value for $\boldsymbol{E}_{\text{out}}$ and carry $\boldsymbol{I}_6$ through the computations as a variable. Thus, letting $\boldsymbol{E}_{\text{out}} = 1$ v and

$I_6 = I$ amperes, we have

$$\begin{aligned} E_1 &= E_{\text{out}} + R_2 I_6 = 1 + R_2 I \\ I_2 &= E_1 s C_3 = s C_3 + R_2 C_3 s I \\ I_1 &= I_2 + I_6 = s C_3 + R_2 C_3 s I + I \\ E_{\text{in}} &= E_1 + R_1 I_1 = 1 + R_2 I + R_1 I + R_1 C_3 s + R_1 R_2 C_3 s I \end{aligned} \tag{22-6}$$

and also

$$\begin{aligned} I_5 &= -I_6 = -I \\ E_2 &= E_{\text{out}} + \frac{I_5}{sC_2} = 1 - \frac{I}{sC_2} \\ I_4 &= \frac{E_2}{R_3} = \frac{1}{R_3} - \frac{I}{R_3 C_2 s} \\ I_3 &= I_4 + I_5 = \frac{1}{R_3} - \frac{I}{R_3 C_2 s} - I \\ E_{\text{in}} &= E_2 + \frac{I_3}{sC_1} = 1 - \frac{I}{sC_2} - \frac{I}{sC_1} + \frac{1}{R_3 C_1 s} - \frac{I}{R_3 C_1 C_2 s^2} \end{aligned} \tag{22-7}$$

Since the value of E_{in} has now been computed by working back along *each* ladder, the two resulting equations may be solved simultaneously. Rewriting these equations as

$$E_{\text{in}} = 1 + R_1 C_3 s + I(R_1 + R_2 + R_1 R_2 C_3 s) \tag{22-8}$$

and

$$E_{\text{in}} = 1 + \frac{1}{R_3 C_1 s} - I\left(\frac{1}{sC_1} + \frac{1}{sC_2} + \frac{1}{C_1 C_2 R_3 s^2}\right) \tag{22-9}$$

they may be solved easily for E_{in}. The ratio of output to input voltage may then be expressed as

$$\frac{E_{\text{out}}}{E_{\text{in}}} = \frac{1 + R_3(C_1 + C_2)s + C_1 C_2 R_3 (R_1 + R_2)s^2 + R_1 R_2 R_3 C_1 C_2 C_3 s^3}{\begin{array}{c} 1 + [R_1 C_3 + R_3(C_1 + C_2) + C_2(R_1 + R_2)]s + [R_1 R_3 C_3 (C_1 + C_2) \\ + C_1 C_2 R_3 (R_1 + R_2) + R_1 R_2 C_2 C_3]s^2 + R_1 R_2 R_3 C_1 C_2 C_3 s^3 \end{array}} \tag{22-10}$$

Although this result appears formidable and the solution involved a considerable amount of algebra, this method is far easier than a straightforward simultaneous equation approach.

There are many other situations in which a similar approach may result in appreciably reducing the labor of computation. The degree of success that may be achieved by such procedures depends largely upon the alertness and ingenuity of the circuit analyst.

22-2. Simplifications Resulting from Symmetry. Practical electric circuits frequently exhibit a symmetry of form which, in many cases, may be exploited to reduce the amount of computation. As an elementary example of such symmetry consider the circuit in Fig. 22–4. As a means of making the symmetry more evident, imagine that the 50-ohm resistor has been removed

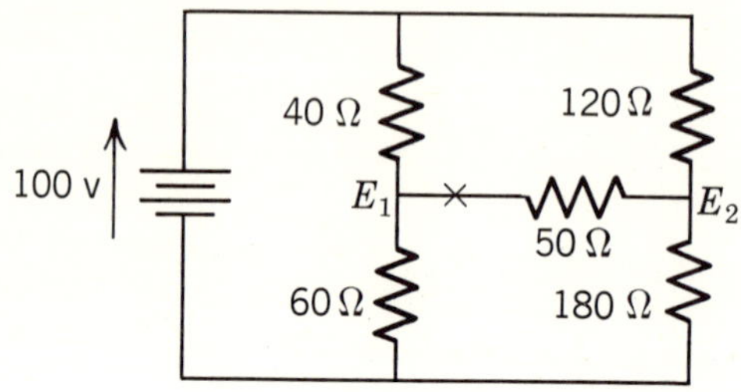

Fig. 22–4. Symmetrical bridge circuit.

by opening the circuit at the point marked ×. With this condition prevailing, compute the values of E_1 and E_2 with respect to the reference node. They are simply

$$E_1 = \frac{60}{40 + 60} \times 100 = 60 \text{ v}$$

and

$$E_2 = \frac{180}{120 + 180} \times 100 = 60 \text{ v}$$

Since these points are at the same potential there will be no current in the 50-ohm resistor when it is connected; hence, its effect on the circuit can be ignored. These points could even be connected together with a short circuit and none of the voltages or currents would be altered.

The foregoing example illustrates the basic ideas concerning symmetry which may be concisely stated as follows:

> If any two nodes in a circuit are at the *same* potential for the specified energy sources, any elements connected directly between these nodes may be removed and the circuit analyzed with either an open circuit or a short circuit between these nodes.

Symmetry conditions are not always immediately obvious. In the circuit of Fig. 22–4 the symmetry could have been recognized without computing voltages by observing the "balanced bridge" relationship existing among the four external resistances. This relationship is a common way for symmetry conditions to occur, and therefore it should be reviewed for general impedance elements. Thus, in Fig. 22–5 the relationship among the impedances which will produce a zero value of $\mathbf{E}_n$ is

$$\frac{\mathbf{Z}_1}{\mathbf{Z}_4} = \frac{\mathbf{Z}_2}{\mathbf{Z}_3} \tag{22–11}$$

If a bridge circuit of this sort is recognized as being part of a more complicated circuit, it becomes a simple matter to check the impedance ratios to see if the bridge is balanced. Generally, however, the most difficult task is that of recognizing the existence of potentially symmetrical circuits. There are no definite rules to guide one in this task other than the necessity of retaining a

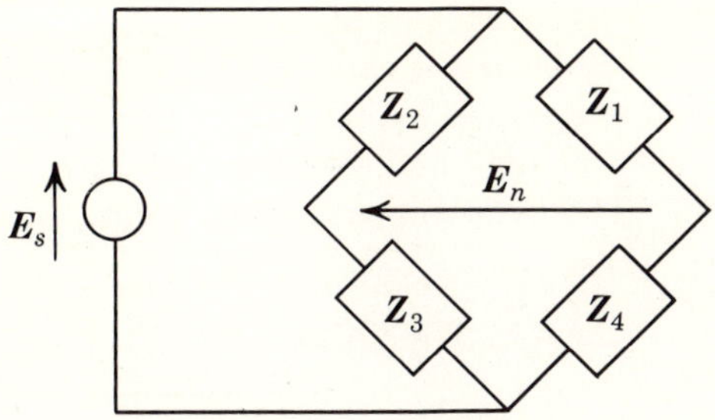

Fig. 22–5. A bridge circuit.

flexible attitude toward circuit analysis and being constantly alert to possibilities for simplification.

As an example of a simple circuit possessing a symmetry which is not immediately obvious, consider the circuit of Fig. 22–6*a*, which is often

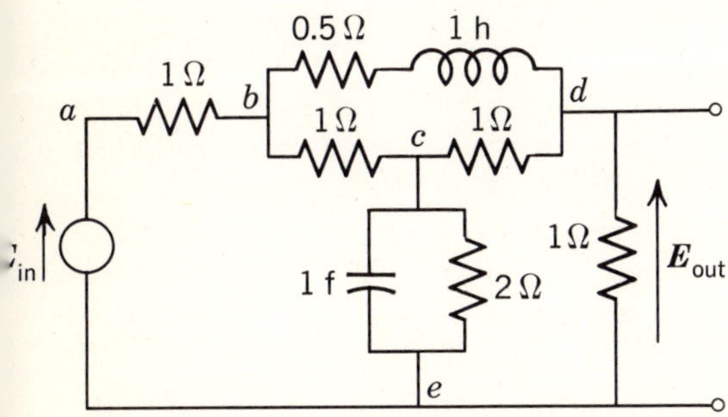

(*a*)

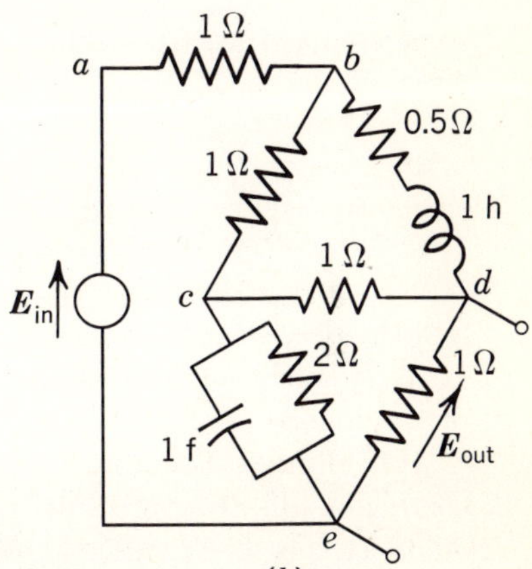

(*b*)

Fig. 22–6. A bridged-T network.

referred to as a *bridged*-T. The potential symmetry of this circuit is made more apparent by redrawing it as shown in Fig. 22–6*b*. In this form the impedances making up the bridge may be identified readily as (using the notation of Fig. 22–5)

$$\mathbf{Z}_1 = s + 0.5$$

$$\mathbf{Z}_2 = 1$$

$$\mathbf{Z}_3 = \frac{2(1/s)}{2 + (1/s)} = \frac{1}{s + 0.5}$$

$$\mathbf{Z}_4 = 1$$

Upon checking the condition for balance, it is seen that

$$\frac{Z_1}{Z_4} = \frac{s + 0.5}{1} = s + 0.5; \qquad \frac{Z_2}{Z_3} = \frac{1}{(1/s) + 0.5} = s + 0.5$$

hence, points c and d will be at the same potential at any frequency.

In order to utilize the symmetry condition to obtain the output voltage, it is convenient to connect points c and d with a short circuit, thus yielding the circuit of Fig. 22–7. The ratio of output to input voltage may then be

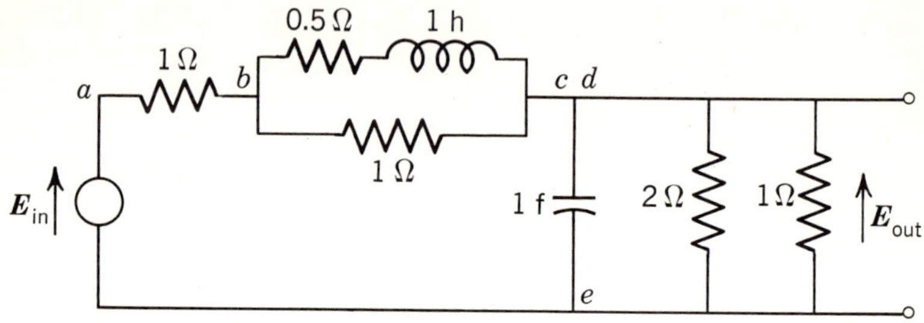

Fig. 22–7. The bridged-T circuit of Fig. 22–6 with points c and d short-circuited.

expressed as the ratio of the impedance between c, d, and e to the total impedance. Thus,

$$\frac{E_{\text{out}}}{E_{\text{in}}} = \frac{\dfrac{1}{0.5 + 1 + s}}{\dfrac{1}{0.5 + 1 + s} + \dfrac{1(s + 0.5)}{s + 1.5} + 1} = \frac{1}{2(s + 1.5)} \qquad (22\text{–}12)$$

and the desired result is obtained with very little effort. To obtain the same result from the equilibrium equations would require solving either three mesh or three node equations and would involve a great deal more labor.

It is of interest to note in connection with this example that the symmetry condition will always result if the impedances designated as Z_1 and Z_3 are duals, that is, if $Z_1 = 1/Z_3$. Since bridged-T circuits are usually designed so as to make these impedances duals (for reasons other than convenience) the simplifications resulting from symmetry are applicable more often than not.

There are many examples of circuits involving a large number of elements which can be handled almost by inspection by observing the existence of symmetry. For the most part these circuits are of academic interest only and do not occur in a practical form. In practical circuits the symmetries are usually more subtle (and, therefore, harder to spot) but the saving in labor that can be achieved makes it worthwhile to inspect almost every circuit carefully before proceeding with a straightforward solution in terms of equilibrium equations.

22–3. Source Transformations. The possibility of transforming from a voltage source to a current source, or vice versa, in order to simplify the writing of equilibrium equations has already been discussed. Situations arise, however, in which a more extensive application of source transformations may reduce the number of equilibrium equations required or even eliminate the need for writing such equations explicitly. In order to exploit this possibility it is advisable to review the concepts of Thévenin's and Norton's theorems.

Thévenin's theorem makes it possible to replace any portion of a circuit with a voltage source and an impedance in series provided the portion replaced has only one pair of terminals. The voltage source in the Thévenin equivalent circuit has a value equal to the open-circuit voltage appearing at the pair of terminals, and the series impedance is the impedance that would be seen looking into the pair of terminals with all energy sources turned off.

The circuit of Fig. 22–8*a* will serve to illustrate how an application of this

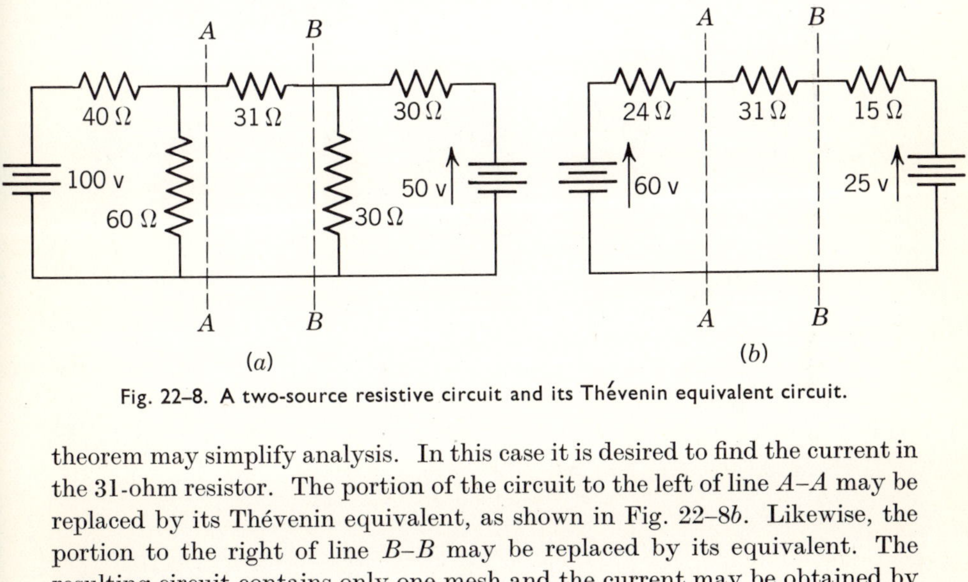

Fig. 22–8. A two-source resistive circuit and its Thévenin equivalent circuit.

theorem may simplify analysis. In this case it is desired to find the current in the 31-ohm resistor. The portion of the circuit to the left of line *A–A* may be replaced by its Thévenin equivalent, as shown in Fig. 22–8*b*. Likewise, the portion to the right of line *B–B* may be replaced by its equivalent. The resulting circuit contains only one mesh and the current may be obtained by inspection as

$$I = \frac{60 - 25}{70} = 0.5 \text{ amp}$$

When utilizing Thévenin's theorem in this manner it is essential that the portions of the circuit that are transformed do not include the particular branch that is being investigated. Once a portion has been transformed, the original voltages and currents within that portion lose their identity and do not appear explicitly in the equivalent circuit. It is always possible, of course, to take the conditions existing at the *terminals* of the transformed portion and work backward to determine the currents and voltages within it. As an example of this, the voltage at the A–A terminals of Fig. 22–8b is easily calculated as

$$E_{A-A} = 60 - 0.5 \times 24 = 48 \text{ v}$$

This same voltage must exist at the A–A terminals of Fig. 22–8a; hence, the current in the 60-ohm resistor must be

$$I_{60} = \tfrac{48}{60} = 0.8 \text{ amp}$$

and the current from the 100-v source is

$$I_s = 0.8 + 0.5 = 1.3 \text{ amp}$$

Norton's theorem provides similar opportunities for simplification by making it possible to replace a portion of the circuit with a current source and

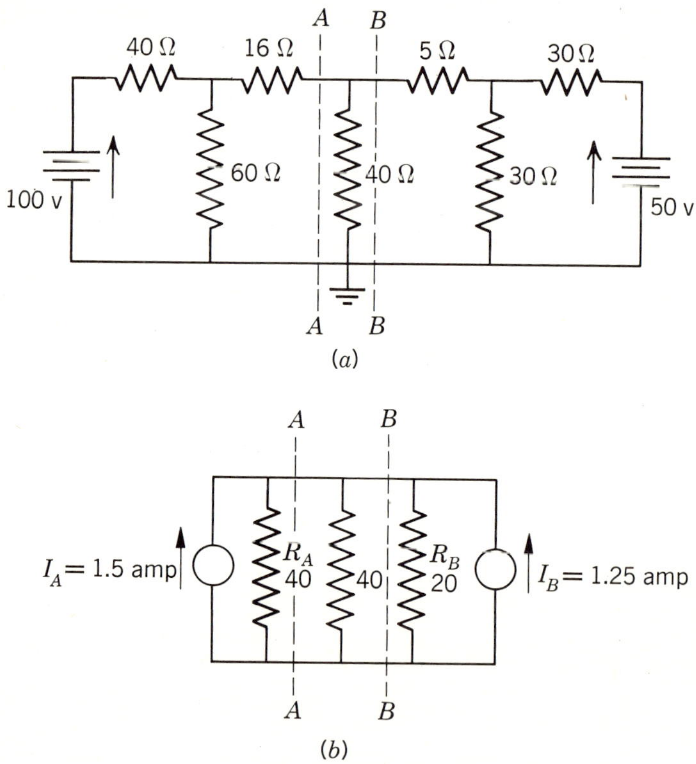

Fig. 22–9. A two-source resistive circuit and its Norton equivalent circuit.

an impedance in parallel. The value of the current source is the short-circuit current that would flow at the terminals, and the impedance is the impedance that would be seen looking into the terminals with all energy sources turned off.

As an example of the application of Norton's theorem, consider the circuit of Fig. 22–9a in which it is desired to find the voltage across the center 40-ohm resistor. Everything to the left of line A–A may be replaced by its Norton equivalent and similarly for everything to the right of line B–B. The resulting equivalent circuit is shown in Fig. 22–9*b*. The current sources and their resistances are calculated as follows:

$$I_A = \frac{100}{40 + \dfrac{60 \times 16}{76}} \times \frac{60}{76} = 1.5 \text{ amp}$$

$$R_A = 16 + \frac{40 \times 60}{100} = 40 \text{ ohms}$$

$$I_B = \frac{50}{30 + \dfrac{5 \times 30}{35}} \times \frac{30}{35} = 1.25 \text{ amp}$$

$$R_B = 5 + 15 = 20 \text{ ohms}$$

From the simplified equivalent circuit it may be seen that

$$E_1 = (1.5 + 1.25) \frac{1}{\frac{1}{40} + \frac{1}{40} + \frac{1}{20}} = 27.5 \text{ v}$$

Although both of the preceding examples have dealt with purely resistive circuits, the same techniques are useful for more general circuits. As a matter of fact, such transformations may be considerably more than just a handy

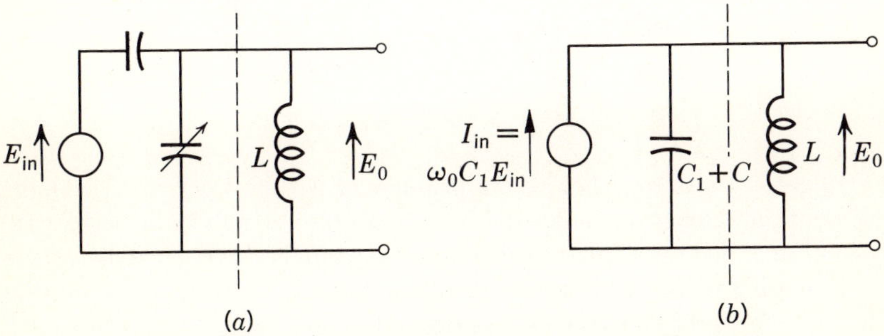

Fig. 22–10. A tuned *L-C* circuit and its Norton equivalent circuit.

computational short cut; they may also be a real aid in visualizing the operation of the circuit. As a very simple example of this aspect assume that in the circuit of Fig. 22–10*a* it is desired to find the value of capacitance C which will make the output voltage a maximum when the input is a steady-state sinusoid of frequency ω_0. At first glance this problem might appear to involve a more general form of resonance than has been previously discussed. However, if the source and the two capacitors are replaced by their Norton equivalent circuit as shown in Fig. 22–10*b*, then it becomes apparent immediately that resonance (and maximum value of $\boldsymbol{E}_0$) will occur when the *sum* of the two capacitors resonates with L. Hence,

$$\omega_0 = \frac{1}{\sqrt{L(C + C_1)}}$$

from which

$$C = \frac{1}{\omega_0^{\,2} L} - C_1 \tag{22–13}$$

Although this application may appear trivial, it actually represents a rather basic approach to many types of problems. When faced with new circuit configurations, any transformation which will convert the circuit to a more familiar form tends to reduce confusion and speed the eventual analysis.

22–4. Circuit Transformations. In accordance with this idea of transforming circuits to more familiar forms, it is often desirable to make transformations that do not involve any sources. Although there are many such transformations possible, only one pair will be discussed here. This pair involves the equivalence of the two circuit forms shown in Fig. 22–11, which are known as the Y and the Δ networks, respectively.

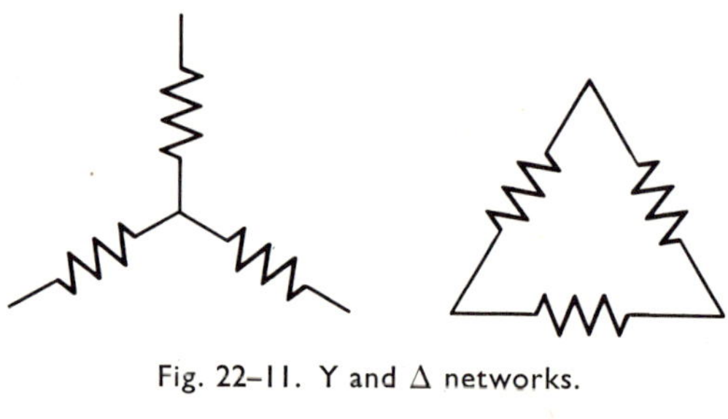

Fig. 22–11. Y and Δ networks.

A straightforward approach to the problem of finding the relations among the elements necessary to make the Y and Δ equivalent would be to write expressions for the impedance at each pair of terminals and equate the expressions for similar pairs. This would result in a set of simultaneous equations which could be solved to yield the desired relationships. An alternative approach would be to write the equilibrium equations for the circuits and then attempt to identify the process of systematic elimination with changes in the circuit geometry. This latter approach will be the one followed here primarily because of the physical interpretation that it provides for systematic elimination.

In the first place, consider that the potential at each of the external nodes is specified as shown in Fig. 22–12 and that the elements are expressed in terms of admittances. The current equilibrium equations pertaining to the

Y circuit (assuming no energy sources connected to the nodes) are

$$\begin{aligned} Y_1E_1 - 0E_2 - 0E_3 - Y_1E_4 &= 0 \\ -0E_1 + Y_2E_2 - 0E_3 - Y_2E_4 &= 0 \\ -0E_1 - 0E_2 + Y_3E_3 - Y_3E_4 &= 0 \\ -Y_1E_1 - Y_2E_2 - Y_3E_3 + (Y_1 + Y_2 + Y_3)E_4 &= 0 \end{aligned} \tag{22-14}$$

and the corresponding matrix representation of these equations is

$$\begin{bmatrix} Y_1 & 0 & 0 & -Y_1 & 0 \\ 0 & Y_2 & 0 & -Y_2 & 0 \\ 0 & 0 & Y_3 & -Y_3 & 0 \\ -Y_1 & -Y_2 & -Y_3 & (Y_1 + Y_2 + Y_3) & 0 \end{bmatrix} \tag{22-15}$$

It is now observed that the Δ circuit contains nodes 1, 2, and 3 but does not contain 4. Hence, if the column pertaining to E_4 can be eliminated from the

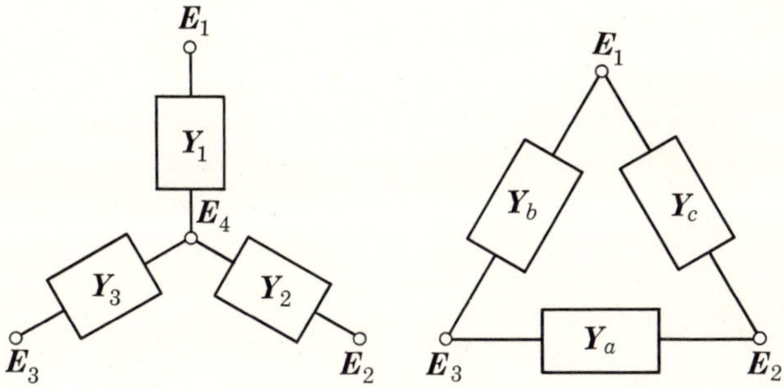

Fig. 22-12. Nodal analysis of Y and Δ networks.

matrix, the remaining columns must represent the equilibrium equations for the Δ circuit and the corresponding branch admittances can be identified easily.

The desired elimination may be carried out by the following steps:

1. Multiply row 4 by $Y_1/Y_1 + Y_2 + Y_3$ and add it to row 1.
2. Multiply row 4 by $Y_2/Y_1 + Y_2 + Y_3$ and add it to row 2.
3. Multiply row 4 by $Y_3/Y_1 + Y_2 + Y_3$ and add it to row 3.

The resulting matrix becomes

$$\begin{bmatrix} Y_1 - \dfrac{Y_1^2}{Y_1+Y_2+Y_3} & -\dfrac{Y_1Y_2}{Y_1+Y_2+Y_3} & -\dfrac{Y_1Y_3}{Y_1+Y_2+Y_3} & 0 & 0 \\ -\dfrac{Y_1Y_2}{Y_1+Y_2+Y_3} & Y_2 - \dfrac{Y_2^2}{Y_1+Y_2+Y_3} & -\dfrac{Y_2Y_3}{Y_1+Y_2+Y_3} & 0 & 0 \\ -\dfrac{Y_1Y_3}{Y_1+Y_2+Y_3} & -\dfrac{Y_2Y_3}{Y_1+Y_2+Y_3} & Y_3 - \dfrac{Y_3^2}{Y_1+Y_2+Y_3} & 0 & 0 \\ -Y_1 & -Y_2 & -Y_3 & (Y_1 + Y_2 + Y_3) & 0 \end{bmatrix} \tag{22-16}$$

The first three rows of this matrix now represent a set of three equations with only three unknowns and thus they can be interpreted as the equilibrium equations for the Δ circuit. Since the terms off the major diagonal represent admittances common to two nodes, it is a simple matter to make the identifications

$$Y_a = \frac{Y_2Y_3}{Y_1 + Y_2 + Y_3} \tag{22-17}$$

$$Y_b = \frac{Y_1Y_3}{Y_1 + Y_2 + Y_3} \tag{22-18}$$

$$Y_c = \frac{Y_1Y_2}{Y_1 + Y_2 + Y_3} \tag{22-19}$$

Thus, it may be seen that the elimination of a node voltage from a set of equilibrium equations corresponds physically to the removal of that node from the circuit.

Equations 22–17, 22–18, and 22–19 may be considered as specifying the transformation of a Y into a Δ. It is of interest to note the symmetry of these equations.

The inverse transformation may be obtained by writing voltage equilibrium equations for the Δ circuit. The mesh currents involved are indicated in Fig. 22–13 and the elements are now specified in terms of impedances.

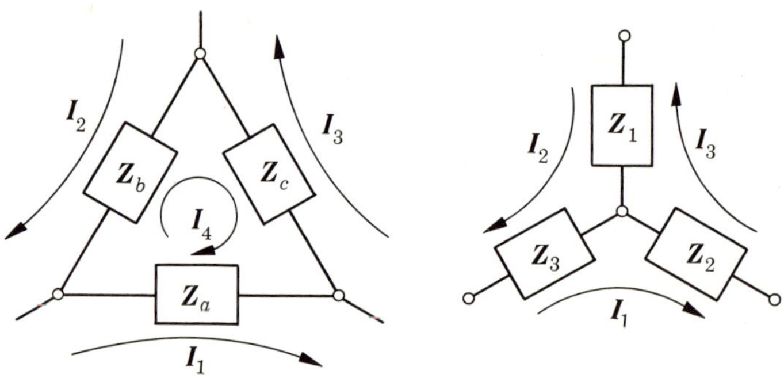

Fig. 22–13. Mesh analysis of Y and Δ networks.

The procedure in this case is to write the four equilibrium equations for the Δ circuit and eliminate the mesh current I_4. The remaining three equations become the equilibrium equations for the Y circuit. The elements in this circuit may be identified easily then and become

$$Z_1 = \frac{Z_bZ_c}{Z_a + Z_b + Z_c} \tag{22-20}$$

$$Z_2 = \frac{Z_aZ_c}{Z_a + Z_b + Z_c} \tag{22-21}$$

$$Z_3 = \frac{Z_aZ_b}{Z_a + Z_b + Z_c} \tag{22-22}$$

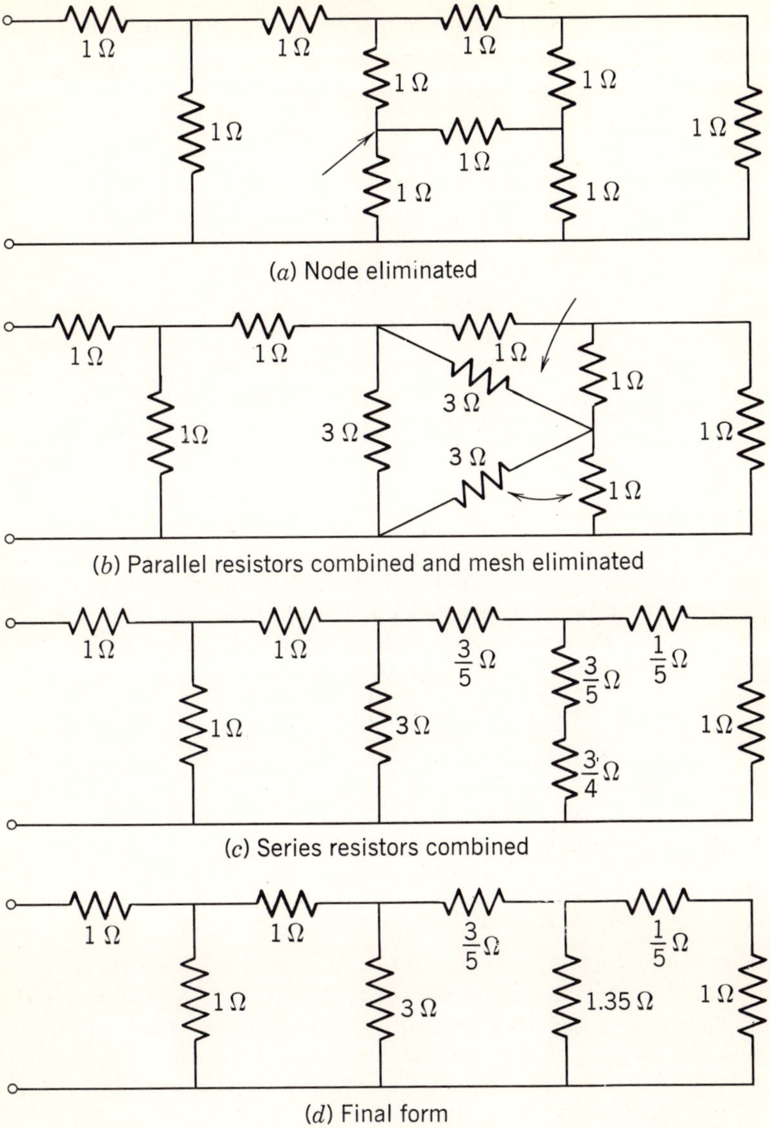

Fig. 22–14. A complex resistance network and its ladder equivalent.

The similarity of these equations to Eqs. 22–17, 22–18, and 22–19 should be noted.

These transformations may be useful in reducing a general network to a form that is more easily handled. For example, assume that it is desired to reduce the resistive network of Fig. 22–14*a* to a ladder form. The successive steps in this reduction are shown and the node or mesh eliminated in each step clearly indicated.

It should be noted that this procedure for reducing the circuit to a simpler form may not be satisfactory if it is desired to compute the potential at one of the nodes to be eliminated or the current in one of the meshes to be eliminated. On the other hand, if only the input and output voltages are of interest, then this type of circuit reduction may result in a considerable saving in labor.

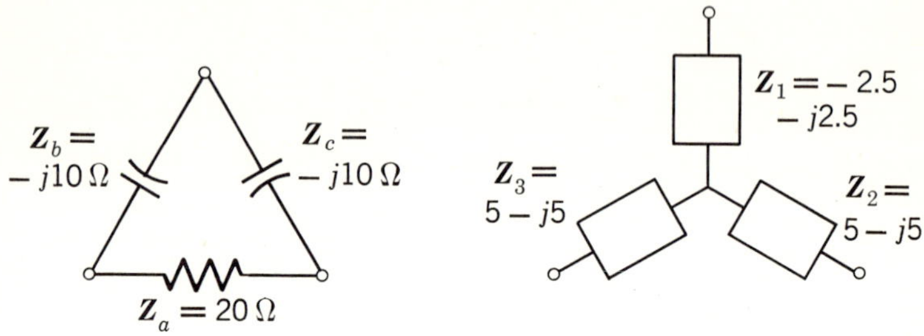

Fig. 22–15. Δ and Y containing reactive elements.

The Y-Δ and Δ-Y transformations may also be applied to circuits involving inductance and capacitance. In this case, it frequently turns out that the resulting circuit contains negative elements and thus is not physically realizable. As a simple illustration of this consider the Δ-Y transformation of Fig. 22–15. The impedance values in the Y circuit may be computed as

$$Z_1 = \frac{Z_b Z_c}{Z_a + Z_b + Z_c} = \frac{(-j10)(-j10)}{20 - j20} = -2.5 - j2.5$$

$$Z_2 = \frac{Z_a Z_c}{Z_a + Z_b + Z_c} = \frac{20(-j10)}{20 - j20} = 5 - j5$$

$$Z_3 = \frac{Z_a Z_b}{Z_a + Z_b + Z_c} = \frac{20(-j10)}{20 - j20} = 5 - j5$$

From this it may be seen that Z_1 contains a negative resistance component, which is not possible in a passive network. The presence of this non-physical element does not in any way prevent us from utilizing the transformation procedure in solving circuit problems; the major purpose of discussing the point here is to warn against becoming alarmed when the situation arises.

22–5. The Reciprocity Theorem. One of the most interesting and useful properties of linear, passive, bilateral networks is that of reciprocity. This may be illustrated in a simple fashion by considering the circuits shown in Fig. 22–16. In Fig. 22–16*a* the indicated current may be computed easily as

$$I = \frac{100}{40 + \dfrac{60 \times 36}{96}} \times \frac{60}{96} = 1 \text{ amp} \tag{22–23}$$

If the points of voltage source and current are interchanged as shown in

Fig. 22-16*b*, the current may again be computed as

$$I = \frac{100}{36 + \dfrac{40 \times 60}{100}} \times \frac{60}{100} = 1 \text{ amp} \tag{22-24}$$

which is seen to be the same value.

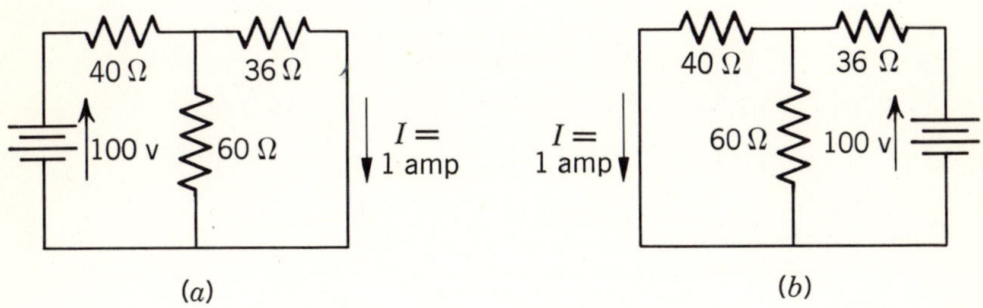

Fig. 22-16. Two-mesh resistive network.

Reciprocity is a general property which may be defined as follows:

> In a circuit with a single excitation applied at one point and a response observed at another point, the ratio of excitation to response is not altered by an interchange of the points of excitation and observation.

A general proof of this property may be obtained rather easily by using the method of determinants to solve for the response in both the original and the

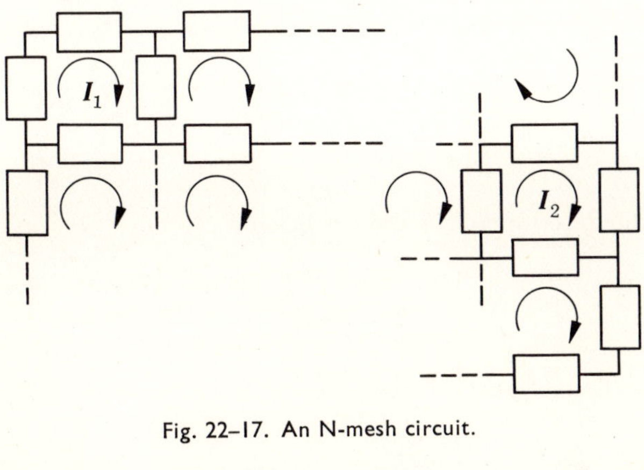

Fig. 22-17. An N-mesh circuit.

interchanged conditions. Since this proof illustrates a number of significant points concerning the nature of general networks, a brief discussion of it will be of value. Either mesh or node equations could be used equally well but the following discussion will be phrased in terms of meshes.

An N-mesh circuit is indicated in Fig. 22-17 and two of the meshes have been singled out for considering the excitation and response. Since the meshes may be numbered in any sequence there is no loss in generality if these meshes are designated as 1 and 2. The voltage equilibrium equations may then be expressed as

$$\begin{aligned}
\mathbf{Z}_{11}\mathbf{I}_1 - \mathbf{Z}_{12}\mathbf{I}_2 - \mathbf{Z}_{13}\mathbf{I}_3 - \cdots - \mathbf{Z}_{1n}\mathbf{I}_n &= \mathbf{E}_1 \\
-\mathbf{Z}_{21}\mathbf{I}_1 + \mathbf{Z}_{22}\mathbf{I}_2 - \mathbf{Z}_{23}\mathbf{I}_3 - \cdots - \mathbf{Z}_{2n}\mathbf{I}_n &= \mathbf{E}_2 \\
-\mathbf{Z}_{31}\mathbf{I}_1 - \mathbf{Z}_{32}\mathbf{I}_2 + \mathbf{Z}_{33}\mathbf{I}_3 - \cdots - \mathbf{Z}_{3n}\mathbf{I}_n &= 0 \\
\cdot \qquad \cdot \qquad \cdot \qquad \cdot \qquad \cdot \qquad & \cdot \\
\cdot \qquad \cdot \qquad \cdot \qquad \cdot \qquad \cdot \qquad & \cdot \\
\cdot \qquad \cdot \qquad \cdot \qquad \cdot \qquad \cdot \qquad & \cdot \\
-\mathbf{Z}_{n1}\mathbf{I}_1 - \mathbf{Z}_{n2}\mathbf{I}_2 - \mathbf{Z}_{n3}\mathbf{I}_3 - \cdots - \mathbf{Z}_{nn}\mathbf{I}_n &= 0
\end{aligned} \tag{22-25}$$

in which $\mathbf{Z}_{11}$, $\mathbf{Z}_{22}$ $\cdots$, $\mathbf{Z}_{nn}$ are the total impedances around each mesh and $\mathbf{Z}_{12}$, $\mathbf{Z}_{13}$, $\cdots$, $\mathbf{Z}_{1n}$, $\mathbf{Z}_{23}$, $\cdots$, are the impedances common to two meshes. Many of these common impedances will, of course, be zero. The voltages $\mathbf{E}_1$ and $\mathbf{E}_2$ are inserted so that they are contained in their respective meshes only and are not common to two meshes. This assumption does not lose generality either, because it is always possible to rearrange the geometry of the network to achieve this condition in any mesh.

The mesh currents $\mathbf{I}_1$ and $\mathbf{I}_2$ may be solved for by Cramer's rule as discussed in Chapter 20. They will be

$$\mathbf{I}_1 = \frac{\begin{vmatrix} \mathbf{E}_1 & -\mathbf{Z}_{12} & \cdots & -\mathbf{Z}_{1n} \\ \mathbf{E}_2 & \mathbf{Z}_{22} & & \cdot \\ 0 & \cdot & & \cdot \\ \cdot & \cdot & & \cdot \\ 0 & -\mathbf{Z}_{n2} & \cdots & \mathbf{Z}_{nn} \end{vmatrix}}{D} \quad \text{and} \quad \mathbf{I}_2 = \frac{\begin{vmatrix} \mathbf{Z}_{11} & \mathbf{E}_1 & \cdots & -\mathbf{Z}_{1n} \\ -\mathbf{Z}_{21} & \mathbf{E}_2 & & \cdot \\ \cdot & 0 & & \cdot \\ \cdot & \cdot & & \cdot \\ -\mathbf{Z}_{n1} & 0 & \cdots & \mathbf{Z}_{nn} \end{vmatrix}}{D} \tag{22-26}$$

where D is the determinant of the coefficients

$$D = \begin{vmatrix} \mathbf{Z}_{11} & \mathbf{Z}_{12} & \cdots & \mathbf{Z}_{1n} \\ -\mathbf{Z}_{21} & -\mathbf{Z}_{22} & \cdots & \mathbf{Z}_{2n} \\ \cdot & \cdot & & \cdot \\ \cdot & \cdot & & \cdot \\ \cdot & \cdot & & \cdot \\ -\mathbf{Z}_{n1} & -\mathbf{Z}_{n2} & \cdots & \mathbf{Z}_{nn} \end{vmatrix} \tag{22-27}$$

In order to prove reciprocity it is necessary to compare the two ratios

$$\left.\frac{\mathbf{I}_2}{\mathbf{E}_1}\right|_{\mathbf{E}_2=0} \quad \text{and} \quad \left.\frac{\mathbf{I}_1}{\mathbf{E}_2}\right|_{\mathbf{E}_1=0}$$

These ratios may be obtained by observing that

$$I_1 \Big|_{E_1=0} = \frac{\begin{vmatrix} 0 & -Z_{12} & \cdot\cdot & -Z_{1n} \\ E_2 & Z_{22} & \cdot\cdot & \\ 0 & \cdot & \cdot\cdot & \cdot \\ \cdot & \cdot & \cdot\cdot & \cdot \\ 0 & \cdot & \cdot\cdot & Z_{nn} \end{vmatrix}}{D} = -E_2 \frac{\begin{vmatrix} -Z_{12} & \cdots & -Z_{1n} \\ -Z_{32} & \cdots & \cdot \\ \cdot & \cdots & \cdot \\ \cdot & \cdots & \cdot \\ -Z_{n2} & \cdots & Z_{nn} \end{vmatrix}}{D} \tag{22–28}$$

and

$$I_2 \Big|_{E_2=0} = \frac{\begin{vmatrix} Z_{11} & E_1 & \cdot\cdot & -Z_{1n} \\ -Z_{21} & 0 & \cdot\cdot & \cdot \\ \cdot & 0 & \cdot\cdot & \cdot \\ \cdot & \cdot & \cdot\cdot & \cdot \\ -Z_{n1} & 0 & \cdot\cdot & Z_{nn} \end{vmatrix}}{D} = -E_1 \frac{\begin{vmatrix} -Z_{21} & -Z_{23} & \cdot\cdot & -Z_{2n} \\ \cdot & \cdot & \cdot\cdot & \cdot \\ \cdot & \cdot & \cdot\cdot & \cdot \\ \cdot & \cdot & \cdot\cdot & \cdot \\ -Z_{n1} & \cdot & \cdot\cdot & Z_{nn} \end{vmatrix}}{D} \tag{22–29}$$

It should be noted that $Z_{12} = Z_{21}$, $Z_{1n} = Z_{n1}, \cdots$, since exactly the same element is common to two meshes regardless of which mesh current is being considered. Upon recalling that interchanging the rows and columns of a determinant does not alter its value, it becomes apparent that the numerator determinants of Eqs. 22–28 and 22–29 are the same. Hence,

$$\frac{I_2}{E_1}\bigg|_{E_2=0} = \frac{I_1}{E_2}\bigg|_{E_1=0} \tag{22–30}$$

and the reciprocity condition is proved.

This proof of reciprocity assumed that the excitation was a voltage source and the response a current. A similar proof could be given for a current source excitation and a voltage response. Reciprocity does *not* apply, however, to a voltage response from a voltage excitation or a current response from a current excitation. In order to illustrate that this is the case, consider the circuit of Fig. 22–18*a* in which the response is considered to be the open-circuit voltage E_2. The ratio of response to excitation is

$$\frac{E_2}{E_1} = \frac{20}{20 + 30} = 0.4 \tag{22–31}$$

If the position of the source and the open-circuit voltage are interchanged, as in Fig. 22–18*b*, the ratio of response to excitation becomes

$$\frac{E_2}{E_1} = 1 \tag{22–32}$$

which is obviously different.

If it is recalled that a voltage source is a generalized short circuit, it may be seen from the example just given that interchanging source and response involves a change in circuit geometry and therefore we cannot expect reciprocity to apply. Interchanging a voltage source and a current involves no such change in geometry and reciprocity will always apply.

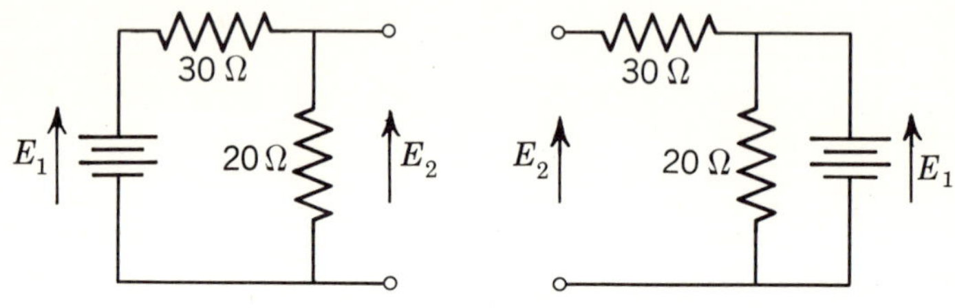

Fig. 22–18. Resistive circuit.

The two-terminal-pair network is a type for which the reciprocity theorem is particularly useful. Such networks may be represented as shown in Fig. 22–19, from which the descriptive nature of the name is clearly apparent.

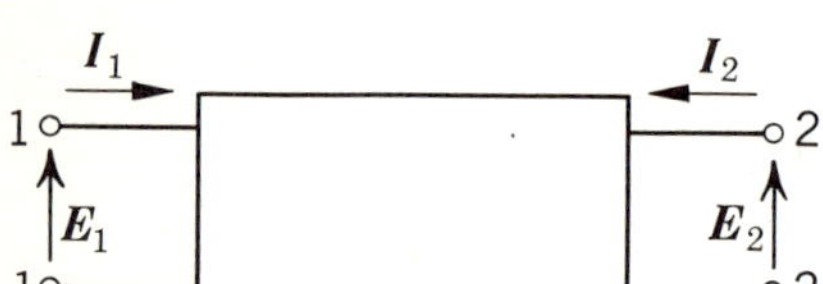

Fig. 22–19. Two-terminal-pair network.

The ladder network of Fig. 22–2, the parallel-T of Fig. 22–3, and the bridged-T of Fig. 22–6 are all examples of two-terminal-pair networks. It is typical of such networks that energy sources may be connected at either end of the network but rarely occur at any place within the network. Since this is the case, a solution for a source at one end will apply equally well for a source at the other end, or, by means of superposition, for sources at both ends. Thus, for Fig. 22–19, the following relationships may be established as a result of reciprocity:

$$\left.\frac{I_2}{E_1}\right|_{E_2=0} = \left.\frac{I_1}{E_2}\right|_{E_1=0} \tag{22–33}$$

$$\left.\frac{E_2}{I_1}\right|_{I_2=0} = \left.\frac{E_1}{I_2}\right|_{I_1=0} \tag{22–34}$$

22–6. Matrix Representation of Networks. In discussing the solution of simultaneous equations by means of systematic elimination, it was found convenient to place the coefficients of the equations into an array called a *matrix*. A modified form of this matrix has many applications other than that of systematic elimination and it is the purpose of this section and the following one to discuss some of these.

By way of review, consider the network in Fig. 22–20 from which all sources have been removed. Using the indicated mesh currents, the voltage equilibrium equations for this network would be

$$\left(1 + \frac{1}{s} + 2s\right) I_1 - 2sI_2 - \frac{1}{s} I_3 = 0 \tag{22-35}$$

$$-2sI_1 + \left(4 + \frac{1}{3s} + 2s\right) I_2 - 4I_3 = 0 \tag{22-36}$$

$$-\frac{1}{s} I_1 - 4I_2 + \left(4 + \frac{1}{s} + 3s\right) I_3 = 0 \tag{22-37}$$

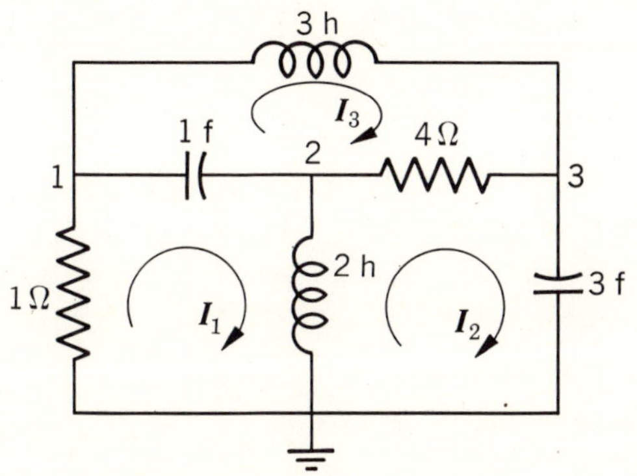

Fig. 22–20. Three-mesh network.

The coefficients of the current variables may be placed in an array which is designated as the *impedance matrix* of the network. This matrix is

$$[Z] = \begin{bmatrix} 1 + \frac{1}{s} + 2s & -2s & -\frac{1}{s} \\ -2s & 4 + \frac{1}{3s} + 2s & -4 \\ -\frac{1}{s} & -4 & 4 + \frac{1}{s} + 3s \end{bmatrix} \tag{22-38}$$

It should be noted that this matrix could have been written directly from the circuit without writing the equilibrium equations. The major diagonal terms are simply the total impedances around each mesh and the terms off the diagonal represent the impedances common to two meshes. This matrix is symmetrical about the major diagonal and this will always be true when it is obtained from mesh currents. It should be recognized, however, that this symmetry is *not* a fundamental property of any set of equilibrium equations but is a direct consequence of writing these equations for the same closed paths that were used for defining loop currents. Therefore, in order to avoid the possibility of ambiguity, it shall be stipulated that the impedance matrix for any network shall always be based on the *mesh equations* and therefore will always be symmetrical.

The impedance matrix may also be expressed in terms of *parameter matrices*. A parameter matrix is written by assembling all of the elements of a single type in an array. Thus, a general network will have three parameter matrices. For the network of Fig. 22–20 these matrices become

$$\text{Resistance matrix} = [R] = \begin{bmatrix} 1 & 0 & 0 \\ 0 & 4 & -4 \\ 0 & -4 & 4 \end{bmatrix} \tag{22–39}$$

$$\text{Inductance matrix} = [L] = \begin{bmatrix} 2 & -2 & 0 \\ -2 & 2 & 0 \\ 0 & 0 & 3 \end{bmatrix} \tag{22–40}$$

$$\begin{array}{r}\text{Inverse capacitance, or} \\ \text{elastance matrix} = [S]\end{array} = \begin{bmatrix} 1 & 0 & -1 \\ 0 & \frac{1}{3} & 0 \\ -1 & 0 & 1 \end{bmatrix} \tag{22–41}$$

The impedance matrix may be expressed as

$$[\mathbf{Z}] = [R] + s[L] + \frac{1}{s}[S] \tag{22–42}$$

It is also possible to represent a network in terms of its *admittance matrix*. This matrix comes from the *current* equilibrium equations but may also be written by inspection of the circuit. Thus, for the network of Fig. 22–20 the admittance matrix is

$$[\mathbf{Y}] = \begin{bmatrix} 1 + s + \dfrac{1}{3s} & -s & -\dfrac{1}{3s} \\ -s & \dfrac{1}{4} + s + \dfrac{1}{2s} & -\dfrac{1}{4} \\ -\dfrac{1}{3s} & -\dfrac{1}{4} & \dfrac{1}{4} + 3s + \dfrac{1}{3s} \end{bmatrix} \tag{22–43}$$

The admittance matrix may also be expressed in terms of parameter matrices. For this example, these would be

$$\text{Conductance matrix} = [G] = \begin{bmatrix} 1 & 0 & 0 \\ 0 & \frac{1}{4} & -\frac{1}{4} \\ 0 & -\frac{1}{4} & \frac{1}{4} \end{bmatrix} \tag{22–44}$$

$$\text{Capacitance matrix} = [C] = \begin{bmatrix} 1 & -1 & 0 \\ -1 & 1 & 0 \\ 0 & 0 & 3 \end{bmatrix} \tag{22–45}$$

$$\text{Inverse inductance matrix} = [\Gamma] = \begin{bmatrix} \frac{1}{3} & 0 & -\frac{1}{3} \\ 0 & \frac{1}{2} & 0 \\ -\frac{1}{3} & 0 & \frac{1}{3} \end{bmatrix} \tag{22–46}$$

The admittance matrix then becomes

$$[Y] = [G] + s[C] + \frac{1}{s}[\Gamma] \tag{22–47}$$

Just as it is possible to obtain the impedance or admittance matrix from the circuit, it is also possible to obtain the circuit from the matrix. As an example, consider the impedance matrix

$$[Z] = \begin{bmatrix} 2 + 3s & -(2 + s) & 0 \\ -(2 + s) & 3 + s + \dfrac{2}{3s} & -\dfrac{1}{3s} \\ 0 & -\dfrac{1}{3s} & 4 + \dfrac{1}{3s} + 2s \end{bmatrix} \tag{22–48}$$

In determining a circuit which corresponds to this matrix it is convenient to consider the terms which are common to two meshes first. The self-impedance terms may then be examined to see how much must be added to each mesh in a branch that is not common to any other mesh. This procedure will yield the network of Fig. 22–21.

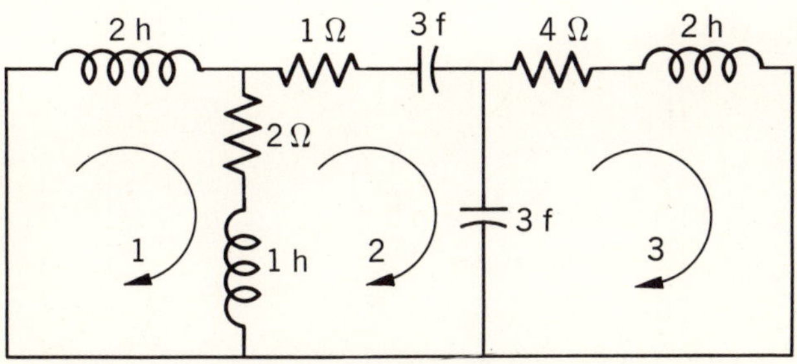

Fig. 22–21. Network determined from the matrix of Eq. 22–48.

22–7. Impedance and Admittance Functions. In the first discussion of impedance and admittance functions in Chapter 17, these functions were determined by making combinations of series and parallel elements. A more general method of determining such functions may be established by considering the impedance or admittance matrix. In order to establish this method, consider first the circuit of Fig. 22–22 for which it is desired to determine the driving point impedance at terminals 1–1′. As an aid in visualizing the procedure it is convenient to assume that a voltage source $\boldsymbol{E}_1$ is connected

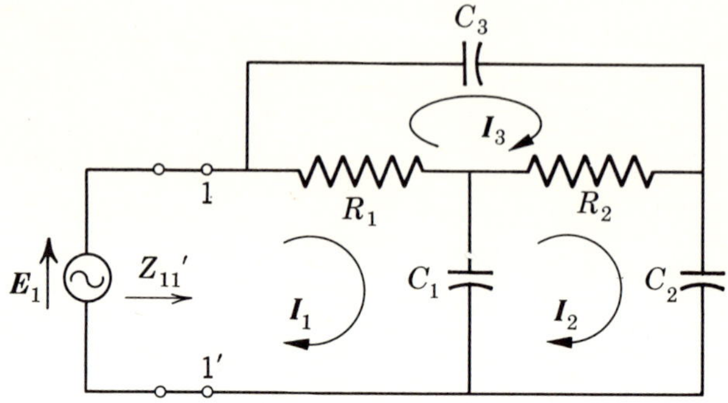

Fig. 22–22. *R-C* network.

to these terminals and designate the mesh currents that would result. The voltage equilibrium equations for the circuit would then be

$$\left(R_1 + \frac{1}{sC_1}\right)\boldsymbol{I}_1 - \left(\frac{1}{sC_1}\right)\boldsymbol{I}_2 - (R_1)\boldsymbol{I}_3 = \boldsymbol{E}_1 \tag{22–49}$$

$$-\left(\frac{1}{sC_1}\right)\boldsymbol{I}_1 + \left(R_2 + \frac{1}{sC_1} + \frac{1}{sC_2}\right)\boldsymbol{I}_2 - (R_2)\boldsymbol{I}_3 = 0 \tag{22–50}$$

$$-(R_1)\boldsymbol{I}_1 - (R_2)\boldsymbol{I}_2 + \left(R_1 + R_2 + \frac{1}{sC_3}\right)\boldsymbol{I}_3 = 0 \tag{22–51}$$

The mesh current $\boldsymbol{I}_1$ may be solved for immediately by application of Cramer's rule. Hence,

$$\boldsymbol{I}_1 = \frac{\begin{vmatrix} \boldsymbol{E}_1 & -\dfrac{1}{sC_1} & -R_1 \\ 0 & R_2 + \dfrac{1}{sC_1} + \dfrac{1}{sC_2} & -R_2 \\ 0 & -R_2 & R_1 + R_2 + \dfrac{1}{sC_3} \end{vmatrix}}{\begin{vmatrix} R_1 + \dfrac{1}{sC_1} & -\dfrac{1}{sC_1} & -R_1 \\ -\dfrac{1}{sC_1} & R_2 + \dfrac{1}{sC_1} + \dfrac{1}{sC_2} & -R_2 \\ -R_1 & -R_2 & R_1 + R_2 + \dfrac{1}{sC_3} \end{vmatrix}} \tag{22–52}$$

which may also be written as

$$
I_1 = \frac{E_1 \begin{vmatrix} R_2 + \dfrac{1}{sC_1} + \dfrac{1}{sC_2} & -R_2 \\ -R_2 & R_1 + R_2 + \dfrac{1}{sC_3} \end{vmatrix}}{\begin{vmatrix} R_1 + \dfrac{1}{sC_1} & -\dfrac{1}{sC_1} & -R_1 \\ -\dfrac{1}{sC_1} & R_2 + \dfrac{1}{sC_1} + \dfrac{1}{sC_2} & -R_2 \\ -R_1 & -R_2 & R_1 + R_2 + \dfrac{1}{sC_3} \end{vmatrix}} \tag{22–53}
$$

by expanding the numerator determinant along the first column. Since I_1 is current flowing into the 1–1′ terminals as a result of applying a voltage E_1, the driving-point impedance is clearly

$$
Z_{11} = \frac{E_1}{I_1} = \frac{\begin{vmatrix} R_1 + \dfrac{1}{sC_1} & -\dfrac{1}{sC_1} & -R_1 \\ -\dfrac{1}{sC_1} & R_2 + \dfrac{1}{sC_1} + \dfrac{1}{sC_2} & -R_2 \\ -R_1 & -R_2 & R_1 + R_2 + \dfrac{1}{sC_3} \end{vmatrix}}{\begin{vmatrix} R_2 + \dfrac{1}{sC_1} + \dfrac{1}{sC_2} & -R_2 \\ -R_2 & R_1 + R_2 + \dfrac{1}{sC_3} \end{vmatrix}} \tag{22–54}
$$

Examination of Eq. 22–54 reveals that the numerator is simply the determinant of the impedance matrix of the network and the denominator is the cofactor of the element in the first row and first column. Thus, it would have been possible to write the driving-point impedance expression directly from the impedance matrix without resorting to the equilibrium equations. Furthermore, the driving-point impedance associated with a pair of terminals in any other mesh could be obtained just as easily by using the appropriate cofactor in the denominator. This cofactor may always be obtained by crossing out the row and column associated with the mesh in which the terminals are located and will always have the same algebraic sign as the minor determinant thus formed since the sum of the row number and column number will always be even.

These results may be stated in more general form as follows:

> If $\boldsymbol{Z}$ is the determinant of the impedance matrix of any linear, passive network and $\boldsymbol{Z}_{kk}$ is a minor determinant formed by crossing out the Kth

row and Kth column, then the driving-point impedance at a pair of terminals in a branch of the Kth mesh, not common to any other mesh, is

$$\boldsymbol{Z}_{kk} = \frac{\boldsymbol{Z}}{\boldsymbol{Z}_{kk}} = \frac{1}{\boldsymbol{Y}_{kk}} \tag{22–55}$$

A very similar result can be expressed in terms of the admittance matrix. To illustrate the derivation of this consider the circuit of Fig. 22–23 in which

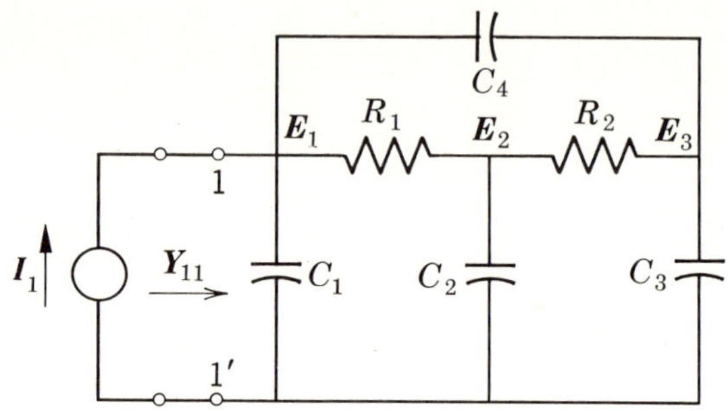

Fig. 22–23. *R-C* network.

a current source and appropriate node voltages have been designated. The resulting current equilibrium equations are

$$\left(\frac{1}{R_1} + sC_1 + sC_4\right)\boldsymbol{E}_1 - \left(\frac{1}{R_1}\right)\boldsymbol{E}_2 - (sC_4)\boldsymbol{E}_3 = \boldsymbol{I}_1 \tag{22–56}$$

$$-\left(\frac{1}{R_1}\right)\boldsymbol{E}_1 + \left(sC_2 + \frac{1}{R_1} + \frac{1}{R_2}\right)\boldsymbol{E}_2 - \left(\frac{1}{R_2}\right)\boldsymbol{E}_3 = 0 \tag{22–57}$$

$$-(sC_4)\boldsymbol{E}_1 - \left(\frac{1}{R_2}\right)\boldsymbol{E}_2 + \left(\frac{1}{R_2} + sC_3 + sC_4\right)\boldsymbol{E}_3 = 0 \tag{22–58}$$

The node voltage $\boldsymbol{E}_1$ may be solved for by application of Cramer's rule and becomes

$$\boldsymbol{E}_1 = \frac{\begin{vmatrix} \boldsymbol{I}_1 & -\dfrac{1}{R_1} & -sC_4 \\ 0 & sC_2 + \dfrac{1}{R_1} + \dfrac{1}{R_2} & -\dfrac{1}{R_2} \\ 0 & -\dfrac{1}{R_2} & \dfrac{1}{R_2} + sC_3 + sC_4 \end{vmatrix}}{\begin{vmatrix} \dfrac{1}{R_1} + sC_1 + sC_4 & -\dfrac{1}{R_1} & -sC_4 \\ -\dfrac{1}{R_1} & sC_2 + \dfrac{1}{R_1} + \dfrac{1}{R_2} & -\dfrac{1}{R_2} \\ -sC_4 & -\dfrac{1}{R_2} & \dfrac{1}{R_2} + sC_3 + sC_4 \end{vmatrix}} \tag{22–59}$$

The driving-point admittance at terminals 1–1′ is simply the ratio of source current $\boldsymbol{I}_1$ to node voltage $\boldsymbol{E}_1$ and therefore may be written as

$$\boldsymbol{Y}_{11} = \frac{\boldsymbol{I}_1}{\boldsymbol{E}_1} = \frac{\begin{vmatrix} \frac{1}{R_1} + sC_1 + sC_4 & -\frac{1}{R_1} & -sC_4 \\ -\frac{1}{R_1} & sC_2 + \frac{1}{R_1} + \frac{1}{R_2} & -\frac{1}{R_2} \\ -sC_4 & -\frac{1}{R_2} & \frac{1}{R_2} + sC_3 + sC_4 \end{vmatrix}}{\begin{vmatrix} sC_2 + \frac{1}{R_1} + \frac{1}{R_2} & -\frac{1}{R_2} \\ -\frac{1}{R_2} & \frac{1}{R_2} + sC_3 + sC_4 \end{vmatrix}} \tag{22–60}$$

after expanding the determinant in the numerator of Eq. 22–59 along its first column. Again it may be recognized that the numerator of Eq. 22–60 is the determinant of the admittance matrix and the denominator is equivalent to the minor formed by crossing out the first row and first column.

The general form of this result may be stated as follows:

> If Y is the determinant of the admittance matrix of any linear, passive network and $\boldsymbol{Y}_{kk}$ is a minor determinant formed by crossing out the Kth row and Kth column, then the driving-point admittance at a pair of terminals connected to the Kth node and the reference node is

$$\boldsymbol{Y}_{kk} = \frac{\boldsymbol{Y}}{\boldsymbol{Y}_{kk}} = \frac{1}{\boldsymbol{Z}_{kk}} \tag{22–61}$$

The preceding results provide very general procedures for finding driving-point impedance and admittance functions. Because of their generality they may be incorrectly used so the following words of caution should be emphasized:

1. The network must be linear and contain *no* energy sources.
2. The driving-point impedance, as obtained from the impedance matrix, applies *only* to a pair of terminals in *series* with a branch containing only *one* mesh current.
3. The driving-point admittance, as obtained from the admittance matrix, applies *only* to a pair of terminals in *parallel* with a node pair, one node of which is the *reference* node.

It is also possible to utilize the impedance and admittance matrices to determine transfer impedances (that is, the ratio of a voltage at one pair of terminals to the current at another pair) and admittances but this application cannot be discussed at this time.

22–8. Matrix Transformations. It is not appropriate at this time to discuss the problem of the *synthesis* of impedance and admittance functions but there is one aspect of that problem which follows so readily from the matrix methods of representation that it seems desirable to introduce it at this stage. The problem of synthesis is that of determining a circuit geometry and element sizes necessary to obtain a given impedance or admittance function. Since there are generally many different networks corresponding to any impedance function, the procedure for finding such a network is usually based on mathematical convenience and as such may not produce the most desirable form of network. For example, it may turn out that certain element sizes are undesirably large (or small) or that the range of element sizes is too great. When such situations arise it would be very convenient to be able to alter the network already determined in order to obtain the more desirable form rather than having to start over again with a different synthesis procedure (and with no assurance that the new procedure will produce any better results). Fortunately, it is often possible to make these changes by means of transformations applied to the matrix of the network and it is the purpose of this section to take an introductory look at such transformations.

In Art. 22–6 it was shown that a matrix could be used as a basis for determining the configuration and elements of a network. Hence, any changes that are made in the matrix for a given network can be interpreted as changes in the network itself. In making such changes, however, there are certain restrictions that must be observed. In the first place, the modified matrix must still be symmetrical about the major diagonal if it is to be interpreted in terms of meshes. In the second place, any changes that occur must be restricted to certain meshes if the desired driving-point impedance or admittance function is to be unaltered. Finally, the changes must not require the insertion of *negative* element values in the modified network.

The procedure is probably best illustrated by means of a simple numerical example. For this purpose, consider the resistive circuit of Fig. 22–24 and assume that it is desired to make both shunt resistances have equal value but retain the same driving-point resistance R_{11}. If the meshes are designated as shown, the resulting resistance matrix is

$$[R] = \begin{bmatrix} 14 & -4 & 0 \\ -4 & 18 & -2 \\ 0 & -2 & 10 \end{bmatrix} \tag{22–62}$$

and the driving-point resistance R_{11} may be expressed as the ratio of the determinant of this matrix to the minor determinant formed by crossing out the first row and first column:

$$R_{11} = \frac{\begin{vmatrix} 14 & -4 & 0 \\ -4 & 18 & -2 \\ 0 & -2 & 10 \end{vmatrix}}{\begin{vmatrix} 18 & -2 \\ -2 & 10 \end{vmatrix}} = 13.09 \text{ ohms} \tag{22–63}$$

The desired matrix transformation may be made by noting the following items:

1. If both determinants in Eq. 22-63 are changed in magnitude by the *same* factor, the value of R_{11} remains unchanged.
2. If all the elements of any row or column are multiplied by a factor K, the value of the determinant is changed by the same factor.
3. If the determinant, hence, the matrix, is to remain symmetrical it is necessary to change both a row and a column. If they are each multiplied by a factor K, the value of the determinant is changed by K^2.
4. If the row and column thus multiplied in the original determinant are *not* crossed out in the minor determinant, then *both* will be altered by K^2 and R_{11} will remain unchanged as mentioned in item 1.

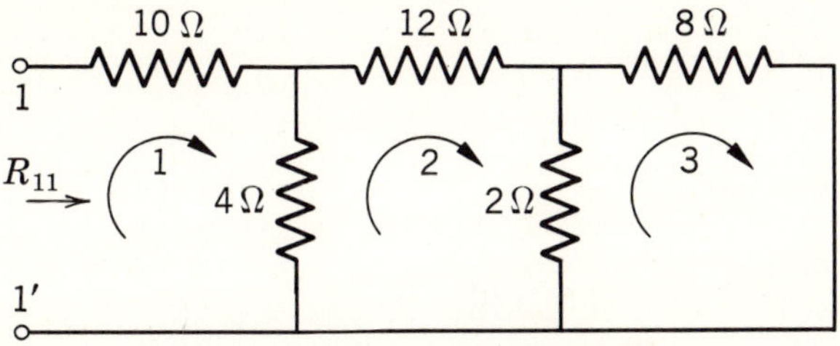

Fig. 22-24. Resistive ladder network.

In the example, it is clear that the 2-ohm resistor common to meshes 2 and 3 appears in both determinants, so multiplying the row and column containing this element by a factor of 2 will make it equal to the 4-ohm resistor between meshes 1 and 2. This change is most easily done in the matrix as follows:

$$[R'] = \begin{bmatrix} 14 & -4 & 0 \\ -4 & 18 & -2 \\ 0 & -2 & 10 \end{bmatrix} \begin{matrix} \\ \\ \leftarrow 2 \end{matrix} = \begin{bmatrix} 14 & -4 & 0 \\ -4 & 18 & -4 \\ 0 & -4 & 40 \end{bmatrix} \qquad (22\text{-}64)$$
$$\qquad\qquad\qquad\uparrow 2$$

Note that the term common to the third row and the third column is multiplied by 2^2 or 4. The modified matrix is easily seen to represent the network of Fig. 22-25 in which the shunt resistances are now equal. A check on the driving-point impedance yields

$$R_{11} = \frac{\begin{vmatrix} 14 & -4 & 0 \\ -4 & 18 & -4 \\ 0 & -4 & 40 \end{vmatrix}}{\begin{vmatrix} 18 & -4 \\ -4 & 40 \end{vmatrix}} = 13.09 \text{ ohms} \qquad (22\text{-}65)$$

which is seen to be unaltered.

Although this example involved only resistance, the same procedure may be used in more general situations. In the circuit of Fig. 22–26a, for example,

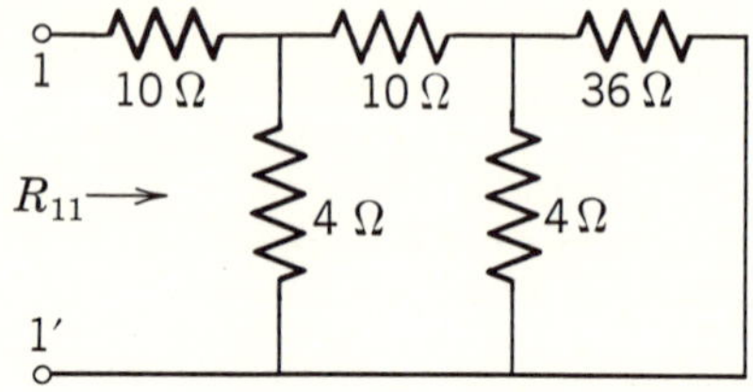

Fig. 22–25. Resistive ladder of Fig. 22–24 with modifications.

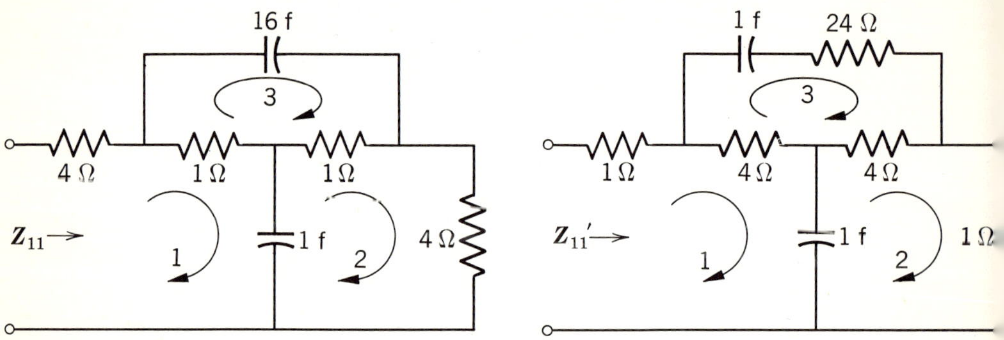

(*a*) (*b*)

Fig. 22–26. *R-C* network.

it is desired to make the two capacitors equal but retain the same driving-point impedance function $\boldsymbol{Z}_{11}$. Using the indicated meshes, the impedance matrix is

$$[\boldsymbol{Z}] = \begin{bmatrix} 5+\dfrac{1}{s} & -\dfrac{1}{s} & -1 \\ -\dfrac{1}{s} & 5+\dfrac{1}{s} & -1 \\ -1 & -1 & 2+\dfrac{1}{16s} \end{bmatrix} \tag{22–66}$$

If the third row and third column are both multiplied by 4, the desired result will be achieved and the resulting matrix will be

$$[\boldsymbol{Z}'] = \begin{bmatrix} 5+\dfrac{1}{s} & -\dfrac{1}{s} & -4 \\ -\dfrac{1}{s} & 5+\dfrac{1}{s} & -4 \\ -4 & -4 & 32+\dfrac{1}{s} \end{bmatrix} \tag{22–67}$$

The circuit corresponding to this matrix is shown in Fig. 22–26*b*. Note that it has been necessary to add an additional resistor in the third mesh.

Although both of the preceding examples have utilized the impedance matrix, exactly the same technique can be used with the admittance matrix. Likewise, there are many other applications of this procedure besides those shown here. To a very large extent, the usefulness of this technique depends upon the ingenuity of the user and his ability to determine the particular combination of multiplying factors which will bring about the desired result.

22–9. Impedance and Admittance Functions for Ladder Networks. It was shown in Art. 22–1 that the driving-point impedance function for a general ladder network could be computed by assuming a voltage or current at the end of the ladder and working back toward the source to determine the required input voltage and current. The ratio of this voltage and current is the driving-point impedance.

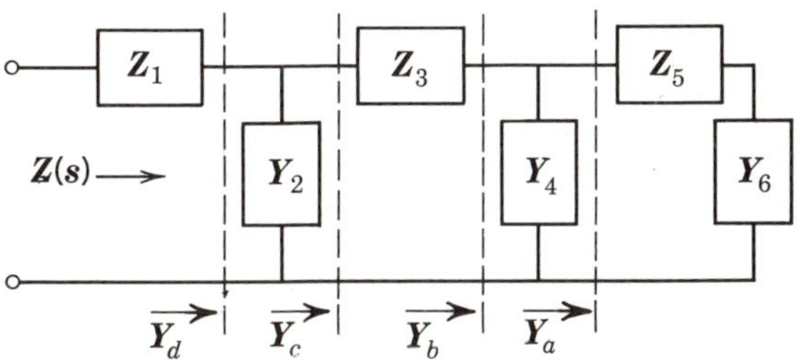

Fig. 22–27. General ladder network.

An alternative procedure, which avoids the necessity of computing any voltages or currents, utilizes a very similar approach. The general form of this procedure may be illustrated by considering the ladder network of Fig. 22–27 in which all series elements are designated as impedances and all shunt elements as admittances. The impedance designated as $\boldsymbol{Z}_a$, which contains the two elements to the right of the dashed line, is simply

$$\boldsymbol{Z}_a = \boldsymbol{Z}_5 + \frac{1}{\boldsymbol{Y}_6} \tag{22–68}$$

Inverting this impedance and adding to $\boldsymbol{Y}_4$ will give the admittance designated as $\boldsymbol{Y}_b$. Hence,

$$\boldsymbol{Y}_b = \boldsymbol{Y}_4 + \frac{1}{\boldsymbol{Z}_a} = \boldsymbol{Y}_4 + \cfrac{1}{\boldsymbol{Z}_5 + \cfrac{1}{\boldsymbol{Y}_6}} \tag{22–69}$$

Similarly, the impedance $\boldsymbol{Z}_c$ is

$$\boldsymbol{Z}_c = \boldsymbol{Z}_3 + \frac{1}{\boldsymbol{Y}_b} = \boldsymbol{Z}_3 + \cfrac{1}{\boldsymbol{Y}_4 + \cfrac{1}{\boldsymbol{Z}_5 + \cfrac{1}{\boldsymbol{Y}_6}}} \tag{22–70}$$

and the admittance Y_d is

$$Y_d = Y_2 + \frac{1}{Z_c} = Y_2 + \cfrac{1}{Z_3 + \cfrac{1}{Y_4 + \cfrac{1}{Z_5 + \cfrac{1}{Y_6}}}} \tag{22–71}$$

Finally, the input impedance $Z(s)$ is given by

$$Z(s) = Z_1 + \frac{1}{Y_d} = Z_1 + \cfrac{1}{Y_2 + \cfrac{1}{Z_3 + \cfrac{1}{Y_4 + \cfrac{1}{Z_5 + \cfrac{1}{Y_6}}}}} \tag{22–72}$$

The mathematical form appearing on the right side of Eq. 22–72 is known as a *continued fraction* and it should be apparent, by comparison with the network, that it can be written by inspection. Furthermore, once the form is

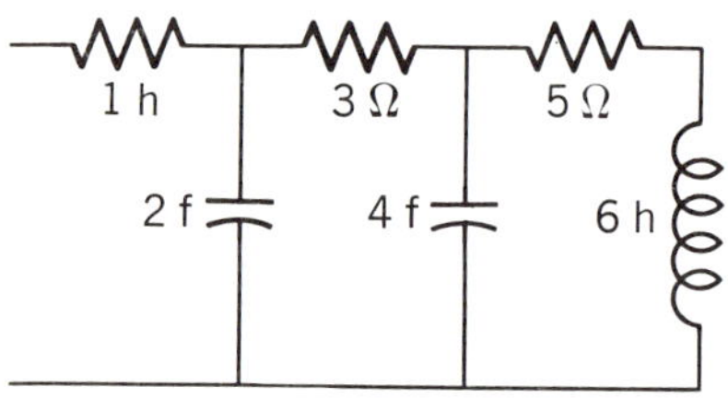

Fig. 22–28. Ladder network.

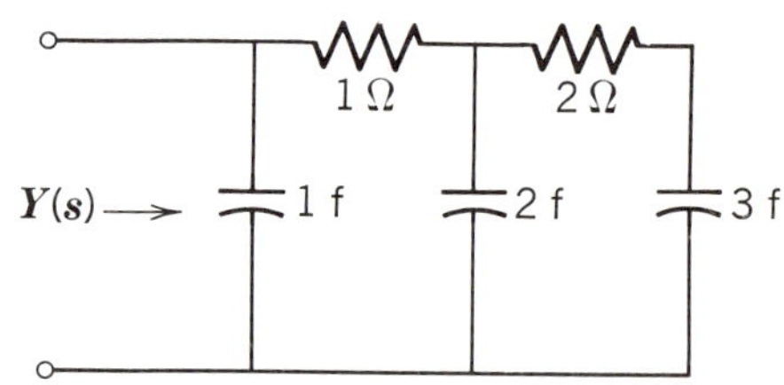

Fig. 22–29. Ladder network.

recognized, it can be written starting from the driving-point end of the ladder rather than the reverse end. Having written the impedance as a continued fraction, the simplification to a more usable form, such as the ratio of polynomials in s, is merely a matter of algebraic manipulation.

As an illustration of this procedure, consider the circuit of Fig. 22–28. The continued-fraction expansion for $Z(s)$ may be written immediately as

$$Z(s) = s + \cfrac{1}{2s + \cfrac{1}{3 + \cfrac{1}{4s + \cfrac{1}{5 + \cfrac{1}{\cfrac{1}{6s}}}}}} \tag{22–73}$$

This may be simplified to

$$Z(s) = \frac{144s^4 + 156s^3 + 108s^2 + 67s + 8}{144s^3 + 156s^2 + 36s + 1} \tag{22–74}$$

If the network starts with a shunt element, as shown in Fig. 22–29, it is more convenient to write the admittance function. This may be done in exactly the same manner. Hence, for this circuit

$$Y(s) = s + \cfrac{1}{1 + \cfrac{1}{2s + \cfrac{1}{2 + \cfrac{1}{3s}}}} \tag{22–75}$$

$$= \frac{12s^3 + 23s^2 + 6s}{12s^2 + 11s + 1} \tag{22–76}$$

The continued-fraction expansion has been presented here as a convenient method of computing the driving-point impedance or admittance of a ladder network. This expansion also provides a method for synthesizing ladder networks. This versatility makes the form a most useful one and deserving of careful consideration.

22–10. Conclusion. This chapter has presented a number of special techniques which are intended to save time and reduce labor. None of these techniques are necessary in the sense that there is no other way to solve the problem because it is always possible to use the equilibrium equations presented in the previous chapters. They do, however, free the engineer from the necessity of following a prescribed pattern in circuit analysis and enable him to capitalize on his ingenuity and his understanding of the fundamental nature of circuit behavior.

It should be emphasized, however, that neither approach is completely adequate within itself. The really competent engineer must be able to use these special techniques with facility where they apply and must also be able to write and solve the necessary equilibrium equations with equal facility in other situations. Mastering this dual approach to circuit analysis is of such tremendous value that the student should not be content with anything less.

PROBLEMS

22–1. Determine all currents and voltages in the ladder network on p. 378.

22–2. Determine the ratio E_2/E_1 of the figure on p. 378.

22–3. Determine the ratio E_{out}/E_{in} of the network illustrated on p. 378 as a function of s. Also determine the input impedance of the ladder network.

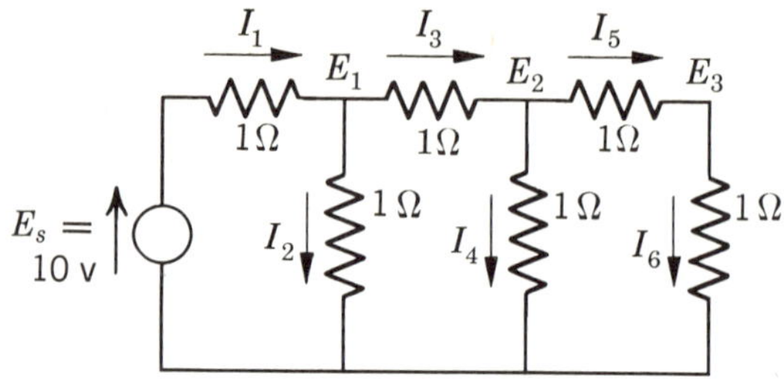

Prob. 22–1.

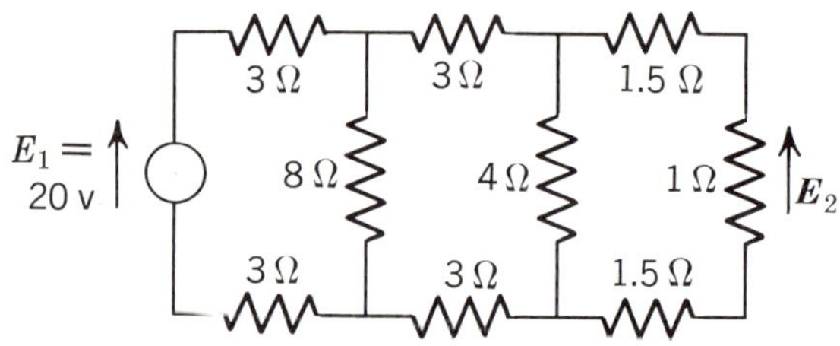

Prob. 22–2.

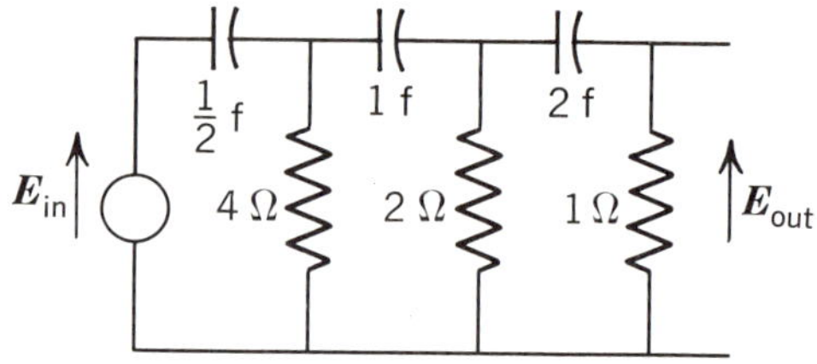

Prob. 22–3.

22–4. Determine $e_{out}(t)$ for the circuit shown and also the input impedance of the ladder network.

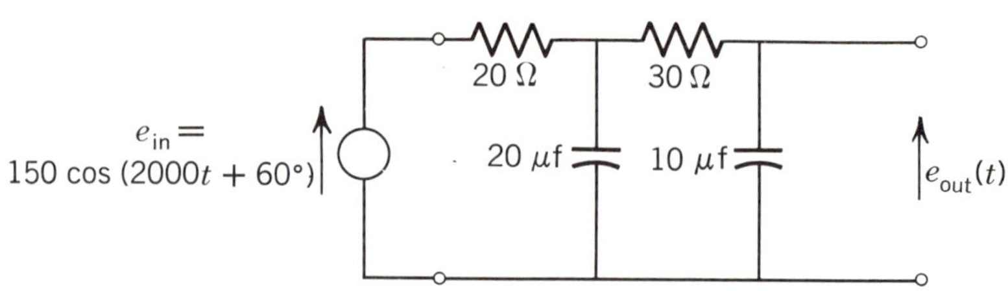

Prob. 22–4.

22–5. Determine E_{out} of the figure.

22–6. Utilize any symmetry exhibited by the network shown and solve for all currents.

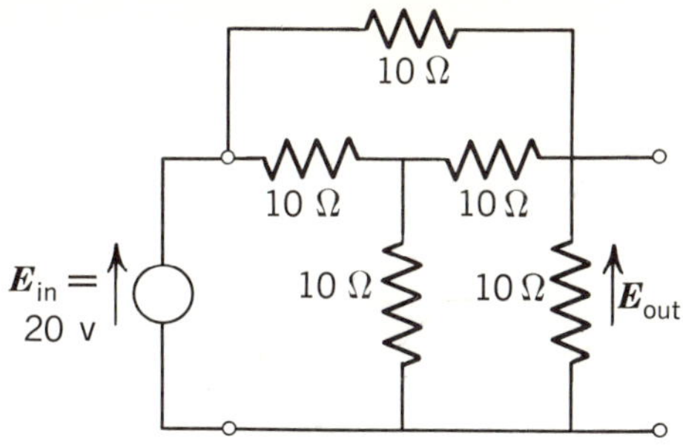

Prob. 22–5.

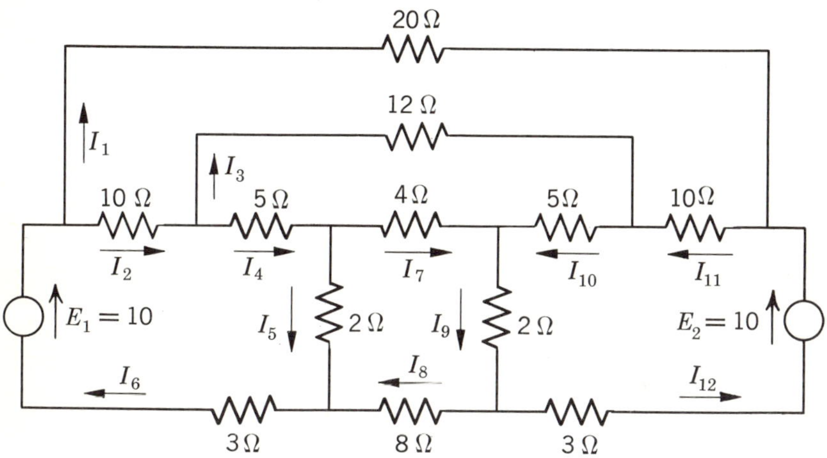

Prob. 22–6.

22–7. Transform each of the circuits shown into an equivalent voltage source.

22–8. Transform each of the circuits of Prob. 22–7 into an equivalent current source.

22–9. Determine $\boldsymbol{E}_{\text{out}}$ of the figure in three different ways: (a) By the techniques applied previously to ladder networks. (b) By replacing the portion of the circuit to the left of the line A–A with the equivalent voltage source determined in (c) of the circuit for Prob. 22–7. (c) By replacing the portion of the circuit to the left of the line A–A with the equivalent current source determined in Prob. 22–8.

22–10. The approximate equivalent circuit of a triode vacuum tube for high frequencies is shown on p. 380. Find the value of ω which will result in the voltage e_0 having a phase angle of 45° relative to the input voltage e_s.

22–11. A network consists of resistances and d-c sources and has two accessible terminals. When the terminals are open-circuited a voltage of 15 v appears. When a resistance of 20 ohms is connected between the terminals a current of $\frac{1}{2}$ amp results. What current will result if the terminals are connected by a short circuit?

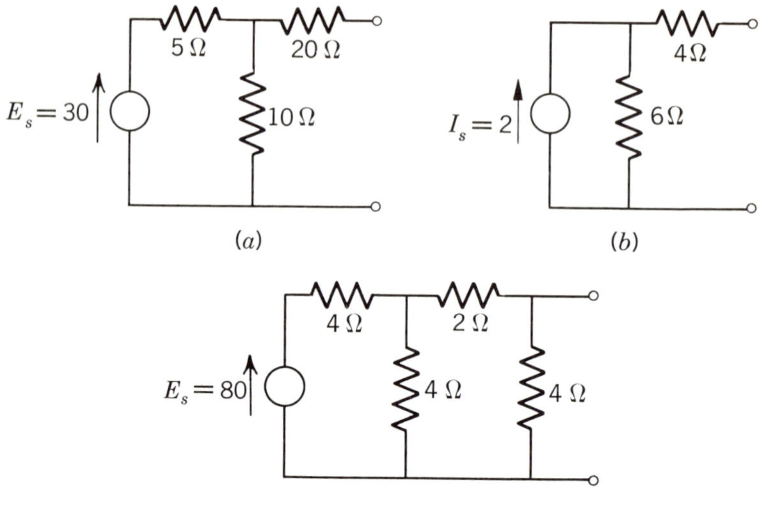

Prob. 22–7.

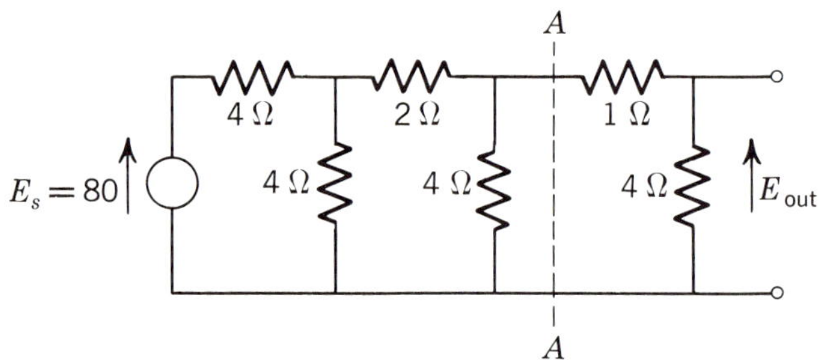

Prob. 22–9.

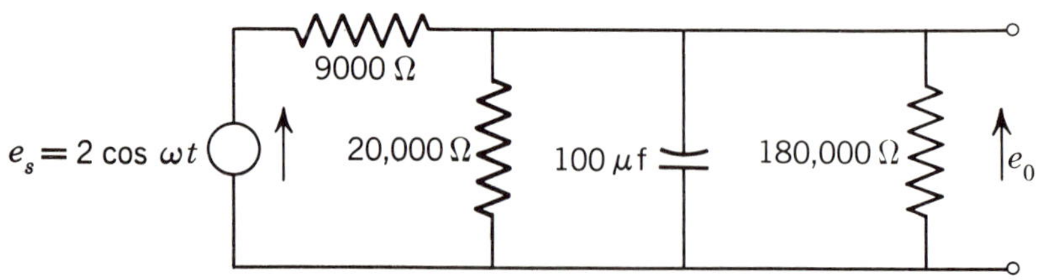

Prob. 22–10.

22–12. Determine the ratio $\boldsymbol{E}_{out}/\boldsymbol{E}_{in}$ of the figure as a function of $\boldsymbol{s}$ by first replacing everything to the left of the line A–A by its Thévenin equivalent.

22–13. Apply Norton's theorem and determine the ratio $\boldsymbol{E}_{out}/\boldsymbol{E}_{in}$ of the figure as a function of $\boldsymbol{s}$. Plot the pole-zero configuration of this ratio and sketch a curve showing how the magnitude of the ratio varies with ω.

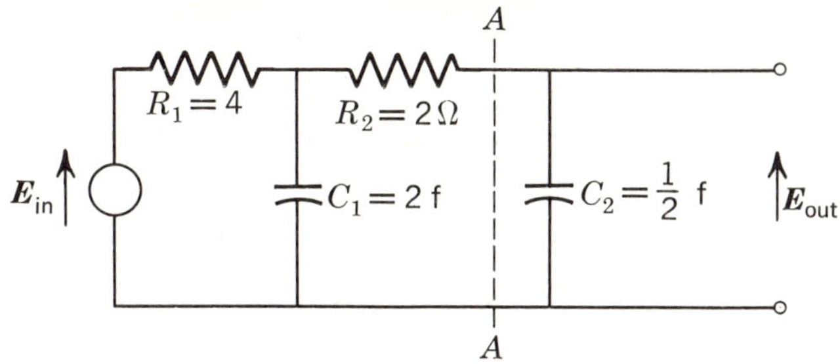

Prob. 22–12.

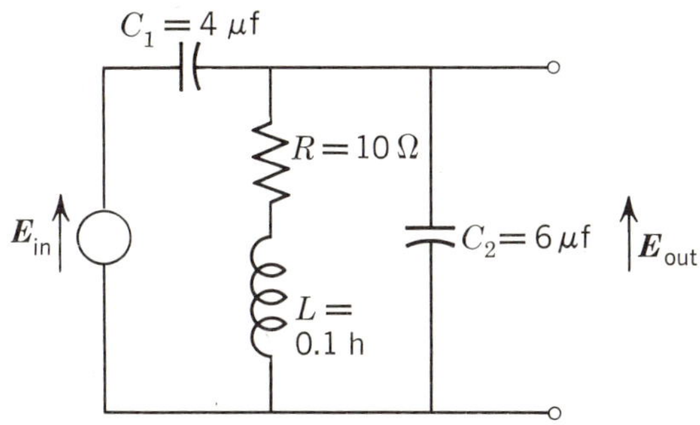

Prob. 22–13.

22–14. Transform each of the given Y-connected circuits into an equivalent Δ.

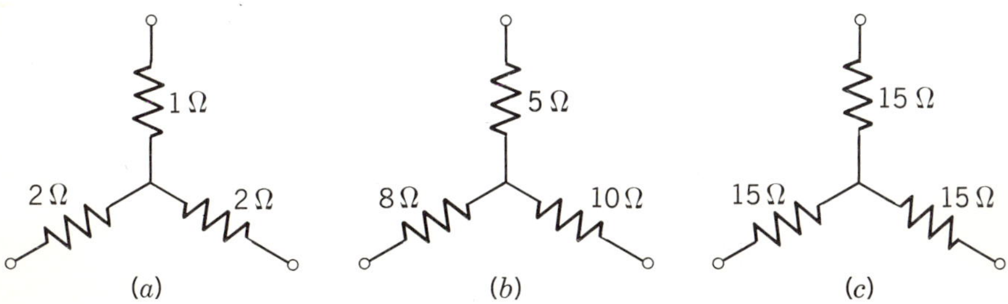

Prob. 22–14.

22–15. Transform each of the given Δ-connected circuits into an equivalent Y.

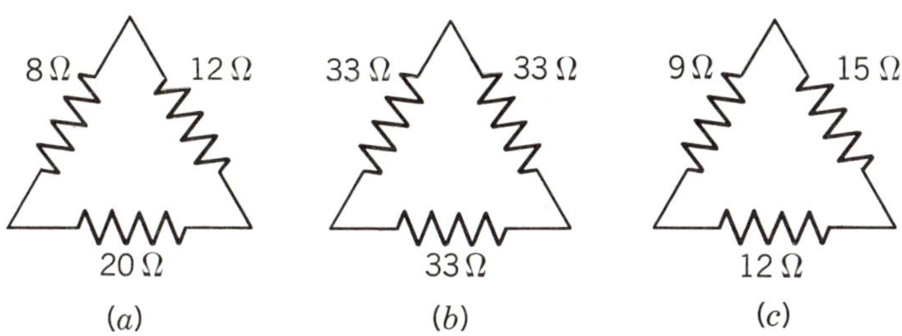

Prob. 22–15.

22–16. Transform the network indicated in (*a*) into one of the forms indicated in (*b*). From these results determine the ratios e_2/i_1 and e_2/e_1.

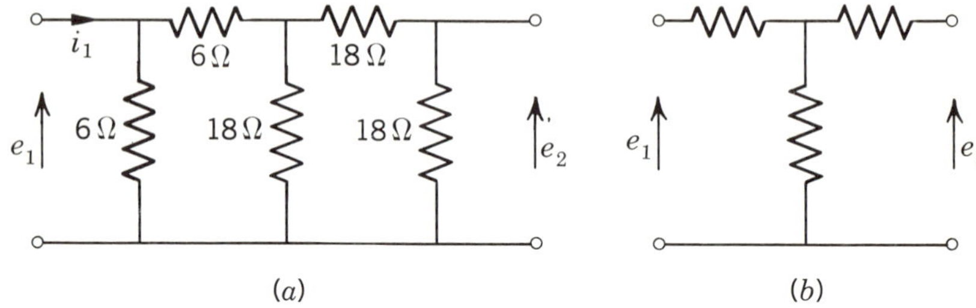

Prob. 22–16.

22–17. Transform this Δ into a Y that will be equivalent in the sinusoidal steady state with $\omega = 2000$. Will this equivalent Y be valid for any other frequency?

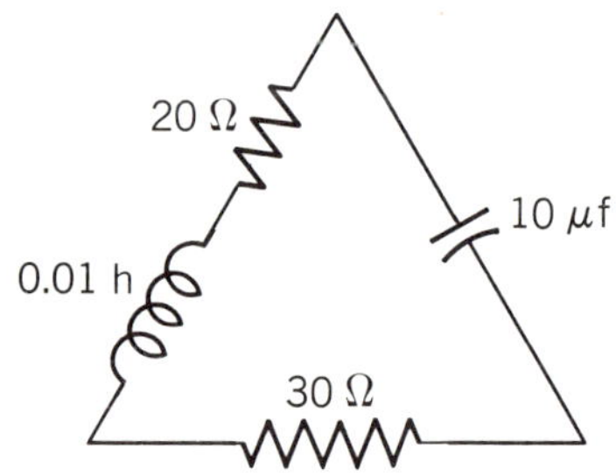

Prob. 22–17.

22–18. Indicate the successive steps that will reduce this network to a single resistance.

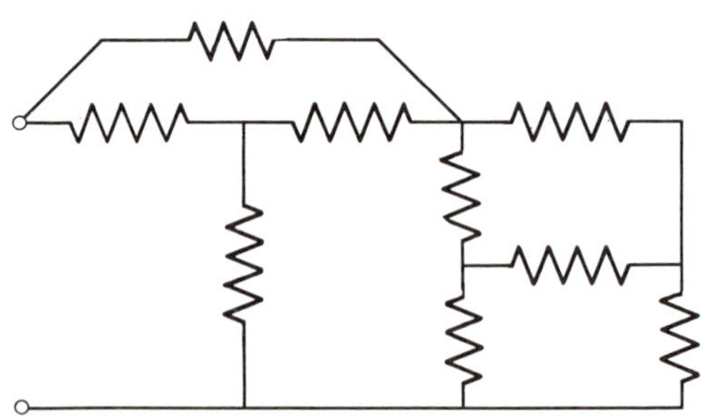

Prob. 22–18.

22–19. Tests are made on a two-terminal-pair resistive network with the results indicated in (*a*) and (*b*). Determine the currents i_1 and i_2 that would result from the conditions indicated in (*c*) and (*d*).

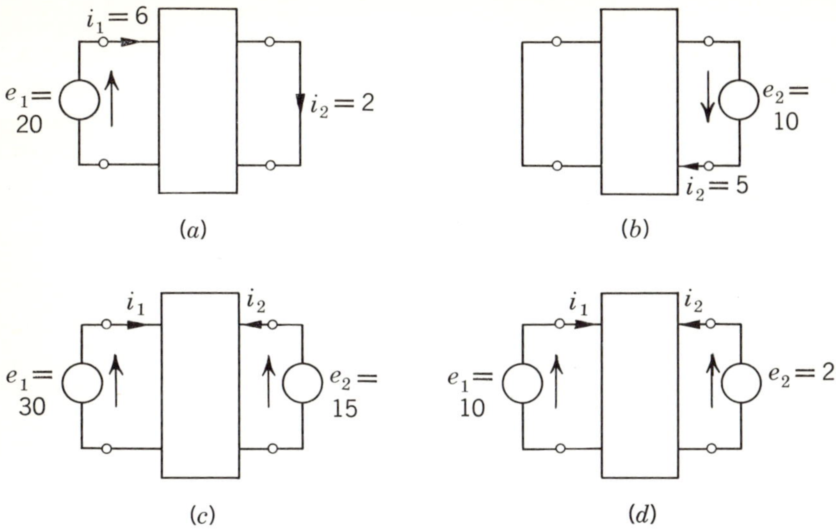

Prob. 22–19.

22–20. Determine an equivalent Δ-connected circuit that could be used to replace the two-terminal-pair network of Prob. 22–19.

22–21. Tests are made on a two-terminal-pair resistive network as indicated in the figure. Determine an equivalent Y-connected circuit that could be used to replace the two-terminal-pair network.

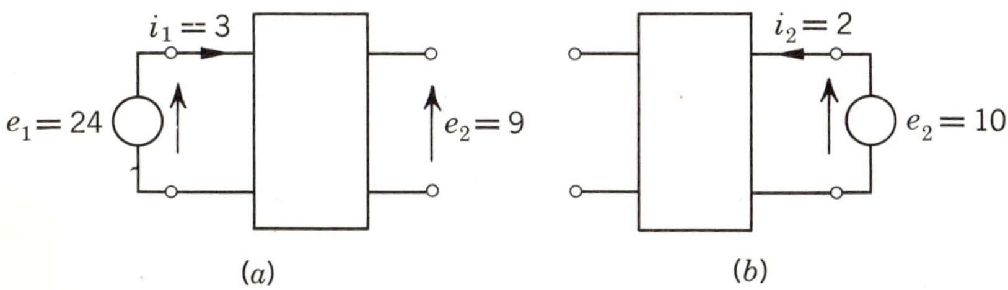

Prob. 22–21.

22–22. Write the impedance matrix for the network shown on p. 384. Express this impedance matrix in terms of the parameter matrices.

22–23. Write the admittance matrix for the network of Prob. 22–22. Express this admittance matrix in terms of the parameter matrices.

22–24. Determine the driving-point impedance at terminals 1–1′ of the figure on p. 384.

22–25. Determine the driving-point admittance at terminals 1–1′ of the figure on the next page.

22–26. Find the driving-point impedance for the network shown on p. 384.

22–27. Find the driving-point admittance for the network illustrated.

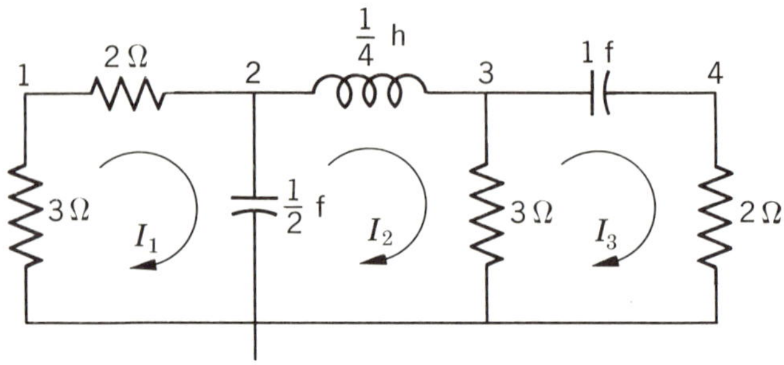

Prob. 22–22.

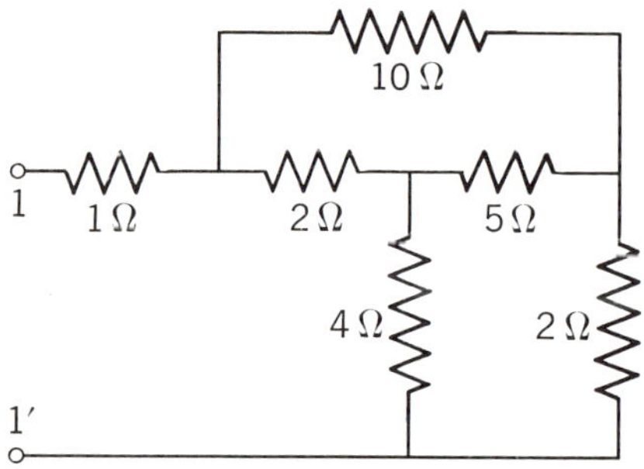

Prob. 22–24.

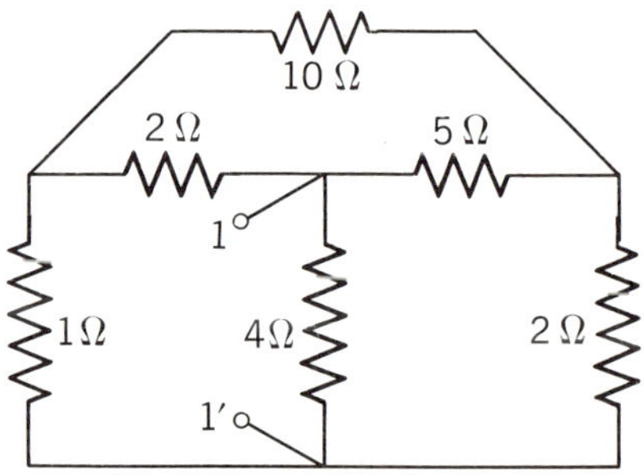

Prob. 22–25.

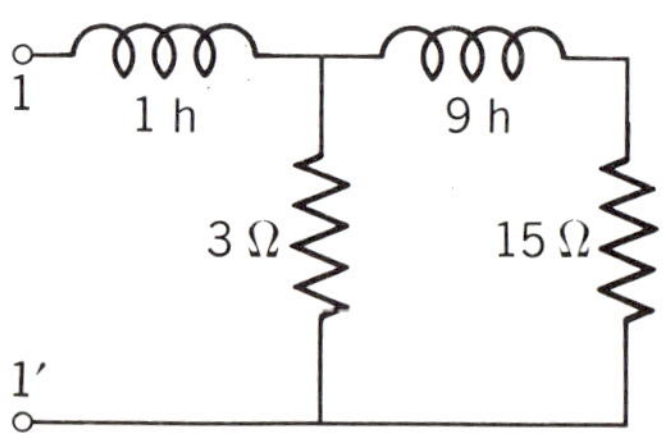

Prob. 22–26.

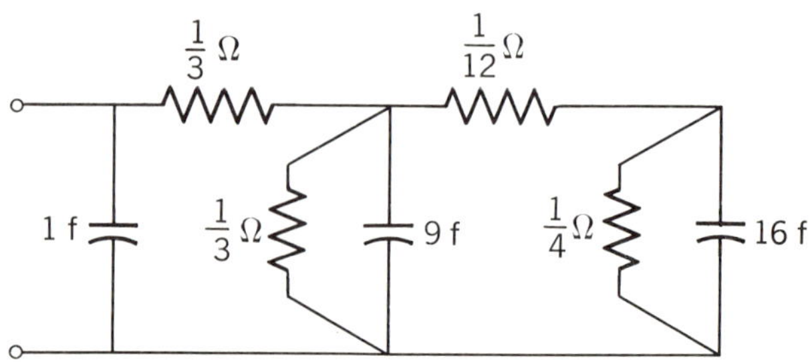

Prob. 22–27.

22–28. Transform the network of Prob. 22–26 into one having the same driving-point impedance and the same number of inductances, but in which the inductances have equal values.

22–29. Transform the network of Prob. 22–27 into one having the same driving-point admittance and the same number of capacitances, but in which all capacitances have the same values.

22–30. Determine the driving-point impedance of the network shown in continued-fraction form and simplify to a ratio of polynomials in s.

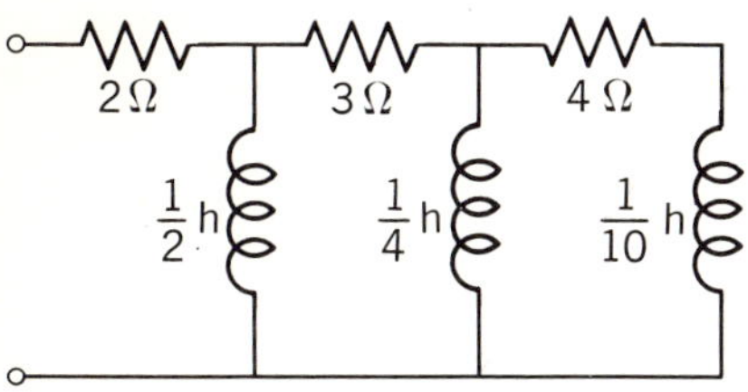

Prob. 22–30.

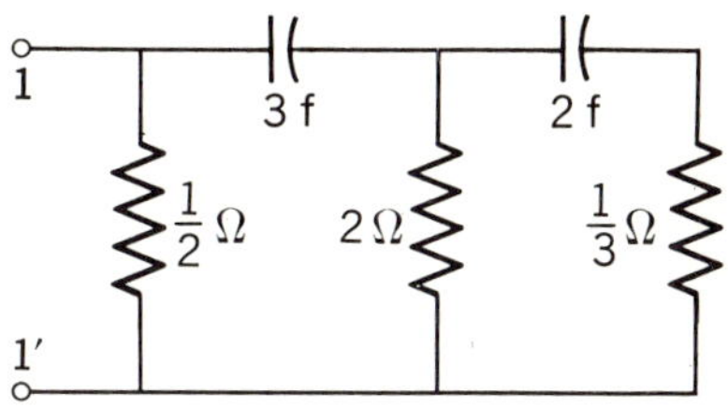

Prob. 22–31.

22–31. Determine the driving-point admittance of the network shown in continued-fraction form and simplify it to a ratio of polynomials in s.

23

THREE-PHASE CIRCUITS

Polyphase circuits can be analyzed by the theories developed in the previous chapters. The superposition and other methods already discussed can be applied to the analysis of polyphase systems. Three-phase circuits, however, are so common that a little study of their behavior is advisable. Furthermore, when the three-phase system is balanced a certain symmetry exists and its analysis becomes an easy problem. In this chapter only the steady-state problem when the three-phase voltages are sine waves of the same frequency will be discussed.

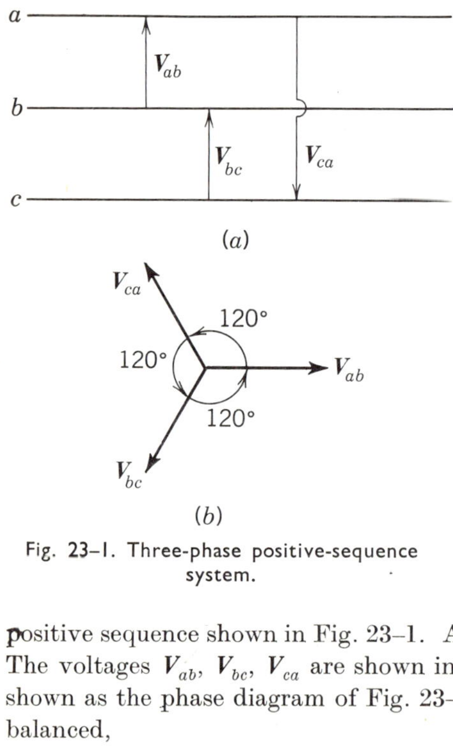

Fig. 23–1. Three-phase positive-sequence system.

23–1. The Three-Phase System. By three-phase balanced systems of voltages (or currents) is meant three voltages (or currents) of equal magnitude and displaced from each other by equal angles. Only one sequence of voltages will be discussed here, namely the so-called positive sequence shown in Fig. 23–1. Assume three conductors *a*, *b*, and *c*. The voltages V_{ab}, V_{bc}, V_{ca} are shown in Fig. 23–1*a*. Their sequence is also shown as the phase diagram of Fig. 23–1*b*. When the system of voltages is balanced,

$$V_{bc} = V_{ab}\epsilon^{j\frac{4\pi}{3}} \qquad \text{and} \qquad V_{ca} = V_{bc}\epsilon^{j\frac{4\pi}{3}}$$

An unbalanced three-phase system of line voltages is a system having voltages unequal in magnitude; also, the phase angles between the phasors are unequal.

23–2. Three-Phase Voltages Applied to Y-Connected Load. Assume that three impedances $\boldsymbol{Z}_A$, $\boldsymbol{Z}_B$, and $\boldsymbol{Z}_C$ are Y-connected as shown in

Fig. 23–2*a*. Let the line applied voltages be V_{ab}, V_{bc}, and V_{ca}, as in Fig. 23–2*b*. Let it be desired to find the currents in the impedances and the voltage drop across them.

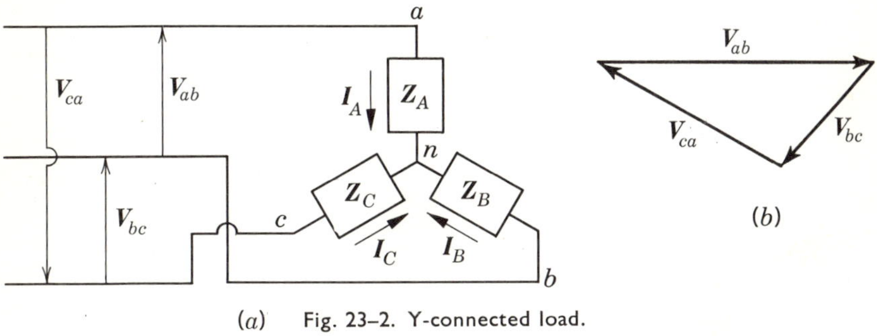

Fig. 23–2. Y-connected load.

The method of superposition can be used to find the currents through and voltages across the impedances. However, it will not be used here. It is a simple matter to assume three phasor currents in the impedances. The direction of the phase currents is toward the common junction n.

By Kirchhoff's second law the sum of the phasor currents is zero, or

$$I_A + I_B + I_C = 0 \tag{23–1}$$

By Kirchhoff's first law the line voltage equations are:

$$V_{ab} = I_A Z_A - I_B Z_B \tag{23–2}$$

$$V_{bc} = I_B Z_B - I_C Z_C \tag{23–3}$$

$$V_{ca} = I_C Z_C - I_A Z_A \tag{23–4}$$

The sum of Eqs. 23–2, 23–3, and 23–4 is zero, or

$$V_{ab} + V_{bc} + V_{ca} = 0 \tag{23–5}$$

From Eq. 23–5 we conclude that the sum of the three *line* voltages of a three-phase system is always zero. This means that there is a restriction on the line voltages of a three-phase system and therefore only two of the line voltages can be independent and the third line voltage is the negative of the sum of the other two phasors.

Similarly, from Eq. 23–1 it can be said that in a three-wire three-phase system the third current is the negative of the sum of the other two currents.

To solve for I_A, I_B, and I_C select Eq. 23–1 and any two of Eqs. 23–2, 23–3, and 23–4. Take, for instance, Eqs. 23–1, 23–2, and 23–3:

$$I_A + I_B + I_C = 0 \tag{23–1}$$

$$V_{ab} = I_A Z_A - I_B Z_B \tag{23–2}$$

$$V_{bc} = I_B Z_B - I_C Z_C \tag{23–3}$$

From these three simultaneous equations the values of I_A, I_B, and I_C can be found. For example, substitute the value of I_C from Eq. 23–1 into Eq. 23–3:

$$V_{bc} = I_B Z_B + (I_A + I_B)Z_C$$
$$= I_B(Z_B + Z_C) + I_A Z_C \qquad (23\text{–}6)$$

Using Eqs. 23–6 and 23–2, we get

$$I_A = \frac{V_{ab}(Z_B + Z_C) + V_{bc}Z_B}{Z_B Z_C + Z_A Z_B + Z_A Z_C} \qquad (23\text{–}7)$$

Other forms in terms of V_{ca} and V_{bc} or V_{ab} and V_{ca} can be found for I_A but Eq. 23–7 is sufficient for the conclusions which are to be drawn.

23–3. Balanced Voltages and Equal Impedances. Equation 23–7 is a general equation for I_A. No restrictions were imposed on the impedances and the voltages V_{ab} and V_{bc}.

First assume that the impedances are equal, namely:

$$Z_A = Z_B = Z_C = Z \qquad (23\text{–}8)$$

The application of Eq. 23–8 to Eq. 23–7 gives:

$$I_A = \frac{2V_{ab} + V_{bc}}{3Z} \qquad (23\text{–}9)$$

Equation 23–7 can be put in another form. From Eq. 23–5 we get

$$V_{ab} + V_{bc} = -V_{ca} \qquad (23\text{–}10)$$

Substitute Eq. 23–10 into Eq. 23–9 and get

$$I_A = \frac{V_{ab} - V_{ca}}{3Z} \qquad (23\text{–}11)$$

Assume further that the line voltages are balanced, namely,

$$V_{bc} = V_{ab}\epsilon^{j240^\circ} \qquad (23\text{–}12)$$
$$V_{ca} = V_{ab}\epsilon^{j120^\circ} \qquad (23\text{–}13)$$

as shown in Fig. 23–3. Then from Eq. 23–13:

$$-V_{ca} = -V_{ab}\epsilon^{j120^\circ}$$
$$= V_{ab}\epsilon^{-j60^\circ} \qquad (23\text{–}14)$$

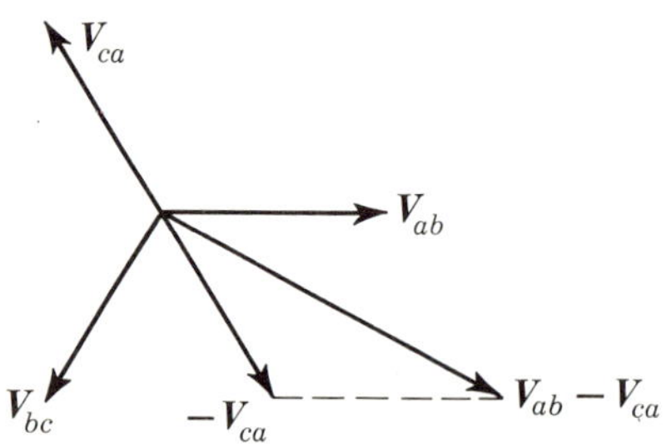

Fig. 23–3. Phasors of balanced line voltages.

Substitute Eq. 23–14 into Eq. 23–11 and obtain:

$$I_A = \frac{\sqrt{3}V_{ab}}{3Z}\epsilon^{-j30^\circ} \qquad (23\text{–}15)$$

In a similar manner the equations for I_B and I_C for balanced voltages and impedances can be found to be:

$$I_B = \frac{\sqrt{3}V_{bc}}{3Z}\epsilon^{-j30^\circ} \qquad (23\text{–}16)$$

$$I_C = \frac{\sqrt{3}V_{ca}}{3Z}\epsilon^{-j30^\circ} \qquad (23\text{–}17)$$

Notice that Eqs. 23–15, 23–16, and 23–17 form a balanced three-phase system of currents. In other words, when the applied three-phase line voltages are balanced and the impedances are equal, and when one of the currents $\boldsymbol{I}_A$, $\boldsymbol{I}_B$, or $\boldsymbol{I}_C$ is known, the other two currents are automatically determined.

From Eq. 23–16, the voltage $\boldsymbol{V}_A$ across $\boldsymbol{Z}_A$ is:

$$\boldsymbol{V}_A = \boldsymbol{I}_A\boldsymbol{Z} \tag{23–18}$$

Comparing Eqs. 23–15 and 23–18, this gives

$$\frac{\boldsymbol{V}_A}{\boldsymbol{Z}} = \frac{\sqrt{3}\boldsymbol{V}_{ab}}{3\boldsymbol{Z}}\,\epsilon^{-j30°}$$

or

$$\boldsymbol{V}_A = \frac{\boldsymbol{V}_{ab}}{\sqrt{3}}\,\epsilon^{-j30°} \tag{23–19}$$

Equation 23–19 states that the phase voltage (voltage drop between line a and point n) is equal to the line voltage $\boldsymbol{V}_{ab}/\sqrt{3}$ shifted backward or through a negative angle of 30°.

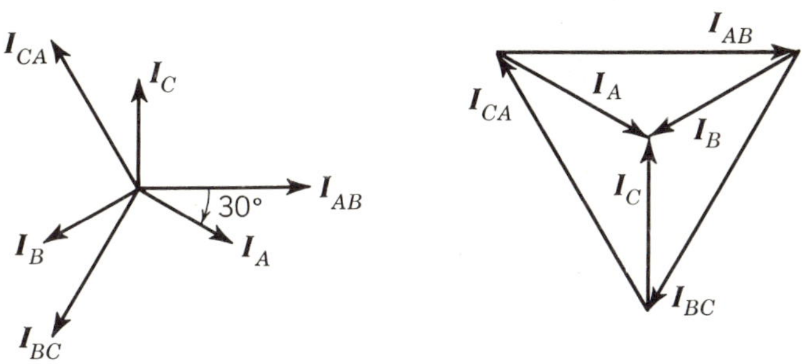

Fig. 23–4. Phase and line voltages of a balanced three-phase system.

Similarly, we find $\boldsymbol{V}_b = \boldsymbol{V}_A\underline{/240°}$ and $\boldsymbol{V}_C = \boldsymbol{V}_A\underline{/120°}$. In other words, for a balanced Y-connected load and balanced line voltages the phase voltage is

$$|\boldsymbol{V}_\phi| = \frac{|\boldsymbol{V}_L|}{\sqrt{3}} \tag{23–20}$$

The line current is given by

$$\boldsymbol{I}_L = \boldsymbol{I}_\phi = \frac{\boldsymbol{V}_\phi}{\boldsymbol{Z}} \tag{23–21}$$

where $\boldsymbol{V}_\phi$ is the phase voltage
$\boldsymbol{V}_L$ is the line voltage
$\boldsymbol{I}_L$ is the line current
$\boldsymbol{I}_\phi$ is the phase current

The phase voltages and line voltages are as shown in Fig. 23–4.

23–4. Three-Phase Currents Flowing in Δ-Connected Loads. Assume in a three-phase system the currents $\boldsymbol{I}_{AB}$, $\boldsymbol{I}_{BC}$, and $\boldsymbol{I}_{CA}$ are flowing toward a Δ-connected load as shown in Fig. 23–5. The current sequence is assumed to be $\boldsymbol{I}_{AB}$ leads $\boldsymbol{I}_{BC}$ and lags $\boldsymbol{I}_{CA}$.

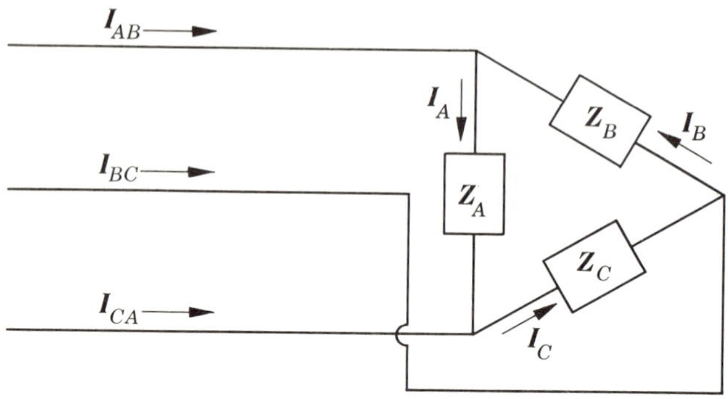

Fig. 23–5. Δ-connected load.

The line currents have two subscripts. $\boldsymbol{I}_{AB}$ enters the load at the junctions of the impedances $\boldsymbol{Z}_A$ and $\boldsymbol{Z}_B$; similarly, $\boldsymbol{I}_{BC}$ enters at junctions of $\boldsymbol{Z}_B$ and $\boldsymbol{Z}_C$ and $\boldsymbol{I}_{CA}$ at junctions of $\boldsymbol{Z}_C$ and $\boldsymbol{Z}_A$. With the phasors as shown,

$$\boldsymbol{I}_{AB} = \boldsymbol{I}_A - \boldsymbol{I}_B \tag{23–22}$$

$$\boldsymbol{I}_{BC} = \boldsymbol{I}_B - \boldsymbol{I}_C \tag{23–23}$$

$$\boldsymbol{I}_{CA} = \boldsymbol{I}_C - \boldsymbol{I}_A \tag{23–24}$$

From Eqs. 23–22, 23–23, and 23–24 we see that

$$\boldsymbol{I}_{AB} + \boldsymbol{I}_{BC} + \boldsymbol{I}_{CA} = 0 \tag{23–25}$$

This means that in a three-phase three-wire system the sum of the line currents is always zero. In other words, the three line currents are not independent of each other. Any one can be obtained from the other two.

In Fig. 23–6 let the voltage drop across the loads be $\boldsymbol{V}_A$, $\boldsymbol{V}_B$, and $\boldsymbol{V}_C$. These three voltages form a closed loop, and

$$\boldsymbol{V}_A + \boldsymbol{V}_B + \boldsymbol{V}_C = 0 \tag{23–26}$$

But

$$\boldsymbol{I}_A = \boldsymbol{V}_A\boldsymbol{Y}_A \tag{23–27}$$

$$\boldsymbol{I}_B = \boldsymbol{V}_B\boldsymbol{Y}_B \tag{23–28}$$

$$\boldsymbol{I}_C = \boldsymbol{V}_C\boldsymbol{Y}_C \tag{23–29}$$

(Although $\boldsymbol{V}_A + \boldsymbol{V}_B + \boldsymbol{V}_C = 0$, this does not mean that $\boldsymbol{I}_A + \boldsymbol{I}_B + \boldsymbol{I}_C = 0$.) From Eqs. 23–22, 23–23, 23–27, 23–28, and 23–29 we get

$$\boldsymbol{I}_{AB} = \boldsymbol{V}_A\boldsymbol{Y}_A - \boldsymbol{V}_B\boldsymbol{Y}_B \tag{23–30}$$

$$\boldsymbol{I}_{BC} = \boldsymbol{V}_B\boldsymbol{Y}_B - \boldsymbol{V}_C\boldsymbol{Y}_C \tag{23–31}$$

where $\boldsymbol{Y}_A$, $\boldsymbol{Y}_B$, and $\boldsymbol{Y}_C$ are the admittances of $\boldsymbol{Z}_A$, $\boldsymbol{Z}_B$, and $\boldsymbol{Z}_C$ respectively.

Substitute the value of V_C from Eq. 23–26 into Eq. 23–31 and obtain

$$I_{BC} = V_B Y_B + (V_A + V_B) Y_C = V_B(Y_B + Y_C) + V_A Y_C \quad (23\text{–}32)$$

From Eqs. 23–30 and 23–32 solve for V_A:

$$V_A = \frac{I_{AB}(Y_B + Y_C) + I_{BC} Y_B}{Y_A Y_B + Y_A Y_C + Y_B Y_C} \quad (23\text{–}33)$$

Similar equations for V_B and V_C can be obtained.

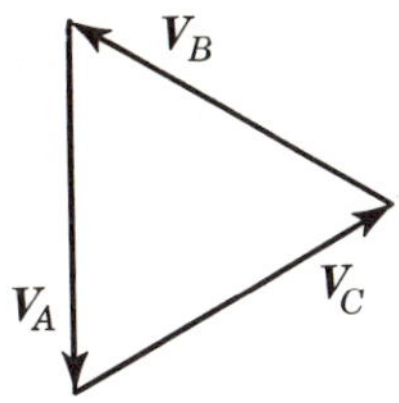

Fig. 23–6. Voltages across Δ load.

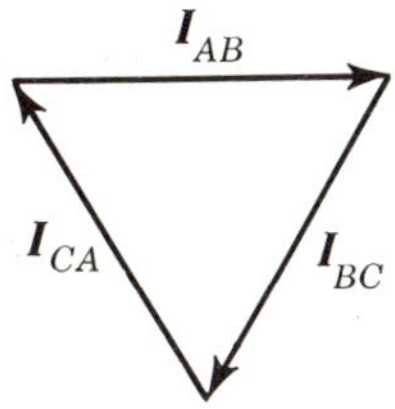

Fig. 23–7. Positive-sequence three-phase line currents.

23–5. Equal Load Admittances and Balanced Line Currents.

If the load admittances are equal, namely,

$$Y_A = Y_B = Y_C = Y \quad (23\text{–}34)$$

Eq. 23–33 reduces to

$$V_A = \frac{2I_{AB} + I_{BC}}{3Y} \quad (23\text{–}35)$$

Equation 23–35 can be put in another form:

$$V_A = \frac{I_{AB} - I_{CA}}{3Y} \quad (23\text{–}36)$$

If it is assumed further that the line currents are balanced, namely,

$$I_{BC} = I_{AB}\epsilon^{j240} \quad (23\text{–}37)$$

and

$$I_{CA} = I_{AB}\epsilon^{j120} \quad (23\text{–}38)$$

as shown in Fig. 23–7, then

$$-I_{CA} = I_{AB}\epsilon^{-j60} \quad (23\text{–}39)$$

and

$$I_{AB} - I_{CA} = I_{AB}(1 + \epsilon^{-j60})$$

$$= \sqrt{3} I_{AB}\epsilon^{-j30} \quad (23\text{–}40)$$

Substitution of Eq. 23–40 into Eq. 23–36 gives

$$V_A = \frac{I_{AB}}{\sqrt{3}Y}\,\epsilon^{-j30} \quad (23\text{–}41)$$

In a similar manner it can be shown that

$$V_B = \frac{I_{BC}}{\sqrt{3}Y} \epsilon^{-j30} \tag{23–42}$$

and

$$V_C = \frac{I_{CA}}{\sqrt{3}Y} \epsilon^{-j30} \tag{23–43}$$

On the other hand,

$$V_A = \frac{I_A}{Y_A} \tag{23–44}$$

$$V_B = \frac{I_B}{Y_B} \tag{23–45}$$

$$V_C = \frac{I_C}{Y_C} \tag{23–46}$$

Equating Eqs. 23–41 and 23–44 results in:

$$I_A = \frac{I_{AB}}{\sqrt{3}} \epsilon^{-j30} \tag{23–47}$$

$$I_B = \frac{I_{BC}}{\sqrt{3}} \epsilon^{-j30} \tag{23–48}$$

$$I_C = \frac{I_{CA}}{\sqrt{3}} \epsilon^{-j30} \tag{23–49}$$

In other words, the phase current is equal to the line current divided by $\sqrt{3}$ and shifted 30° in the negative direction, or

$$I_\phi = \frac{I_L}{\sqrt{3}} \epsilon^{-j30} \tag{23–50}$$

The three line currents and the three phase currents are shown in Fig. 23–8.

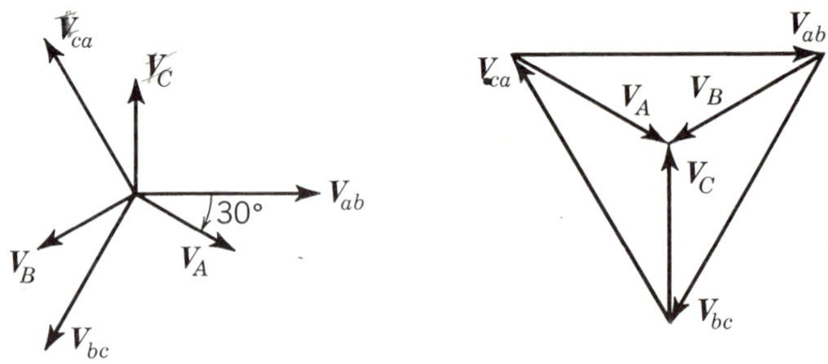

Fig. 23–8. Phase and line currents of a balanced three-phase Δ-connected load.

From the foregoing we see that, if in a three-phase system the line currents are balanced and known and the impedances of the load are equal, the current in the phase or impedance is equal to the line current divided by $\sqrt{3}$. Its angle is shifted 30° negative with respect to the line current. The phase voltages are equal to the line voltages and are equal to:

$$\mathbf{V}_\phi = \mathbf{I}_\phi \mathbf{Z} \tag{23–51}$$

PROBLEMS

23–1. In this chapter it was proved that the phasor sum of the three line voltages of a three-phase system is zero. In the circuit shown, measurements were made with a voltmeter between three lines labeled a, b, and c. The following readings were given:

$$V_{ab} = 120 \text{ v}, \qquad V_{bc} = 100 \text{ v}, \qquad V_{ca} = 110 \text{ v}$$

It is desired to express these three voltages as phasors with $\mathbf{V}_{ab}$ as reference and it is known that the sequence of voltages is ab, bc, ca. Write the three line voltages as phasors, namely, give their magnitude and the angle they make with each other.

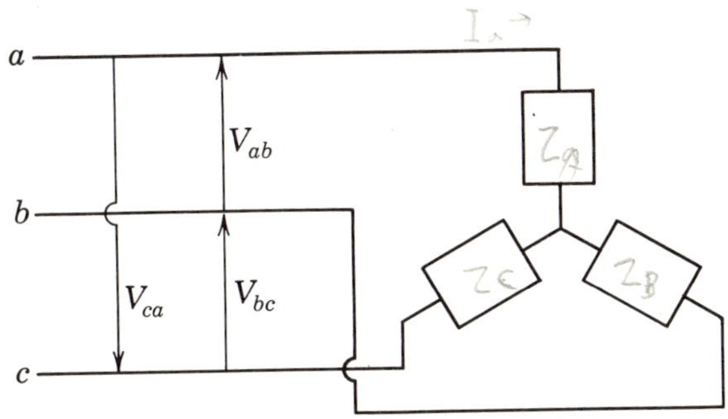

Prob. 23–1.

23–2. The voltages in Prob. 23–1 are applied to three impedances connected Y as shown in the figure for Prob. 23–1. If

$$\mathbf{Z}_A = 3 + j4 \text{ ohms}$$

$$\mathbf{Z}_B = 8 + j6 \text{ ohms}$$

and

$$\mathbf{Z}_C = 5 + j0 \text{ ohms}$$

find the currents (magnitude and angle) in lines a, b, and c.

23–3. In Prob. 23–2, what are the voltages $\mathbf{V}_A$, $\mathbf{V}_B$, and $\mathbf{V}_C$ across the three impedances? Is the sum of $\mathbf{V}_A$, $\mathbf{V}_B$, and $\mathbf{V}_C$ equal to zero?

23–4. Assume that the line voltages of Prob. 23–1 are to be applied across three impedances connected Δ as shown in the diagram. (a) Find the current in each impedance and the currents in lines a, b, and c, knowing that:

$$\mathbf{Z}_A = 30 + j40 \text{ ohms}$$

$$\mathbf{Z}_B = 40 + j30 \text{ ohms}$$

and

$$\mathbf{Z}_C = 50 + j0 \text{ ohms}$$

(b) Is the sum of the currents through $\boldsymbol{Z}_A$, $\boldsymbol{Z}_B$, and $\boldsymbol{Z}_C$ zero? (c) Is the sum of the currents in lines a, b, and c zero?

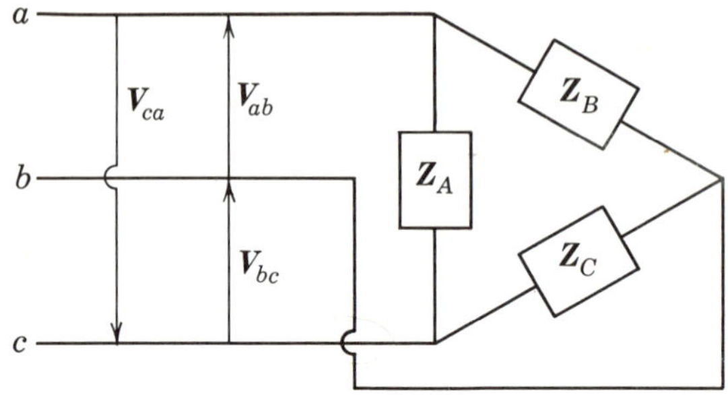

Prob. 23–4.

23–5. An a-c ammeter was used to measure the line currents in a three-wire three-phase system as illustrated. The ammeter gave the following readings:

$$I_{AB} = 30 \text{ amp}, \qquad I_{BC} = 25 \text{ amp}, \qquad I_{CA} = 35 \text{ amp}$$

If the sequence of the currents is AB, BC, and CA, write these currents in phasor forms with $\boldsymbol{I}_{AB}$ as reference.

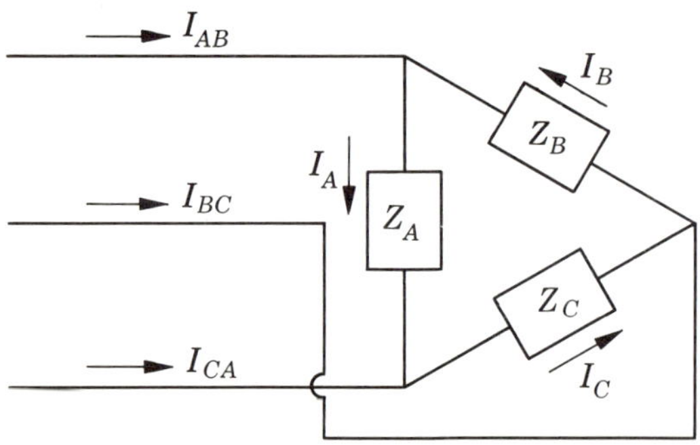

Prob. 23–5.

23–6. The impedances $\boldsymbol{Z}_A$, $\boldsymbol{Z}_B$, and $\boldsymbol{Z}_C$ shown in the figure for Prob. 23–5 have the following values:

$$\boldsymbol{Z}_A = 5 + j0 \text{ ohms}$$

$$\boldsymbol{Z}_B = 3 + j4 \text{ ohms}$$

and

$$\boldsymbol{Z}_C = 4 + j3 \text{ ohms}$$

(a) Find the currents $\boldsymbol{I}_A$, $\boldsymbol{I}_B$, and $\boldsymbol{I}_C$. (b) Is the sum of the currents $\boldsymbol{I}_A$, $\boldsymbol{I}_B$, and $\boldsymbol{I}_C$ equal to zero? (c) Is the sum of the voltages $\boldsymbol{V}_A$, $\boldsymbol{V}_B$, and $\boldsymbol{V}_C$ equal to zero?

23–7. A three-phase balanced system of line voltages of 120 v is applied across three equal impedances of $10\epsilon^{-j20}$ each connected Y. Find the voltage per phase and the currents in the three phases. Draw a phasor diagram of line voltages, phase voltages, and phase currents.

23–8. A three-phase balanced line current of 50 amp each feeds three equal loads, connected Δ. If the impedance of each load is $4\epsilon^{+j20}$ find the current in each impedance and the voltage between the lines. Draw a three-phase phasor diagram for the line currents, the phase currents, and the line voltages.

23–9. A three-phase balanced line voltage of sequence a, b, and c is connected across two incandescent lamps and one inductance connected Y as shown in the diagram. The lamps have pure resistance impedances and the inductance has pure reactance. The impedances of the lamps and reactances are equal. Which lamp, 1 or 2, will show more brightness?

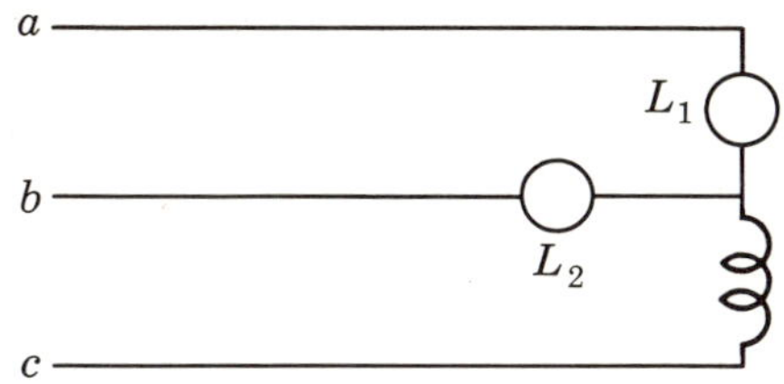

Prob. 23–9.

23–10. The method of Prob. 23–9 is used to find the phase sequence of a three-phase system. How is this done?

23–11. In Prob. 23–9 a capacitor is substituted for the inductor. How can you find the sequence by this method?

23–12. It is claimed that the total power in a three-phase balanced system is given by

$$P = \sqrt{3}VI \cos \theta$$

where P is the total power in watts
V is the line voltage (effective) in volts
I is the line current (effective) in amperes
θ is the angle between *phase* voltage and *phase* current

Prove or refute the foregoing statement.

23–13. Someone suggests that the power in a three-phase three-wire system can be given by means of two wattmeters connected as shown. Can you show that the sum of the two wattmeter readings gives the total power in a three-phase three-wire system?

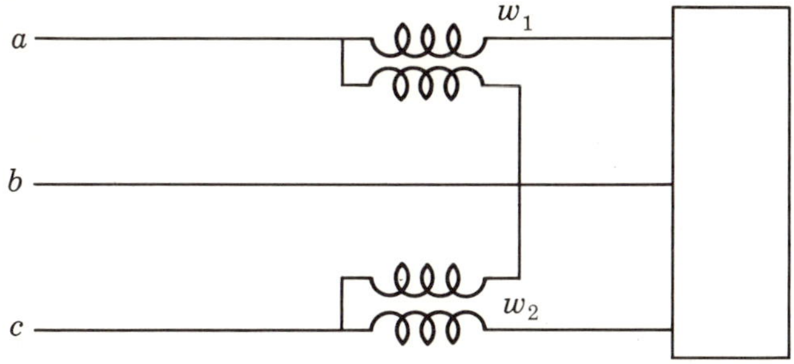

Prob. 23–13.

24

MAGNETICALLY COUPLED CIRCUITS

The discussions in the previous chapters concerning methods of writing equilibrium equations did not mention methods of handling circuits containing magnetic coupling. There are many practical circuits, however, which do involve magnetic coupling so a knowledge of how to deal with this type of circuit is essential. This chapter will discuss methods for writing equilibrium equations for magnetically coupled circuits and introduce the term *mutual inductance*.

24–1. Self- and Mutual Inductance. Assume two coils 1 and 2 as shown in Fig. 24–1. When current flows in coil 1, a magnetic field is produced. This magnetic field links with the turns of coil 1. Let the total flux linkages of coil 1 set by current i_1 in coil 1 be:

$$\lambda_{11} = N_1\phi_{11} \tag{24–1}$$

where λ_{11} is the total flux linkages in coil 1 set by current i_1 in weber-turns
N_1 is the turns of coil 1 linked by ϕ_{11}, the flux set by i_1
ϕ_{11} is the flux linking coil 1

The voltage drop in coil 1 caused by a change of the flux linkage is

$$e_1 = \frac{d\lambda_{11}}{dt} = N_1 \frac{d\phi_{11}}{dt} \tag{24–2}$$

But $d\phi_{11}/dt$ may be written as

$$\frac{d\phi_{11}}{dt} = \frac{d\phi_{11}}{di_1}\frac{di_1}{dt} \tag{24–3}$$

Therefore

$$e_1 = N_1 \frac{d\phi_{11}}{di_1}\frac{di_1}{dt} \quad \text{volts} \tag{24–4}$$

$$= L_{11}\frac{di_1}{dt}$$

where

$$L_{11} = N_1 \frac{d\phi_{11}}{di_1} \tag{24–5}$$

L_{11} is called the *self-inductance*.

ϕ_{21}, a part of flux ϕ_{11}, links N_2 turns of coil 2. Then the flux linkage in coil 2 caused by a current in coil 1 is $N_2\phi_{21}$. The voltage induced in coil 2 due to a change in the flux ϕ_{21} is:

$$e_2 = N_2 \frac{d\phi_{21}}{dt} \tag{24–6}$$

But

$$N_2 \frac{d\phi_{21}}{dt} = N_2 \frac{d\phi_{21}}{di_1}\frac{di_1}{dt} \tag{24–7}$$

Therefore

$$e_2 = N_2 \frac{d\phi_{21}}{di_1}\frac{di_1}{dt}$$

$$= M_{21}\frac{di_1}{dt} \tag{24–8}$$

where

$$M_{21} = N_2 \frac{d\phi_{21}}{dt} \tag{24–9}$$

From Eq. 24–5 we obtain the definition of self-inductance for a linear system as the flux linkages in a coil caused by the current in the same coil per ampere.

Equation 24–9 gives the definition of a mutual inductance between two coils as the flux linkages in coil 2 caused by a current in coil 1. Similarly,

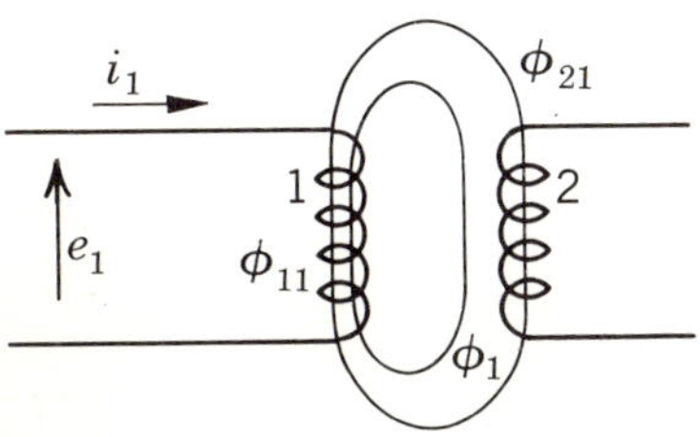

Fig. 24–1. Self- and mutual flux linkages.

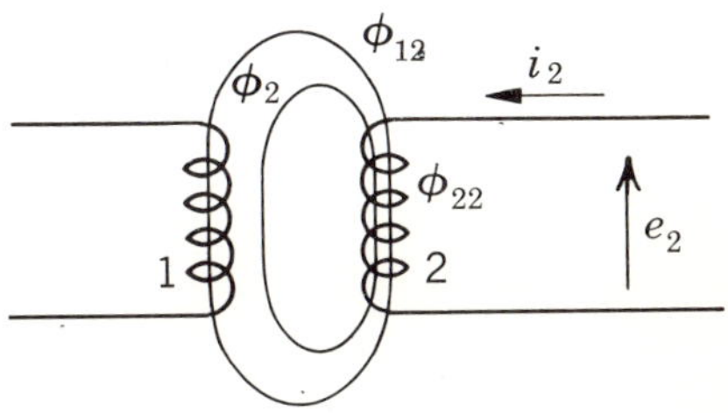

Fig. 24–2. Self- and mutual flux linkages.

if coil 2 was excited and coil 1 was not, the current in coil 2 sets up a field which links with the turns of coil 2. The flux linkage of coil 2 is:

$$\lambda_{22} = N_2\phi_{22} \tag{24–10}$$

The self-inductance of coil 2 is:

$$L_{22} = N_2 \frac{\phi_{22}}{i_2} = N_2 \frac{d\phi_{22}}{di_2} \tag{24–11}$$

The flux ϕ_{12} links coil 1, as shown in Fig. 24–2, and produces a flux linkage in coil 1 of value $N_1\phi_{12}$. This flux linkage when changing induces a voltage in coil 1 equal to

$$e_1 = N_1 \frac{d\phi_{12}}{dt}$$

$$= N_1 \frac{d\phi_{12}}{di_2} \frac{di_2}{dt}$$

$$= M_{12} \frac{di_2}{dt} \tag{24–12}$$

where

$$M_{12} = N_1 \frac{d\phi_{12}}{di_2} \tag{24–13}$$

Since the reciprocity theorem can be shown to hold for any linear, passive system, the two mutual inductances defined by Eqs. 24–9 and 24–13 must be equal. Therefore:

$$M_{12} = M_{21} \tag{24–14}$$

Another definition of mutual inductance between two coils is obtained from Eq. 24–8 or 24–12. From these equations mutual inductance is defined as the voltage induced in one coil per ampere per second of the rate of change of current in another coil.

The concept of mutual inductance may be extended to cases in which there are more than two coils involved. For example, consider the situation illustrated in Fig. 24–3 in which there are three coils so positioned that flux from any one of them links both of the other two. If a current is assumed to flow in coil 1, the open-circuit voltage produced in the other two coils may be written as:

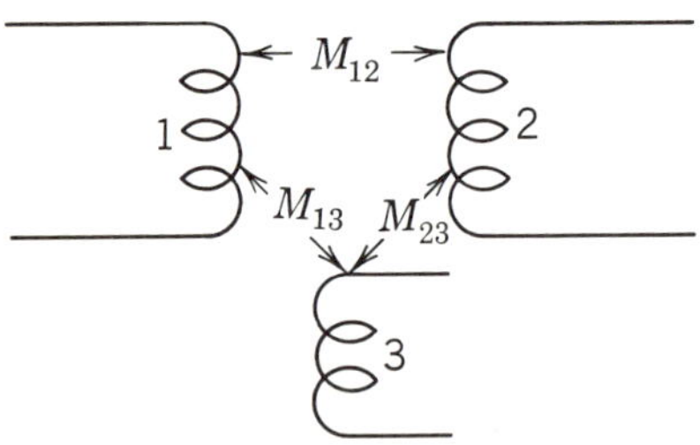

Fig. 24–3. Three coils with mutual inductances.

$$e_2 = M_{21} \frac{di_1}{dt} \tag{24–15}$$

$$e_3 = M_{31} \frac{di_1}{dt} \tag{24–16}$$

Likewise, for a current in coil 2 the voltages in the other coils are

$$e_1 = M_{12} \frac{di_2}{dt} \tag{24–17}$$

$$e_3 = M_{32} \frac{di_2}{dt} \tag{24–18}$$

and for a current in coil 3

$$e_1 = M_{13}\frac{di_3}{dt} \tag{24–19}$$

$$e_2 = M_{23}\frac{di_3}{dt} \tag{24–20}$$

Here too

$$M_{12} = M_{21}, \qquad M_{13} = M_{31}, \qquad M_{23} = M_{32} \tag{24–21}$$

The significant item to be noted in these results is that the mutual inductance between any pair of coils is the same whether other coils are present or not.

24–2. Coupling Coefficients. An additional concept which is often convenient when dealing with coupled circuits is that of the *coupling coefficient.*

From Eqs. 24–9 and 24–5, we get

$$\frac{M_{21}}{L_{11}} = \frac{N_2\,d\phi_{21}}{N_1\,d\phi_{11}} \tag{24–22}$$

From Eqs. 24–11 and 24–13 we obtain

$$\frac{M_{12}}{L_{22}} = \frac{N_1\,d\phi_{12}}{N_2\,d\phi_{22}} \tag{24–23}$$

But $M_{21} = M_{12} = M$. Therefore:

$$\frac{M^2}{L_{11}L_{22}} = \frac{d\phi_{21}}{d\phi_{11}}\frac{d\phi_{12}}{d\phi_{22}} \tag{24–24}$$

But from Figs. 24–1 and 24–2

$$\phi_{21} < \phi_{11} \qquad \text{and} \qquad \phi_{12} < \phi_{22} \tag{24–25}$$

Therefore:

$$\frac{M}{\sqrt{L_{11}L_{22}}} = K < 1 \tag{24–26}$$

Although in Figs. 24–1 and 24–2 $\phi_{12} < \phi_{11}$ and $\phi_{21} < \phi_{22}$, it is possible to have all the flux ϕ_{11} link coil 2 and all the flux ϕ_{22} link coil 1. Under this condition $\phi_{12} = \phi_{11}$ and $\phi_{21} = \phi_{22}$ and therefore Eq. 24–26 becomes

$$\frac{M}{\sqrt{L_{11}L_{22}}} = K = 1 \tag{24–27}$$

K is called the coupling coefficient of the coils. Its value lies between 1 and 0.

When $K = 1$ there is perfect coupling in the system and it represents a situation in which all the flux links all the windings of both coils. This condition is actually never attained but it may be closely approached in carefully designed transformers with iron cores. In any event, circuits in which a high value of K is desired are said to be *closely coupled*. Such circuits are generally desired when the efficient transfer of energy is of primary concern.

There are many other cases, however, in which a low value of coupling is desired. This is generally the case when the coupled circuit is to be used as a frequency selective device. Circuits having a low value of K are said to be *loosely coupled*.

24–3. Polarity. The question of polarity of the induced voltage and the algebraic sign of the mutual inductance remain to be settled. The schematic representation used in Figs. 24–1, 24–2, and 24–3 gives no indication as to the relative direction of winding in the coils, and thus makes it impossible to determine the polarity of the induced voltage for a given current direction. One method of resolving this ambiguity is to designate corresponding ends of any pair of coils by means of plus signs, dots, or other suitable distinguishing marks. The corresponding ends are defined as those which will always go positive together or negative together regardless of which winding is producing the flux. When more than two coils are magnetically coupled the use of markings becomes much more involved and, in the case of a large number of coils, may become prohibitively complicated.

Since each *pair* of coils must be assigned a separate mutual inductance, a reaonable solution to the problem of specifying relative polarities can be achieved by letting the algebraic sign of this mutual inductance carry such information. In order to do this, however, it is necessary to assign a positive current direction in *each* coil regardless of which coil is producing flux, and to adopt a convention with regard to the sign on the mutual inductance. This convention is arbitrary, so long as one is consistent, but the one most often used is as follows: If currents *in* the *assumed positive direction* of a given coil pair produce fluxes that *add*, the mutual inductance between that pair is assumed to be *positive*; if these currents produce fluxes that *subtract*, the mutual inductance is considered *negative*. This is the convention that will be adhered to throughout the remainder of this chapter. A current is assumed in a *positive direction* if it *enters* the coil at the *dotted end*.

24–4. Equilibrium Equations with Mutual Inductance. The discussion so far has expressed the volt-ampere relationship for mutual inductance as an instantaneous time function. It is known, however, that in the case of *self-inductance* with *sinusoidal excitation*, the introduction of complex frequencies and the impedance concept resulted in a much more compact notation; hence, it should not be surprising if a similar result is achieved in the case of mutual inductance. The justification for this is identical with that used for self-inductance and will not be repeated here. Thus the sinusoidal volt-ampere relationships for Fig. 24–4 may be expressd as:

$$\boldsymbol{E}_1 = s L_{11} \boldsymbol{I}_1 \tag{24–28}$$

$$\boldsymbol{E}_2 = s M_{21} \boldsymbol{I}_1 \tag{24–29}$$

in which I_1, E_1, and E_2 are complex numbers specifying the magnitude and phase of the corresponding sinusoidal function and $s = jw$ is the complex frequency of this function.

One additional point of notation needs clarification before we proceed to the writing of equilibrium equations. This has to do with designating current directions and an artificial sign for the mutual inductance.

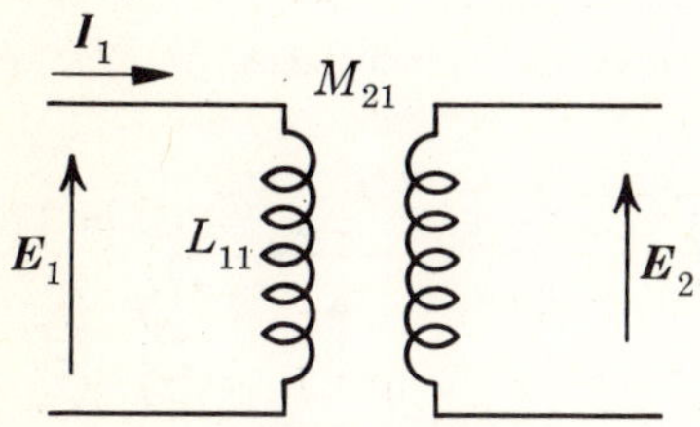

Fig. 24–4. Sinusoidal voltages and equilibrium equations.

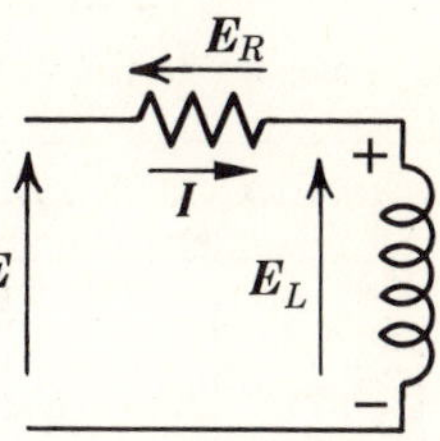

Fig. 24–5. Current in direction of terminal voltage.

Assume a circuit as shown in Fig. 24–5. If the terminal voltage is as shown and the current is in the direction of the terminal voltage, the equation of the circuit is:

$$E = iR + L_{11}\frac{di}{dt} \tag{24–30}$$

If the current is against the terminal voltage as shown in Fig. 24–6, the equation of the circuit is:

$$E = -iR - L_{11}\frac{di}{dt} \tag{24–31}$$

Assume now two magnetically coupled circuits as shown in Fig. 24–7; a current I_1 entering coil 1 at the dotted end of coil 1 will produce a positive voltage $sM_{21}I_1$ at the dotted end of coil 2 if the mutual inductance M_{21} is positive as shown in the diagram.

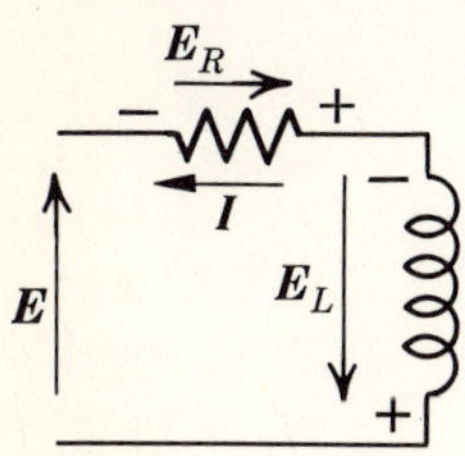

Fig. 24–6. Current in opposite direction to terminal voltage.

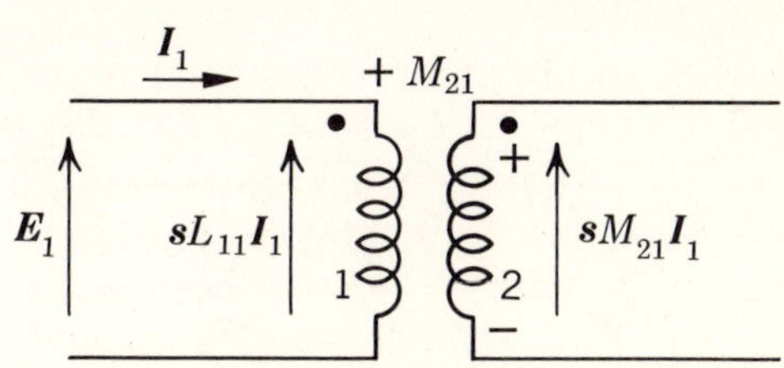

Fig. 24–7. The dotted ends and positive mutual inductance.

On the other hand, a current I_1 entering the dotted end of coil 1 induces a negative voltage $sM_{21}I_1$ at the dotted end of coil 1 if the mutual inductance is negative. This is shown in Fig. 24–8.

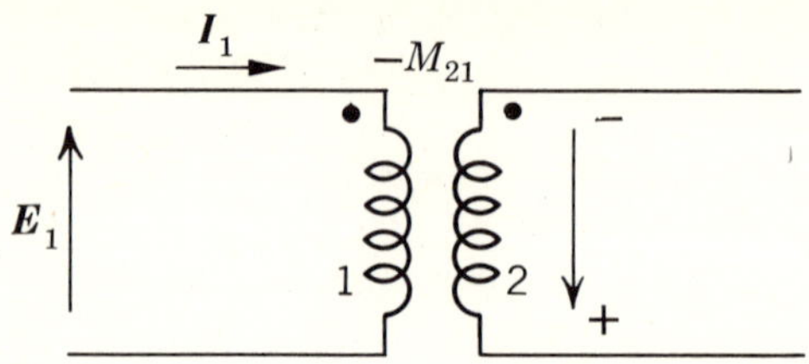

Fig. 24–8. The dotted ends and negative mutual inductance.

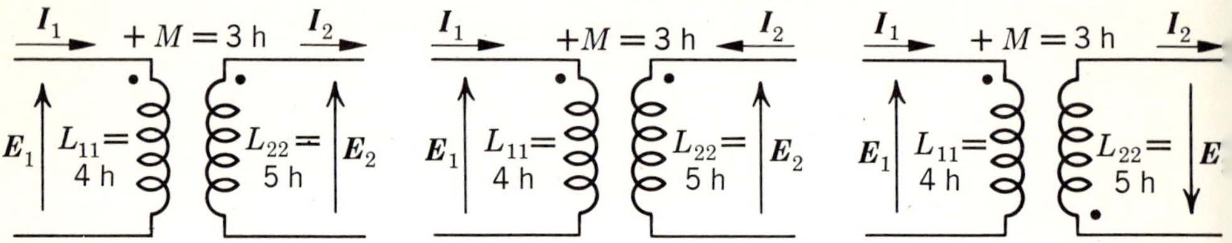

$$E_1 = 4sI_1 - 3sI_2 \qquad E_1 = 4sI_1 + 3sI_2 \qquad E_1 = 4sI_1 + 3sI_2$$

$$E_2 = 3sI_1 - 5sI_2 \qquad E_2 = 3sI_1 + 5sI_2 \qquad E_2 = 3sI_1 + 5sI_2$$

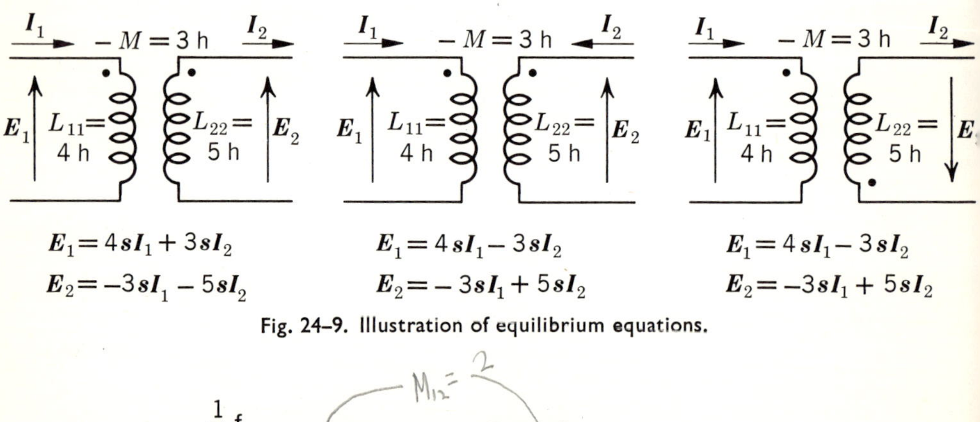

$$E_1 = 4sI_1 + 3sI_2 \qquad E_1 = 4sI_1 - 3sI_2 \qquad E_1 = 4sI_1 - 3sI_2$$

$$E_2 = -3sI_1 - 5sI_2 \qquad E_2 = -3sI_1 + 5sI_2 \qquad E_2 = -3sI_1 + 5sI_2$$

Fig. 24–9. Illustration of equilibrium equations.

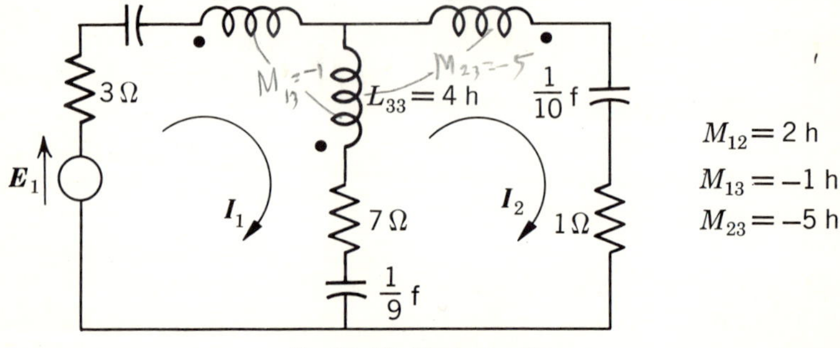

Fig. 24–10. Mesh circuits.

Similarly, a current I_2 entering the dotted end of coil 2 induces a voltage $sM_{12}I_2$ in coil 1. The polarity of this voltage is positive at the dotted end of coil 1 if the mutual inductance is positive and is negative at the dotted end of coil 1 if the mutual inductance is negative.

In order to illustrate the procedure for writing complete equilibrium equations when mutual inductance is present, consider Fig. 24–9. Under each combination the equilibrium equations are written.

24–5. Mesh Equation with Mutual Inductances. Assume a circuit as shown in Fig. 24–10 with the indicated self- and mutual inductances. This circuit shows three coils with self- and mutual inductances. In the equilibrium equations to follow, note that in a given mesh *each* of the coils contained in that mesh contributes one self-inductance term and *two* mutual inductance terms. Consider the voltage equilibrium around mesh 1:

$$E_1 = \left(3 + \frac{5}{s} + 2s\right)I_1 + \left(4s + 7 + \frac{9}{s}\right)(I_1 - I_2) + 1s(I_1 - I_2) + 1sI_1 - 2sI_2 - 5sI_2 \quad (24\text{–}32)$$

The voltage equilibrium around mesh 2 is:

$$0 = \left(8s + \frac{10}{s} + 1\right)I_2 + \left(4s + 7 + \frac{9}{s}\right)(I_1 - I_1) - 1sI_1 + 5sI_2 - 2sI_1 + 5s(I_2 - I_1) \quad (24\text{–}33)$$

These two equations may be written as:

$$E_1 = \left(10 + 8s + \frac{14}{s}\right)I_1 - \left(7 + 12s + \frac{9}{s}\right)I_2 \quad (24\text{–}34)$$

$$0 = -\left(7 + 12s + \frac{9}{s}\right)I_1 + \left(8 + 22s + \frac{19}{s}\right)I_2 \quad (24\text{–}35)$$

The loop inductance matrix for this circuit may now be written and is:

$$[L] = \begin{bmatrix} 8 & -12 \\ -12 & 22 \end{bmatrix} \quad (24\text{–}36)$$

The other parameter matrices are no different in this circuit than they would be if there were no mutual inductance.

24–6. Node Equilibrium Equations. Recall that in a circuit containing no mutual inductances for node equations it was convenient to utilize branch *admittances* and *consequently reciprocal* inductances. Since *mutual inductance* is now being considered, it will again be found convenient to introduce coefficients which have the nature of reciprocal inductances. Unfortunately, these coefficients are no longer just the reciprocals of the circuit inductances but involve a more complicated relation among these. The following generalized example will serve to illustrate the nature of this relationship and a method of calculating the proper reciprocal inductance coefficients.

Consider the three inductances shown in Fig. 24–11 with their corresponding mutual inductances. Therefore:

$$e_1 = L_{11}\frac{di_1}{dt} + M_{12}\frac{di_2}{dt} + M_{13}\frac{di_3}{dt} \tag{24–37}$$

$$e_2 = M_{12}\frac{di_1}{dt} + L_{22}\frac{di_2}{dt} + M_{23}\frac{di_3}{dt} \tag{24–38}$$

$$e_3 = M_{13}\frac{di_1}{dt} + M_{32}\frac{di_2}{dt} + L_{33}\frac{di_3}{dt} \tag{24–39}$$

In order to use these relations in setting up current equilibrium equations, it is necessary to be able to express the currents explicitly in terms of the

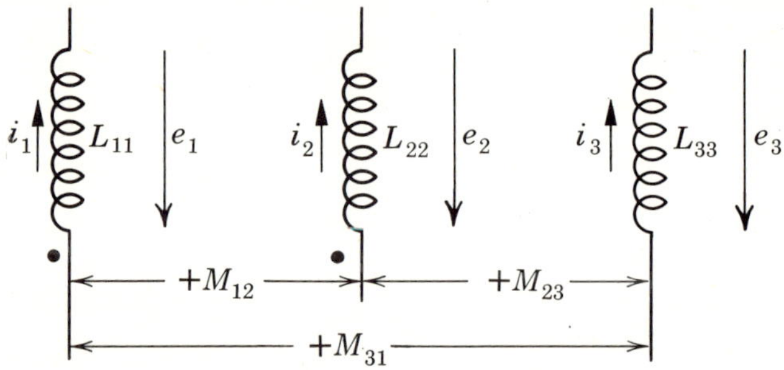

Fig. 24–11. Currents in coupled coils.

voltages. This may be done by solving Eqs. 24–37, 24–38, and 24–39 simultaneously to yield the currents. The first step in carrying out such a solution is to integrate each equation with respect to time. This results in:

$$\int e_1\,dt = L_{11}i_1 + M_{12}i_2 + M_{13}i_3 = \lambda_1 \tag{24–40}$$

$$\int e_2\,dt = M_{21}i_1 + L_{22}i_2 + M_{23}i_3 = \lambda_2 \tag{24–41}$$

$$\int e_3\,dt = M_{31}i_1 + M_{32}i_2 + L_{33}i_3 = \lambda_3 \tag{24–42}$$

Equations 24–40, 24–41, and 24–42 are the flux linkages in coils 1, 2, and 3 respectively. Solving these equations for i_1, i_2, and i_3 yields:

$$i_1 = \frac{\lambda_1 D_{11} + \lambda_2 D_{21} + \lambda_3 D_{31}}{D} \tag{24–43}$$

where

$$D = \begin{vmatrix} L_{11} & M_{12} & M_{13} \\ M_{12} & L_{22} & M_{23} \\ M_{13} & M_{23} & L_{33} \end{vmatrix} \tag{24–44}$$

$$D_{11} = \begin{vmatrix} L_{22} & M_{23} \\ M_{23} & L_{22} \end{vmatrix} \qquad D_{21} = -\begin{vmatrix} M_{12} & M_{13} \\ M_{23} & L_{33} \end{vmatrix}$$
$$D_{31} = \begin{vmatrix} M_{12} & M_{13} \\ L_{22} & M_{23} \end{vmatrix} \tag{24–45}$$

Similar expressions can be written for i_2 and i_3.

Let:

$$\Gamma_{sk} = \frac{D_{ks}}{D} \tag{24–46}$$

Then:

$$\begin{aligned} i_1 &= \Gamma_{11}\lambda_1 + \Gamma_{12}\lambda_2 + \Gamma_{13}\lambda_3 \\ i_2 &= \Gamma_{21}\lambda_1 + \Gamma_{22}\lambda_2 + \Gamma_{23}\lambda_3 \\ i_3 &= \Gamma_{31}\lambda_1 + \Gamma_{32}\lambda_2 + \Gamma_{33}\lambda_3 \end{aligned} \tag{24–47}$$

Note the interchange of subscript sequence used in Eq. 24–46 for purposes of preserving notational symmetry. This will be of no consequence in all situations of interest because the original symmetry in D will always result in:

$$\Gamma_{sk} = \Gamma_{ks} \tag{24–48}$$

At this point it should be noted that Γ_{sk} is the reciprocal inductances that are required for writing general current equilibrium equations. Furthermore, it is *not* necessary to solve equations of the form of Eqs. 24–42, 24–41, and 24–40 in order to obtain them. Instead, it may be observed that the Γ_{sk}'s are uniquely obtained from the inductance matrix 24–44.

Equations 24–47 are written in terms of instantaneous values of current and flux linkages. If the circuit is to be operated in the sinusoidal steady state, these equations may be rewritten in complex notation by noting first that

$$\lambda = \int e\,dt \tag{24–49}$$

and if

$$e = \boldsymbol{E}\epsilon^{st} \tag{24–50}$$

then

$$\lambda = \int \boldsymbol{E}\epsilon^{st}\,dt = \frac{\boldsymbol{E}}{s}\,\epsilon^{st} \tag{24–51}$$

If λ is expressed in similar form as

$$\lambda = \Lambda\epsilon^{st} \tag{24–52}$$

then it becomes apparent that

$$\Lambda = \frac{\boldsymbol{E}}{s}$$

Therefore, Eqs. 24–47 may be written for the sinusoidal state as:

$$\begin{aligned} \boldsymbol{I}_1 &= \frac{\Gamma_{11}}{s}\boldsymbol{E}_1 + \frac{\Gamma_{12}}{s}\boldsymbol{E}_2 + \frac{\Gamma_{13}}{s}\boldsymbol{E}_3 \\ \boldsymbol{I}_2 &= \frac{\Gamma_{21}}{s}\boldsymbol{E}_1 + \frac{\Gamma_{22}}{s}\boldsymbol{E}_2 + \frac{\Gamma_{23}}{s}\boldsymbol{E}_3 \\ \boldsymbol{I}_3 &= \frac{\Gamma_{31}}{s}\boldsymbol{E}_1 + \frac{\Gamma_{32}}{s}\boldsymbol{E}_2 + \frac{\Gamma_{33}}{s}\boldsymbol{E}_3 \end{aligned} \tag{24–53}$$

24–7. Illustration. To illustrate the result of Art. 24–6, consider the two-node-pair circuit shown in Fig. 24–12 along with the specified mutual inductances and dotted references.

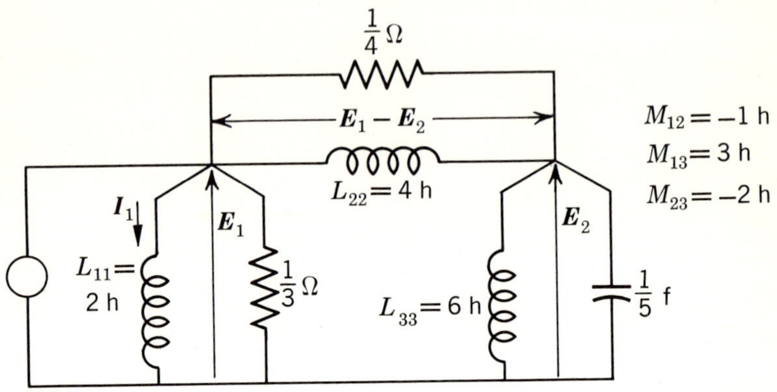

Fig. 24–12. Two-node-pair circuit.

First, the inductance matrix may be written from the specified self- and mutual inductance as:

$$[L] = \begin{bmatrix} 2 & -1 & 3 \\ -1 & 4 & -2 \\ 3 & -2 & 6 \end{bmatrix} \tag{24–54}$$

The determinant and its cofactors may now be written:

$$D = \begin{vmatrix} 2 & -1 & 3 \\ -1 & 4 & -2 \\ 3 & -2 & 6 \end{vmatrix} = 10$$

$$D_{11} = \quad (1)\begin{vmatrix} 4 & -2 \\ -2 & 6 \end{vmatrix} = 20 \qquad D_{12} = (-1)\begin{vmatrix} -1 & 3 \\ -2 & 6 \end{vmatrix} = 0$$

$$D_{13} = \quad (1)\begin{vmatrix} -1 & 3 \\ 4 & -2 \end{vmatrix} = -10 \qquad D_{22} = \quad (1)\begin{vmatrix} 2 & 3 \\ 3 & 6 \end{vmatrix} = 3$$

$$D_{23} = (-1)\begin{vmatrix} 2 & 3 \\ -1 & -2 \end{vmatrix} = 1 \qquad D_{33} = \quad (1)\begin{vmatrix} 2 & -1 \\ -1 & 4 \end{vmatrix} = 7$$

From these the reciprocal inductance matrix may be formed by recalling that:

$$\Gamma_{sk} = \frac{D_{ks}}{D}$$

Since $D = 10$ in this case, the desired matrix is:

$$|\Gamma| = \begin{bmatrix} 2 & 0 & -1 \\ 0 & 0.3 & 0.1 \\ -1 & 0.1 & 0.7 \end{bmatrix} \tag{24-55}$$

The current equations in the inductances for the sinusoidal state are:

$$\begin{aligned} \boldsymbol{I}_1 &= \frac{2}{s}\boldsymbol{E}_1 + \frac{0}{s}(\boldsymbol{E}_1 - \boldsymbol{E}_2) - \frac{1}{s}\boldsymbol{E}_2 \\ \boldsymbol{I}_2 &= \frac{0}{s}\boldsymbol{E}_1 + \frac{0.3}{s}(\boldsymbol{E}_1 - \boldsymbol{E}_2) + \frac{0.1}{s}\boldsymbol{E}_2 \\ \boldsymbol{I}_3 &= -\frac{1}{s}\boldsymbol{E}_1 + \frac{0.1}{s}(\boldsymbol{E}_1 - \boldsymbol{E}_2) + \frac{0.7}{s}\boldsymbol{E}_2 \end{aligned} \tag{24-56}$$

The total current equation is:

$$\boldsymbol{I}_s = \left(3 + \frac{2}{s}\right)\boldsymbol{E}_1 - \frac{1}{s}\boldsymbol{E}_2 + \left(4 + \frac{0.3}{s}\right)(\boldsymbol{E}_1 - \boldsymbol{E}_2) + \frac{0.1}{s}\boldsymbol{E}_2 \tag{24-57}$$

$$\begin{aligned} 4(\boldsymbol{E}_1 - \boldsymbol{E}_2) + \frac{0.3}{s}(\boldsymbol{E}_1 - \boldsymbol{E}_2) + \frac{0.1}{s}\boldsymbol{E}_2 = 5s\boldsymbol{E}_2 - \frac{1}{s}\boldsymbol{E}_1 \\ + \frac{0.1}{s}(\boldsymbol{E}_1 - \boldsymbol{E}_2) + \frac{0.7}{s}\boldsymbol{E}_2 \end{aligned} \tag{24-58}$$

Simplify by collecting coefficients and obtain:

$$\boldsymbol{I}_s = \left(7 + \frac{2.3}{s}\right)\boldsymbol{E}_1 - \left(4 + \frac{1.2}{s}\right)\boldsymbol{E}_2 \tag{24-59}$$

$$0 = -\left(4 + \frac{1.2}{s}\right)\boldsymbol{E}_1 + \left(4 + 5s + \frac{0.8}{s}\right)\boldsymbol{E}_2 \tag{24-60}$$

The reciprocal matrix for the circuit may now be written as:

$$|\Gamma| = \begin{bmatrix} 2.3 & -1.2 \\ -1.2 & 0.8 \end{bmatrix} \tag{24-61}$$

where again the symmetry serves as a partial check on the accuracy of the numerical calculations.

When we compare the mesh and node methods of writing equilibrium equations, we note that the node method generally requires the additional step of determining the reciprocal inductance matrix. However, the procedure for doing this is perfectly straightforward, although it might become tedious if the number of inductances were large.

24–8. Simple Circuits with One Mutual Inductance. Consider first the situation in which two inductances with magnetic coupling are connected in series so that exactly the same current flows in each. Since they may be connected so that the fluxes produced by this current may be

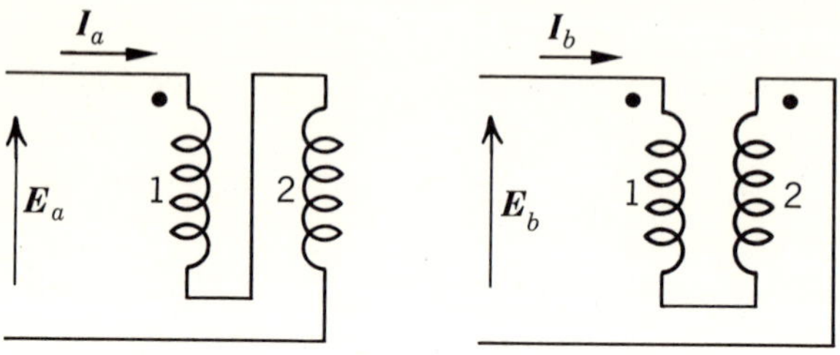

(*a*) Series aiding (*b*) Series opposing

Fig. 24–13. Series-connected coils with mutual inductance.

either aiding or opposing, there are two situations to consider as shown in Fig. 24–13. In the series-aiding case (Fig. 24–13*a*) the voltage equilibrium equation is simply:

$$\boldsymbol{E}_a = sL_{11}\boldsymbol{I}_a + sL_{22}\boldsymbol{I}_a + sM\boldsymbol{I}_a + sM\boldsymbol{I}_a \tag{24–62}$$

This equation may be written:

$$\boldsymbol{E}_a = s(L_{11} + L_{22} + 2M)\boldsymbol{I}_a \tag{24–63}$$

The impedance offered to the current is:

$$\boldsymbol{Z}_a = \frac{\boldsymbol{E}_a}{\boldsymbol{I}_a} = s(L_{11} + L_{22} + 2M) \tag{24–64}$$

The equivalent inductance is therefore:

$$L_{aa} = L_{11} + L_{22} + 2M \tag{24–65}$$

In the series-opposing case (Fig. 24–13*b*) the voltage equilibrium equation is:

$$\boldsymbol{E}_b = sL_{11}\boldsymbol{I}_b + sL_{22}\boldsymbol{I}_b - sM\boldsymbol{I}_b - sM\boldsymbol{I}_b \tag{24–66}$$

The impedance offered to the current is:

$$\boldsymbol{Z}_b = \frac{\boldsymbol{E}_b}{\boldsymbol{I}_b} = s(L_{11} + L_{22} - 2M) \tag{24–67}$$

The equivalent inductance is:

$$L_{bb} = L_{11} + L_{22} - 2M \tag{24–68}$$

Parallel connection of inductances having magnetic coupling is also of some interest. Again there are two situations to consider as shown in Fig. 24–14. The details of carrying out the analysis of the parallel-connected coils with mutual inductance are left to the student. He can carry out the analysis in a mesh or node equation. The end results, however, are:

$$L_{aa} = \frac{L_{11}L_{22} - M^2}{L_{11} + L_{22} - 2M} \qquad (24\text{–}69)$$

for the parallel-aiding case, and

$$L_{bb} = \frac{L_{11}L_{22} - M^2}{L_1 + L_2 + 2M} \qquad (24\text{–}70)$$

for the parallel-opposing case.

Some additional insight into the characteristics of series and parallel combinations of inductances having magnetic coupling may be obtained by

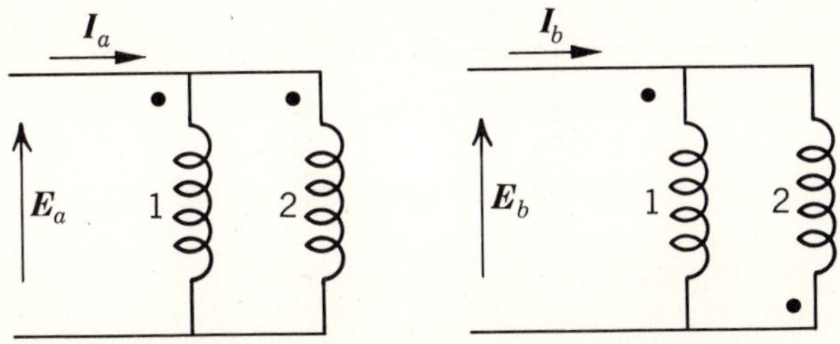

(*a*) Parallel aiding (*b*) Parallel opposing

Fig. 24–14. Parallel-connected coils with mutual inductance.

expressing the preceding results in terms of the coupling coefficient rather than the mutual inductance. This may be done by rewriting

$$M = K\sqrt{L_{11}L_{22}} \qquad (24\text{–}71)$$

and considering only positive values of M. Substitution of this expression into Eqs. 24–65, 24–68, 24–69, and 24–70 results in the forms shown in Table 24–1.

TABLE 24–1

Connection	General	Equivalent Inductance: Loose Coupling	Equivalent Inductance: Close Coupling
Series aiding . . .	$L_{11} + L_{22} + 2K\sqrt{L_{11}L_{22}}$	$L_{11} + L_{22}$	$(\sqrt{L_{11}} + \sqrt{L_{22}})^2$
Series opposing . .	$L_{11} + L_{22} - 2K\sqrt{L_{11}L_{22}}$	$L_{11} + L_{22}$	$(\sqrt{L_{11}} - \sqrt{L_{22}})^2$
Parallel aiding . . .	$\dfrac{(1 - K^2)L_{11}L_{22}}{L_{11} + L_{22} - 2K\sqrt{L_{11}L_{22}}}$	$\dfrac{L_{11}L_{22}}{L_{11} + L_{22}}$	0 if $L_{11} \neq L_2$ L_{11} if $L_{11} = L_{22}$
Parallel opposing . .	$\dfrac{(1 - K^2)L_{11}L_{22}}{L_{11} + L_{22} + 2K\sqrt{L_{11}L_{22}}}$	$\dfrac{L_{11}L_{22}}{L_{11} + L_{22}}$	0

PROBLEMS

24–1. Three coils having magnetic coupling are shown. In order to determine the self- and mutual inductance a sinusoidal voltage source having a maximum value of 100 v and an angular frequency of $\omega = 100$ is applied to each coil in *turn*. The resulting current in *that* coil and the open-circuit voltage of *one other* coil are recorded as shown. Compute all self- and mutual inductances.

Current:

$I_1 = 1$ amp	$E_2 = 10$ v
$I_2 = 0.5$ amp	$E_3 = 15$ v
$I_3 = 2$ amp	$E_1 = 20$ v

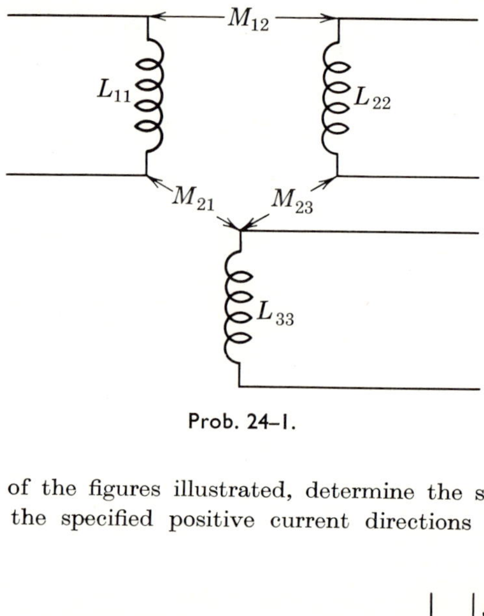

Prob. 24–1.

24–2. In each of the figures illustrated, determine the sign of all mutual inductances from the specified positive current directions and coil winding directions.

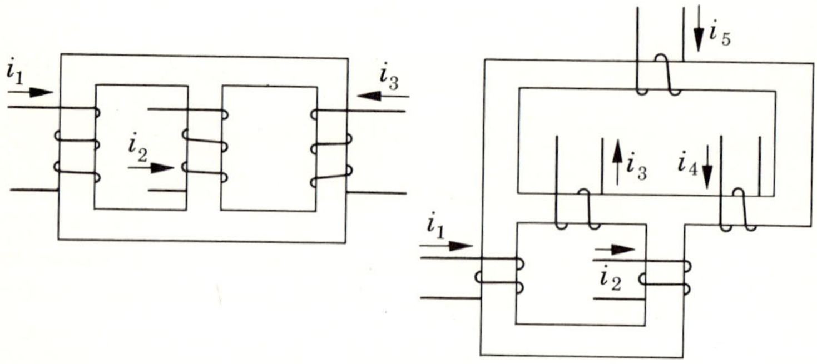

Prob. 24–2.

24–3. Fill in the unspecified items in Table 24–2.

TABLE 24–2

	L_{11}	L_{22}	M	K
a	10	5		1
b		6	8	0.5
c	5	4	3	
d	12		6	0.1

24–4. Write the voltage equilibrium equations for each of the circuits shown.

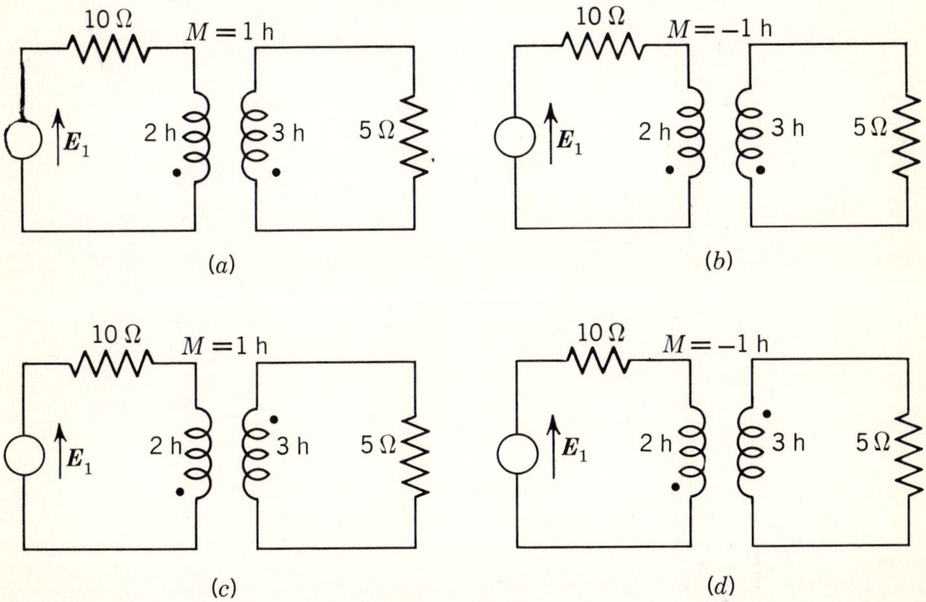

Prob. 24–4.

24–5. Write the voltage equilibrium equations for the circuit illustrated.

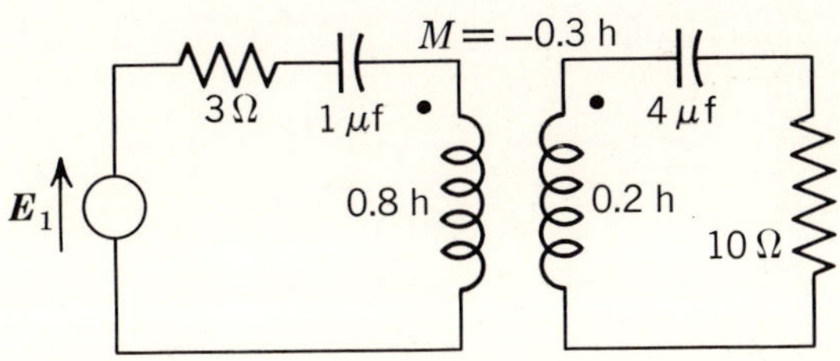

Prob. 24–5.

24–6. Write the voltage equilibrium equations for the circuit on p. 412.

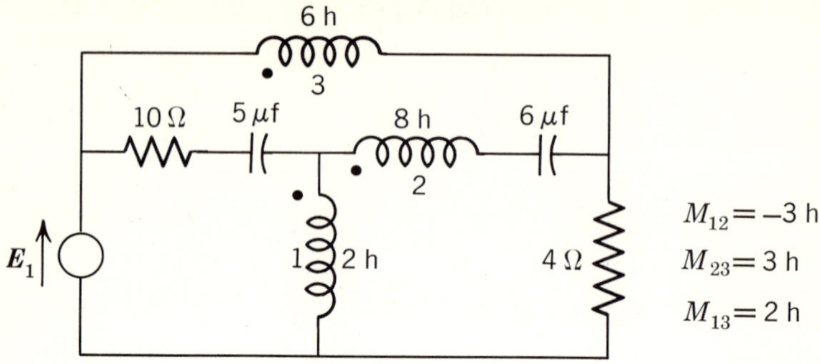

Prob. 24–6.

24–7. Write the current equilibrium equations for each of the circuits illustrated.

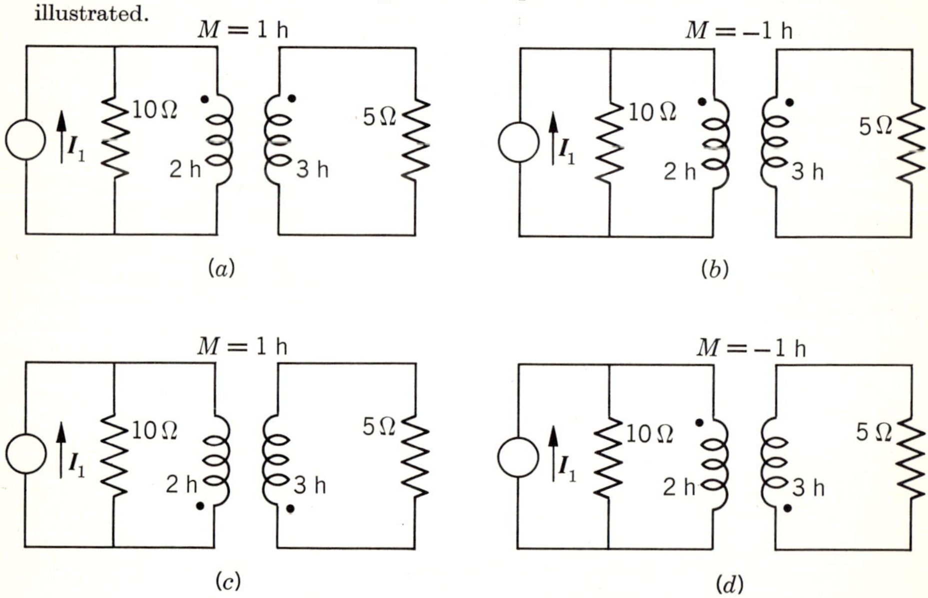

Prob. 24–7.

24–8. Write the current equilibrium equations for the adjacent circuit.

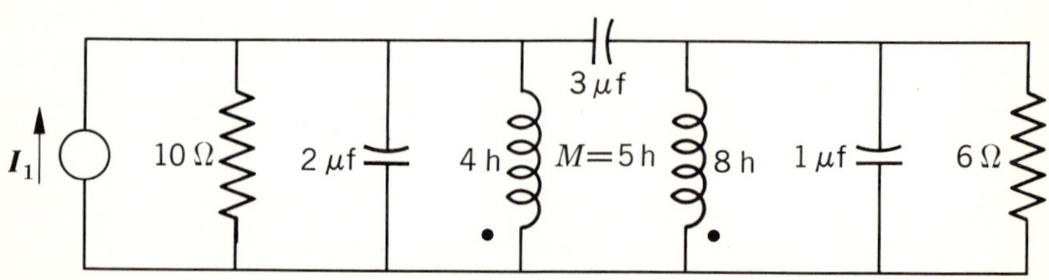

Prob. 24–8.

24–9. Write the current equilibrium equations for the circuit shown.

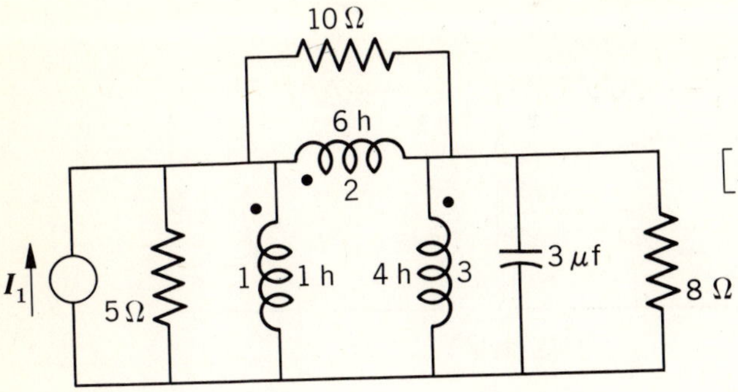

$$[L'] = \begin{bmatrix} 1 & -1 & 1 \\ -1 & 6 & -2 \\ 1 & -2 & 4 \end{bmatrix}$$

Prob. 24–9.

24–10. Derive the equivalent inductance of two coils in the figure illustrated.

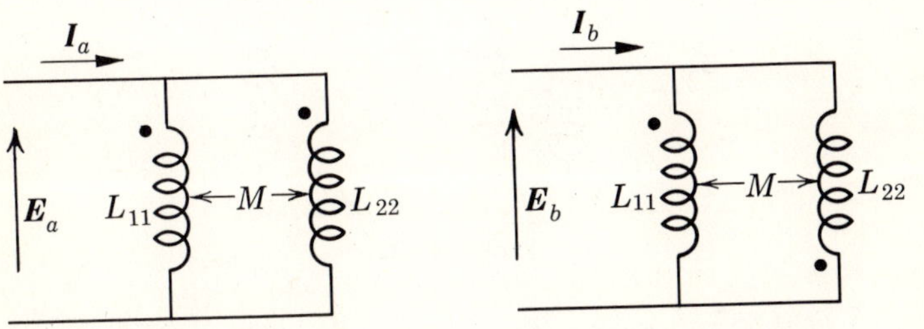

(*a*) Parallel aiding (*b*) Parallel opposing

Prob. 24–10.

24–11. Calculate the input impedance of the circuit at frequencies of $\omega = 0$; 100; 1000.

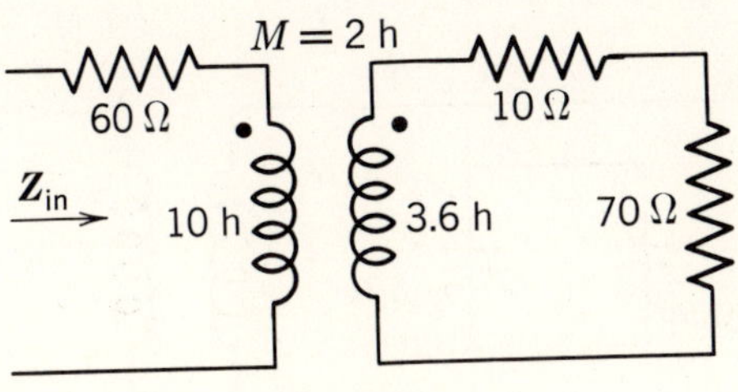

Prob. 24–11.

24–12. In the circuit illustrated show that: (a)

$$I_1 = \frac{E_s}{Z_s + R_1 + sL_1 - \dfrac{s^2M^2}{Z_L + R_2 + sL_2}}$$

(b) What is the impedance of the circuit if circuit 2 is open-circuited? (c) What conclusion do you draw from parts (a) and (b)?

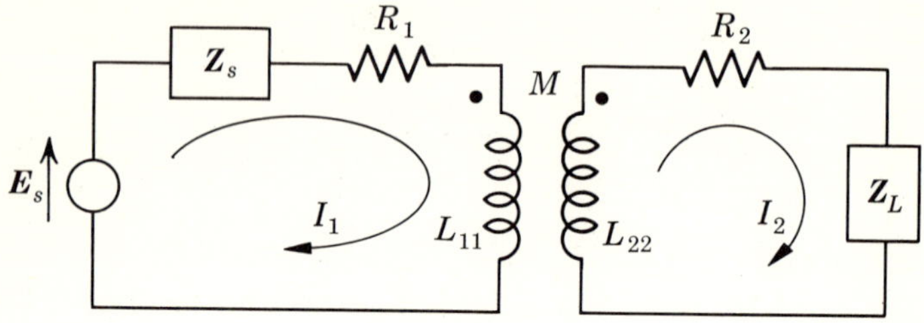

Prob. 24–12.

24–13. In the circuit illustrated is there any element value in circuit 2 that makes Z_{in} capacitive?

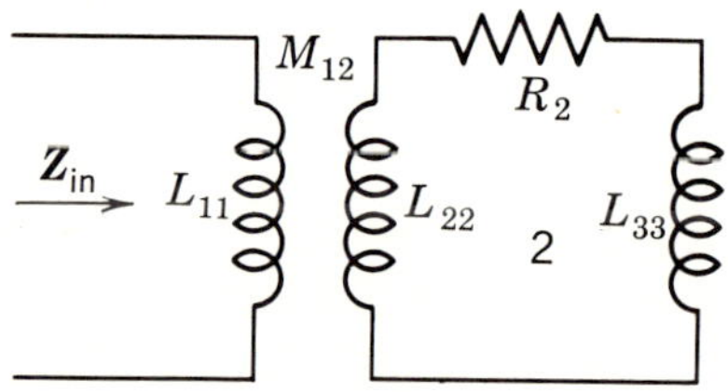

Prob. 24–13.

24–14. A method to measure the mutual inductance M between two coils is shown in the diagram. (a) With switch 1 open the capacitor is changed until Z_{in} is maximum and the circuit is in resonance. Assume for this condition that the capacitance has a value C_0. (b) When switch 1 is closed resonance is lost. Change the capacitance until resonance is established, namely Z_{in} is maximum again. Assume that the capacitance to give this condition is now C_s. Show that $K^2 = (C_s - C_0)/C_s$ where $K^2 = M^2/(L_{11}L_{22})$.

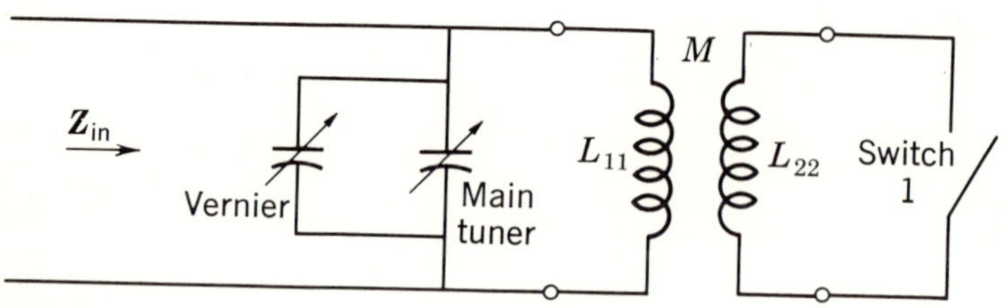

Prob. 24–14.

24–15. A system is as shown with $K = M/\sqrt{L_{11}L_{22}}$. Both meshes of the system are tuned to a frequency ω_0. (a) Find the value of ω_0. (b) Find the ratio of $|E_2/E_1|$ in terms of K. (c) For what value of K will $|E_2/E_1|$ be maximum? At what frequency? (d) What will this maximum value be?

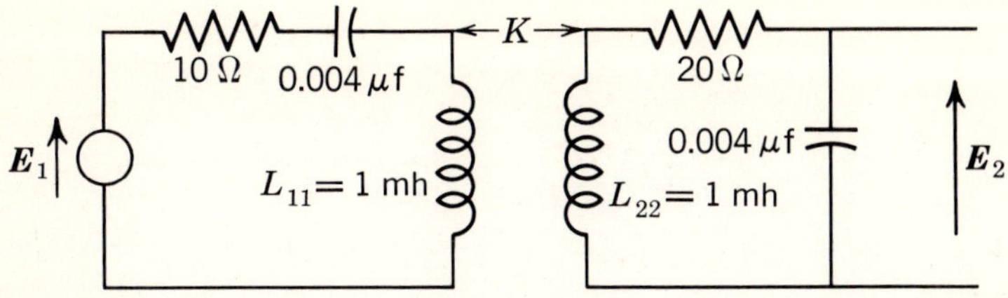

Prob. 24–15.

24–16. For the value of K found in (c) of Prob. 24–15, plot $|E_2/E_1|$ versus ω for the ω's around the value found in (c) of Prob. 24–15.

24–17. If the coupling coefficient in Prob. 24–15 is 0.01, find: (a) The voltage ratio $|E_2/E_1|$ at ω_0. (b) The voltage ratio at $1.2\omega_0$. (c) Sketch a curve of $|E_2/E_1|$ as a function of ω.

24–18. Repeat Prob. 24–17 for a coupling coefficient of 0.05.

25

FOURIER SERIES

25–1. Introduction. In the preceding chapters the response of circuits to singular functions of time, to sinusoidal forcing functions of time, and to exponential forcing functions of form ϵ^{st} was studied in the time and frequency domains. In the present chapter the response of circuits to periodic but non-sinusoidal functions of time is discussed.

If we take a harmonic function of time of any one discrete frequency f_1 and plot it versus time, we obtain a sine or cosine wave as a function of time.

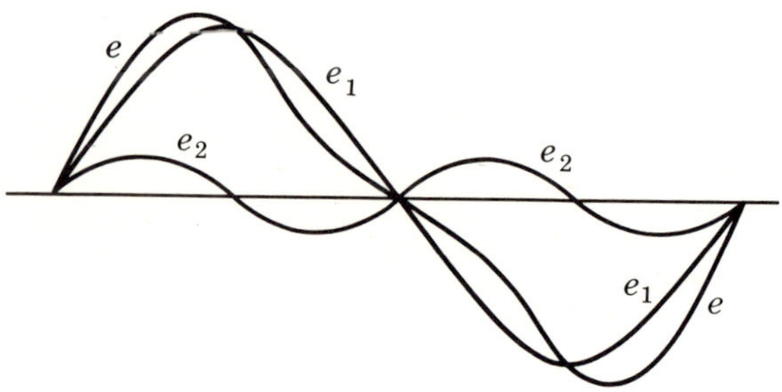

Fig. 25–1. Sum of a sine wave and its second harmonic.

Assume the plot to be a sine wave given by

$$e_1 = E_{s1} \sin \omega_1 t \tag{25–1}$$

as shown in Fig. 25–1. On the same abscissa let us plot a second harmonic of the original wave and let us further assume that the instantaneous value of the second harmonic is zero and ascending at $t = 0$, namely, let

$$e_2 = E_{s2} \sin 2\omega_1 t \tag{25–2}$$

This also is shown in Fig. 25–1. If we add the two functions of time we get (Fig. 25–1)

$$e = e_1 + e_2 = E_{s1} \sin \omega_1 t + E_{s2} \sin 2\omega_1 t \tag{25–3}$$

Notice that e is not a sinusoid, but is periodic.

If on the same abscissa we plot the third, fourth, and higher harmonics and if we add all the instantaneous values, we obtain a function of time which is not sinusoidal, but which is periodic. The shape of the resulting wave will in general depend upon:

1. The number of harmonics present.
2. The maximum value of the fundamental and each harmonic.
3. The value and direction of the fundamental and each harmonic at zero time, in other words, on the phase angle of the fundamental and each harmonic.

Thus by changing the magnitudes of the fundamental and harmonics, and by changing the number of harmonics and their phase angles, we can obtain all sorts of periodic functions. These functions, however, will have the same period as the fundamental. The functions will also have an average value of zero over a period. On the other hand, each of these periodic functions will be made up of discrete sinusoidal waves having discrete frequencies which are multiples of the fundamental frequency ω_1.

If a d-c or constant value wave is added to the periodic non-sinusoidal wave, the resulting wave will also be periodic, but will be translated upward or downward by the added direct current. The average value of the newly formed wave is equal to the d-c wave.

In the following text such periodic functions will be analyzed.

25-2. Fourier Series. In the introduction we have seen that the sum of a fundamental sinusoid and its harmonics produces a non-sinusoidal periodic function of time. We shall now reverse the order and decompose a periodic function into a fundamental and harmonics. This we shall state without proof.

> Any periodic function of time which is single-valued, continuous, or if discontinuous has a finite number of finite discontinuities and which at most has a finite number of maxima and minima in a finite time interval can be expressed as a series of sine and cosine terms thus:

$$f(t) = a_0 + a_1 \cos \omega_1 t + a_2 \cos 2\omega_1 t + a_3 \cos 3\omega_1 t + \cdots + b_1 \sin \omega_1 t + b_2 \sin 2\omega_1 t + b_3 \sin 3\omega_1 t + \cdots \tag{25-4}$$

$$= a_0 + \sum_{n=1}^{n=\infty} a_n \cos n\omega_1 t + \sum_{n=1}^{n=\infty} b_n \sin n\omega_1 t \tag{25-5}$$

where $n = 1, 2, 3, 4, \cdots$.

This series is known as Fourier series in honor of J. L. Fourier, a French physicist who first developed the idea in connection with his studies on heat flow. Here the period is:

$$T = \frac{2\pi}{\omega_1} \tag{25-6}$$

where ω_1 is the fundamental angular frequency in radians per second.

$$\omega_1 = 2\pi f_1 \tag{25-7}$$

and

$$f_1 = \frac{\omega_1}{2\pi} \tag{25-8}$$

where f_1 is the fundamental frequency in cycles per second. Notice that $\cos 2\omega_1 t$, $\sin 2\omega_1 t \cdots \cos n\omega_1 t$, $\sin n\omega_1 t \cdots$ are cosines and sines of the harmonics of the fundamentals.

If Eq. 25–5 is multiplied by dt and integrated between $-(\pi/\omega_1)$ and π/ω_1, we get:

$$\int_{-\frac{\pi}{\omega_1}}^{\frac{\pi}{\omega_1}} f(t)\,dt = \int_{-\frac{\pi}{\omega_1}}^{\frac{\pi}{\omega_1}} a_0\,dt + \sum_{n=1}^{n=\infty} \int_{-\frac{\pi}{\omega_1}}^{\frac{\pi}{\omega_1}} a_n \cos n\omega_1 t\,dt + \sum_{n=1}^{n=\infty} \int_{-\frac{\pi}{\omega_1}}^{\frac{\pi}{\omega_1}} b_n \sin n\omega_1 t\,dt = a_0 \frac{2\pi}{\omega_1} \tag{25-9}$$

because

$$\int_{-\frac{\pi}{\omega_1}}^{\frac{\pi}{\omega_1}} a_n \cos n\omega_1 t\,dt = a_n \frac{\sin n\omega_1 t}{n\omega_1}\Bigg|_{-\frac{\pi}{\omega_1}}^{\frac{\pi}{\omega_1}} = 0$$

and

$$\int_{-\frac{\pi}{\omega_1}}^{\frac{\pi}{\omega_1}} b_n \sin n\omega_1 t\,dt = -b_n \frac{\cos n\omega_1 t}{n\omega_1}\Bigg|_{-\frac{\pi}{\omega_1}}^{\frac{\pi}{\omega_1}} = 0$$

From Eq. 25–9 we get the value of the constant a_0 thus:

$$a_0 = \frac{\omega_1}{2\pi} \int_{-\frac{\pi}{\omega_1}}^{\frac{\pi}{\omega_1}} f(t)\,dt \tag{25-10}$$

From Eq. 25–10 we see that a_0 is the average value of $f(t)$ over a period. In other words, it is the value of the d-c component in the function of time $f(t)$.

Similarly, if Eq. 25–5 is multiplied by $\cos m\omega_1 t\,dt$ and integrated over a period, we get:

$$\int_{-\frac{\pi}{\omega_1}}^{\frac{\pi}{\omega_1}} f(t) \cos m\omega_1 t\,dt = \int_{-\frac{\pi}{\omega_1}}^{\frac{\pi}{\omega_1}} a_0 \cos m\omega_1 t\,dt + \sum_{n=1}^{n=\infty} \int_{-\frac{\pi}{\omega_1}}^{\frac{\pi}{\omega_1}} a_n \cos n\omega_1 t \cos m\omega_1 t\,dt + \sum_{n=1}^{n=\infty} \int_{-\frac{\pi}{\omega_1}}^{\frac{\pi}{\omega_1}} b_n \sin n\omega_1 t \cos m\omega_1 t\,dt \tag{25-11}$$

But

$$\int_{-\frac{\pi}{\omega_1}}^{\frac{\pi}{\omega_1}} a_0 \cos m\omega_1 t\,dt = a_0 \frac{\sin m\omega_1 t}{m\omega_1}\Bigg|_{-\frac{\pi}{\omega_1}}^{\frac{\pi}{\omega_1}} = 0$$

$$\begin{aligned}\int_{-\frac{\pi}{\omega_1}}^{\frac{\pi}{\omega_1}} a_n \cos n\omega_1 t \cos m\omega_1 t\,dt &= \frac{a_n}{2} \int_{-\frac{\pi}{\omega_1}}^{\frac{\pi}{\omega_1}} [\cos (n-m)\omega_1 t + \cos (n+m)\omega_1 t]\,dt \\ &= \frac{a_n}{2}\left[\frac{\sin (n-m)\omega_1 t}{n-m} + \frac{\sin (n+m)\omega_1 t}{n+m}\right]\Bigg|_{-\frac{\pi}{\omega_1}}^{\frac{\pi}{\omega_1}} \\ &= 0 \qquad \text{if } n \neq m\end{aligned}$$

If $n = m$,

$$\int_{-\frac{\pi}{\omega_1}}^{\frac{\pi}{\omega_1}} a_n \cos^2 n\omega_1 t\, dt = \frac{a_n}{2} \int_{-\frac{\pi}{\omega_1}}^{\frac{\pi}{\omega_1}} (1 + \cos 2n\omega_1 t)\, dt$$

$$= \frac{a_n}{2} \left(t + \frac{\sin 2n\omega_1 t}{2n\omega_1} \right) \Bigg|_{-\frac{\pi}{\omega_1}}^{\frac{\pi}{\omega_1}}$$

$$= a_n \frac{\pi}{\omega_1} \tag{25–12}$$

From Eqs. 25–11 and 25–12 we get:

$$a_n = \frac{\omega_1}{\pi} \int_{-\frac{\pi}{\omega_1}}^{\frac{\pi}{\omega_1}} f(t) \cos n\omega_1 t\, dt \tag{25–13}$$

In like manner it can be shown that:

$$b_n = \frac{\omega_1}{\pi} \int_{-\frac{\pi}{\omega_1}}^{\frac{\pi}{\omega_1}} f(t) \sin n\omega_1 t\, dt \tag{25–14}$$

From Eq. 25–10 we notice that a_0 is the average value of $f(t)$ during a period. It is the d-c component in the function. Equation 25–13 states that a_n is twice the average value of $f(t) \cos n\omega_1 t$ for a period. Equation 25–14 shows that b_n is twice the average value of $f(t) \sin n\omega_1 t$ for a period.

Notice that $a_0, a_1, a_2, a, \cdots, b_1, b_2, b, \cdots$ are not functions of time. After integration and substitution of the limiting values of t, these values are given as functions of ω_1. The a's are the coefficients of the cosine terms and therefore are the maximum values of the cosine terms. The b's are the maximum values of the sine terms. We can plot $a_0, a_1, a_2, a, \cdots, b_1, b_2, b, \cdots$ versus ω as shown in Fig. 25–2. We further notice that there are no a's or b's for values of ω between 0 and and ω_1, between ω_1 and $2\omega_1$, between $2\omega_1$ and $3\omega_1$, and so forth. This means that there are only cosine and sine functions of discrete frequencies ω_1, $2\omega_1$, $3\omega_1$, and so on.

Example 25–1. What is the Fourier series of the function whose graph is shown in Fig. 25–3? Here the periodic function is expressed as:

$$f(t + 2\pi m) = f(t) \qquad m = 0, 1, 2, 3, \cdots$$

For this example

$$f(t) = E \qquad 0 \leqslant t \leqslant \frac{\pi}{\omega_1}$$

$$f(t) = 0 \qquad \frac{\pi}{\omega_1} \leqslant t \leqslant \frac{2\pi}{\omega_1}$$

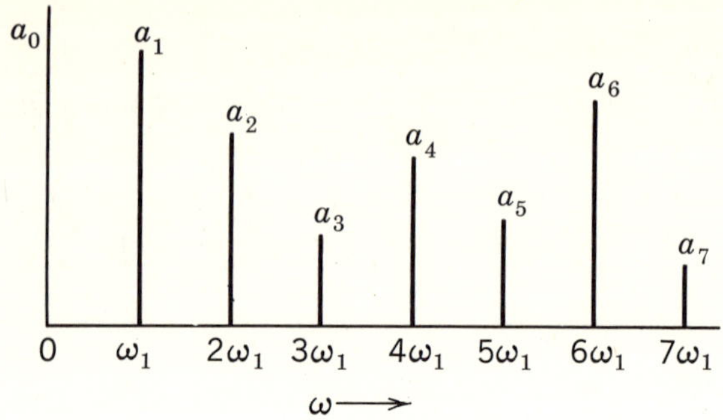

(a) Cosine terms

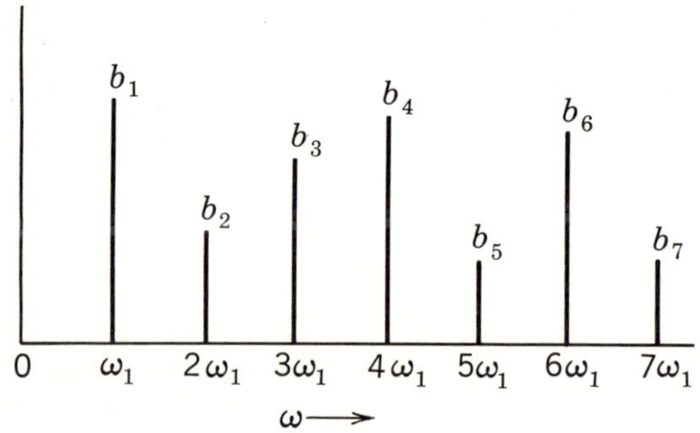

(b) Sine terms

Fig. 25–2. Plots of a_0, a_n, and b_n versus ω.

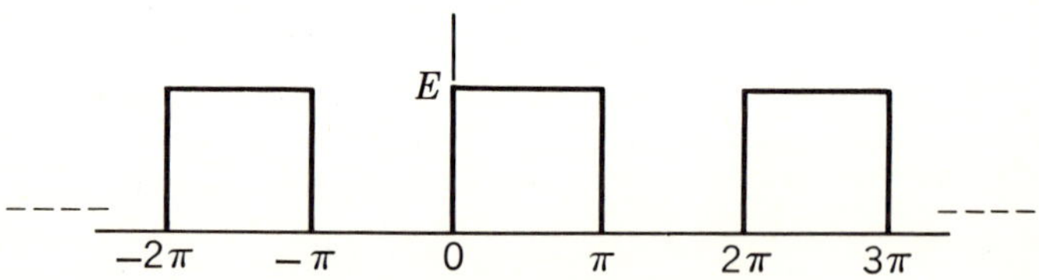

Fig. 25–3. Wave of equal periodic pulses.

Therefore:

$$a_0 = \frac{\omega_1}{2\pi}\int_0^{\frac{2\pi}{\omega_1}} f(t)\,dt$$

$$= \frac{\omega_1}{2\pi}\int_0^{\frac{\pi}{\omega_1}} E\,dt$$

$$= \frac{E}{2}$$

$$a_n = \frac{\omega_1}{\pi}\int_0^{\frac{2\pi}{\omega_1}} E\cos n\omega_1 t\,dt = \frac{\omega_1}{\pi}\int_0^{\frac{\pi}{\omega_1}} E\cos n\omega_1 t\,dt$$
$$= \frac{\omega_1}{\pi}\left(\frac{E\sin n\omega_1 t}{n\omega_1}\right)\Bigg|_0^{\frac{\pi}{\omega_1}}$$
$$= 0$$

This function has no cosine terms.

$$b_n = \frac{\omega_1}{\pi}\int_0^{\frac{\pi}{\omega_1}} E\sin n\omega_1 t\,dt$$
$$= -\frac{\omega_1}{\pi}\frac{E}{n\omega_1}\cos n\omega_1 t\Bigg|_0^{\frac{\pi}{\omega_1}}$$
$$= -\frac{E}{\pi n}(\cos n\pi - 1)$$
$$= 0 \qquad \text{if } n \text{ is even}$$
$$= \frac{2E}{\pi n} \qquad \text{if } n \text{ is odd}$$

Thus the series is:

$$f(t) = \frac{E}{2} + \frac{2E}{\pi}\left(\sin\omega_1 t + \frac{\sin 3\omega_1 t}{3} + \frac{\sin 5\omega_1 t}{5} + \cdots\right)$$

The function of time just given has no cosine terms and no even sine terms. It has a d-c term and the odd sine terms. Notice that b_n is the maximum value of the nth harmonic. The magnitude of each harmonic depends upon the harmonic and not upon time. Thus b_1, the value of the first harmonic or fundamental, is given by:

$$b_1 = \frac{2E}{\pi}$$

and b_3, the value of the third harmonic or three times the fundamental frequency, is given by:

$$b_3 = \frac{2E}{3\pi}$$

Similarly, b_n is given by:

$$b_n = \frac{2E}{\pi n}$$

Let us now plot the constant a_0 and the b's versus ω. The plot gives Fig. 25–4.

The following observations are important:

1. The b's are functions of ω.
2. There are no b's between 0 and ω_1, between ω_1 and $3\omega_1$, between $3\omega_1$ and $5\omega_1$, and so forth.

3. The maximum value of the harmonic depends upon the harmonic.
4. The b's form line spectra versus frequency.

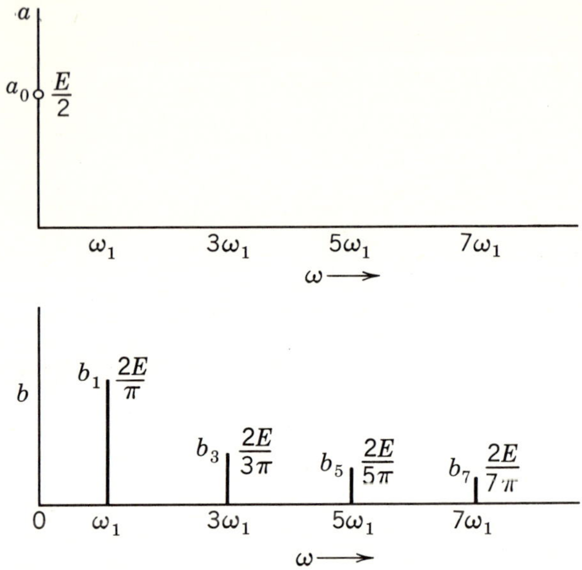

Fig. 25–4. Line spectra of Fig. 25–3.

25–3. Symmetry. Very often not all the a's and b's are found in the Fourier series, as was seen in the previous example. Some functions possess certain symmetries which result in the a's or b's having zero value. There are two main symmetries which often characterize a system.

EVEN SYMMETRY. If

$$f(t) = f(-t) \tag{25–15}$$

the function is said to be *even*. Under this condition the Fourier series has no sine terms and the b's are zero. To prove this, write:

$$f(t) = a_0 + \sum_{n=1}^{n=\infty} a_n \cos n\omega_1 t + \sum_{n=1}^{n=\infty} b_n \sin n\omega_1 t \tag{25–16}$$

Also

$$f(-t) = a_0 + \sum_{n=1}^{n=\infty} a_n \cos(-n\omega_1 t) + \sum_{n=1}^{n=\infty} b_n \sin(-n\omega_1 t) \tag{25–17}$$

From Eqs. 25–15, 25–16, and 25–17 we get:

$$a_0 + \sum_{n=1}^{n=\infty} a_n \cos n\omega_1 t + \sum_{n=1}^{n=\infty} b_n \sin n\omega_1 t$$
$$= a_0 + \sum_{n=1}^{n=\infty} a_n \cos(-n\omega_1 t) + \sum_{n=1}^{n=\infty} b_n \sin(-n\omega_1 t) \tag{25–18}$$

But:

$$\cos(-n\omega_1 t) = \cos n\omega_1 t \tag{25–19}$$

and

$$\sin(-n\omega_1 t) = -\sin n\omega_1 t \tag{25–20}$$

From Eqs. 25–18, 25–19, and 25–20 we get:

$$a_0 + \sum_{n=1}^{n=\infty} a_n \cos n\omega_1 t + \sum_{n=1}^{n=\infty} b_n \sin n\omega_1 t$$
$$= a_0 + \sum_{n=1}^{n=\infty} a_n \cos n\omega_1 t - \sum_{n=1}^{n=\infty} b_n \sin n\omega_1 t \quad (25\text{–}21)$$

or

$$2 \sum_{n=1}^{n=\infty} b_n \sin n\omega_1 t = 0 \quad (25\text{–}22)$$

Therefore:

$$b_n = 0 \quad (25\text{–}23)$$

When $f(t) = f(-t)$, the series can be written as:

$$f(t) = a_0 + \sum_{n=1}^{n=\infty} a_n \cos n\omega_1 t \quad (25\text{–}24)$$

Example 25–2. Find the Fourier series of the function shown in Fig. 25–5. Here $f(t) = f(-t)$ and b_n should be zero. Let us verify this first and then find the Fourier series.

From Eq. 25–14 we have:

$$b_n = \frac{\omega_1}{\pi} \int_{-\frac{\pi}{2\omega_1}}^{\frac{\pi}{2\omega_1}} E \sin n\omega_1 t \, dt$$
$$= -\frac{E\omega_1}{\pi} \frac{\cos n\omega_1 t}{n\omega_1} \Bigg|_{-\frac{\pi}{2\omega_1}}^{\frac{\pi}{2\omega_1}}$$
$$= -\frac{E\omega_1}{\pi n\omega_1} \left[\cos \frac{\pi n}{2} - \cos \left(-\frac{\pi n}{2} \right) \right]$$
$$= 0$$

as expected.

Now let us find the Fourier series:

$$a_0 = \frac{\omega_1}{2\pi} \int_{-\frac{\pi}{2\omega_1}}^{\frac{\pi}{2\omega_1}} E \, dt$$
$$= \frac{\omega_1}{2\pi} Et \Bigg|_{-\frac{\pi}{2\omega_1}}^{\frac{\pi}{2\omega_1}}$$
$$= \frac{E}{2}$$

$$a_n = \frac{\omega_1}{\pi} \int_{-\frac{\pi}{2\omega_1}}^{\frac{\pi}{2\omega_1}} E \cos n\omega_1 t \, dt = \frac{\omega_1}{\pi} E \frac{\sin n\omega_1 t}{n\omega_1} \Bigg|_{-\frac{\pi}{2\omega_1}}^{\frac{\pi}{2\omega_1}}$$
$$= \frac{2E}{\pi n} \sin \frac{\pi n}{2} \qquad \text{if } n \text{ is odd}$$
$$= 0 \qquad \text{if } n \text{ is even}$$

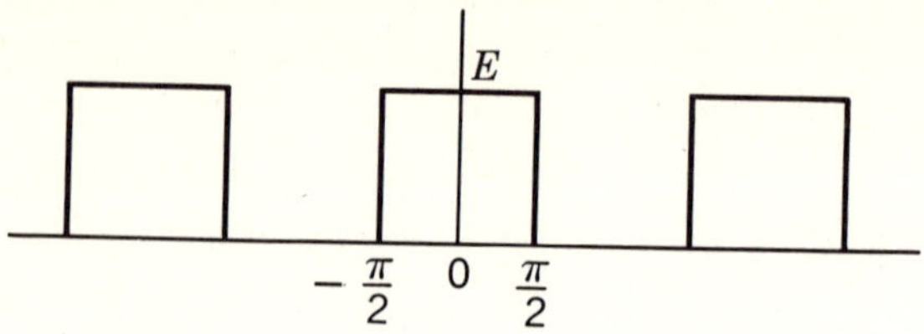

Fig. 25–5. Change of reference on pulses.

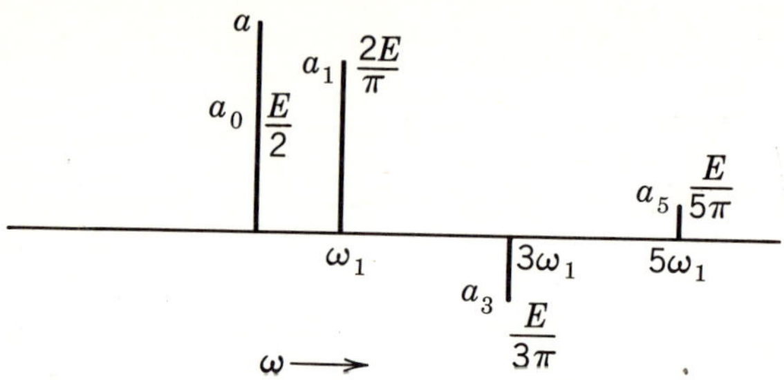

Fig. 25–6. Line spectra of Fig. 25–5.

Plotting a_n versus ω, we get Fig. 25–6. The Fourier series is then:

$$f(t) = \frac{E}{2} + \frac{2E}{\pi} \cos \omega_1 t - \frac{2E}{3\pi} \cos 3\omega_1 t + \frac{2E}{5\pi} \cos 5\omega_1 t \cdots$$

$$= \frac{E}{2} + \frac{2E}{\pi} \left(\cos \omega_1 t - \frac{\cos 3\omega_1 t}{3} + \frac{\cos 5\omega_1 t}{5} - \frac{\cos 7\omega_1 t}{7} + \cdots \right)$$

Notice that when the function is even or has the symmetry $f(t) = f(-t)$, a_n can be found by integration over half the function and doubling the value, thus:

$$a_n = \frac{2\omega_1}{\pi} \int_0^{\frac{\pi}{2\omega_1}} E \cos n\omega_1 t \, dt$$

$$= \frac{2\omega_1}{\pi} E \frac{\sin n\omega_1 t}{n\omega_1} \Bigg|_0^{\frac{\pi}{2\omega_1}}$$

$$= \frac{2\omega_1 E}{\pi n \omega_1} \sin \frac{\pi n}{2}$$

$$= \frac{2E}{\pi n} \sin \frac{\pi n}{2} \qquad \text{if } n \text{ is odd}$$

$$= 0 \qquad \text{if } n \text{ is even}$$

as previously found.

Odd Symmetry. If

$$f(t) = -f(-t) \tag{25–25}$$

the function is said to be *odd*. Under this condition no cosine terms exist, namely, a_n is zero. The function $f(t)$ can then be written:

$$f(t) = \sum_{n=1}^{n=\infty} b_n \sin n\omega_1 t \tag{25–26}$$

To prove Eq. 25–26, let:

$$f(t) = a_0 + \sum_{n=1}^{n=\infty} a_n \cos n\omega_1 t + \sum_{n=1}^{n=\infty} b_n \sin n\omega_1 t \tag{25–27}$$

and

$$f(-t) = a_0 + \sum_{n=1}^{n=\infty} a_n \cos(-n\omega_1 t) + \sum_{n=1}^{n=\infty} b_n \sin(-n\omega_1 t) \tag{25–28}$$

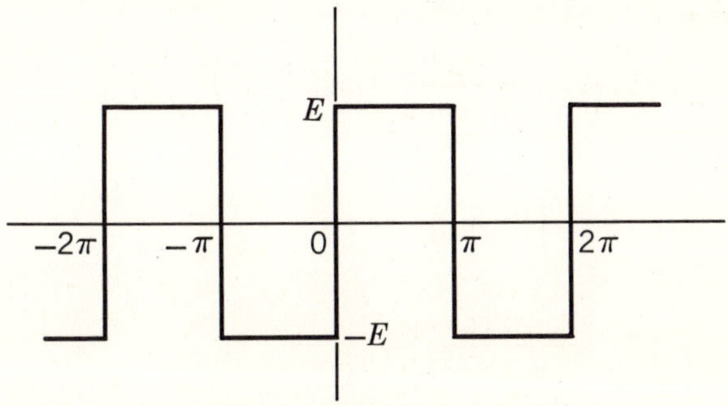

Fig. 25–7. Rectangular wave.

From Eqs. 25–25, 25–27, and 25–28 we get:

$$a_0 + \sum_{n=1}^{n=\infty} a_n \cos n\omega_1 t + \sum_{n=1}^{n=\infty} b_n \sin n\omega_1 t = -a_0 - \sum_{n=1}^{n=\infty} a_n \cos n\omega_1 t + \sum_{n=1}^{n=\infty} b_n \sin n\omega_1 t \tag{25–29}$$

From Eq. 25–20 we get:

$$2a_0 + 2\sum_{n=1}^{n=\infty} a_n \cos n\omega_1 t = 0 \tag{25–30}$$

under all conditions and for all values of t. This is possible only if

$$a_0 = 0 \qquad \text{and} \qquad a_n = 0 \tag{25–31}$$

Therefore, when $f(t) = -f(-t)$, then

$$f(t) = \sum_{n=1}^{n=\infty} b_n \sin n\omega_1 t \tag{25–32}$$

Here, too, b_n can be found by integration over half of the wave only and doubling the result. Thus:

$$b_n = \frac{2\omega_1}{\pi}\int_0^{\frac{\pi}{\omega_1}} f(t) \sin n\omega_1 t \, dt \tag{25–33}$$

EXAMPLE 25–3. Find the Fourier series of the function of time shown in Fig. 25–7. Here $f(t) = -f(-t)$ and $a_n = 0$. This can be verified as follows:

$$a_n = \frac{\omega_1}{\pi} \int_{-\frac{\pi}{\omega_1}}^{\frac{\pi}{\omega_1}} f(t) \cos n\omega_1 t \, dt$$

$$= \frac{\omega_1}{\pi}\left(-E \int_{-\frac{\pi}{\omega_1}}^{0} \cos n\omega_1 t \, dt + E \int_{0}^{\frac{\pi}{\omega_1}} \cos n\omega_1 t \, dt \right)$$

$$= \frac{\omega_1}{\pi}\left(-E \left. \frac{\sin n\omega_1 t}{n\omega_1} \right|_{-\frac{\pi}{\omega_1}}^{0} + E \left. \frac{\sin n\omega_1 t}{n\omega_1} \right|_{0}^{\frac{\pi}{\omega_1}} \right)$$

$$= \frac{\omega_1}{\pi}\left(-E \frac{\sin n\pi}{n\omega_1} + E \frac{\sin n\pi}{n\omega_1} \right)$$

$$= 0$$

$$b_n = \frac{2\omega_1}{\pi} \int_{0}^{\frac{\pi}{\omega_1}} E \sin n\omega_1 t \, dt$$

$$= -\frac{2\omega_1 E}{\pi} \left. \frac{\cos n\omega_1 t}{n\omega_1} \right|_{0}^{\frac{\pi}{\omega_1}}$$

$$= -\frac{2E}{\pi n}(\cos n\pi - 1)$$

$$= \frac{2E}{\pi n}(1 - \cos n\pi)$$

$$= \frac{4E}{\pi n} \qquad \text{if } n \text{ is odd}$$

$$= 0 \qquad \text{if } n \text{ is even}$$

If we plot b_n versus ω, we get Fig. 25–8.

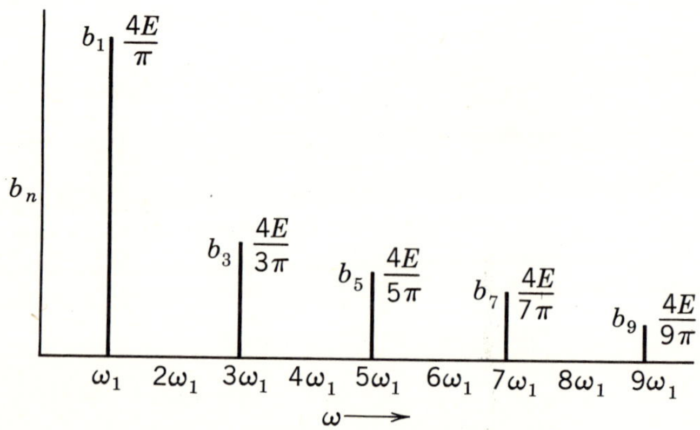

Fig. 25–8. Harmonics of a rectangular wave.

25-4. Other Symmetries. There are other symmetries which can be investigated, but we shall give them as exercises at the end of the chapter.

25-5. Fourier Series in Terms of Cosines. The Fourier series can be written in terms of cosines, thus:

$$f(t) = a_0 + a_1 \cos \omega_1 t + a_2 \cos 2\omega_1 t + \cdots + b_1 \sin \omega_1 t + b_2 \sin 2\omega_1 t + \cdots$$

But a_1 and b_1 are two sinusoids of the same frequency but 90° apart, the a_1 being 90° ahead of b_1. If we let a_1 and b_1 form the legs of a right-angled triangle as shown in Fig. 25-9, then:

$$a_1 = r_1 \cos \phi_1 \qquad (25\text{-}34)$$

$$b_1 = r_1 \sin \phi_1 \qquad (25\text{-}35)$$

with

$$r_1 = \sqrt{a_1^{\,2} + b_1^{\,2}} \qquad (25\text{-}36)$$

$$\phi_1 = \tan^{-1} \frac{b_1}{r_1} \qquad (25\text{-}37)$$

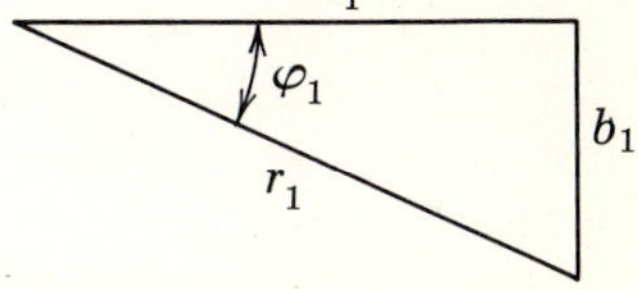

Fig. 25-9. How *a* and *b* are added.

Therefore:

$$a_1 \cos \omega_1 t + b_1 \sin \omega_1 t = r_1(\cos \omega_1 t \cos \phi_1 + \sin \omega_1 t \sin \phi_1)$$

$$= r_1 \cos (\omega t - \phi_1) \qquad (25\text{-}38)$$

Similarly,

$$a_n \cos n\omega_1 t + b_n \sin n\omega_1 t = r_n \cos (n\omega_1 t - \phi_n) \qquad (25\text{-}39)$$

where

$$r_n = \sqrt{a_n^{\,2} + b_n^{\,2}} \qquad (25\text{-}40)$$

$$\phi_n = \tan^{-1} \frac{b_n}{a_n} \qquad (25\text{-}41)$$

The series then becomes:

$$f(t) = a_0 + \sum_{n=1}^{n=\infty} r_n \cos (n\omega_1 t - \phi_n) \qquad (25\text{-}42)$$

25-6. Fourier Series in Terms of Sines. If we let

$$b_n = r_n \cos \theta_n \qquad (25\text{-}43)$$

$$a_n = r_n \sin \theta_n \qquad (25\text{-}44)$$

as in Fig. 25-10, then

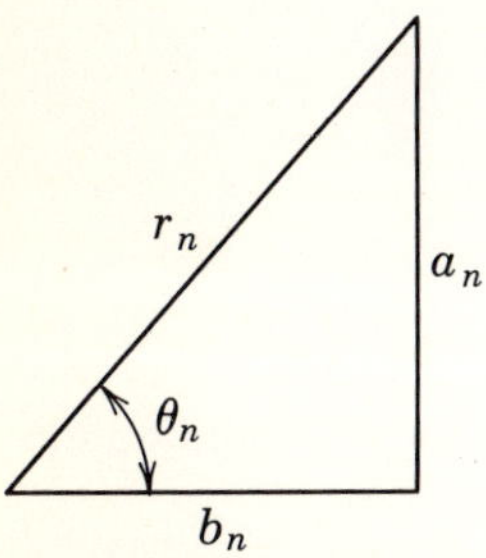

Fig. 25-10. Another method of addition.

$$a_n \cos n\omega_1 t + b_n \sin n\omega_1 t$$

$$= r_n(\sin \theta_n \cos n\omega_1 t + \cos \theta_n \sin n\omega_1 t)$$

$$= r_n \sin (n\omega_1 t + \theta_n) \qquad (25\text{-}45)$$

Here

$$r_n = \sqrt{a_n^{\,2} + b_n^{\,2}} \tag{25–46}$$

$$\theta_n = \tan^{-1} \frac{a_n}{b_n} \tag{25–47}$$

The series is then written as:

$$f(t) = a_0 + \sum_{n=1}^{n=\infty} r_n \sin (n\omega_1 t + \theta_n) \tag{25–48}$$

25–7. Fourier Series of a Graph. If the function of time is given graphically but if it cannot be expressed explicitly as a simple function of time, its Fourier series can be obtained in the following manner.

The base or abscissa of the function can be divided into n equal parts. Let n be 36. Take the ordinate corresponding to each division. Each division corresponds to 10°. Tabulate the angle and the ordinate corresponding to each angle, thus:

	y
0	y_1
10	y_2
20	y_3
30	y_4
40	y_5
.	.
.	.
.	.

To find a_0 add the positive ordinates and also the negative ordinates. Subtract the negative sum from the positive sum. Let the answer be Y. Then:

$$a_0 = \frac{Y}{36}$$

To find a_1 multiply each ordinate by the cosine of its angle and tabulate thus:

$$\begin{aligned} y_1 \cos 0 &= y'_1 \\ y_2 \cos 10 &= y'_2 \\ y_3 \cos 20 &= y'_3 \\ y_4 \cos 30 &= y'_4 \\ y_5 \cos 40 &= y'_5 \\ \cdot \quad & \quad \cdot \\ \cdot \quad & \quad \cdot \\ \cdot \quad & \quad \cdot \end{aligned}$$

Add all the positive values and negative values of $y'_1, y'_2, y'_3, \cdots$ and obtain Y_1. Divide Y_1 by 18 and get:

$$a_1 = \frac{Y_1}{18}$$

To find b_1 multiply each ordinate by the sine of its angle and tabulate thus:

$$\begin{aligned} y_1 \sin 0 &= x_1 \\ y_2 \sin 10 &= x_2 \\ y_3 \sin 20 &= x_3 \\ y_4 \sin 30 &= x_4 \\ \cdot \quad &\quad \cdot \\ \cdot \quad &\quad \cdot \\ \cdot \quad &\quad \cdot \end{aligned}$$

Add the $x''s$, taking the signs into consideration, and obtain X. The value of b_1 is then given by:

$$b_1 = \frac{X}{18}$$

The resulting component of the fundamental is:

$$r_1 = \sqrt{a_1^2 + b_1^2}$$

The angle ϕ_1 is:

$$\phi_1 = \tan^{-1} \frac{b_1}{a_1}$$

To find a_2 multiply each ordinate by the cosine of twice its angle and tabulate thus:

$$\begin{aligned} y_1 \cos 2 \times 0 &= y''_1 \\ y_2 \cos 2 \times 10 &= y''_2 \\ y_3 \cos 2 \times 20 &= y''_3 \\ y_4 \cos 2 \times 30 &= y''_4 \\ y_5 \cos 2 \times 40 &= y''_5 \\ y_6 \cos 2 \times 50 &= y''_6 \\ y_7 \cos 2 \times 60 &= y''_7 \\ \cdot \quad &\quad \cdot \\ \cdot \quad &\quad \cdot \\ \cdot \quad &\quad \cdot \end{aligned}$$

Add all the $y'''s$ and obtain Y_2. Then

$$a_2 = \frac{Y_2}{18}$$

To find b_2 multiply each ordinate by the sine of twice its angle and tabulate thus:

$$\begin{aligned} y_1 \sin 2 \times 0 &= 0 \\ y_2 \sin 2 \times 10 &= x'_2 \\ y_3 \sin 2 \times 20 &= x'_3 \\ y_4 \sin 2 \times 30 &= x'_4 \\ \cdot \quad &\quad \cdot \\ \cdot \quad &\quad \cdot \\ \cdot \quad &\quad \cdot \end{aligned}$$

Add all the $x''s$, taking the signs into consideration, and obtain X_2. The value of b_2 is then found thus:

$$b_2 = \frac{X_2}{18}$$

and

$$r_1 = \sqrt{a_2^2 + b_2^2}$$

$$\phi_2 = \tan^{-1} \frac{b_2}{a_2}$$

The other harmonics are found in the same manner. The Fourier series is then:

$$f(t) = a_0 + r_1 \cos (\omega_1 t - \phi_1) + r_2 \cos (2\omega_1 t - \phi_2) + r_3 \cos (3\omega_1 t - \phi_3) \cdots$$

When the graphical function has certain symmetries the number of divisions can often be reduced and the amount of work reduced.

25–8. Application. The Fourier series is used to find the steady-state response of a circuit to non-sinusoidal periodic functions and to explain why the response does not have the same form as the actuating function.

Assume the forcing function to be a voltage

$$e_c = a_n \cos n\omega_1 t$$

Let this function be applied to an R-L series circuit. The steady-state response is found by the following steps:

1. Find a_n, the maximum value of e_c.
2. Find the impedance of the circuit as a function of frequency $n\omega_1$ thus:

$$|Z| = \sqrt{R^2 + (n\omega_1 L)^2}$$

3. Find θ_z, the phase angle of $\mathbf{Z}$, as a function of frequency thus:

$$\theta_z = \tan^{-1} \frac{n\omega_1 L}{R}$$

4. Find the maximum response of the system as a function of $n\omega_1$

$$\mathbf{I}(n\omega_1) = \frac{a_n}{|Z(n\omega_1)|}$$

Notice that $\mathbf{I}(n\omega_1)$ is a function of frequency.

5. Find the instantaneous value of the current

$$i(t) = \mathbf{I}(n\omega_1) \cos (n\omega_1 t - \theta_z)$$

Steps 1–4 are shown on the ω domain in Fig. 25–11.

If the forcing function is a sine wave, then

$$e_s = b_n \sin n\omega_1 t$$

The corresponding steps are:

1. Find b_n, the maximum voltage of e_s.
2. Find the impedance of the circuit as a function of frequency $n\omega_1$ thus:

$$\mathbf{Z}(n\omega_1) = \sqrt{R^2 + (n\omega_1 L)^2}$$

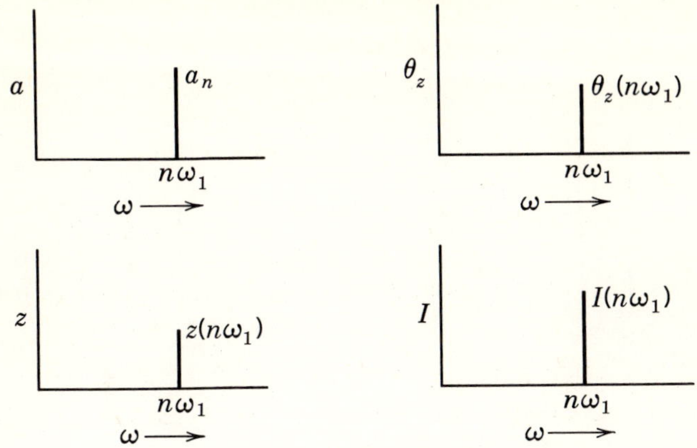

Fig. 25–11. Steps to follow in solution.

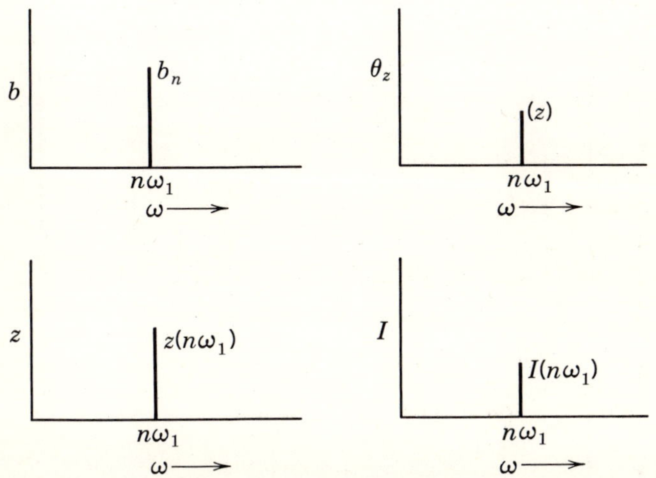

Fig. 25–12. More steps to follow.

3. Find θ_z, the phase angle of $\mathbf{Z}(n\omega_1)$ thus:

$$\theta_z = \tan^{-1} \frac{n\omega_1 L}{R}$$

4. Find the maximum response of the system to frequency $n\omega_1$

$$\mathbf{I}(n\omega_1) = \frac{b_n}{\mathbf{Z}(n\omega_1)}$$

5. Find the instantaneous value of the response

$$i(t) = \mathbf{I}(n\omega_1) \sin (n\omega_1 t - \theta_z)$$

On the frequency domain steps 1–4 are shown in Fig. 25–12.

Frequently, instead of using $\mathbf{Z}(n\omega_1)$ it is preferable to use $\mathbf{Y}(n\omega_1)$ where $\mathbf{Y}(n\omega_1)$ is the admittance of the circuit and is equal to $1/[\mathbf{Z}(n\omega_1)]$.

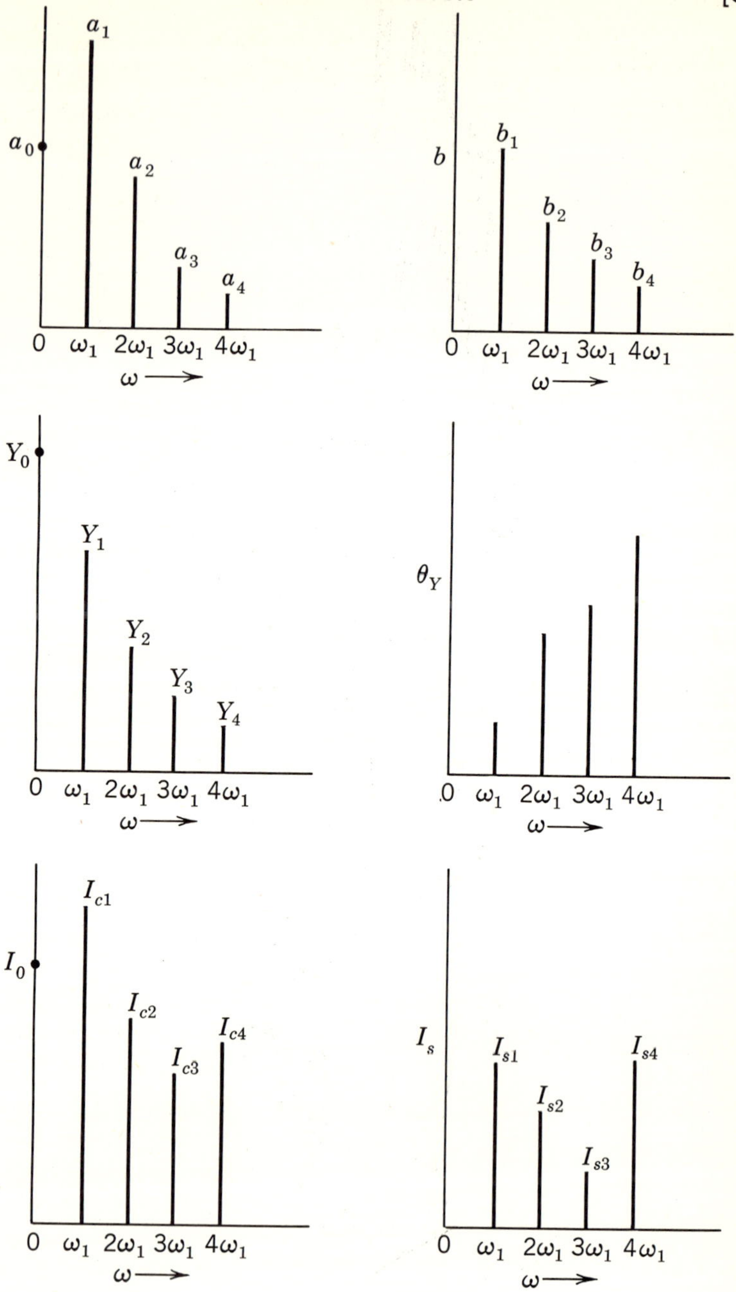

Fig. 25–13. Line spectra.

25-9. Response of Systems. If a periodic but non-sinusoidal force, meeting the conditions stated in Art. 25-8, is applied to a linear system, the steady-state response of the system can be found by the following steps:

1. Find the a's and b's (the maximum value of the harmonics) in the periodic function and plot them versus ω.
2. Find the admittance or impedance of the system in terms of frequency and plot it versus frequency ω.
3. Find the phase angle of the admittance or impedance of the system in terms of frequency and plot it versus ω.
4. Find the value of the maximum response of the system to each harmonic by multiplying a_n by $\boldsymbol{Y}_n$ and b_n by $\boldsymbol{Y}_n$.
5. Find the instantaneous response of each harmonic thus:

$$i_n(t) = a_n|Y_n| \cos (n\omega_1 t \pm \theta_n) + b_n|Y_n| \sin (n\omega_1 t \pm \theta_n)$$

6. The total response of the system is then:

$$i(t) = a_0 Y_0 + \sum_{n=1}^{n=\infty} a_n|Y_n| \cos (n\omega_1 t \pm \theta_n) + \sum_{n=1}^{n=\infty} b_n|Y_n| \sin (n\omega_1 t \pm \theta_n)$$

$$= \boldsymbol{I}_0 + \sum_{n=1}^{n=\infty} \boldsymbol{I}_{cn} \cos (n\omega_1 t \pm \theta_n) + \sum_{n=1}^{n=\infty} \boldsymbol{I}_{sn} \sin (n\omega_1 t \pm \theta_n)$$

Notice that the preceding steps can be stated in the following way:

(a) Convert the periodic function of time into a function of frequency.
(b) Find the response of the circuit to each existing frequency.
(c) Reconvert the responses from terms of frequency to terms of time.

The first five steps are depicted in Fig. 25-13. Notice that the admittance or impedance of the circuit is plotted for 0, ω_1, $2\omega_1$, $3\omega_1$, $\cdots$, only because these are the values to be used with the a's and the b's of the forcing function. Although the periodic function is made up of discrete frequencies forming line spectra in the ω domain, the $\boldsymbol{Y}$'s and $\boldsymbol{Z}$'s are continuous functions of frequency and form a continuous spectrum in the ω domain. We could have plotted them as such and used in our operation the values corresponding to the line spectra of the forcing function.

Example 25-4. Assume that a periodic voltage wave as shown in Fig. 25-3 is applied to an R-L series circuit having $R = 10$ ohms and $L = 1$ h. Find the steady-state current, knowing that the frequency is 0.5 cps and $\boldsymbol{E} = 100$ v.

The solution for this problem is carried out as follows:

$$\omega_1 = 2\pi f = 2\pi 0.5 = \pi$$

We have seen that $f(t)$ can be written as:

$$f(t) = \frac{\boldsymbol{E}}{2} + \frac{2\boldsymbol{E}}{\pi}\left(\sin \omega_1 t + \frac{\sin 3\omega_1 t}{3} + \frac{\sin 5\omega_1 t}{5} + \cdots\right)$$

Therefore:

$$f(t) = 50 + 63.7 \sin \pi t + 21.2 \sin 3\pi t + 12.7 \sin 5\pi t + \cdots$$

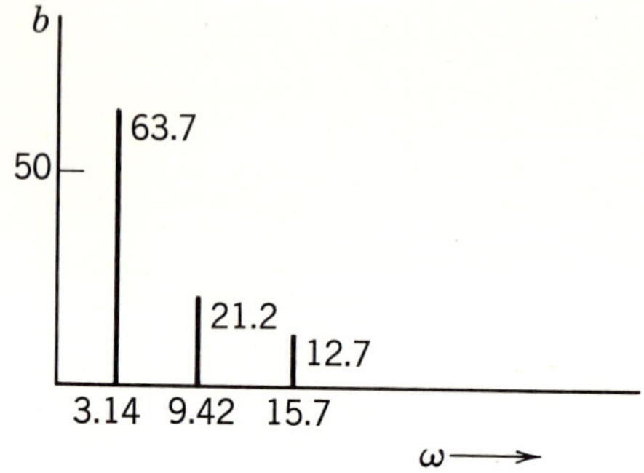

Fig. 25–14. Voltage spectrum of Fig. 25–3.

The frequency spectrum is as shown in Fig. 25–14. The impedance of the circuit is:

$$Z = \sqrt{R^2 + (n\omega L)^2}$$

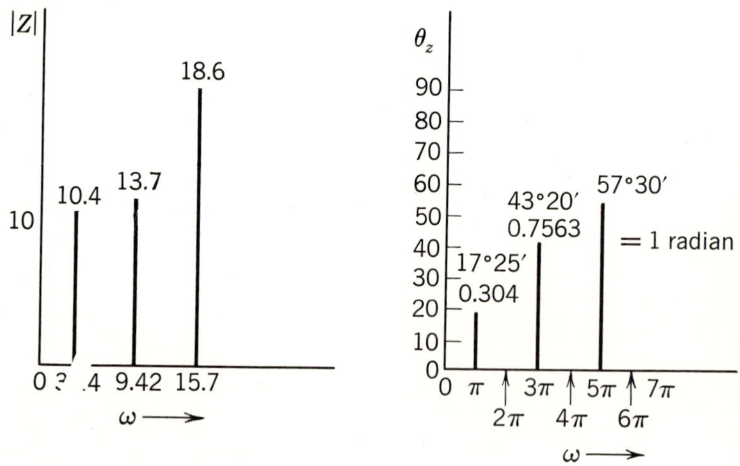

Fig. 25–15. $|Z|$ and θ_z versus ω.

Here we make a table of $|Z|$ and the phase angle θ_z, as shown in Table 25–1 and Fig. 25–15. The phase angle θ_z of the impedance is given by:

$$\theta_z = \tan^{-1} \frac{n\omega L}{R}$$

TABLE 25–1

ω	$\lvert Z \rvert$	θ_z (degrees)	θ_z (radians)
0	10	0	1
3.14	10.4	17°25′	0.304
9.42	13.7	43°20′	0.7563
15.7	18.6	57°30′	1.00

The response at different frequencies is given by

$$I(n\omega) = \frac{E(n\omega)}{|Z(n\omega)|}$$

and is shown in Table 25–2.

TABLE 25–2

ω	$I(n\omega)$
0	5
3.14	6.12
9.42	1.54
15.7	0.68

The steady-state response of the circuit to the foregoing voltage wave is:

$$i(t) = 5 + 6.12 \sin(\pi t - 0.304) + 1.54 \sin(3\pi t - 0.7563) + 0.68 \sin(5\pi t - 1)$$

This can be plotted point by point for a complete cycle and the shape of the current wave versus time can be obtained.

EXAMPLE 25–5. Assume the voltage applied to an R-L-C series circuit to have the form given in Fig. 25–7. $E = 100$ v, $\omega = 377$ radians/sec, $R = 5$ ohms, $L = 1$ h, and $C = 1$ μf. Find the current caused by the fundamental and third harmonic.

To solve this problem, proceed as follows (see Fig. 25–16):

1. b_n when n is odd is given by:

$$b_n = \frac{4E}{\pi n}$$

Therefore:

$$b_1 = \frac{400}{\pi}$$

$$b_3 = \frac{400}{3\pi}$$

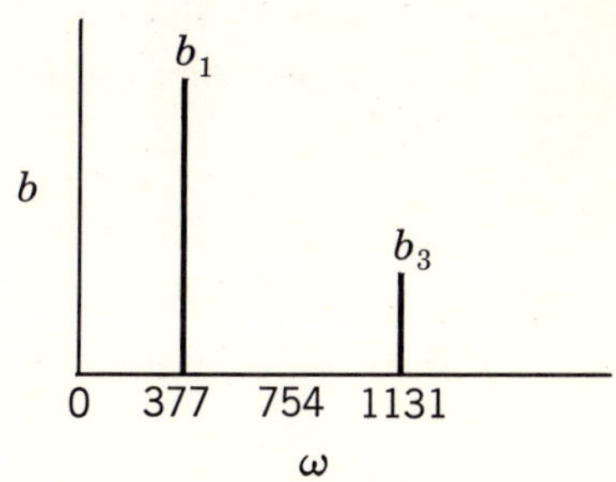

Fig. 25–16 (contd. on p. 436).

2. The impedance is:

$$Z = \sqrt{R^2 + \left(\omega L - \frac{1}{\omega c}\right)^2}$$

$$|Z_1(\omega)| \simeq 2775$$

$$|Z_3(\omega)| \simeq 247$$

3. The phase angle of Z is given by:

$$\theta_z = \tan^{-1} \frac{\omega L - \dfrac{1}{\omega c}}{R}$$

Therefore:

$$\theta_{z1} \simeq -90^\circ$$

$$\theta_{z3} \simeq +90^\circ$$

4. The magnitude of $I_{n\omega}$ is given by:

$$I_{n\omega} = \frac{b_{n\omega}}{|Zn\omega|}$$

Therefore:

$$I_{1\omega} = \frac{400}{\pi \times 2775} = 0.046$$

$$I_{3\omega} = \frac{400}{3\pi \times 247} = 0.172$$

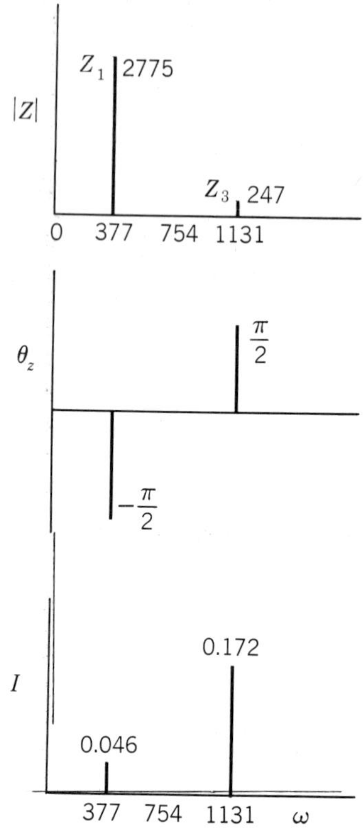

Fig. 25–16. Steps in solving Example 25–5.

5. The instantaneous value of the nth harmonic is

$$i_n(t) = I_{n\omega} \sin (n\omega t \pm \theta_z)$$

Therefore:

$$i_1(t) = 0.046 \sin \left(377t + \frac{\pi}{2}\right) = 0.046 \cos 377t$$

$$i_3(t) = 0.172 \sin \left(1131t - \frac{\pi}{2}\right) = -0.172 \cos 1131t$$

6. The total instantaneous current (fundamental and third harmonic only) is

$$i(t) = 0.046 \cos 377t - 0.172 \cos 1131t$$

In this example the third harmonic current is larger than the fundamental although the third harmonic voltage is only one-third as large as the fundamental voltage, The third harmonic impedance, however, is approximately one-eleventh of the fundamental impedance.

25-10. Exponential Form of Fourier Series. The steady-state response of circuits when excited with sinusoidal forcing functions is usually found by means of complex algebra. This being the case, one wonders if the same method can be applied to circuits having periodic but non-sinusoidal excitation applied to them. The answer is yes. Complex algebra methods can be used even in this case. To show how this can be done, we shall express the Fourier series in exponential form in the following way.

We have seen that a periodic function can be expressed thus:

$$f(t) = a_0 + \sum_{n=1}^{n=\infty} a_n \cos n\omega_1 t + \sum_{n=1}^{n=\infty} b_n \sin n\omega_1 t \tag{25-49}$$

But:

$$\cos n\omega_1 t = \frac{\epsilon^{jn\omega_1 t} + \epsilon^{-jn\omega_1 t}}{2} \tag{25-50}$$

and

$$\sin n\omega_1 t = \frac{\epsilon^{jn\omega_1 t} - \epsilon^{-jn\omega_1 t}}{2j} \tag{25-51}$$

Therefore we get from Eqs. 25-49, 25-50, and 25-51:

$$f(t) = a_0 + \sum_{n=1}^{n=\infty} \frac{a_n}{2}(\epsilon^{jn\omega_1 t} + \epsilon^{-jn\omega_1 t}) + \frac{b_n}{2j}(\epsilon^{jn\omega_1 t} - \epsilon^{-jn\omega_1 t})$$

$$= a_0 + \sum_{n=1}^{n=\infty} \frac{a_n - jb_n}{2}\epsilon^{jn\omega_1 t} + \sum_{n=1}^{n=\infty} \frac{a_n + jb_n}{2}\epsilon^{-jn\omega_1 t} \tag{25-52}$$

Let:

$$C_n = \frac{a_n - jb_n}{2} \tag{25-53}$$

$$C_n{}^* = \frac{a_n + jb_n}{2} \tag{25-54}$$

Notice that C^* is the conjugate of C. Therefore:

$$C_n = \frac{a_n - jb_n}{2} = \frac{\omega_1}{2\pi}\left[\int_{-\frac{\pi}{\omega_1}}^{\frac{\pi}{\omega_1}} f(t)\cos n\omega_1 t\,dt - j\int_{-\frac{\pi}{\omega_1}}^{\frac{\pi}{\omega_1}} f(t)\sin n\omega_1 t\,dt\right]$$

$$= \frac{\omega_1}{2\pi}\int_{-\frac{\pi}{\omega_1}}^{\frac{\pi}{\omega_1}} f(t)\epsilon^{-jn\omega_1 t}\,dt \tag{25-55}$$

and

$$C_n{}^* = \frac{a_n + jb_n}{2} = \frac{\omega_1}{2\pi}\left[\int_{-\frac{\pi}{\omega_1}}^{\frac{\pi}{\omega_1}} f(t)\cos n\omega_1 t\,dt + j\int_{-\frac{\pi}{\omega_1}}^{\frac{\pi}{\omega_1}} f(t)\sin n\omega_1 t\,dt\right]$$

$$= \frac{\omega_1}{2\pi}\int_{-\frac{\pi}{\omega_1}}^{\frac{\pi}{\omega_1}} f(t)\epsilon^{jn\omega_1 t}\,dt \tag{25-56}$$

Notice that by making n negative in Eq. 25–55 we obtain C_n^*, the conjugate of C_n, in Eq. 25–56 because $\cos(-n\omega_1 t) = \cos n\omega_1 t$ and $\sin(-n\omega_1 t) = -\sin n\omega_1 t$. From Eqs. 25–53 and 25–54 we can write Eq. 25–52 as:

$$f(t) = a_0 + \sum_{n=1}^{n=\infty} C_n \epsilon^{jn\omega_1 t} + \sum_{n=1}^{n=\infty} C_n^* \epsilon^{-jn\omega_1 t} \tag{25–57}$$

Notice also that:

$$\sum_{n=1}^{n=\infty} C_n^* \epsilon^{-jn\omega_1 t} = \sum_{n=-\infty}^{n=-1} C_n \epsilon^{jn\omega_1 t} \tag{25–58}$$

By means of Eq. 25–58 we can write Eq. 25–57 in the following form:

$$f(t) = a_0 + \sum_{n=-\infty \text{ to } -1}^{n=1 \text{ to } \infty} C_n \epsilon^{jn\omega_1 t} \tag{25–59}$$

Observe that in the limits of summation of Eq. 25–59, $n = 0$ is skipped. However, if $n = 0$ is used in Eq. 25–55 we get:

$$C_0 = \underset{n=0}{C_n} = \frac{\omega_1}{2\pi} \int_{-\frac{\pi}{\omega_1}}^{\frac{\pi}{\omega_1}} f(t)\, dt = a_0$$

and

$$C_0 \epsilon^{j0\omega_1 t} = C_0 = a_0$$

Therefore Eq. 25–59 can be put as, including $n = 0$,

$$f(t) = \sum_{n=-\infty}^{n=\infty} C_n \epsilon^{jn\omega_1 t} \tag{25–60}$$

with

$$C_n = \frac{\omega_1}{2\pi} \int_{-\frac{\pi}{\omega_1}}^{\frac{\pi}{\omega_1}} f(t)\epsilon^{-jn\omega_1 t}\, dt \tag{25–61}$$

Equations 25–60 and 25–61 are the Fourier series pair in exponential form. Equation 25–61 changes the function of time into a function of frequency. Equation 25–60 changes a function of frequency into a function of time. Notice that C_n is a complex number or a phasor with magnitude and angle. Its magnitude and its angle extend to cover the negative as well as the positive frequencies. When C_n is extended in the negative frequencies we really are getting C_n^*.

EXAMPLE 25–6. Find C_n of Fig. 25–7.

The function $f(t)$ is mathematically expressed by:

$$f(t) = E \qquad 0 < t < \frac{\pi}{\omega_1}$$

$$f(t) = -E \qquad -\frac{\pi}{\omega_1} < t < 0$$

Then:

$$C_n = \frac{\omega_1}{2\pi}\left(\int_{-\frac{\pi}{\omega_1}}^{0} -E\epsilon^{-jn\omega_1 t}\,dt + \int_{0}^{\frac{\pi}{\omega_1}} E\epsilon^{-jn\omega_1 t}\,dt\right)$$

$$= \frac{\omega_1}{2\pi}\left(E\,\frac{\epsilon^{-jn\omega_1 t}}{jn\omega_1}\Bigg|_{-\frac{\pi}{\omega_1}}^{0} + E\,\frac{\epsilon^{-jn\omega_1 t}}{-jn\omega_1}\Bigg|_{0}^{\frac{\pi}{\omega_1}}\right)$$

$$= \frac{\omega_1}{2\pi}\,E\left(\frac{1-\epsilon^{-jn\pi}}{jn\omega_1} + \frac{\epsilon^{-jn\pi}-1}{-jn\omega_1}\right)$$

$$= \frac{\omega_1}{\pi}\,E\left(\frac{1-\epsilon^{-jn\pi}}{jn\omega_1}\right)$$

$$= 0 \qquad \text{for } n \text{ even}$$

$$= \frac{2E}{j\pi n} \qquad \text{for } n \text{ odd}$$

$$= \frac{2E}{\pi n}\,\epsilon^{-j(\pi/2)} \qquad \text{for } n \text{ odd and positive}$$

$$= \frac{2E}{\pi n}\,\epsilon^{j(\pi/2)} \qquad \text{for } n \text{ odd and negative}$$

Notice that for n odd and negative the magnitude of C_n is the same as for n odd and positive. The angle, however, is the negative of the angle for n odd and positive. The magnitude and phase of C_n are shown in Fig. 25–17.

25–11. Application. Assume that a non-sine wave periodic function is applied to a circuit and that the steady-state response is desired. To find the steady-state response the following steps are suggested. Find:

1. C_n in both magnitude and angle for positive and negative n.
2. The impedance or the admittance of the circuit, in phase and magnitude, for positive and negative n.
3. The response of the circuit in terms of frequency. The response is found in magnitude and angle for both positive and negative n.
4. The steady-state response as a function of time as obtained by the following formula:

$$f(t) = \sum_{n=-\infty}^{n=\infty} I_n \epsilon^{jn\omega_1 t}$$

EXAMPLE 25–7. Assume the periodic function of Fig. 25–7 to be applied to an R-L series circuit. If $E = 10$ v, $R = 1$ ohm, $L = 1$ h, and $\omega = 1$ radian, find the equation of the steady-state current in the circuit.

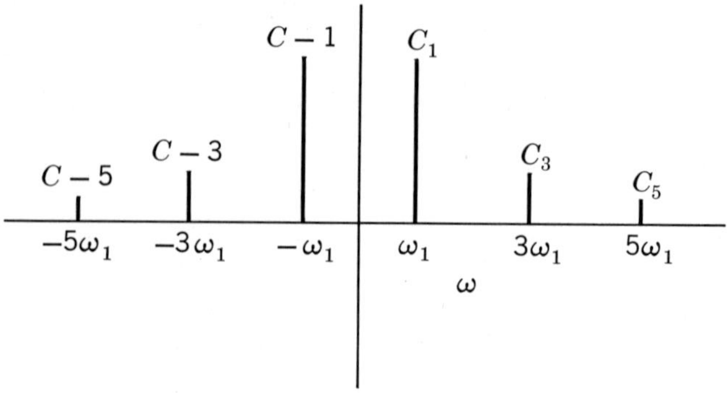

(*a*) Magnitude of $\boldsymbol{C}_n$ versus ω

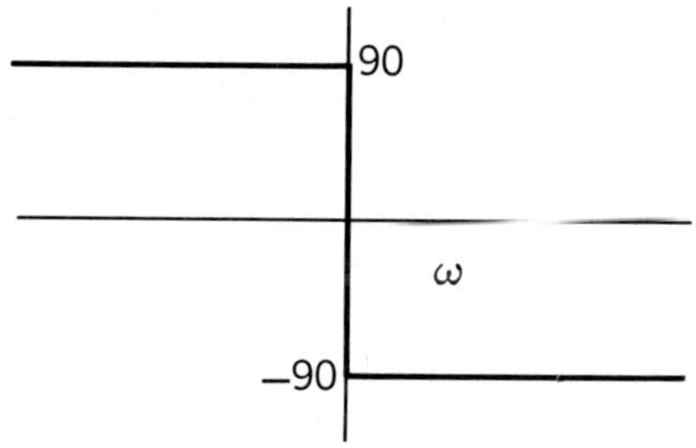

(*b*) Angle of $\boldsymbol{C}_n$ versus ω

Fig. 25–17. Magnitude and phase of $\boldsymbol{C}_n$ versus ω.

To solve this problem, follow the four steps that are suggested in Art. 25–11.

1. $\boldsymbol{C}_n$ is found to be

$$\boldsymbol{C}_n = \frac{2\boldsymbol{E}}{j\pi n} \qquad \text{for } n = \text{odd number}$$

   $\boldsymbol{C}_n$ is plotted versus ω for positive and negative ω both in magnitude and angle. See Fig. 25–18.

2. The impedance of the circuit is:

$$\boldsymbol{Z}_n = R + jn\omega L = 1 + jn$$

   The magnitude of $\boldsymbol{Z}_n$ as well as its angle are plotted versus ω for both positive and negative n.

3. $\boldsymbol{I}_n = \dfrac{\boldsymbol{C}_n}{\boldsymbol{Z}_n}$

4. The equation of the instantaneous response is:

$$i(t) = \sum_{n=-\infty}^{n=\infty} \frac{\boldsymbol{C}_n}{\boldsymbol{Z}_n} \epsilon^{jn\omega_1 t}$$

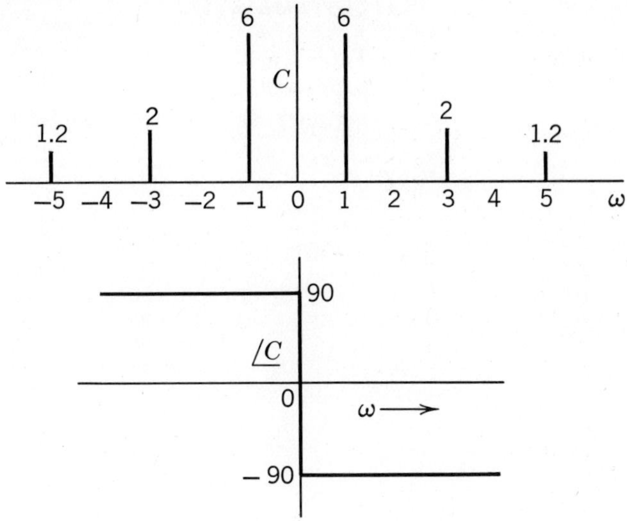

(*a*). C versus ω and $\underline{/C}$ versus ω for Example 25–7.

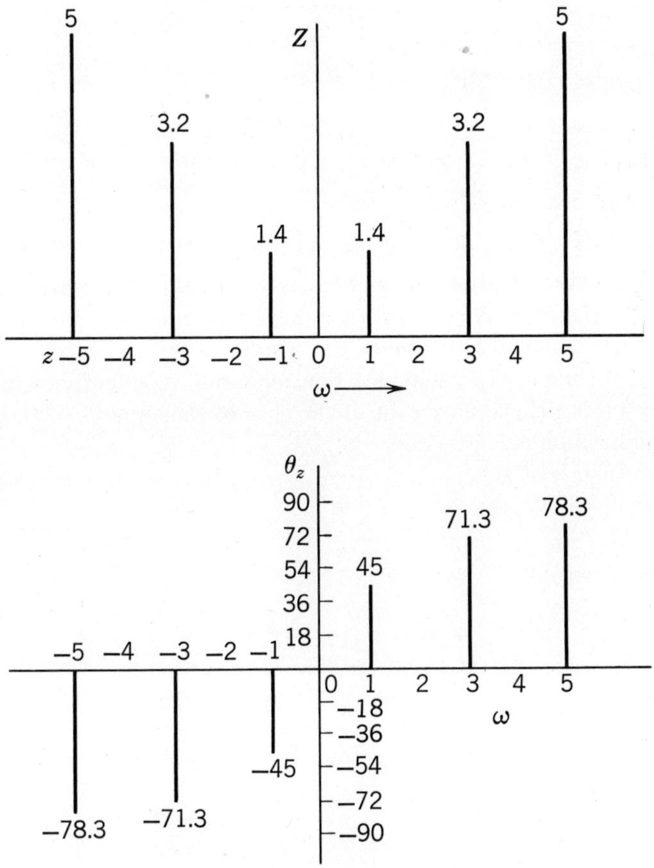

(*b*). Z versus ω and θ_z versus ω.
Fig. 25–18.

A table is made as shown in Table 25–3.

TABLE 25–3

n	Z	θ_z	$\|C_n\|$	$\angle C_n$	$\dfrac{C_n}{Z_n}$
1	1.414	45	6	−90	$4.2\epsilon^{-j(90+45)}$
3	3.2	71.3	2	−90	$0.6\epsilon^{-j(90+71.3)}$
5	5.0	78.3	1.2	−90	$0.22\epsilon^{-j(90+78)}$
−1	1.414	−45	6	+90	$4.2\epsilon^{j(90+45)}$
−3	3.2	−71.3	2	+90	$0.6\epsilon^{j(90+71.3)}$
−5	5	−78.3	1.2	+90	$0.22\epsilon^{j(90+78)}$

$$\begin{aligned} i(t) &= \sum_{n=-\infty}^{n=\infty} \frac{C_n}{Z_n} \epsilon^{jn\omega_1 t} \\ &= 4.2\epsilon^{-j(90+45)}\epsilon^{jt} + 4.2\epsilon^{j(90+45)}\epsilon^{-jt} \\ &\quad + 0.6\epsilon^{-j(90+71.3)}\epsilon^{j3t} + 0.6\epsilon^{j(90+71.3)}\epsilon^{-j3t} \\ &\quad + 0.22\epsilon^{-j(90+78.3)}\epsilon^{j5t} + 0.22\epsilon^{j(90+78.3)}\epsilon^{-j5t} \\ &= 4.2[\epsilon^{j(t-90-45)} + \epsilon^{-j(t-90-45)}] \\ &\quad + 0.6[\epsilon^{j(3t-90-71.3)} + \epsilon^{-j(3t-90-71.3)}] \\ &\quad + 0.22[\epsilon^{j(5t-90-78.3)} + \epsilon^{-j(5t-90-78.3)}] \\ &= 8.4 \cos(t - 45^\circ - 90^\circ) + 1.2 \cos(3t - 71.3^\circ - 90^\circ) \\ &\quad + 0.44 \cos(5t - 78.3^\circ - 90^\circ) \\ &= 8.4 \sin(t - 45^\circ) + 1.2 \sin(3t - 71.3^\circ) + 0.44 \sin(5t - 78.3^\circ) \end{aligned}$$

25–12. Effective Value of a Non-sinusoidal Periodic Function. In Chapter 13 the effective value of a function of time was found by taking the square root of the mean value of the squares of the function of time. If the function of time is expressed as a Fourier series, its effective value can be obtained by taking the square root of the sum of the squares of the effective value of the harmonics.

To prove this, let

$$f(t) = \sum_{n=-\infty}^{n=\infty} C_n \epsilon^{jn\omega_1 t} \tag{25–60}$$

This can be written as

$$\begin{aligned} f(t) &= C_0 + \sum_{n=1}^{n=\infty} C_n \epsilon^{jn\omega_1 t} + \sum_{n=1}^{n=\infty} C_n{}^* \epsilon^{-jn\omega_1 t} \qquad (25\text{–}57) \\ &= C_0 + \sum_{n=1}^{n=\infty} |C_n| \epsilon^{j(n\omega_1 t - \phi_n)} + \sum_{n=1}^{n=\infty} |C_n| \epsilon^{-j(n\omega_1 t - \phi_n)} \\ &= C_0 + \sum_{n=1}^{n=\infty} |C_n| [\epsilon^{j(n\omega_1 t - \phi_n)} + \epsilon^{-j(n\omega_1 t - \phi_n)}] \\ &= C_0 + 2 \sum_{n=1}^{n=\infty} |C_n| \cos(n\omega_1 t - \phi_n) \end{aligned}$$

The square of the function $f(t)$ is

$$f(t)^2 = \left[C_0 + 2\sum_{n=1}^{n=\infty} |C_n| \cos(n\omega_1 t - \phi_n)\right]^2$$

$$= [C_0^2 + 8|C_0|\,|C_1| \cos(\omega_1 t - \phi_1) + 8|C_0|\,|C_2| \cos(2\omega_1 t - \phi_2) + \cdots$$

$$+ 4|C_1|^2 \cos^2(\omega_1 t - \phi_1) + 8|C_1|\,|C_2| \cos(\omega_1 t - \phi_1)$$

$$(\cos 2\omega_1 t - \phi_2) + \cdots$$

$$+ 4|C_2|^2 \cos^2(2\omega_1 t - \phi_2) + 8|C_2|\,|C_3| \cos(2\omega_1 t - \phi_2)$$

$$(\cos 3\omega_1 t - \phi_3) + \cdots$$

$$+ 4|C_3|^2 \cos^2(3\omega_1 t - \phi_3) + 8|C_3|\,|C_4| \cos(3\omega_1 t - \phi_3)$$

$$(\cos 4\omega_1 t - \phi_4) + \cdots]$$

But

$$\cos(n\omega_1 t - \phi_n)(\cos m\omega_1 t - \phi_m) = \tfrac{1}{2}\{\cos[(n+m)\omega_1 t - (\phi_n + \phi_m)] + \cos[(n-m)\omega_1 t - (\phi_n - \phi_m)]\} \tag{25–62}$$

when $n \neq m$

$$\int_{-\frac{\pi}{\omega_1}}^{\frac{\pi}{\omega_1}} \cos[(n+m)\omega_1 t - (\phi_n + \phi_m)]\,dt = -\left.\frac{\sin[(n+m)\omega_1 t - (\phi_n + \phi_m)]}{n+m}\right|_{-\frac{\pi}{\omega_1}}^{\frac{\pi}{\omega_1}} = 0$$

Similarly,

$$\int_{-\frac{\pi}{\omega_1}}^{\frac{\pi}{\omega_1}} \cos[(n-m)\omega_1 t - (\phi_n - \phi_m)]\,dt = -\left.\frac{\sin[(n-m)\omega_1 t - (\phi_n - \phi_m)]}{n-m}\right|_{-\frac{\pi}{\omega_1}}^{\frac{\pi}{\omega_1}} = 0$$

On the other hand, when $n = m$ the integral of Eq. 25–62 gives

$$\int_{-\frac{\pi}{\omega_1}}^{\frac{\pi}{\omega_1}} C_0^2 + 4|C_n|^2 \cos^2(n\omega_1 t - \phi_n)\,dt = C_0^2 t + 4\left.\frac{|C_n^2|}{2}\,t\right|_{-\frac{\pi}{\omega_1}}^{\frac{\pi}{\omega_1}}$$

Therefore

$$\int_{-\frac{\pi}{\omega_1}}^{\frac{\pi}{\omega_1}} f^2(t)\,dt = \frac{2\pi}{\omega_1}[C_0^2 + 2|C_1|^2 + 2|C_2|^2 + 2|C_3|^2 + \cdots] \tag{25–63}$$

The root mean square of $f(t)$ is

$$\sqrt{\frac{\omega_1}{2\pi}\int_{-\frac{\pi}{\omega_1}}^{\frac{\pi}{\omega_1}} f^2(t)\,dt} = \sqrt{C_0^2 + 2|C_1|^2 + 2|C_2|^2 + \cdots} \tag{25–64}$$

But $C_0 = a_0$. Therefore

$$|C_1|^2 = \frac{a_1^2 + b_1^2}{4}$$

and

$$|C_n|^2 = \frac{a_n^2 + b_n^2}{4} \qquad n = 1, 2, \cdots$$

The effective value therefore is equal to

$$\sqrt{a_0^2 + \frac{a_1^2 + b_1^2}{2} + \frac{a_2^2 + b_2^2}{2} + \cdots} = \sqrt{a_0^2 + \frac{r_1^2}{2} + \frac{r_2^2}{2} + \cdots}$$

$$= \sqrt{a_0^2 + |r_{1\text{eff}}|^2 + |r_{2\text{eff}}|^2 + \cdots}$$

PROBLEMS

25–1. If

$$f(\omega_1 t) = -f(\pi + \omega_1 t)$$

show that the Fourier series is given by:

$$f(\omega_1 t) = \Sigma\, a_n \cos n\omega_1 t + \Sigma\, b_n \sin n\omega_1 t$$

where n is odd.

25–2. If

$$f(\omega_1 t) = -f(\pi - \omega_1 t)$$

show that

$$f(\omega_1 t) = \Sigma\, a_n \cos n\omega_1 t + \Sigma\, b_m \sin m\omega_1 t$$

where n is odd and m is even.

25–3. If

$$f(\omega_1 t) = f(\pi + \omega_1 t)$$

show that

$$f(\omega_1 t) = a_0 + \Sigma\, a_m \cos m\omega_1 t + \Sigma\, b_m \sin m\omega_1 t$$

where m is even.

25–4. If

$$f(\omega_1 t) = f(\pi - \omega_1 t)$$

show that

$$f(\omega_1 t) = a_0 + \Sigma\, a_m \cos m\omega_1 t + \Sigma\, b_n \sin n\omega_1 t$$

where n is odd and m is even.

25–5. Find the Fourier series of the curve in the figure shown and plot its a's and b's versus ω.

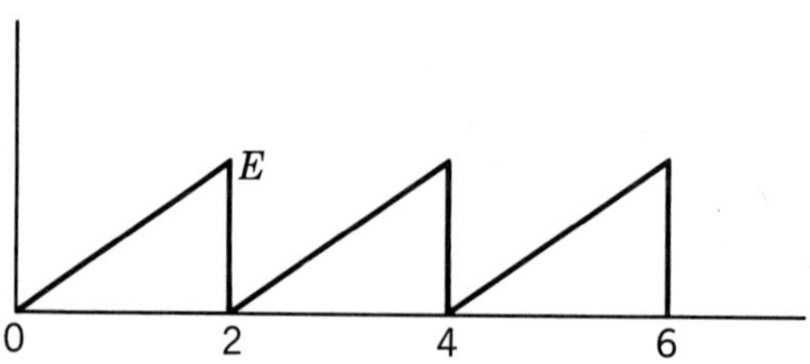

Prob. 25–5.

25–6. Repeat Prob. 25–5 for the figure illustrated.

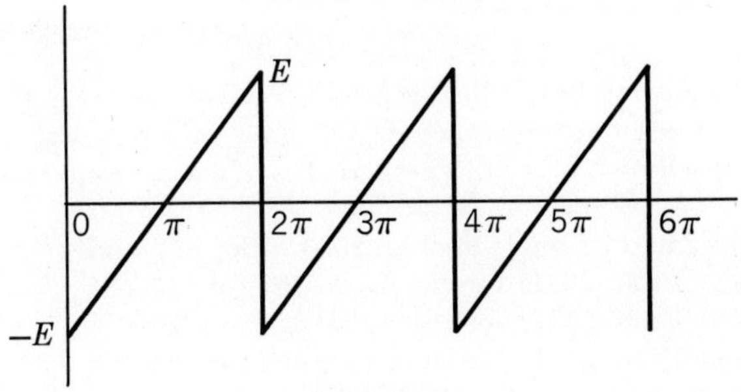

Prob. 25–6.

25–7. Repeat Prob. 25–5 for the waveform shown.

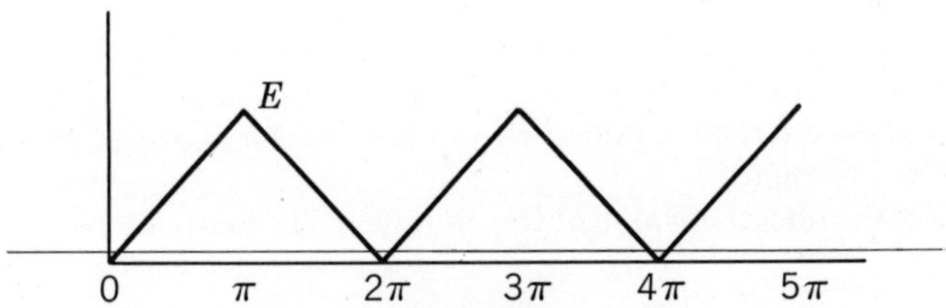

Prob. 25–7.

25–8. The voltage wave per cycle on a 400-cycle source is measured and found to be as given in Table 25–4. Find its Fourier series.

TABLE 25–4

ωt (degrees)	$f(t)$	ωt (degrees)	$f(t)$
0	0	190	−15
10	10	200	−30
20	15	210	−50
30	20	220	−70
40	25	230	−69
50	30	240	−68
60	40	250	−67
70	50	260	−66
80	60	270	−65
90	65	280	−60
100	66	290	−50
110	67	300	−40
120	68	310	−30
130	69	320	−25
140	70	350	−20
150	50	340	−15
160	30	350	−10
170	15	360	0
180	0		

25–9. The figure of Prob. 25–5 represents a current flowing into a circuit made up of a resistance and capacitance in parallel. If $G = 1$ mho and $C = 100\ \mu$f and if $f = 400$ cycles and $\boldsymbol{I} = 0.1$ amp, find the shape of the voltage across the parallel circuit.

25–10. The figure of Prob. 25–6 represents a voltage with $\boldsymbol{E} = 10$ v. This voltage is applied across a resistance of 100 ohms in series with an inductance of 1 h and a capacitance of 1 μf. The period is $\frac{1}{400}$ sec. Find the shape of the current.

25–11. Find C_n and its angle for Fig. 25–3. Plot $|C_n|$ and $\angle C_n$ versus ω.

25–12. Repeat Prob. 25–11 for the figure of Prob. 25–5.

25–13. Repeat Prob. 25–11 for the figure of Prob. 25–6.

25–14. Repeat Prob. 25–11 for the figure of Prob. 25–7.

25–15. If the function of Prob. 25–11 is a current flowing into an R-C parallel circuit and $R = 1$ ohm, $C = 100\ \mu$f, and the frequency is 400 cycles, find the shape of the voltage by the use of Fourier exponential form. $\boldsymbol{E} = 1$ amp.

26–16. The figure of Prob. 25–6 represents a voltage with $\boldsymbol{E} = 10$ v. This voltage is applied across a resistance of 100 ohms in series with an inductance of 1 h and a capacitance of 1 μf. The period is $\frac{1}{400}$ sec. Find by the use of the exponential form of the Fourier series the shape of the current.

25–17. Find the effective value of the voltage in the figure of Prob. 25–5 by two methods and compare.

25–18. Find the effective value of the voltage in the figure of Prob. 25–7.

INDEX

1 12 1/2 cents